Proceedings of GeoShanghai 2018 International
Conference: Geoenvironment and Geohazard

Arvin Farid · Hongxin Chen
Editors

Proceedings of GeoShanghai 2018 International Conference: Geoenvironment and Geohazard

 Springer

Editors
Arvin Farid
Civil Engineering
Boise State University
Boise, ID
USA

Hongxin Chen
Tongji University
Shanghai
China

ISBN 978-981-13-0127-8   ISBN 978-981-13-0128-5   (eBook)
https://doi.org/10.1007/978-981-13-0128-5

Library of Congress Control Number: 2018939621

Printed on acid-free paper

This Springer imprint is published by the registered company Springer Nature Singapore Pte Ltd. part of Springer Nature
The registered company address is: 152 Beach Road, #21-01/04 Gateway East, Singapore 189721, Singapore

Preface

The 4th GeoShanghai International Conference was held on May 27–30, 2018, in Shanghai, China. GeoShanghai is a series of international conferences on geotechnical engineering held in Shanghai every four years. The conference was inaugurated in 2006 and was successfully held in 2010 and 2014, with more than 1200 participants in total. The conference offers a platform of sharing recent developments of the state-of-the-art and state-of-the-practice in geotechnical and geoenvironmental engineering. It has been organized by Tongji University in cooperation with the ASCE Geo-Institute, Transportation Research Board, and other cooperating organizations.

The proceedings of the 4th GeoShanghai International Conference include eight volumes of over 560 papers; all were peer-reviewed by at least two reviewers. The proceedings include Volumes 1: Fundamentals of Soil Behavior edited by Dr. Annan Zhou, Dr. Junliang Tao, Dr. Xiaoqiang Gu, and Dr. Liangbo Hu; Volume 2: Multi-physics Processes in Soil Mechanics and Advances in Geotechnical Testing edited by Dr. Liangbo Hu, Dr. Xiaoqiang Gu, Dr. Junliang Tao, and Dr. Annan Zhou; Volume 3: Rock Mechanics and Rock Engineering edited by Dr. Lianyang Zhang, Dr. Bruno Goncalves da Silva, and Dr. Cheng Zhao; Volume 4: Transportation Geotechnics and Pavement Engineering edited by Dr. Xianming Shi, Dr. Zhen Liu, and Dr. Jenny Liu; Volume 5: Tunneling and Underground Construction edited by Dr. Dongmei Zhang and Dr. Xin Huang; Volume 6: Advances in Soil Dynamics and Foundation Engineering edited by Dr. Tong Qiu, Dr. Binod Tiwari, and Dr. Zhen Zhang; Volume 7: Geoenvironment and Geohazards edited by Dr. Arvin Farid and Dr. Hongxin Chen; and Volume 8: Ground Improvement and Geosynthetics edited by Dr. Lin Li, Dr. Bora Cetin, and Dr. Xiaoming Yang. The proceedings also include six keynote papers presented at the conference, including "Tensile Strains in Geomembrane Landfill Liners" by Prof. Kerry Rowe, "Constitutive Modeling of the Cyclic Loading Response of Low Plasticity Fine-Grained Soils" by Prof. Ross Boulanger, "Induced Seismicity and Permeability Evolution in Gas Shales, CO_2 Storage and Deep Geothermal Energy" by Prof. Derek Elsworth, "Effects of Tunneling on Underground Infrastructures" by Prof. Maosong Huang, "Geotechnical Data Visualization and Modeling of Civil

Infrastructure Projects" by Prof. Anand Puppala, and "Probabilistic Assessment and Mapping of Liquefaction Hazard: from Site-specific Analysis to Regional Mapping" by Prof. Hsein Juang. The Technical Committee Chairs, Prof. Wenqi Ding and Prof. Xiong Zhang, the Conference General Secretary, Dr. Xiaoqiang Gu, the 20 editors of the 8 volumes and 422 reviewers, and all the authors contributed to the value and quality of the publications.

The Conference Organizing Committee thanks the members of the host organizations, Tongji University, Chinese Institution of Soil Mechanics and Geotechnical Engineering, and Shanghai Society of Civil Engineering, for their hard work and the members of International Advisory Committee, Conference Steering Committee, Technical Committee, Organizing Committee, and Local Organizing Committee for their strong support. We hope the proceedings will be valuable references to the geotechnical engineering community.

Shijin Feng
Conference Chair
Ming Xiao
Conference Co-chair

Organization

International Advisory Committee

Herve di Benedetto	University of Lyon, France
Antonio Bobet	Purdue University, USA
Jean-Louis Briaud	Texas A&M University, USA
Patrick Fox	Penn State University, USA
Edward Kavazanjian	Arizona State University, USA
Dov Leshchinsky	University of Illinois, USA
Wenhao Liang	China Railway Construction Corporation Limited, China
Robert L. Lytton	Texas A&M University, USA
Louay Mohammad	Louisiana State University, USA
Manfred Partle	KTH Royal Institute of Technology, Switzerland
Anand Puppala	University of Texas at Arlington, USA
Mark Randolph	University of Western Australia, Australia
Kenneth H. Stokoe	University of Texas at Austin, USA
Gioacchino (Cino) Viggiani	Université Joseph Fourier, France
Dennis T. Bergado	Asian Institute of Technology, Thailand
Malcolm Bolton	Cambridge University, UK
Yunmin Chen	Zhejiang University, China
Zuyu Chen	Tsinghua University, China
Jincai Gu	PLA, China
Yaoru Lu	Tongji University, China
Herbert Mang	Vienna University of Technology, Austria
Paul Mayne	Georgia Institute of Technology, USA
Stan Pietruszczak	McMaster University, Canada
Tom Papagiannakis	Washington State University, USA
Jun Sun	Tongji University, China

Scott Sloan   University of Newcastle, Australia
Hywel R. Thomas   Cardiff University, UK
Atsashi Yashima   Gifu University, Japan

Conference Steering Committee

Jie Han   University of Kansas, USA
Baoshan Huang   University of Tennessee, USA
Maosong Huang   Tongji University, China
Yongsheng Li   Tongji University, China
Linbin Wang   Virginia Tech, USA
Lianyang Zhang   University of Arizona, USA
Hehua Zhu   Tongji University, China

Technical Committee

Wenqi Ding (Chair)   Tongji University, China
Charles Aubeny   Texas A&M University, USA
Rifat Bulut   Oklahoma State University, USA
Geoff Chao   Asian Institute of Technology, Thailand
Jian Chu   Nanyang Technological University, Singapore
Eric Drumm   University of Tennessee, USA
Wen Deng   Missouri University of Science and Technology,
                                  USA
Arvin Farid   Boise State University, Idaho, USA
Xiaoming Huang   Southeast University, China
Woody Ju   University of California, Los Angeles, USA
Ben Leshchinsky   Oregon State University, Oregon, USA
Robert Liang   University of Dayton, Ohio, USA
Hoe I. Ling   Columbia University, USA
Guowei Ma   Hebei University of Technology, China
Roger W. Meier   University of Memphis, USA
Catherine O'Sullivan   Imperial College London, UK
Massimo Losa   University of Pisa, Italy
Angel Palomino   University of Tennessee, USA
Krishna Reddy   University of Illinois at Chicago, USA
Zhenyu Yin   Tongji University, China
ZhongqiYue   University of Hong Kong, China
Jianfu Shao   Université des Sciences et Technologies
                                  de Lille 1, France
Jonathan Stewart   University of California, Los Angeles, USA

Wei Wu	University of Natural Resources and Life Sciences, Austria
Jianhua Yin	The Hong Kong Polytechnic University, China
Guoping Zhang	University of Massachusetts, USA
Jianmin Zhang	Tsinghua University, China
Xiong Zhang (Co-chair)	Missouri University of Science and Technology, USA
Yun Bai	Tongji University, China
Jinchun Chai	Saga University, Japan
Cheng Chen	San Francisco State University, USA
Shengli Chen	Louisiana State University, USA
Yujun Cui	École Nationale des Ponts et Chaussees (ENPC), France
Mohammed Gabr	North Carolina State University, USA
Haiying Huang	Georgia Institute of Technology, USA
Laureano R. Hoyos	University of Texas at Arlington, USA
Liangbo Hu	University of Toledo, USA
Yang Hong	University of Oklahoma, USA
Minjing Jiang	Tongji University, China
Richard Kim	North Carolina State University, USA
Juanyu Liu	University of Alaska Fairbanks, USA
Matthew Mauldon	Virginia Tech., USA
Jianming Ling	Tongji University, China
Jorge Prozzi	University of Texas at Austin, USA
Daichao Sheng	University of Newcastle, Australia
Joseph Wartman	University of Washington, USA
Zhong Wu	Louisiana State University, USA
Dimitrios Zekkos	University of Michigan, USA
Feng Zhang	Nagoya Institute of Technology, Japan
Limin Zhang	Hong Kong University of Science and Technology, China
Zhongjie Zhang	Louisiana State University, USA
Annan Zhou	RMIT University, Australia
Fengshou Zhang	Tongji University, China

Organizing Committee

Shijin Feng (Chair)	Tongji University, China
Xiaojqiang Gu (Secretary General)	Tongji University, China
Wenqi Ding	Tongji University, China
Xiongyao Xie	Tongji University, China

Yujun Cui	École Nationale des Ponts et Chaussees (ENPC), France
Daichao Sheng	University of Newcastle, Australia
Kenichi Soga	University of California, Berkeley, USA
Weidong Wang	Shanghai Xian Dai Architectural Design (Group) Co., Ltd., China
Feng Zhang	Nagoya Institute of Technology, Japan
Yong Yuan	Tongji University, China
Weimin Ye	Tongji University, China
Ming Xiao (Co-chair)	Penn State University, USA
Yu Huang	Tongji University, China
Xiaojun Li	Tongji University, China
Xiong Zhang	Missouri University of Science and Technology, USA
Guenther Meschke	Ruhr-Universität Bochum, Germany
Erol Tutumluer	University of Illinois, Urbana—Champaign, USA
Jianming Zhang	Tsinghua University, China
Jianming Ling	Tongji University, China
Guowei Ma	Hebei University of Technology, Australia
Hongwei Huang	Tongji University, China

Local Organizing Committee

Shijin Feng (Chair)	Tongji University, China
Zixin Zhang	Tongji University, China
Jiangu Qian	Tongji University, China
Jianfeng Chen	Tongji University, China
Bao Chen	Tongji University, China
Yongchang Cai	Tongji University, China
Qianwei Xu	Tongji University, China
Qingzhao Zhang	Tongji University, China
Zhongyin Guo	Tongji University, China
Xin Huang	Tongji University, China
Fang Liu	Tongji University, China
Xiaoying Zhuang	Tongji University, China
Zhenming Shi	Tongji University, China
Zhiguo Yan	Tongji University, China
Dongming Zhang	Tongji University, China
Jie Zhang	Tongji University, China
Zhiyan Zhou	Tongji University, China
Xiaoqiang Gu (Secretary)	Tongji University, China
Lin Cong	Tongji University, China
Hongduo Zhao	Tongji University, China

Fayun Liang	Tongji University, China
Bin Ye	Tongji University, China
Zhen Zhang	Tongji University, China
Yong Tan	Tongji University, China
Liping Xu	Tongji University, China
Mengxi Zhang	Tongji University, China
Haitao Yu	Tongji University, China
Xian Liu	Tongji University, China
Shuilong Shen	Tongji University, China
Dongmei Zhang	Tongji University, China
Cheng Zhao	Tongji University, China
Hongxin Chen	Tongji University, China
Xilin Lu	Tongji University, China
Jie Zhou	Tongji University, China

Contents

Behavior of Biotreated Geomaterials and Foundations

About the Editors

Farid received his PhD from Northeastern University, Boston, MA, USA, with a focus on Geoenvironmental Engineering. He is currently an Associate Professor and the Graduate Coordinator of the Civil Engineering Department at Boise State University.

Chen received his PhD from the Hong Kong University of Science and Technology, HKSAR, China. He is currently an Assistant Professor in the Department of Geotechnical Engineering at Tongji University, China.

Geohazards

Prediction Method of Unsaturated Slope Stability Under Rainfall and Fluctuation of Reservoir Water Level

X. Xiong[1], Z. M. Shi[2,3(✉)], Y. L. Xiong[4], X. L. Ma[5], and F. Zhang[1]

[1] Nagoya Institute of Technology, Nagoya, Japan
[2] Department of Geotechnical Engineering, Tongji University, Shanghai, China
shi_tongji@tongji.edu.cn
[3] Ministry of Education Key Laboratory of Geotechnical and Underground Engineering, Tongji University, Shanghai, China
[4] Institute of Geotechnical Engineering, Ningbo University, Ningbo, Zhejiang, China
[5] CCCC Second Highway Consultants Co., Ltd, Wuhan, Hubei, China

Abstract. As a sudden catastrophic geological disaster, landslides always cause a lot of casualties and property damage. Landslides in reservoir areas are not caused by a single factor, such as rainfall or fluctuation of reservoir water level, but the result of the combined actions. In order to accurately describe landslide mechanism and eventually be able to forecast a landslide accurately, it is necessary to select suitable constitutive model and rational parameters for numerical calculation of slope stability. In this paper, an unsaturated soil constitutive model was selected to describe the soil mechanical properties, in which the parameters of the model for the unsaturated soil S1 were obtained by element tests. With these parameters, numerical analyses were conducted to calculate the boundary value problem of slope model tests, in which the stability of model slope with the same soil S1 was tested under rainfall and fluctuation of reservoir water level. By comparing the test and calculation results, it was found that the numerical method proposed in this paper has satisfactory accuracy, being able to describe the mechanism of landslide and make a rational prediction of slope failure.

Keywords: Slope stability · Unsaturated soil · Rainfall
Fluctuation of reservoir water level

1 Introduction

In recent years, due to rapid development of economy, infrastructure in mountainous area of China develops very quickly, which involves the construction of large water power stations. Number of landslides in the reservoir areas happened frequently. Landslides often occur during the period of fluctuation of reservoir water level, and this period is often accompanied by large-scale rainfall. Therefore, landslides in reservoir areas are caused by the combined actions. In addition, the natural condition of the slopes is mostly unsaturated. In order to accurately describe landslide mechanism, which is important in geological catastrophes mitigation, it is necessary to select

© Springer Nature Singapore Pte Ltd. 2018
A. Farid and H. Chen (Eds.): GSIC 2018, *Proceedings of GeoShanghai 2018*
International Conference: Geoenvironment and Geohazard, pp. 3–10, 2018.
https://doi.org/10.1007/978-981-13-0128-5_1

suitable constitutive model and rational parameters of soils to estimate the slope stability under rainfall and fluctuation of reservoir water level.

Many researches on the slope stability under rainfall and fluctuation of reservoir water level can be found in literature. Field monitoring data have been used to analyze the deformation of slopes due to rainfall and fluctuation, indicating that these two input factors have different influence on the slope stability (Wang et al. 2016; Sun et al. 2016; Huang et al. 2016). Numerical method has also been utilized, e.g., probabilistic models (Zhou et al. 2016) and limited equilibrium (Zhang et al. 2014; Liang et al. 2015; Huang et al. 2016). However, in these calculations, the mechanical properties, such as the stress-strain relation of unsaturated soil, was not described by rational constitutive model. Moreover, the determination of the parameters involved in constitutive model was lack of explicit description. As the results, these numerical calculations cannot be used to predict landslides, but only fit the landslides behavior to some extent that have already happened.

In this paper, an unsaturated soil constitutive model, proposed by Zhang and Ikariya (2011), was selected to describe the soil mechanical properties, and the parameters were determined from element tests. Based on the model and using these parameters, soil-water-air full coupling analyses with a finite element-finite deformation (FE-FD) scheme were conducted to calculate the boundary value problem (BVP) of slope failure model tests. By comparing the results of the model test and the calculation, the method was verified if it could predict the landslide due to rainfall and fluctuation of reservoir water level.

2   Constitutive Model and Parameters

To properly describe the changes of slope soil saturation under rainfall and fluctuation of reservoir water level, which lead to the reduction of soil strength and landslides, unsaturated soil constitutive model was considered in the numerical calculation.

Basing on experimental results, Zhang and Ikariya (2011) proposed an unsaturated soil constitutive model, using skeleton stress and degree of saturation as independent variables. In the model, it is assumed that normally consolidated line in unsaturation state (*N.C.L.S.*) is parallel to the normally consolidated line in saturated state (*N.C.L.*) but in a higher position than *N.C.L.* Skeleton stress is a kind of Bishop effective stress, defined as,

$$\sigma_{ij}'' = \sigma_{ij}^t - u_a \delta_{ij} + S_r(u_a - u_w)\delta_{ij} = \sigma_{ij}^n + S_r s \delta_{ij} \tag{1}$$

where σ_{ij}'' is skeleton stress tensor, σ_{ij}' is total stress tensor, σ_{ij}^n is net stress tensor, S_r is degree of saturation, u_a is air pressure, u_w is water pressure and s is suction. The constitutive model is able to describe not only the behavior of unsaturated soil but also saturated soil because the skeleton stress can smoothly shift to effective stress from unsaturated condition to saturated condition. The constitutive model includes nine material parameters.

In the unsaturated soil constitutive model, a moisture characteristics curve (MCC) considering moisture hysteresis of unsaturated soil is also proposed. Depending on the state of the moisture, the moisture characteristics are given in three different tangential and arc-tangential functions as primary drying curve, secondary drying curve and wetting curve, which include eight parameters.

There have been some studies on the application of the unsaturated soil constitutive model (Xiong et al. 2014). Thus, it is thought to be proper to select the constitutive model in the stability analysis of unsaturated slope under rainfall and fluctuation of reservoir water level.

In this paper, to assure the accuracy of the calculation of BVP, all the parameters of the unsaturated soil constitutive model are determined based on element tests. A geo-material, named as S1 whose grading curve is shown in Fig. 1, was used as the model ground of slope failure model tests. The values of the parameters involved in the moisture characteristics curve, obtained from the water retention test of S1, are listed in Table 1.

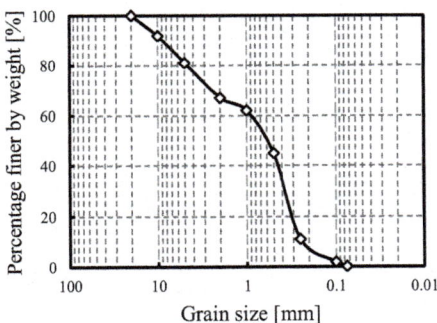

Fig. 1. Gradation curve of S1.

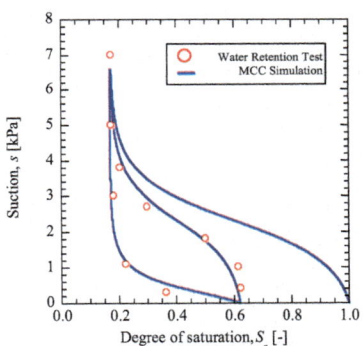

Fig. 2. Water retention test and numerical simulation results of unsaturated S1.

Table 1. Parameters of MCC of S1.

Saturated degrees of saturation S_r^s	0.62
Residual degrees of saturation S_r^r	0.17
Parameter corresponding to drying air entry value (kPa) S_d	2.30
Parameter corresponding to wetting air entry value (kPa) S_w	0.10
Initial stiffness of scanning curve (kPa) k_{sp}^e	500
Parameter of shape function c_1	1.20
Parameter of shape function c_2	1.70
Parameter of shape function c_3	24

Figure 2 shows the simulation result that fits the experimental result well. Triaxial tests of S1 have been done under consolidated-undrained conditions with the void ratio $e = 0.70$ (Zuo 2013). In the triaxial tests, confining stresses were 50, 100 and 200 kPa, and the shear rate is 0.4 mm/min. Based on the simulation of the triaxial tests, the material parameters of S1 were calibrated and their values are listed in Table 2. Compared results of test and simulation shown in Fig. 3 revealed that the magnitude and trend of shear stress and pore water pressure are basically the same, implying that the constitutive model can properly describe the mechanical properties of S1.

Table 2. Material parameters of S1.

Compression index λ	0.123
Swelling index κ	0.024
Critical state parameter R_{cs}	3.55
Void ratio N (p' = 10 kPa on $N.C.L.$)	0.57
Poisson's ratio v	0.25
Parameter of overconsolidation a	0.85
Parameter of suction b	0.00
Parameter of overconsolidation β	1.00
Void ratio N_r (p' = 10 kPa on $N.C.L.S.$)	0.77

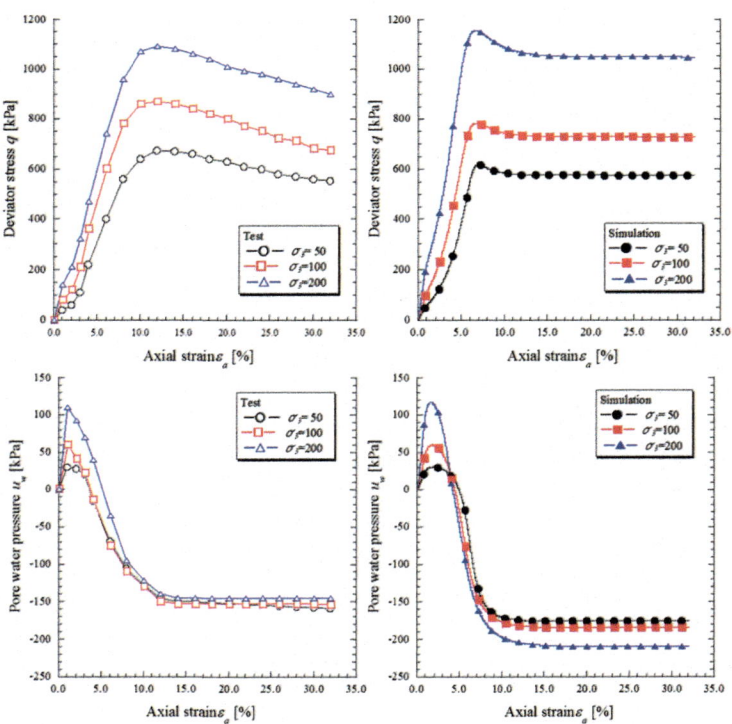

Fig. 3. Comparison between CU triaxial tests and numerical simulation of S1.

3   Brief Introduction of Model Test

Taking the Three Gorges Reservoir area as an example, slope failure model test under rainfall and fluctuation of reservoir water was designed, used S1 as the model ground material. As shown in Fig. 4, the slope model was made of a homogeneous ground. When preparing the slope model, layered compaction method was used to control the void ratio ($e = 0.70$). The boundary conditions are given as: (a) the top and left surfaces are permeable, while the front, back, bottom and right surfaces are impermeable; (b) the artificial rainfall could cover the entire top surface; (c) the initial reservoir water level of left surface is 35 cm. The measuring points are arranged as shown in Fig. 4.

The characteristics of the reservoir water level operation and seasonal rainfall in the area are summarized and simplified for the test as shown in Fig. 5. The test result shows that under the rise of the reservoir water level and repeated rainfall, the shallow landslide occurred at 451 min, just after the third rainfall.

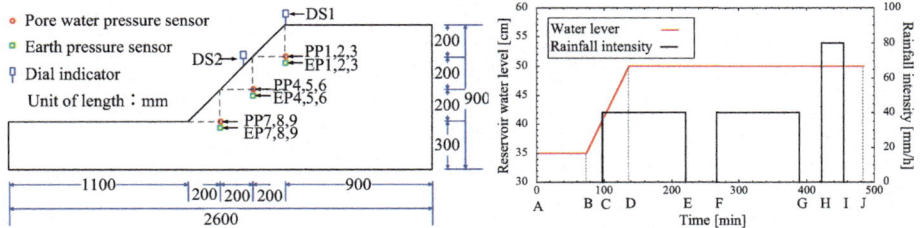

Fig. 4. Measuring points of slope model.   **Fig. 5.** Model testing process.

4   Simulation of Model Test

4.1   Numerical Model and Initial Conditions

The slope failure model test was simulate with finite element method (FEM) with the program named as SOFT (Xiong et al. 2014), using a finite element-finite difference scheme (FE-FD) for soil-water-air three-phase coupling problems. In the analysis, FEM is used for spatial discretization of solid, liquid and gas phases, while finite difference method (FDM) is used for discretization of time domain (Fig. 6), a typical hybrid scheme of FD-FE. The values of the parameters were the same as Tables 1 and 2. Figure 7 shows the finite element mesh used in the simulation. The size of the FEM mesh, composed of 1778 nodes and 1680 4-node isoparametric elements, is the same as that of the model test under plane strain conditions. Hydraulic and air ventilation boundary conditions, as shown in Fig. 8(a), are set the same as the model test. For the displacement boundary condition, as shown in Fig. 8(b), surface E-F is fixed in both directions; surface A-F and D-E are fixed at the vertical displacement in the x direction; the other surfaces are free in the x and y directions.

In the numerical simulation of BVP, the settings of the initial conditions are very important, which would greatly affect the accuracy of the calculation. Based on the moisture characteristics curve, the initial degree of saturation of S1 is $S_{r0} = 0.177$, and the initial pore pressure is $p_0 = -6$ kPa ($s_0 = 6$ kPa). Since the model slope was prepared by layered compaction method, the S1 was overconsolidated at the initial condition. For simplicity, an extra mean effective stress of 5 kPa was added to the whole area to consider the compaction effect. Figure 9 shows the initial vertical effective stress field in the calculation.

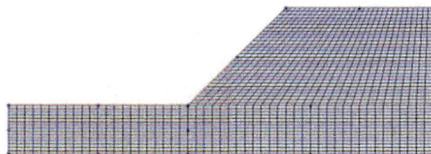

Fig. 6. FEM mesh for model test.

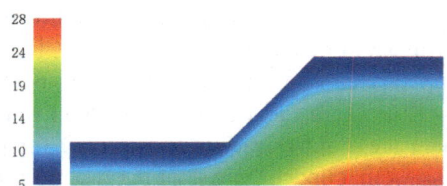

Fig. 7. Initial effective stress field (Unit: kPa).

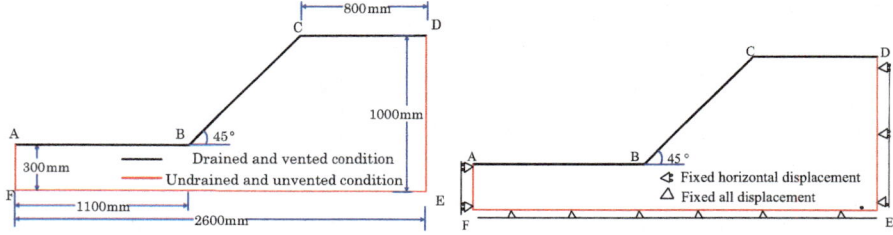

Fig. 8. Boundary condition of numerical modeling.

In FE–FD analysis related to soil-water-air full coupling problems, time interval of each stage could influence the calculation accuracy. After comparing different time intervals, time interval was set to be 0.2 s/step in all stages. At initial stage, the reservoir water level of 35 cm was given, and the reservoir water level and rainfall conditions at other stages are set in the way as shown in Fig. 5.

4.2 Numerical Simulation and Discussion

As shown in Fig. 9, the degree of saturation at the bottom of slope increased obviously with the rise of the reservoir water level. While the degree of saturation changed little at the top of the slope. The rainfall increased the degree of saturation at the upper part of slope and after three rainfall stages, the slope was almost full saturated.

Figure 10 shows the comparison between the test and the simulation results. The excess pore water pressure (EPWP) at PP7 occurred earlier than those at PP1 and PP4, which are in good agreement with the test results. In addition, the numerical simulation results can clearly reflect the influence of the boundary conditions on the slope. The EPWP of the measuring points increased obviously under the short heavy rainfall, and the earth pressures decreased at the same time. At the end of short heavy rainfall, the vertical displacement increment was distinct, indicating the occurrence of a landslide. Though the test and simulation results are not completely coincident with each other in its value, some physical quantities, such as EPWP, earth pressure and displacement, are qualitatively in the same changing trends. Therefore, it is reasonable to say that the numerical calculation used in this paper can describe the slope stability rationally.

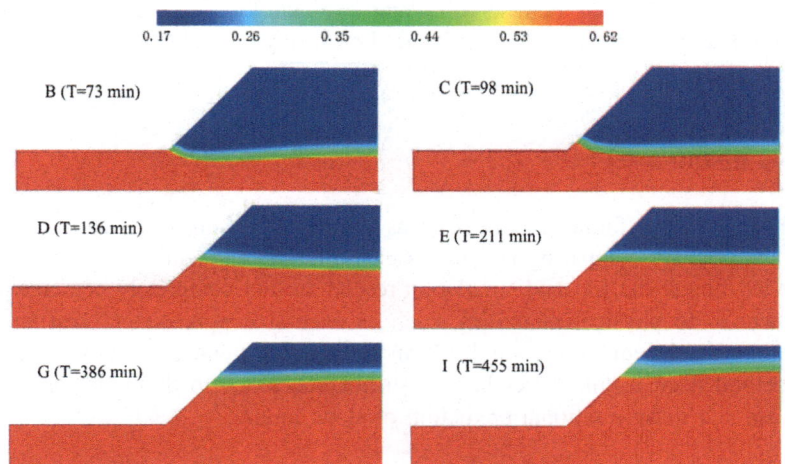

Fig. 9. Simulation results of degree of saturation distributions at different stages.

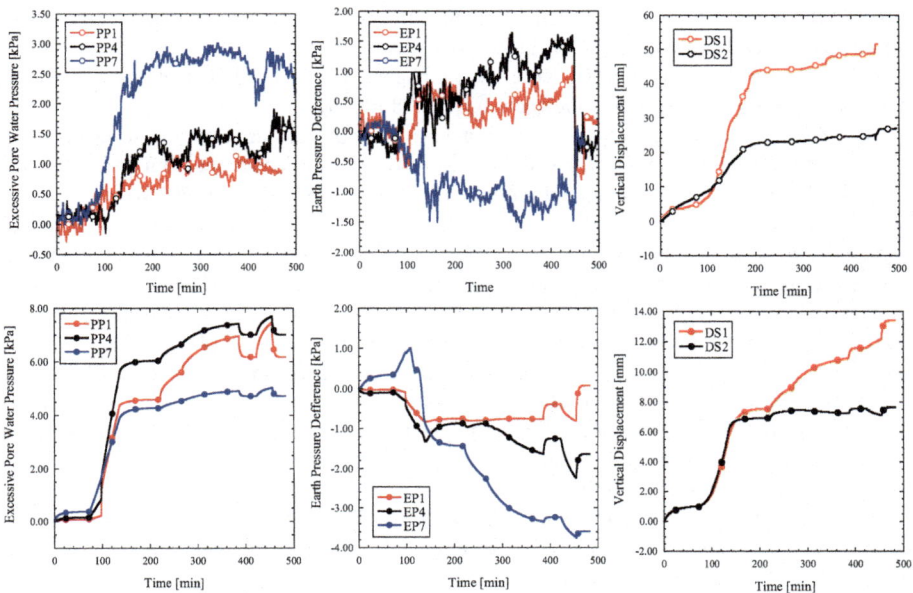

Fig. 10. Comparison between test (upper) and simulation (lower) results.

Figure 10 also shows that the rising of reservoir water level, long weak rainfall and short heavy rainfall have different influence on the slope stability. The increment of reservoir water level (73–98 min) led to the slow increase of EPWP and displacement, while combined with the long weak rainfall (98–136 min), EPWP and displacement increased dramatically. Slope stability was more influenced by the short heavy rainfall than long weak rainfall, since short heavy rainfall caused sudden decrease of earth pressure and increase of vertical displacement in the upper part of the slope, resulting in a shallow landslide. During the whole process, an important reason of slope stability reduction is the increase of the reservoir water level and long weak rainfall, even these do not completely destroy the slope. It is the short heavy rainfall that triggered the final shallow landslide.

5 Conclusion

In this paper, a slope failure model test was simulated by soil-water-air full coupling FE-FD analysis. By comparing the results of the model test and the calculation, it is known that though the test and simulation results are not completely coincident with each other, some physical quantities, such as excessive pore water pressure, earth pressure and displacement, are similar in their changing trends. Therefore, the calculation method used in this paper has satisfactory accuracy to describe the landslide mechanism and make a rational prediction of slope failure.

References

Huang, Q.X., Wang, J.L., Xue, X.: Interpreting the influence of rainfall and reservoir infilling on a landslide. Landslides 13(5), 1139–1149 (2016)

Liang, C., Jaksa, M.B., Ostendorf, B., Kuo, Y.L.: Influence of river level fluctuations and climate on riverbank stability. Comput. Geotech. 63, 83–98 (2015)

Sun, G., Zheng, H., Huang, Y., Li, C.: Parameter inversion and deformation mechanism of Sanmendong landslide in the Three Gorges Reservoir region under the combined effect of reservoir water level fluctuation and rainfall. Eng. Geol. 205, 133–145 (2016)

Wang, J., Su, A., Xiang, W., Yeh, H.F., Xiong, C., Zou, Z., Cheng, Z., Liu, Q.: New data and interpretations of the shallow and deep deformation of Huangtupo No. 1 riverside sliding mass during seasonal rainfall and water level fluctuation. Landslides 13(4), 795–804 (2016)

Xiong, Y., Bao, X., Ye, B., Zhang, F.: Soil-water-air fully coupling finite element analysis of slope failure in unsaturated ground. Soils Found. 54(3), 377–395 (2014)

Zhou, A., Li, C.Q., Huang, J.: Failure analysis of an infinite unsaturated soil slope. Proc. Inst. Civ. Eng.-Geotech. Eng. 169(5), 410–420 (2016)

Zhang, F., Ikariya, T.: A new model for unsaturated soil using skeleton stress and degree of saturation as state variables. Soils Found. 51(1), 67–81 (2011)

Zhang, S., Xiang, L., Wang, L.: Influence of reservoir water level' s fluctuation rate under rainfall on landslide stability. Zhejiang Hydrotech. 42(2), 69–72 (2014)

Zuo, Z.: Investigation on rainfall-induced colluvium landslides using laboratory model tests. PhD thesis, Shanghai Jiao Tong University, China (2013)

Generation of Complex Slope Geometries by DEM for Modeling Landslides: A Case Study of Tangjiashan Landslide

Tao Zhao[1,2(✉)], Giovanni B. Crosta[3], Feng Dai[1], and Nu-wen Xu[1]

[1] State Key Laboratory of Hydraulics and Mountain River Engineering,
College of Water Resource and Hydropower, Sichuan University,
Chengdu 610065, China
zhaotao@scu.edu.cn
[2] Key Laboratory of Geotechnical and Underground Engineering,
Tongji University, Ministry of Education, Shanghai 200092, China
[3] Department of Earth and Environmental Sciences,
Università degli Studi di Milano Bicocca, Piazza della Scienza 4,
20126 Milan, Italy

Abstract. The generation of complex slope geometries by discrete element method (DEM) for modelling landslides is difficult, among other issues, due to the lack of an efficient technique to manipulate and pack a large number of particles within a predefined slope domain. The existing numerical techniques generally use some regular and simple slope geometries, which can lead to some unreliable results. The current study proposes a flexible and simple technique to generate complex slope samples in DEM by feeding the slope domain with rigid and frictionless spherical particles. The slope profile and failure surface are imported from available site investigation data, which can be used to construct the slide boundary in DEM. Based on the geometry of the slide boundary, a virtual hopper is placed atop of the profile, feeding granular particles continuously into the slide bounding space, until it is fully filled. The layered structure of the slope can be attained in this model by generating each sub-layer separately. The slope mass generated by the proposed technique in this study has been used to simulate the failure and subsequent valley damming of Tangjiashan landslide, from which some mechanisms of slope motion and deformation, as controlled by the complex layering geometries, are clearly illustrated. The model is very flexible and requires only minor corrections to fully describe the slope or other complex geometries with realistic stress-strain state in DEM models.

Keywords: Discrete element method · Landslides · Complex slope geometry Sample generation · Layered structure

1 Introduction

Landslides can be associated with an almost instantaneous collapse and spreading, posing a significant hazard to human lives and lifeline facilities worldwide. In addition, slope failures occurring near river valleys with steep bank slopes could potentially

© Springer Nature Singapore Pte Ltd. 2018
A. Farid and H. Chen (Eds.): GSIC 2018, *Proceedings of GeoShanghai 2018*
International Conference: Geoenvironment and Geohazard, pp. 11–19, 2018.
https://doi.org/10.1007/978-981-13-0128-5_2

create landslide dams blocking the river channel. These landslide dams may frequently fail catastrophically, leading to serious downstream inundation and flooding, often with very large social and economic consequences [1]. For instance, rockslides and landslides induced by the 2008 Wenchuan earthquake destroyed many villages and killed more than 20,000 people. These phenomena are under intensive research due to their significant destructive power as well as the still unexplained mechanisms of long travelling distances [2].

For many years, myriads of efforts have been devoted to study the failure of slope mass and the subsequent runout. However, less attention has been paid in analyzing the size and geometry of landslides resulting from given geological and topographical settings which are essentially the characteristics preserved in geologic records [3]. For numerical investigations via the Discrete Element Method (DEM) [4], it is always difficult to generate realistic initial slope geometries due to a lack of highly efficient techniques to manipulate and pack a large number of particles within a predefined complex slope domain. Alternatively, many researchers focus only on the failure and runout of simplified regular granular blocks, attempting to extract some useful information from their numerical analyses. To study the propagation of landslides, the slope mass was commonly generated by gravitational deposition of granular particles within a prismatic box until the desired solid fraction is reached, while the slope sliding surface was represented by rigid planes. Then, the slope mass was released to fall down the slope under gravity [5]. In these studies, the slope size and geometry were significantly simplified, which are sometimes problematic or even misleading. Bonilla-Sierra et al. [6] attempted to generate a complex slope with predefined topography and discontinuity using photogrammetric data and open source DEM platform YADE. In their modeling, the slope mass was generated via cloning a brick of bonded particles to fill the entire model volume. Even though each small granular element can be densely packed, the whole slope mass cannot represent the same stress and strain states as real slope, because the profile of the whole granular packing would change when gravity is applied on each particle during the simulations. Thus, both the slope geometry and its consolidation state should be considered during the generation of slope models.

Since the slope motion and deformation are closely related to the initial slope geometry, it is important to generate realistic slope models in the DEM simulations. In this regard, the current research has focused on proposing an efficient numerical technique for generating complex slope geometries using DEM, aiming to provide new insights into the detailed DEM modeling process.

2   Generation of the Tangjiashan Slide by the DEM

The 2008 Ms 8.0 Wenchuan earthquake triggered lots of catastrophic landslides, distributed along the Longmenshan seismic fault zone within a 300 km long and 10 km wide region in Sichuan Province, China [7] (see Fig. 1(a)). In this region, intense weathering leads to the formation of large unstable weak rock masses and thick deposits covering the mountain slopes. As a result, slopes in this area are very susceptible to mass movement. Among these slides, the Tangjiashan landslide is one of the

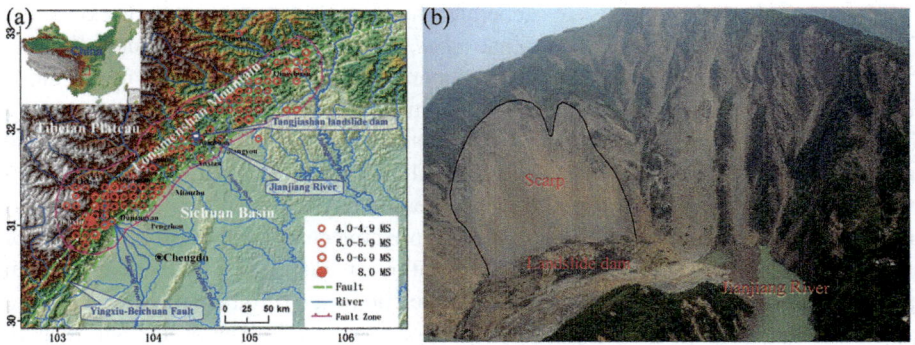

Fig. 1. (a) Location of the seismic zone and Tangjiashan landslide dam (after Chen *et al.* [9]), (b) Aerial view of Tangjiashan landslide dam

largest, with a total displaced mass of approximately 2.04×10^7 m^3 [8]. This landslide occurred immediately after the earthquake and it moved atop of the fragmented bedrock scouring the bank of Jianjiang River for 2400 m and subsequently forming a large landslide dam [7] (see Fig. 1(b)).

According to the site investigation by Xu *et al.* [8], the Tangjiashan landslide dam was approximately 800 m long, 611 m wide and 82 to 124 m high. The landslide dam, as described by Xu et al. (2009), in a first approximation, consists of three layers with a slight increase in grain size with depth. As a back analysis, the initial slope geometry is assumed to be composed of three layers, as shown in Fig. 2.

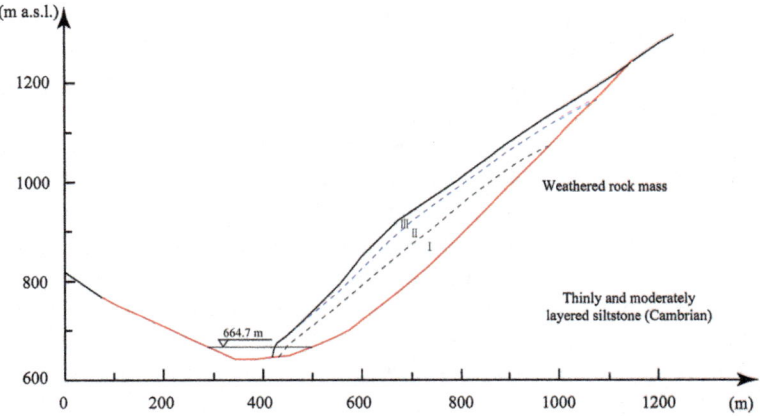

Fig. 2. Profile of the layered slope mass, with layers I: weakly weathered rocks; II: boulders and blocks; III: fragmented rocks with soil. The slope profile is modified after Xu *et al.* [8]

The current study intends to generate the complex slope geometries of the Tangjiashan landslide by the DEM. Zhao *et al.* [10] has described a simple 'hopper discharge technique', by which the spherical particles were allowed to fill up the slope domain. From the Tangjiashan landslide case study, we aim to illustrate how to use a simple but useful numerical technique to generate complex slope models for numerical analyses. As the failed slope mass suffers mainly the translating and partially rotating movements along the failure plane, it can be reasonably assumed that the slope mass probably suffered a contained disintegration, more concentrated at the front and toe of the slide mass. To consider the layered deposition characteristics of the Tangjiashan landslide, the initial slope mass is assumed to be composed of three layers which align approximately parallel to the slope failure surface (see Fig. 2). The profile and relative size of each layer is estimated by the profile of final deposit identified in Xu *et al.* [8]. The contact stiffness of particles located within each layer are set differently, with low stiffness for fragmented rocks and soils, moderate stiffness for the large boulders and blocks and high stiffness for the non-completely disintegrated rock strata.

According to the geometrical data extracted from Fig. 2, the slope profile can be generated as an enclosed space. The slope geometry consists of the upper slope profiles (*e.g.* profiles of layer I, II and III) and the lower slope sliding surface, as represented by smooth and rigid walls in DEM. The performed simulation of Tangjiashan rockslide has a plane strain boundary condition in which the out of plane direction of the model is set as a periodic boundary. In the DEM model, the size of the periodic dimension is set as 22 m, which is at least ten times larger than the size of the effective particle size. In this framework, a number of frictionless particles will be used to fill up this unit periodic cell which can be regarded as one fraction of the real Tangjiashan slope. Any particle with its centroid moving out of the periodic cell through one side of the boundary will be mapped back into the cell domain at a corresponding location on the opposite side of the cell. Particles with only one part of the volume lying outside the cell can interact with particles near the boundary and one image particle will be introduced into the opposite side at a corresponding location, so that it can interact with other particles near the opposite boundary [10].

The 'hopper discharge technique' has been employed in Zhao *et al.* [10] to generate complex slope models containing uniformly distributed granular particles. The generation procedures can be summarized briefly as follows: (1) Create a hole on the upper part of the slope profile, and then, place a large hopper just onto the hole. (2) Continuously generate particles with predefined properties within the hopper. These particles can flow into the slope space through the hopper, and pile up gradually within the slope domain. (3) The generation process stops once the slope domain is filled up with particles. Part of the granular packing in the upper region will be trimmed to get the aimed pre-failure slope geometry.

This 'hopper discharge technique' has two obvious limitations. Firstly, it fails to consider the layered structure of real slopes, and only uniform solid properties can be assigned to the granular sample. As a result, it cannot reproduce correct soil/rock properties of the slope mass. Secondly, the slope mass generated by gravitational deposition would lead to high initial stress concentration near the slope surface region, because of the interaction between the boundary rigid walls and particles. In the current DEM model, this boundary confining stress can be as large as 14 MPa (as measured from the model). However, this boundary stress should not exist in the model, since it is a free surface boundary for a real slope mass. The boundary confining stress would cause the boundary particles to jump out of the slope domain for a sudden removal of the boundary wall. Consequently, the confining wall should be released properly before landslide simulations, so that the slope mass can become as geostatic. Thus, special numerical manipulation should be taken in DEM modeling to avoid the boundary stress concentration problem. The current study will focus on the two above mentioned problem and provide solutions in the following sections.

In the current study, the Tangjiashan slope is assumed to be composed of three major layers, and these layers are generated successively from the bottom to the top region. As shown in Fig. 3(a), a hopper hole was created at the upper part of the layer I profile, which was then connected to a virtual hopper. A particle generator (coded as a C++ module in the DEM program) was placed near the top of the hopper, generating layers of polydispersed frictionless spherical particles with predefined granular properties (in this section, the inter-particle friction is zero, while other properties are set the same as those listed in Table 1). The particle diameters are set randomly within the predefined range, and the generated particles can drop into the slope domain under gravitational forces. In the generation stage, all the solid particles are set frictionless and the width of the hopper hole is at least 10 times larger than the mean particle diameter, such that particle jamming can be avoided during the simulation. In addition, the use of random particle size and zero friction can effectively avoid potential particle size segregation (as discussed by Ketterhagen et al. [11]).

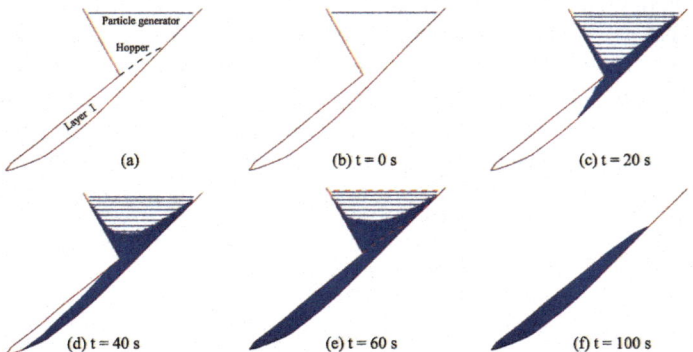

Fig. 3. (a) Configuration of layer I, (b)–(g) generation of the slope sample by 'hopper discharge', and (h) the final trimmed slope sample.

The detailed generation process of the layer I is illustrated by the successive snapshots shown in Fig. 3(b)–(g) (also see the animation in the supplementary material). It can be observed that the granular particles were generated and assigned with initial vertical velocities (5 m/s) within the virtual hopper layer by layer. These particles can drop into the bounding space of layer I through the hopper hole and gradually deposit starting at the slope toe region (t = 40 s). The granular mass slowly piles up

Table 1. Input parameters of the DEM model

Parameters	Value	Parameters	Value
Number of grains, N	100,389	Particle-wall[*] contact stiffness (N/m)	10^9
Particle diameter, D (m)	[0.8, 2]	Inter-particle friction angle, θ (°)	30
Particle density, ρ_s (kg/m^3)	2650	Basal friction angle, θ_b (°)	3
Young's modulus, E (GPa)	1	Viscous damping coefficient, β	0.01
Poisson ratio, υ	0.25	Gravity, g, (m/s^2)	9.81
Bond modulus, E_b (GPa)	1	DEM time step, Δt, (s)	5.0×10^{-5}
Cohesion of bonds, c (MPa)	[1, 4]		

[*]Note a "Wall" is a rigid boundary of the slope profile in DEM.

and fills up the space of slope layer I. The final granular deposit can be obtained at t = 60 s, when the whole space of layer I and part of the hopper volume are filled up with particles. The additional particles fallen into the hopper can be removed, so that the correct slope geometry can be achieved, as shown in Fig. 3(h). After generating the layer I, layer II and III can be constructed similarly atop of layer I, by releasing its boundary and trimming the extra particles for the packed granular assembly. Then, the final Tangjiashan slope mass can be obtained. For each layer, the total number of particles is 16,890, 35,384 and 54,384, respectively.

As discussed before, high initial stresses may concentrate at the upper boundary of the slope mass generated by the above mentioned 'hopper discharge technique'. Thus, it is necessary to release the initial boundary stress before the landslide simulation. During the simulation, the horizontal freedom of the particle motion is fixed, such that all particles can only move upwards/downwards and backwards/forwards when the upper boundary of the slope mass is removed. After removing the upper boundary, the slope mass will expand due to the releasing of boundary stresses, as shown in Fig. 4(b). As a result, the final static deposit is slightly thicker than the aimed slope mass, and the displacement of the boundary particles can be as large as 22.5 m. After releasing the boundary stress, particles above the real slope profile can be trimmed to the real slope

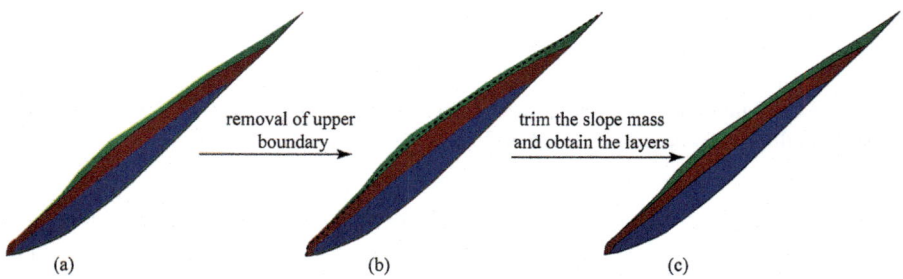

Fig. 4. Expansion of slope mass after the removal of upper boundary (a) layered slope mass from "hopper discharge", (b) stable sample after releasing the boundary stress, and (c) sample with the accurate slope profile and layered structure. The dashed line in (b) represents the geometry of the upper boundary of the slope.

geometry at the end of the simulation. It should be noted that the layered structure of the final slope is only slightly different from the one depicted in Fig. 2. This is due to the combined effects of granular mass compression and boundary stress release. Thus, the layered structure should be re-constructed according to the original slope geometry (see Fig. 4(c)), such that the final slope after stress releasing can match the structure of the real slope.

3 Tangjiashan Slope Failure Modeling

Using the previously generated slope, a simple simulation of the Tangjiashan landslide can be performed, as shown in Fig. 5 (also see the animation in the supplementary material). The input parameters for the slope failure simulations are listed in Table 1, where the inter-particle friction and slope basal friction are set to the aimed values. In the current study, the granular assembly is bonded together to represent the initial integrate slope mass. The values of bonding stiffness for each pair of particles are proportional to their initial centroid distance, while the bonding strength of layer I, II and III are set as 10^6 Pa, 2×10^6 Pa and 4×10^6 Pa, respectively. Thus, the strength of each slope layer can be clearly differentiated. However, the authors are also aware that different values of input parameters can be used in the modeling for a preliminary study. Since the current study only focus on elucidating the detailed sample generation technique and validation, the parametric analyses are ignored. According to Fig. 5, the slope mass moves as a whole along the sliding surface once the landslide is triggered. During the sliding process, the solid materials with a layered structure are translated and partially rotated along the failure surface, as suggested in snapshots at successive time steps. Some cracks occur at the surface of the slope (see Fig. 5(b) and (c)) due to large tensile stresses induced by slope disintegration. The bottom layer is wrapped at the top by the middle layer, especially in the frontal part, moving a short distance. The sliding front region consists of fine and medium sized grains. The granular deposit

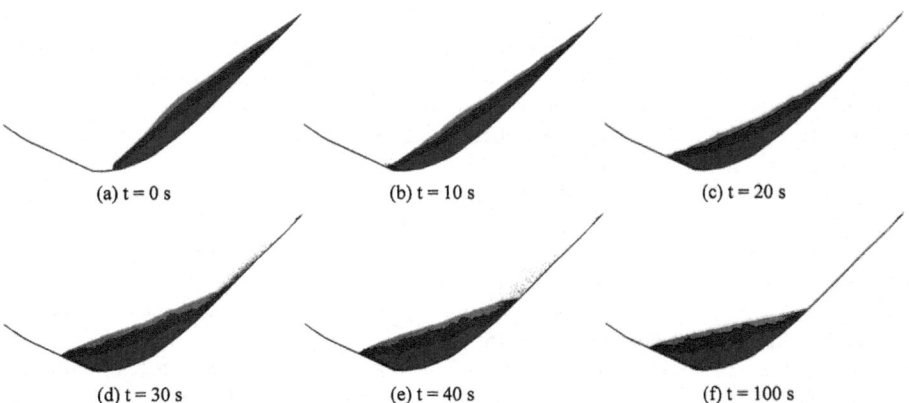

Fig. 5. DEM modeling of the Tangjiashan landslide showing the preserved stratigraphic sequence with the exception of the frontal part of the landslide.

profile changes little after 40 s of sliding, and this sliding duration time can match approximately that reported by eye witness of Tangjiashan landslide (50 s). The final deposit has a relatively steeper inclination angle (11°), compared to the real deposition (10° in Xu *et al.* [8]). The initial stratigraphy is maintained in the final deposit apart for the frontal part where the reversal is partially observed for the occurrence of a sort of conveyor belt movement of the particles from the upper layer towards the front. This motion and the recycling below the front can also be partially controlled by the local geometry at the valley bottom.

4 Conclusions

This paper presents an efficient numerical technique for generating complex slope geometries via the 3D DEM modeling. The slope profile is represented by smooth and rigid walls, while the layered structure of slope mass is reproduced via the 'hopper discharge technique' with boundary stress release and particle trimming for each layer of packed spherical particles. This technique has the advantage of generating realistic complex slope structure without the inverse influence of boundary stress concentration. Since only the slope profile and granular properties are needed in the simulation, this technique is very flexible and efficient to generate slope masses with complex geometries and geological structures. By employing the slope sample generated in this research, the failure and motion of Tangjiashan landslides has been analyzed. Further research on landslide modeling can employ the correct slope failure triggering mechanism, e.g. ground vibration data according to the recorded earthquake wave information.

References

1. Wu, C.-H., Chen, S.-C., Feng, Z.-Y.: Formation, failure, and consequences of the Xiaolin landslide dam, triggered by extreme rainfall from Typhoon Morakot Taiwan. Landslides **11** (3), 357–367 (2014)
2. Legros, F.: The mobility of long-runout landslides. Eng. Geol. **63**(3–4), 301–331 (2002)
3. Katz, O., et al.: Controls on the size and geometry of landslides: insights from discrete element numerical simulations. Geomorphology **220**, 104–113 (2014)
4. Cundall, P.A., Strack, O.D.L.: A discrete numerical model for granular assemblies. Géotechnique **29**(1), 47–65 (1979)
5. Mollon, G., et al.: Discrete modelling of rock avalanches: sensitivity to block and slope geometries. Granul. Matter **17**(5), 645–666 (2015)
6. Bonilla-Sierra, V., et al.: Rock slope stability analysis using photogrammetric data and DFN-DEM modelling. Acta Geotech. **10**(4), 497–511 (2015)
7. Yin, Y., Wang, F., Sun, P.: Landslide hazards triggered by the 2008 Wenchuan earthquake, Sichuan, China. Landslides **6**(2), 139–152 (2009)
8. Xu, Q., et al.: Landslide dams triggered by the Wenchuan Earthquake, Sichuan Province, South West China. Bull. Eng. Geol. Environ. **68**(3), 373–386 (2009)
9. Chen, X.Q., et al.: Emergency response to the Tangjiashan landslide-dammed lake resulting from the 2008 Wenchuan Earthquake, China. Landslides **8**(1), 91–98 (2011)

10. Zhao, T., Utili, S., Crosta, G.B.: Rockslide and impulse wave modelling in the Vajont reservoir by DEM-CFD analyses. Rock Mech. Rock Eng. **49**, 2437–2456 (2015)
11. Ketterhagen, W.R., et al.: Modeling granular segregation in flow from quasi-three-dimensional, wedge-shaped hoppers. Powder Technol. **179**(3), 126–143 (2008)

Investigation of Dry Debris Flow Impact Against a Rigid Barrier via a Discrete Element Approach

Weigang Shen[1], Tao Zhao[1,2(✉)], Feng Dai[1], and Lu Jing[3]

[1] State Key Laboratory of Hydraulics and Mountain River Engineering,
College of Water Resource and Hydropower, Sichuan University,
Chengdu 610065, China
zhaotao@scu.edu.cn
[2] Key Laboratory of Geotechnical and Underground Engineering,
Tongji University, Ministry of Education, Shanghai 200092, China
[3] Department of Civil Engineering, The University of Hong Kong,
Pokfulam Road, Hong Kong, China

Abstract. This study investigates the interaction between dry debris flows and a rigid barrier through numerical flume tests based on a discrete element method. The initial deposition material is modelled as an assembly of spherical particles, and the rigid barrier is modelled as a layer of fixed particles. The numerical model is used to investigate the effect of the flume inclination on the debris-barrier interactions. Based on the numerical results, three interaction mechanisms are identified, namely frontal impact, run up and pile up. The impact force is also detected. In addition, the energy transformation is investigated by analyzing the evolution of the kinetic energy, gravitational potential energy and the energy dissipated by the flume, barrier and inter-particle interactions.

Keywords: Debris flow · Rigid barrier · Discrete element method
Impact force · Energy transformation

1 Introduction

Debris flow is a form of rapid mass movement in which a combination of loose soil, rock, organic matter, air, and water are mobilized to flow downslope [1]. Due to their high mobility and large entrainment solid volume, they can always pose significant hazards to human lives, structures and infrastructures, and lifeline facilities worldwide, threatening populated areas located far away from the slope source. Thus, protective structures, such as rigid barriers, are widely constructed to mitigate such destructive flows.

The design of a rigid barrier requires the knowledge of the maximum impact force of the debris flow. In the literature, the debris-barrier interaction has been addressed by analytical models [2], numerical simulations [3, 4], and laboratory flume tests [5, 6]. Up to now, several semi-empirical methods have been developed to determine the impact force, such as hydrostatic approach, shock wave approach and hydrodynamic approach [7]. These approaches have been widely used in engineering practices. Nevertheless, these available formulas aiming at estimating impact loading stresses have been shown

© Springer Nature Singapore Pte Ltd. 2018
A. Farid and H. Chen (Eds.): GSIC 2018, *Proceedings of GeoShanghai 2018*
International Conference: Geoenvironment and Geohazard, pp. 20–27, 2018.
https://doi.org/10.1007/978-981-13-0128-5_3

to result in high discrepancies [8], and none of them have been acknowledged as a universal formula, notably because these models were obtained in specific impact and boundary conditions that cannot be generalized, and they are usually based on strong assumptions. In addition, these methods do not consider the effect of debris-barrier interaction on the flow velocity and flow thickness. Tiberghien et al. [9] found that the deposited materials behind the barrier due to debris-barrier interaction play an important role in the dynamic load transfer process. Thus, it is worthwhile to highlight the importance of debris-barrier interaction. However, the debris-barrier interaction is still a complicated problem for an effective design of a rigid barrier.

The dynamic interactions between debris flow and rigid barrier are highly complex because it depends on the kinematics of debris flow (in particular solid mass and velocity), mechanical characteristics of soil and geometrical characteristics of rigid wall [5, 10, 11]. The current approaches have limitations in predicting the debris-barrier interactions in the sense that they cannot allow for all the mechanisms occurring in the debris during impact. These mechanisms lead to large deformation, displacement, as well as energy transfer and dissipation, which modify the debris-barrier interaction. These mechanisms and their results may be conveniently addressed using numerical simulations [12]. The energy evolutions can be tracked through numerical simulations, such as Discrete Element Method (DEM), while it is nearly impossible to obtain such information from measurements during experiments. The DEM has been widely used for numerical modeling of avalanches or debris flows [13, 14], and it is an appropriate tool to model debris flows because of the granular nature of these phenomena.

In the present study, a numerical flume model is established using a three-dimensional discrete element method. This model is validated and employed to investigate the debris-barrier interactions. The basic objective is to reveal the detailed mechanism of debris-barrier interaction. The remainder of this paper is organized as follows: in Sect. 2, the details of the numerical flume model are illustrated. Section 3 presents a parametric study conducted to investigate the influence of flume inclination on the debris-barrier interaction. Finally, some conclusions on the capability of the DEM to model debris-barrier interactions are provided.

2 Numerical Model Configuration

Experimental flume tests carried out by Jiang and Towhata [5] have been used to define the geometry and the initial conditions of the numerical model in this study. During the experimental campaign, a cubic granular deposition was released from the top of a flume. The granular mass slid in the flume confined by two side walls. At the end of slide, the granular mass impacted on a rigid barrier which was installed at the bottom end of the flume and perpendicular to the flume base. A sketch of the flume is shown in Fig. 1. The flume is 2.93 m in length, 0.3 m in width, and 0.35 m in height. The granular deposition is an assembly of limestone particles with a length of L = 44 cm, height of H = 15 cm, and width of 0.3 m. The travel distance of the deposition is L_2, which is calculated as the sum of L_1 and half of L.

The numerical flume model established via an open source DEM code ESyS-Particle [15] is illustrated in Fig. 2. In the DEM model, the flume base and the

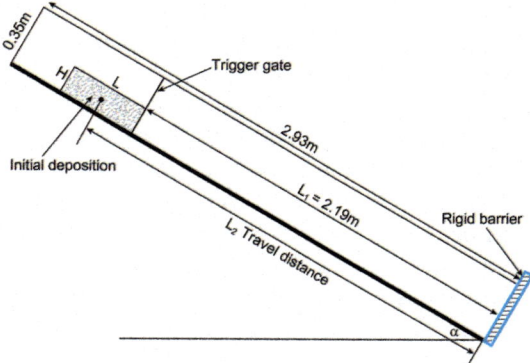

Fig. 1. Sketch of the experimental flume: α is the slope angle.

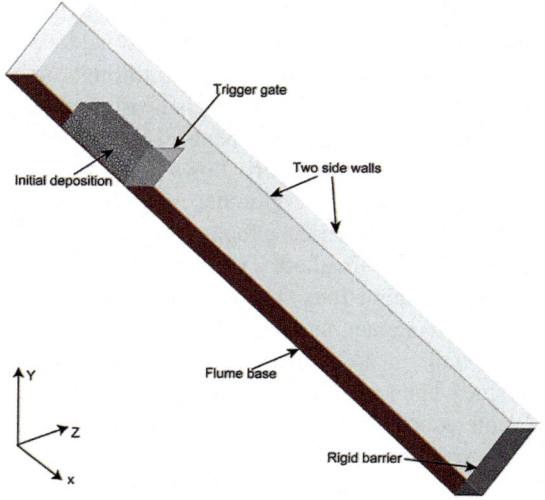

Fig. 2. Numerical model configurations

rigid barrier are respectively represented by a layer of fixed spherical particles with a radius of 0.2 cm, which are used to replicate the friction of the flume base and rigid barrier. The friction of the two side walls are ignored, and the two side walls are represented by two rigid frictionless walls. The trigger gate is also represented by a rigid frictionless wall, which is movable. In all the simulations, the granular flow is initiated by the instantaneous removal of this wall.

The initial deposition is modelled as an aggregate of spherical particles obtained by gravitational deposition. The value of gravitational acceleration adopted in the simulation is 9.81 m/s^2. The initial granular deposition consists of 6, 993 randomly distributed particles, and the ratio of largest to smallest radius equal to 2.5. Due to the constraint of computational time, the particle size distribution (PSD) of the deposition

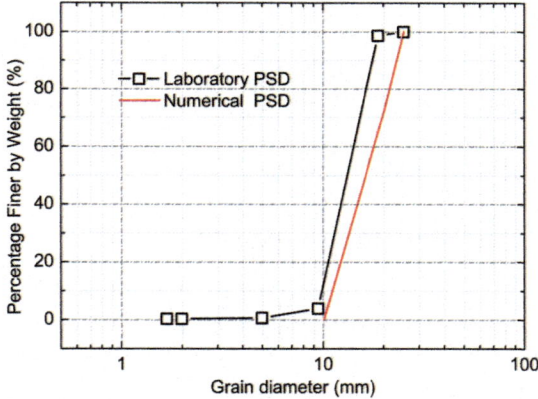

Fig. 3. Particle size distribution of the debris particles, the PSD adopted in the simulations is plotted as the red curve.

used by Jiang and Towhata [5] cannot be reproduced. Therefore, a narrow PSD is adopted in the simulations with sizes ranging from 10 mm to 25 mm (see Fig. 3).

During the granular flow, the particles can contact with other deposited particles, the flume base and the rigid barrier. The same contact model of Cohesiveness Frictional Model (CFM) is used in the three kinds of potential contacts. The corresponding contact model parameters are listed in Table 1. The three friction angles are chosen referring to the work of Jiang and Towhata [5]. All local Young's modulus is chosen to be 10^8 Pa, the mass density of the particles is 2500 kg/m^3, and the Poisson's ratio of the particles is 0.25, which are commonly used in modeling granular medium [4]. The time step size is set as 10^{-6} s in the simulations. The damping of the particles is employed to dissipate a small amount of energy due to elastic wave propagation or particle asperities being sheared off.

Table 1. Contact model parameters used in the simulations

Input parameter	Value
Deposition-deposition contact:	
Local Young's modulus, E (Pa)	10^8
Friction angle, φ (°)	53
Damping coefficient, β	0.05
Deposition-flume base contact:	
Local Young's modulus, E (Pa)	10^8
Friction angle, φ (°)	25
Damping coefficient, β	0.0
Deposition-barrier contact:	
Local Young's modulus, E (Pa)	10^8
Friction angle, φ (°)	21
Damping coefficient, β	0.5

3   Results

The model developed in the previous section is used to numerically investigate the effect of the flume inclination on the debris-barrier interactions. The flume inclined angle is varied from 30° to 45° to simulate different topographic conditions. The rigid barrier is kept perpendicular to the flume base.

3.1   Mechanisms of Granular Flow Impact

Figure 4 shows a side view of the dry granular flow impacting on a rigid barrier ($\alpha = 45°$). The flow direction is from the upper left to the lower right. The attention is focused on the interaction between the granular flow and the barrier. A small static (deposition) zone was retained behind the barrier after the first stage of frontal impact (see Fig. 4(a1) and (b1)). The subsequent flow then ran over the deposited particles to hit the barrier at a higher elevation, and the zone of stationary granular material extended to the upstream direction (see Fig. 4(b2−b4)), hereafter called run-up stage. Gradually, the approaching granular flow cannot run over the deposition zone (see Fig. 4(b5)). Meanwhile, the run-up height of the granular flow on the barrier increased to a maximum value (see Fig. 4a4 and a5), hereafter called pile-up stage. After that, the approaching materials only impacted and piled up on the existing deposits. The development of the deposition zone, the frontal impact, run-up and pile-up mechanisms have also been observed in laboratory studies [6, 8, 10, 16] and in numerical simulations [4]. The formation of deposition zone after frontal impact is acknowledged as the main reason of the energy dissipation in the subsequent granular flow stages [6, 8]. The energy transformation during the deposition process will be discussed in Sect. 3.3.

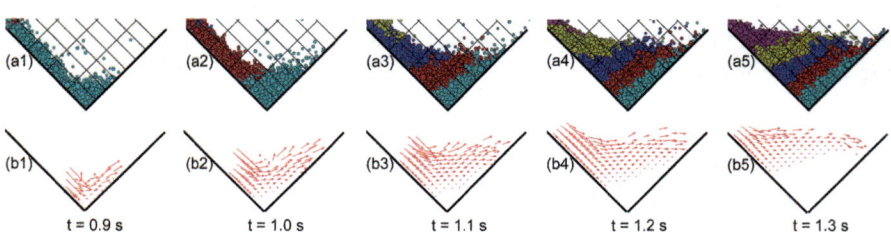

Fig. 4. Observed interaction kinematics for test $\alpha = 45°$: (a1–a5) is the flow profile from t = 0.9 s to t = 1.3 s and (b1–b5) is the corresponding velocity fields.

3.2   Impact Force

Figure 5 shows the measured time history of impact force for different flume inclinations. The approximate time at which the run-up stage transforms into pile-up stage is indicated by colored five-pointed stars (hereafter called "transformation time"). As expected, the steeper the flume inclination is, the higher the flow velocity will be, such that the duration time of the pile-up stage decreases with increasing flume inclination. It is clear that the impact forces have peak values during the impact process, which indicates that the impact has a significant influence on the rigid barrier. In addition, the peak impact force increases rapidly with increasing value of α. As shown in Fig. 5, the time interval between the critical time of peak impact force and the transformation time decreases with the increase of flume inclination angle. Particularly, the peak impact force is reached before the transformation time for α equal to 45°. This result indicates that, for gentle slopes (e.g., α < 45°), the critical time occurs in the pile-up stage, while for steep slopes (e.g., α = 45°), the critical time occurs in the run-up stage.

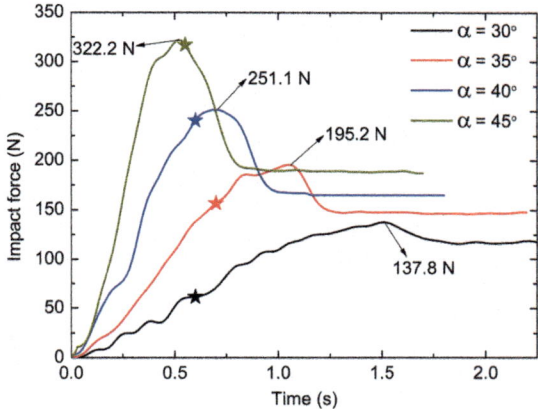

Fig. 5. Measured time histories of impact force at different flume inclinations. The time at which run-up stage transformed into pile-up stage is indicated by colored five-pointed stars.

3.3   Energy Dissipation

The total system energy (E_T) includes gravitational potential energy (E_p), kinetic energy (E_k), strain energy (E_S) and energy loss due to friction (E_F) and local contact damping (E_D). All of these forms of energy can be tracked during simulations to see how the energy is transformed and how the energy transformation is related to the impact mechanism of granular flow. The energy dissipated by the flume base $\left(E_d^f\right)$, rigid barrier $\left(E_d^b\right)$ and inter-particle interaction $\left(E_d^p\right)$ are separately recorded. Figure 6 shows the system energy transformation and energy loss for different flume inclinations. Zero time corresponds to the time when the initial deposition is released. The strain energy (E_S) is not plotted since its value is extremely small (close to nil). As shown in Fig. 6, the granular flow motion involves a cascade of energy that begins with incipient slope movement and ends with deposition. As the granular flow moves downslope, the gravitational potential energy decreases gradually, a portion of the reduced energy transforms into the kinetic energy of the granular material, and the remainder is dissipated by the flume friction, grain contact friction and inelastic collisions. After the granular flow arrives at the barrier, a small part of the system energy is dissipated due to friction and damping of the barrier. With the increase of flume inclination, the granular material run-up height increases and more grains can interact with the barrier (see Fig. 4), hence the energy dissipated by barrier increases from 1.1%

to 2.5%. Meanwhile, the energy dissipated by grain interaction increases from 36.0% to 46.2%. The numerical results indicate that the vast majority of the energy is not dissipated by the barrier. Instead, the debris-barrier interaction facilitates the dissipation of energy in the flow mass, because the debris deposited behind the barrier plays as a cushion layer which absorbs the kinetic energy of the subsequent approaching granular flows. This finding is in agreement with the research outcomes of Ng et al. [6] and Koo et al. [8].

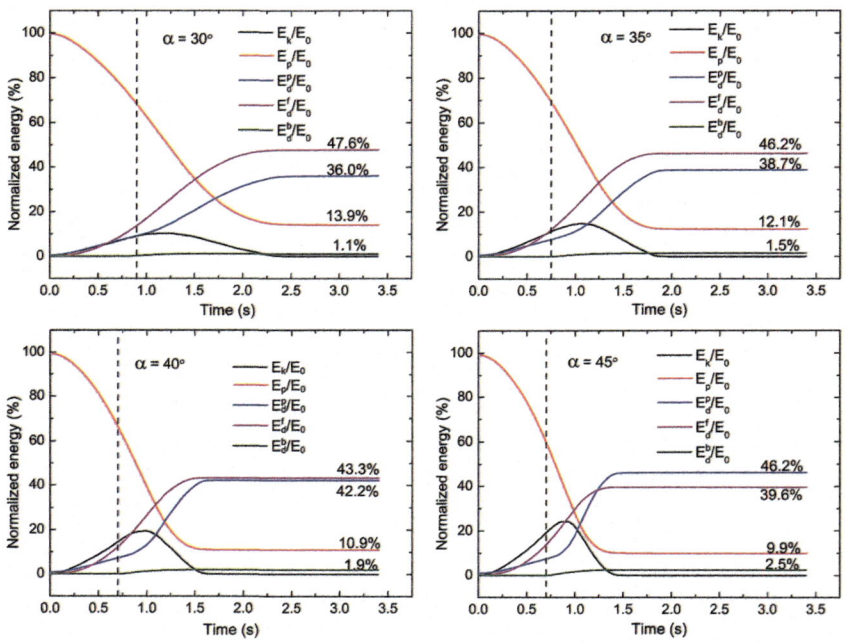

Fig. 6. Energy transformation and energy loss during simulations of various flume inclinations. Vertical dashed lines indicate the arrival time of granular flow at barriers. All of the energy is normalized by the initial gravitational potential energy (E_0).

4   Conclusions

A numerical flume model of the impact of a dry granular flow on a rigid barrier is developed using the open-source code ESyS-Particle. The model is used to investigate the influence of the flume inclination on the debris-barrier interaction. The obtained results clarified the characteristics of flow-barrier interaction.

To mimic various natural slope topographies, simulations for a range of slope angles are performed with the barrier perpendicular to the slope. The side profile of the granular materials impacting on the barrier, the impact force, and energy transformation are recorded and analyzed. Three key interaction mechanisms, namely frontal impact stage, run-up stage and pile-up stage are observed in various testing conditions. The results indicate that the duration of run-up stage is transient which shows negligible dependence on the slope inclination, whereas the duration of pile-up stage decreases with an increase of the slope inclination. For a gentle slope, the peak impact force occurs in the pile-up stage, nevertheless for a steep slope, that occurs in the run-up stage. As expected, the maximum impact force increases as the slope inclination increases. In addition, the analysis of energy transformation shows that the kinetic energy of a granular flow is mainly dissipated by the inter-particle interactions but not by the barrier.

References

1. Cruden, D.M., Varnes, D.J.: Landslide types and processes. Spec. Rep. – Natl. Res. Counc. Transp. Res. Board **247**, 20–47 (1958)
2. Mancarella, D., Hungr, O.: Analysis of run-up of granular avalanches against steep, adverse slopes and protective barriers. Can. Geotech. J. **47**(8), 827–841 (2010)
3. Moriguchi, S., Borja, R.I., Yashima, A., Sawada, K.: Estimating the impact force generated by granular flow on a rigid obstruction. Acta Geotech. **4**(1), 57–71 (2009)
4. Law, P.H.: Computational Study of Granular Debris Flow Impact on Rigid Barriers and Baffles. The Hong Kong University of Science and Technology, Hong Kong (2015)
5. Jiang, Y.J., Towhata, I.: Experimental study of dry granular flow and impact behavior against a rigid retaining wall. Rock Mech. Rock Eng. **46**(4), 713–729 (2013)
6. Ng, C.W.W., Song, D., Choi, C.E., Liu, L.H.D., Kwan, J.S.H., Koo, R.C.H., et al.: Impact mechanisms of granular and viscous flows on rigid and flexible barriers. Can. Geotech. J. **54** (2), 188–206 (2016)
7. Proske, D., Suda, J., Hübl, J.: Debris flow impact estimation for breakers. Georisk: Assessment and Management of Risk for Engineered Systems and Geohazards, p. 143–155 (2011)
8. Koo, R.C.H., Kwan, J.S.H., Ng, C.W.W., Lam, C., Choi, C.E., Song, D., et al.: Velocity attenuation of debris flows and a new momentum-based load model for rigid barriers. Landslides **14**, 1–13 (2016)
9. Tiberghien, D., Laigle, D., Naaim, M., Thibert, E., Ousset, F.: Experimental investigations of interaction between mudflow and an obstacle. Debris Flow Hazard s Mitigation: Mechanics, Prediction, and Assessment; Chengdu, p. 681–687 (2007)
10. Choi, C.E., Au-Yeung, S.C.H., Ng, C.W.W., Song, D.: Flume investigation of landslide granular debris and water runup mechanisms. Geotechn. Lett. **5**(1), 28–32 (2015)
11. Song, D., Ng, C.W.W., Choi, C., Zhou, G.D., Kwan, J.S.H., Koo, R.C.H.: Influence of debris flow solid fraction on rigid barrier impact. Can. Geotech. J. **5**(2), 1421–1434 (2016)
12. Morenoatanasio, R.: Energy dissipation in agglomerates during normal impact. Powder Technol. **223**, 12–18 (2012)
13. Zhao, T., Utili, S., Crosta, G.B.: Rockslide and impulse wave modelling in the vajont reservoir by DEM-CFD analyses. Rock Mech. Rock Eng. **49**(6), 2437–2456 (2015)
14. Zhao, T., Dai, F., Xu, N.: Coupled DEM-CFD investigation on the formation of landslide dams in narrow rivers. Landslides **14**, 1–13 (2016)
15. Wang, Y., Mora, P.: The ESyS_particle: a new 3-D discrete element model with single particle rotation. In: Advances in Geocomputing, pp. 183–228. Springer, Heidelberg (2009)
16. Ng, C.W.W., Choi, C.E., Liu, L.H.D., Wang, Y., Song, D., Yang, N.: Influence of particle size on the mechanism of dry granular run-up on a rigid barrier. Geotechn. Lett. **7**(1), 79–89 (2017)

A Simple THC Coupled Model for Assessing Stability of a Submarine Infinite Slope with Methane Hydrates

Lin Tan[1,2], Fang Liu[1,2(✉)], Giovanni Crosta[3], Paolo Frattini[3], and Mingjing Jiang[1,2]

[1] State Key Laboratory of Disaster Reduction in Civil Engineering, Tongji University, Shanghai 200092, China
liufang@tongji.edu.cn
[2] Key Laboratory of Geotechnical and Underground Engineering (Tongji University), Ministry of Education, Shanghai 200092, China
[3] Department of Earth and Environmental Science, University Milano, Bicocca, Italy

Abstract. Methane hydrate (MH) is an ice-like clathrate compound of methane molecules trapped in cages of water molecules under high-pressure and low-temperature. In shallow marine region, MHs tend to dissociate due to seafloor temperature rise and massive submarine slope failure could be triggered or primed as a result of excess pore pressure buildup. This study develops a simple thermo-hydro-chemically (THC) coupled model to quantify the excess pore pressure buildup due to hydrate dissociation, and incorporates this model into the limit equilibrium method in order to analyze the stability of an idealized infinite slope embedding a hydrate layer. The results show that the presence of the overburden layer above the hydrate-bearing layer plays two opposite roles on the stability of the slope. It serves as a barrier that hampers excess pore pressure dissipation and therefore endangers the slope stability. Meanwhile it has beneficial effects by providing overburden pressure that mobilizes additional shear resistance in the slope. For the circumstances under consideration, the potential failure surface of the slope is constrained within a narrow band at to the top of the hydrate layer.

Keywords: Submarine landslide · Methane hydrate dissociation
Thermo-hydro-chemically coupled model · Excess pore pressure

1 Introduction

Methane hydrate (MH) is an ice-like crystalline compound widely spreading in the sediments on marine continental margins where low temperature and high pressure are present. Natural or anthropologic perturbations could cause massive hydrate dissociation that releases a large amount of methane gas, thereby triggering or priming seafloor instability due to pore pressure buildup in addition to reduction in shear strength of the sediments. The connection of oceanic hydrates dissociation and submarine landslides has been a scientific concern that attracts increasing attentions.

© Springer Nature Singapore Pte Ltd. 2018
A. Farid and H. Chen (Eds.): GSIC 2018, *Proceedings of GeoShanghai 2018*
International Conference: Geoenvironment and Geohazard, pp. 28–36, 2018.
https://doi.org/10.1007/978-981-13-0128-5_4

An accurate quantification of expected excess pore pressure levels is necessary to properly predict the occurrence and position of the failure surface of submarine landslides triggered or primed by hydrate dissociation. Different models have been proposed for this purpose [1–3]. For instance, Xu and Germanovich [1] related the volume expansion during hydrate dissociation to density difference and compressibility of the system, and theoretically formulated the excess pore pressure for confined or interconnected pore space. Nixon and Grozic [2] quantified the excess pore pressure assuming undrained conditions for a given amount of dissociating hydrate, and proposed a conservative approach for assessing seafloor instability by neglecting pressure diffusion during progressive hydrate dissociation. Kwon and Cho [3] established a sequentially coupled model that decouples hydrate dissociation (with pressure buildup) from consolidation (with pressure diffusion) in each time step. Since oceanic hydrate reservoirs are usually seated in marine sediments with more or less interconnected pore space, as addressed by Xu and Germanovich [1], the magnitude of excess pore pressure primarily depends on dissociation rate and pressure diffusion rate (controlled by the sediment permeability) at similar or miscellaneous time scales. Meanwhile, the excess pore pressure impacts the pressure-dependent dissociation process and in turn affects pore connectivity. It is still challenging and however desired to quantify excess pore pressure associated with hydrate dissociation under different submarine settings where multi-physics coupling process can be considered.

This paper develops a simple thermo-hydro-chemically (THC) coupled model to quantify evolving excess pore pressure during hydrate dissociation by considering thermodynamic chemical action, heat transfer, pressure diffusion, and their interplay at different time scales. The model is incorporated into limit equilibrium analysis in order to assess the stability of an idealized submarine slope with a hydrate layer under rising temperature.

2 A Theoretical Model

Figure 1 illustrates a simplified infinite submarine slope with a MH-bearing layer. The stability of the slope can be quantified within the framework of the limit equilibrium method using a safety factor computed from:

$$F_s = \frac{c}{\gamma' H \sin \beta \cos \beta} + (1 - \frac{u_e}{\gamma' H}) \frac{\tan \varphi}{\tan \beta} \tag{1}$$

where H is the depth of the potential slip surface below the seafloor; c and φ are the cohesion and the friction angle of the soil at the potential slip surface, respectively; γ' is the submerged unit weight of the soil; β is the slope angle; and u_e is the excess pore pressure:

$$u_e = P - P_{static} \tag{2}$$

where P_{static} is the hydrostatic pressure; and P is the total pore pressure weighted by the pore water pressure P_w and pore gas pressure P_g [4]:

$$P = \frac{S_w}{S_w + S_g} P_w + \frac{S_g}{S_w + S_g} P_g \qquad (3)$$

where S_w and S_g are water and gas saturations, respectively.

This slope could become unstable (i.e. $F_s < 1$) because of hydrate dissociation. For simplicity, we assume that the cohesion is correlated with hydrate saturation through a linear relationship once MHs dissociate and focus on the effect of excess pore pressure on stability of the slope. To quantify the excess pore water pressure, we develop a THC coupled model to be elaborated on below.

In general, a mass of MH-bearing sediment can be viewed as a porous system composed of a soil skeleton and three components (i.e. water, methane and hydrate) in pores of the skeleton. For simplification, we assume: (1) temperature is the same cross all phases at the same locality; that is, thermal equilibrium between phases is assumed; (2) each phase contains only one single component, that is, water, methane and hydrate are present as liquid, gas, and solid phase, respectively; (3) hydrates are immobile in pores and only dissociate in the same locality; (4) pore water and hydrate are incompressible; (5) soil skeleton does not deform.

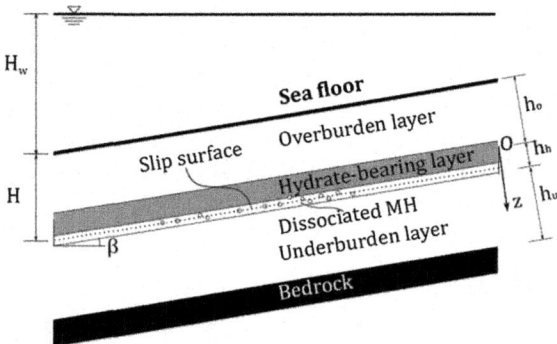

Fig. 1. Schematic illustration of an infinite submarine slope with a MH bearing layer

Under one-dimensional conditions as illustrated in Fig. 1, the mass and energy conservation equations of the slope are written as:

$$\frac{\partial}{\partial t}(\phi S_\alpha \rho_\alpha) = -\frac{\partial q_\alpha}{\partial z} + \frac{\partial m_\alpha^d}{\partial t}, \quad \alpha \equiv w, g, h \qquad (4)$$

$$\left[(1-\phi)\rho_s c_s + \sum_{\alpha \equiv w,g,h}(\phi S_a \rho_\alpha c_\alpha)\right]\frac{\partial \theta}{\partial t} = -\frac{\partial h}{\partial z} - \frac{\partial}{\partial z}\sum_{\alpha \equiv w,g}(q_\alpha c_\alpha \theta) + \frac{\partial Q_h^d}{\partial t} \qquad (5)$$

where ϕ is the porosity; S_α, ρ_α, q_α and c_α are the saturation, density, mass flux and specific heat capacity of the component α (i.e., w: water; g: methane gas; h: hydrate), respectively; θ and h are temperature and heat flux, respectively; the subscript

s represents soil; $\partial m_\alpha^d / \partial t$ is the mass rate of component α per unit soil volume; and $\partial Q_h^d / \partial t$ is the dissociation heat rate.

By introducing Darcy's law, Eq. (4) is re-written as

$$\frac{\partial}{\partial t}(\phi S_\alpha \rho_\alpha) = -\frac{\partial}{\partial z}\left[-\frac{k k_{r\alpha}\rho_\alpha}{\mu_\alpha}\left(\frac{\partial P_\alpha}{\partial z} - \rho_\alpha g \cos \beta\right)\right] + \frac{\partial m_\alpha^d}{\partial t}, \quad \alpha \equiv w, g, h \quad (6)$$

where $k_{r\alpha}$ and μ_α are the relative permeability and viscosity of component α respectively. Note that k_{rh} is zero since hydrate is immobile and the relative permeability k_{rw} and k_{rg} depend on water and gas saturations (see Table 1).

By introducing Fourier's law, Eq. (5) is re-written as

$$\begin{aligned}
&\left[(1 - \phi)\rho_s c_s + \sum_{\alpha \equiv w, g, h}(\phi S_\alpha \rho_\alpha c_\alpha)\right]\frac{\partial \theta}{\partial t} = -\frac{\partial}{\partial z}\left(-\lambda_\Theta \frac{\partial \theta}{\partial z}\right) \\
&- \frac{\partial}{\partial z}\sum_{\alpha \equiv w, g}\left[-\frac{k k_{r\alpha}\rho_\alpha}{\mu_\alpha}\left(\frac{\partial P_\alpha}{\partial z} - \rho_\alpha g \cos \beta\right)c_\alpha \theta\right] + \frac{\partial Q_h^d}{\partial t}
\end{aligned} \quad (7)$$

where λ_Θ is the composite thermal conductivity of the system, of which the calculating model is given in Table 1.

The mass rate in Eq. (4) is computed as:

$$\frac{\partial m_w^d}{\partial t} = -\eta M_w \chi; \quad \frac{\partial m_g^d}{\partial t} = -M_g \chi; \quad \frac{\partial m_h^d}{\partial t} = M_h \chi \quad (8)$$

where M_w, M_g and M_h are the molar masses of water, methane and hydrate, respectively; η represents the hydrate number; χ (mol/s/m^3) is the reaction rate given by Kim-Bishnoi model [5]:

$$\chi = -k_0 \exp(-\frac{\Delta E_a}{R\theta})A_s(P_e - P_g) \quad (9)$$

where k_0 is the intrinsic dissociation constant; ΔE_a is the activation energy of hydrate dissociation; R is the universal gas constant; A_s is the hydrate action area; P_g is pressure of gas phase; P_e is the equilibrium pressure at temperature θ, given by Moridis's model [6].

The dissociation heat rate in Eq. (5) is computed as:

$$\frac{\partial Q_h^d}{\partial t} = \chi \Delta H \quad (10)$$

where ΔH is the latent heat, given by the Kamath equation [7]:

$$\begin{aligned}
&\Delta H = A + \frac{B}{\theta} \\
&\begin{cases} A = 56552, \ B = -16.8137; & 273.15 < \theta \le 298.15 \\ A = 27329, \ B = -48.2748; & 248.151 \le \theta < 273.15 \end{cases}
\end{aligned} \quad (11)$$

In summary, Eqs. (4) and (5) form the governing equations of the system that involve four primary variables: gas pressure P_g, temperature θ, water saturation S_w and gas saturation S_g. These equations are closed by the following auxiliary equations:

$$S_w + S_g + S_h = 1 \tag{12}$$

$$P_g = P_w + P_c \tag{13}$$

where P_c is the capillary pressure to be determined by the van Genuchten function of S_w and S_g [8], as shown in Table 1.

No analytical solution is available for Eqs. (6) and (7). Instead, a numerical solution was coded in MATLAB by linearizing the governing equations with the implicit finite-difference method. The central and forward difference approximation was used for the spatial and time derivatives, respectively. Then the linearized difference equation set was solved with the Newton-Raphson iteration method. The proposed model and the numerical implementation have been validated via experimental data [9] and have not been included in this paper due to length limit.

3   A Case Study of an Overburden-Free Slope T

The proposed model is applied to a case study addressed by Reagan and Moridis [10] that represents a scenario of gentle submarine slopes with a shallow hydrate-bearing layer (hydrate layer for short here). Below a water depth of 570 m, the hydrate layer extends vertically from the seafloor to a depth of 16 m, and the initial hydrate saturation is 3%. Initially, the temperature is 6 °C at the seafloor and linearly increases with depth at a geothermal gradient of 28 °C/km. It is assumed that a temperature rise occurs at the seafloor at a rate of 0.03 °C/year due to climate change. This triggers thermal dissociation in the hydrate layer.

To verify our model and code, we adopted the boundary and initial conditions, and parameters as the same as those in [10]. Note that salt and methane dissolution in water are ignored in our simulation. Table 1 lists the major parameters.

In this setting-up, the dissociation front propagates from the top to the bottom of the hydrate layer. Figures 2a and b compare our results with those given by Reagan and Moridis. Our simulation agrees with the published data except that the dissociation lags for about 60 years. This lag is due to the fact that the effect of salinity on the phase equilibrium, considered by Reagan and Moridis, is ignored in our simulation. Under a pressure of 5.7 MPa, the equilibrium temperature for MH in pure water is 1.7 °C higher than that in sea water with a salinity of 0.035 (considered by Reagan and Moridis). Given a temperature rising rate of 0.03 °C/year, hydrate dissociate 60 years later in our simulation than that provided by Reagan and Moridis. Nevertheless, the consistent profiles of hydrate and gas saturation as illustrated in Figs. 2a and b verified our numerical code.

Figure 2c plots the excess pore pressure against time obtained from our simulation. The magnitude of the induced excess pore pressure is in the order of several kilopascals, and is insufficient to cause slope instability (assuming a gentle slope with an inclination angle of several degrees) because of low hydrate saturation and absence of a relatively impermeable overburden above the hydrate layer. The next section will show more vulnerable cases with an overburden.

Table 1. Parameters of the hydrate layer

Parameters	Value
Absolute permeability k (m^2)	1×10^{-15}
Porosity ϕ (%)	30
Initial hydrate saturation S_{h0} (%)	3
Initial gas saturation S_{g0} (%)	0
Composite thermal conductivity model [11]	$\lambda_\Theta = \left(\sqrt{S_h} + \sqrt{S_w}\right)(\lambda_{sw} - \lambda_{sd}) + \lambda_{sd}$
Dry thermal conductivity λ_{sd} (W/m/K)	1.0
Wet thermal conductivity λ_{sw} (W/m/K)	3.3
Capillary pressure model [8]	$P_c = P_0\left[(S*)^{-1/\xi} - 1\right]^{\xi}$ $S^* = \dfrac{S_r - S_{irw}}{1 - S_{irw}};\ S_r = \dfrac{S_w}{S_w + S_g}$
S_{irw}	0.19
P_0 (Pa)	2000
ξ	0.45
Relative permeability model [12]	$k_{r\alpha} = \left(\dfrac{S_\alpha - S_{ir\alpha}}{1 - S_{ir\alpha}}\right)^n,\quad \alpha \equiv w, g$
n	4
S_{irg}	0.02
S_{irw}	0.20

4 Effect of Overburden Layer

Here we consider the effect of an overburden layer with a thickness ranging from 0 to 80 m. The permeability of the overburden and underburden layers is 1×10^{-17} m^2. The thickness of the hydrate layer is 20 m with a hydrate saturation of 30%. The water depth from the sea level to the top of the hydrate layer is constant at 570 m. The temperature at the bottom of hydrate layer is constant at 8 °C. The slope angle is assumed to be 3°, and the cohesion and the internal frictional angle of all the sediments are 200 kPa and 30°, respectively. The other parameters remain the same as the overburden-free case in the preceding section.

Figure 3 illustrates the profile of S_h, u_e, and F_s against the time elapsed from the first onset of hydrate dissociation under different overburden thicknesses. As shown in Fig. 3a, as the dissociation front propagates from the top to the bottom of the hydrate layer, u_e continuously builds up and F_s decreases. Meanwhile, we assume the cohesion decreases linearly from 200 kPa to zero with the MHs dissociation. As a result, a peak develops on the profile of F_s near the dissociation front (see Fig. 3). The failure onset time refers to the time when F_s first reaches one at a specific depth, where the failure surface is recognized.

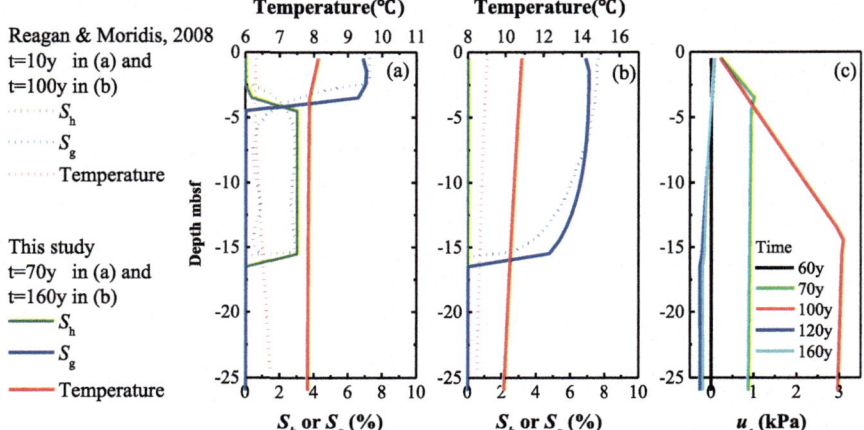

Fig. 2. Profiles of hydrate saturation, gas saturation, and temperature in comparison with results by Reagan and Moridis [10] at different times (a–b), and profile of the excess pore pressure evolution (c).

Figure 4 shows the failure onset time and failure surface position against different overburden thicknesses h_o. The presence of the overburden plays two different roles. On one hand, the overburden serves as a barrier that hampers dissipation of the excess pore pressure accumulating in the hydrate layer during dissociation [1]. This tends to endanger the slope. Even though hydrate dissociation is faster at 5 m thick overburden

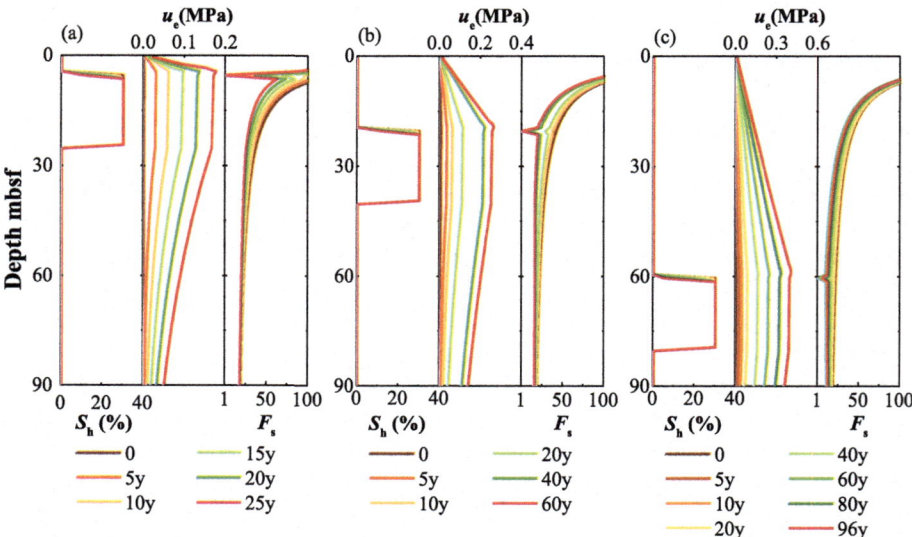

Fig. 3. Profiles of hydrate saturation, excess pore pressure, safety factor at different time periods under a overburden thickness of (a) 5 m; (b) 20 m; and (c) 60 m. The elapsed time starts from the onset of hydrate dissociation at the top of the hydrate layer.

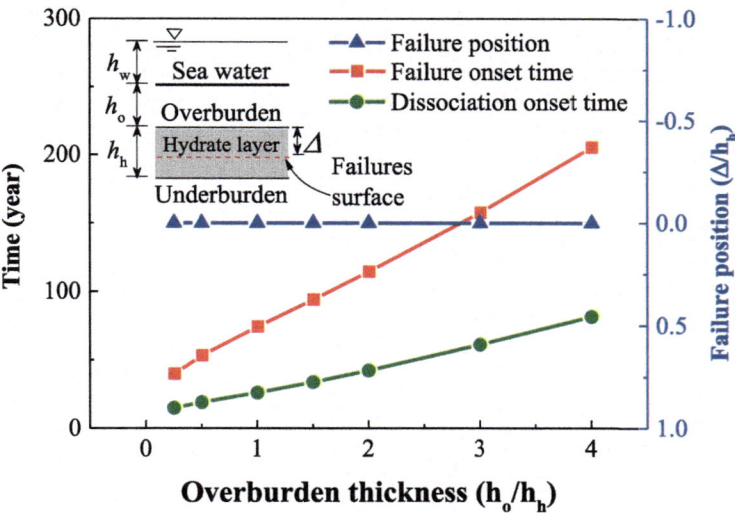

Fig. 4. Failure onset time and failure surface depth against different overburden thicknesses.

(Fig. 3a) than that at 20 m (Fig. 3b), the pressure buildup is more remarkable for the same amount of dissociated hydrate under the thicker overburden, because the pressure dissipates more slowly. On the other hand, the overburden layer provides overburden pressure that mobilizes additional shear resistance at the potential failure surface [13]. This is beneficial to stabilize the slope. The second mechanism seems dominant in the circumstance under consideration. Figure 4 also provides the position of the failure surface against different overburden thickness. In general, the failure surface is constrained within a narrow band at the top of the hydrate layer. This weak zone could move upwards in the overburden layer if this layer is less cohesive.

5   Conclusions

This paper presents a simple THC coupled model for quantifying the excess pore pressure caused by MH dissociation, and incorporates this model to the stability analysis of an idealized submarine slope with a MH-bearing layer. Through an overburden-free case study, the proposed model is demonstrated to be able to capture the multi-physics coupled processes involved in MH dissociation in porous sediments. The stability analysis of the gentle slope with a relatively impervious overburden layer shows that the presence of the overburden plays two opposite roles in impacting the stability of the slope. It serves as a barrier that hampers excess pore pressure dissipation and therefore endangers the slope. Meanwhile it has beneficial effects by providing overburden pressure that mobilizes additional shear resistance in the slope. For the circumstances under consideration, the potential failure surface of the slope is constrained within a narrow band at the top of the hydrate layer. This could be affected by the strength properties of the overburden layer, which will be further investigated in our ongoing work.

Acknowledgement. The work is supported by the Chinese National Natural Science Foundation (with grant No. 41572267, 51639008, and 51239010), and the Fundamental Research Funds for the Central Universities. The first author would like to appreciate the International Exchange Program (with grant No. 2018XKJC-015) funded by the Graduate School of Tongji University that enables the collaboration with the University of Milano-Bicocca.

References

1. Xu, W.Y., Germanovich, L.N.: Excess pore pressure resulting from methane hydrate dissociation in marine sediments: a theoretical approach. J. Geophys. Res. Solid Earth **111**(B1), B01104 (2006)
2. Nixon, M.F., Grozic, J.L.H.: A simple model for submarine slope stability analysis with gas hydrates. Norw. J. Geol. **86**(3), 309–316 (2006)
3. Kwon, T.H., Cho, G.C.: Submarine slope failure primed and triggered by bottom water warming in oceanic hydrate-bearing deposits. Energies **5**(8), 2849–2873 (2012)
4. Kimoto, S., Oka, F., Fushita, T.: A chemo–thermo–mechanically coupled analysis of ground deformation induced by gas hydrate dissociation. Int. J. Mech. Sci. **52**(2), 365–376 (2010)
5. Kim, H.C., Bishnoi, P.R., Heidemann, R.A., Rizvi, S.S.H.: Kinetics of methane hydrate decomposition. Chem. Eng. Sci. **42**(7), 1645–1653 (1987)
6. Moridis, G.J.: TOUGH + v1.5: A Code for the Simulation of System Behavior in Hydrate-Bearing Geologic Media. Lawrence Berkeley National Laboratory (2014)
7. Kamath, V.A., Holder, G.D.: Dissociation heat transfer characteristics of methane hydrates. AIChE J. **33**(2), 347–350 (1987)
8. Van Genuchten, M.T.: A closed-form equation for predicting the hydraulic conductivity of unsaturated soils. Soil Sci. Soc. Am. J. **44**(5), 892–898 (1980)
9. Tang, L.G., Li, X.S., Feng, Z.P., Li, G., Fan, S.S.: Control mechanisms for gas hydrate production by depressurization in different scale hydrate reservoirs. Energy Fuels **21**(1), 227–233 (2007)
10. Reagan, M.T., Moridis, G.J.: Dynamic response of oceanic hydrate deposits to ocean temperature change. J. Geophys. Res. **113**(C12), C12023 (2008)
11. Moridis, G.J., Seol Y., Kneafsey, T.J.: Studies of Reaction Kinetics of Methane Hydrate Dissocation in Porous Media. Lawrence Berkeley National Laboratory (2005)
12. Stone, H.L.: Probability model for estimating three-phase relative permeability. J. Pet. Technol. **22**(2), 214–218 (1970)
13. Crutchley, G.J., Mountjoy, J.J., Pecher, I.A., Gorman, A.R., Henrys, S.A.: Submarine slope instabilities coincident with shallow gas hydrate systems: insights from New Zealand examples. In: Lamarche, G., et al. (eds.) Submarine Mass Movements and their Consequences. Advances in Natural and Technological Hazards Research, vol. 41, pp. 401–409. Springer, Cham (2016)

Managing Information of Gas Hydrate Reservoirs of the South China Sea in an Integrated GIS Database

Fang Liu[1,2(✉)], Yuxuan Che[3], Shixin Xu[3], Qiushi Chen[4],
and Lin Tan[3]

[1] Guangxi Key Laboratory of Geomechanics
and Geotechnical Engineering, Guilin, China
liufang@tongji.edu.cn
[2] State Key Laboratory of Disaster Reduction in Civil Engineering,
Tongji University, Shanghai 200092, China
[3] Key Laboratory of Geotechnical and Underground Engineering
(Tongji University), Ministry of Education, Shanghai 200092, China
[4] Glenn Department of Civil Engineering,
Clemson University, Clemson, SC 29634, USA

Abstract. The seabed of the South China Sea (SCS) is a vast reserve of oceanic gas hydrates, a potential future energy attracting global interest. Gas production from oceanic hydrates poses a significant concern of impending disasters of massive seabed instability. Data inventory of hydrate reservoirs becomes essential to optimize hydrate production strategies and to minimize potential risks. To facilitate the data mining for future risk assessment of gas production in SCS, this study compiles a GIS-driven database for managing and analyzing information of gas hydrate reservoirs in SCS. The information, collected by different agents with various techniques at various scales and accuracies, is overlaid and geo-referenced. The possible bounds of hydrate reservoir are roughly deduced at the regional scale from the data of water depth and the phase equilibrium equation of hydrate. The presented GIS-driven database is adaptable for data accumulating over time. This study is a preliminary and significant step towards a regional risk analysis of hydrate production in SCS.

Keywords: Methane hydrate reservoir · Information management
Spatial database · GIS · South China Sea

1 Introduction

Methane hydrate (MH) is ice-like crystalline compound extensively found in the sediments of continental margins and permafrost regions at low temperature and high pressure. Referred to as "combustible ice", MHs are viewed as potential sources of future energy with a global reserve about twice as much as that of conventional fossil fuels combined [1, 2]. However, utilization of oceanic MHs remains debatable due to high risk, since extracting MHs from marine sediments may prime seafloor instability and trigger severe geohazards such as massive submarine landslides among others.

© Springer Nature Singapore Pte Ltd. 2018
A. Farid and H. Chen (Eds.): GSIC 2018, *Proceedings of GeoShanghai 2018*
International Conference: Geoenvironment and Geohazard, pp. 37–44, 2018.
https://doi.org/10.1007/978-981-13-0128-5_5

Assessment of geohazards associated with MH production is of importance for the decision-makers. To serve this goal, information of MH reservoirs (e.g., geometric boundaries, geotechnical and geochemical characteristics of the sediments, bathymetric map etc.) is indispensable.

The South China Sea (SCS) is recognized as the most important reserve of oceanic MHs in China with favorable settings for MHs formation. Intensive efforts have been made for characterizing the hydrate reservoirs. The collected data are of great potential to engine regional-scale risk assessments associated with MH production. However, these data are partially accessible via research reports and articles. Extreme efforts are required to dig out useful information in order to accommodate a specific scientific research. To ease the procedure for data mining, this study initiates an integrated database in a geographic information system (GIS) for systematically compiling and managing data of MH reservoirs in SCS. This work aims to enable future regional-scale applications such as hydrate production optimization, and relevant hazards assessment and control.

2   The Study Area

SCS is estimated to be home to over 6.8 billion tons of oil equivalent MH, equaling half of the China's onshore oil and gas reserve and able to sustain 200 years' energy consumption of China [3]. As part of Western Pacific Ocean, SCS covers an area of 3.5×10^6 km^2, where three tectonic plates intersect: the Eurasian plate, the Pacific plate and the India Ocean plate. The water depth is 1212 m in average with a maximum depth of 5377 m. Even seabed accompanied with widespread seamounts appears ladder-like form from edge to center [4]. The geological structures (including submarine slumps, diapiric structures, accretionary wedges, tectonic slope-break zones and polygonal faults) are well-developed and widespread in SCS, providing good gas-bearing fluid migration sub-systems. Bottom simulating reflectors manifest extensive existence of MHs in SCS. As listed in Table 1, five expeditions have been accomplished since 1999 for characterizing MH reservoirs in SCS [5–13]. These data are particularly useful for regional-scale risk assessments associated with MH production.

3   Data Management Strategy

The desired database is geohazard-oriented, geo-referenced, extensible, and supports visualization and spatial analysis. Although a database can serve multiple purposes, we aim at a data house that particularly supports geohazard assessment associated with MH dissociation (due to natural and/or anthropologic perturbation) at the regional scale. To serve this purpose, we stress on the datasets including (but not limited to) geometrical boundaries and characteristics of MH reservoirs, thermal conditions, bathymetry data, and stratigraphy. All datasets should be consistently overlain in a unified coordinate system. The data structure should be adequately flexible in order to accommodate growing data. The data can be visualized in spatial context so that data inconsistency can be visually identified. In addition, the database should provide necessary functions of spatial analysis to facilitate further applications.

Taking account of these considerations, we employ ArcGIS [14] to store, manage, visualize and analyze the collected datasets. Figure 1 illustrates the workflow. First, the useful raw data are identified and collected based on literature survey and online resources search. Second, the selected data are compiled into GIS. Hardcopy data need to be digitized, and data without attached geographic information have to be geo-referenced to a predefined coordinate system. Spatial analyses are often necessary to derive relevant information from raw data, for instance, obtaining the seafloor terrain model from the bathymetric data originally archived in a raster file. These can be facilitated with the tools offered by ArcGIS. The processed and geo-referenced data in proper formats enter the GIS database, being ready for various applications such as modeling hydrate reservoirs, MH-related geohazards assessment, and MH production risk evaluation.

Table 1. Major expeditions to SCS [5–13]

Expedition	Period	Location	Mission	Drilling #
ODP Leg 184	2/19–4/12 1999	Jianfengbei Basin	To understand climate changes on a variety of time scales ranging from millennial to tectonic	6
GMGS-1	4/12–6/12 2007	Shenhu area	To determine the distribution of gas hydrate at as many different sites as possible	8
GMGS-2	5/28–9/8 2013	Eastern part of the Pearl River Mouth Basin	To further investigate the resource potential of gas hydrates in SCS	13
GMGS-3	6/1–9/1 2015	Shenhu area	To identify and evaluate gas hydrate reservoirs for future development	19
GMGS-4	Apr–Aug, 2016	Shenhu and Xisha Sea areas	Ditto	21

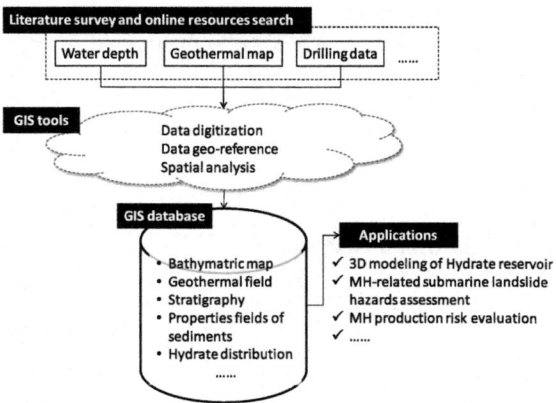

Fig. 1. Workflow for constructing and applying the database

4   Basic Datasets

All information is geo-referenced to the World Geodetic System of 1984 (WGS 84). We report the principal sources for our scope as follows.

(1) Geographic lineaments (such as line coast). These data are directly imported from the base map collections provided by ArcGIS.
(2) Bathymetric maps. The bathymetric maps of SCS are obtained from a subset of ETOPO1 database provided by NOAA [15, 16]. ETOPO1 is a one arc-minute global relief model of Earth's surface that integrates land topography and ocean bathymetry. ETOPO1 is vertically referenced to sea level, and horizontally referenced to WGS 84. The data is imported into our database after removing outliers with raster calculator. Figures 2a and b present the water depth map and the landscape view of the seafloor derived from the bathymetric map, respectively. The bathymetric map is of particular value to derive a terrain model of the sea bottom (required in a submarine landslide hazards assessment) and characterize the pressure conditions of the hydrate reservoirs (necessary for delineating hydrate stabilization zones).
(3) Geothermal data. Spatial and temporal variation of temperature in the marine sediments is of significance to characterize hydrate reservoirs. The relevant data include the geothermal gradient and heat flux. These data are widespread in the literature, but efforts are needed to digitize them (most presented as image maps) and geo-reference them to the proper coordinate system.
(4) Stratification and properties of marine sediments. This information is crucial to reach a reasonable reservoir model required for relevant geo-hazard assessment. The rational procedure is to extract necessary information from available drilling data, and then construct the stratification profiles and properties fields (e.g., porosity, hydraulic conductivity, deformation modulus, thermal conductivity, strength etc.) by employing geostatistical interpolation or stochastic simulations, which are readily available techniques intensively applied for constructing geological systems from limited data. Currently very limited drilling activities have been conducted at SCS (see Table 1). The available drilling data were extracted from the literature, digitized and imported into our database. Apparently, the current dataset is too scarce to reach a well-constrained reservoir model. Nevertheless, we expect more dense drilling data available in the near future, since the success of the first MH extraction trial at SCS is encouraging intensive follow-up expeditions at SCS.

5   Database Application: Estimating GHSZ

This section illustrates the usefulness of the constructed database via a simple application for roughly estimating gas hydrate stability zone (GHSZ). The database is able to accommodate an increasing number of applications as data accumulate with time.

GHSZ indicates the possible maximum bound of MH reservoirs dependent on water depth, geothermal temperature and pore water salinity; however, the actual geometric boundaries of the reservoirs are also affected by many other factors such as geologic setting, pore connectivity in the sediments, gas origin and composition [17]. Phase equilibrium method is prevailing to theoretically estimate the lower bound of GHSZ [18]. The base of the GHSZ locates where water pressure reaches the equilibrium value. According to [19–21], given temperature T (in K) at a certain depth, the equilibrium pressure (in Pa) is computed as:

$$P \le c \, \exp\!\left(a - \frac{b}{T}\right) \tag{1}$$

where a, b and c are constants (as suggested in [22], $a = 49.3185$, $b = 9459$ K, $c = 1$ Pa used in this study). The temperature profile with depth is estimated from the bottom water temperature (i.e. temperature at the seafloor) and the geothermal gradient. The bottom water temperature in SCS is estimated from [23]:

$$T_o = 373.41 \times h_w^{-0.6269} \tag{2}$$

where h_w is the water depth. The geothermal gradient is not uniform in SCS. However, as a rough estimate here, the averaged value 35 °C/km [18] is used for the entire region. Assuming a hydrostatic pressure distribution and a linearly increasing geothermal temperature, the depth of GHSZ base can be calculated from Eq. (1).

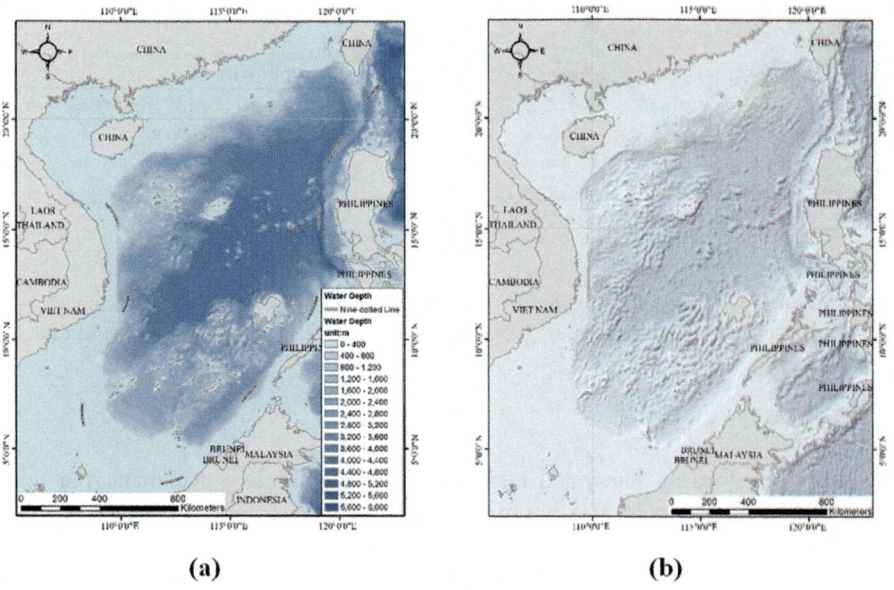

(a)   (b)

Fig. 2. Maps derived from ETOPO1 dataset: (a) water depth and (b) seafloor landscape view

Figure 3a presents the map of the estimated thickness of GHSZ. It indicates that GHSZ thickness is more than 100 m in most areas of SCS. Relatively thicker GHSZ can be found in central and eastern SCS. Areas where water depth is less than 500 m are unfavorable for MH formation. To roughly validate the estimation, we compare the estimated base of GHSZ with available drilling data [4–13]. As shown in Fig. 3b, the estimated base of GHSZ is below the MH-bearing deposits detected at all drilling boreholes except GMG2-08, where dual MH-bearing layers are detected. The estimated base of GHSZ is 45 m below the seafloor, slightly above the bottom MH-bearing layer extending from 69 to 86 m below the seafloor. This slight discrepancy arises from the simplification of the model for estimating GHSZ without considering the secondary factors other than pressure and temperature.

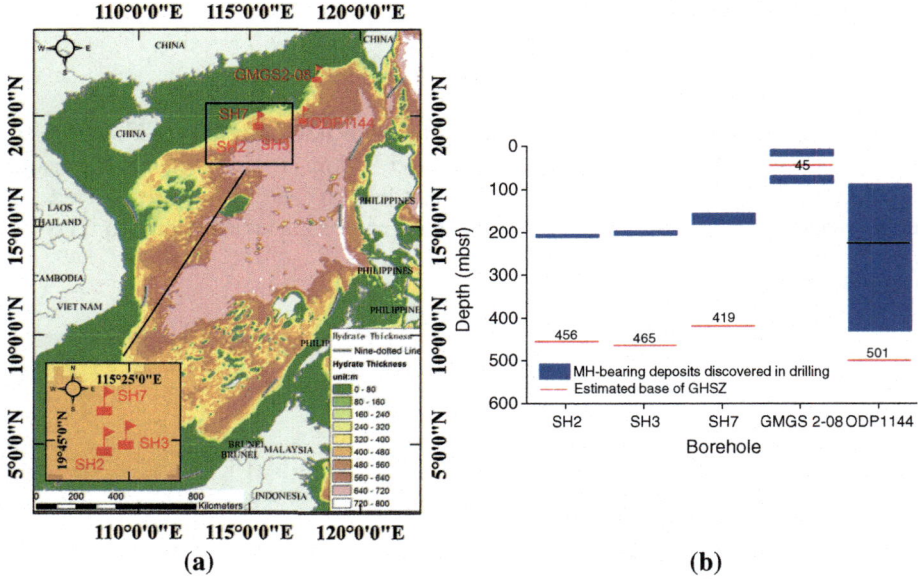

Fig. 3. A map of estimated thickness of GHSZ (a) compared with hydrate-bearing deposits discovered at five available boreholes (b), of which the spatial locations are shown in (a)

6 Summary

This study develops an integrated GIS database for compiling and managing information of MH reservoirs in SCS in order to accommodate regional-scale analyses such as MH production optimization and related geohazard assessment. We target a geohazard-oriented, geo-referenced, and extensible database that supports visualization and spatial analysis. Geographic lineaments, bathymetric maps, geothermal data, stratification and properties of marine sediments are collected from the literature and online resources, digitized, geo-referenced and imported into the database. Useful information is derived from the database to outline the geometry boundaries and geologic properties of the MH reservoirs. The usefulness of the database is demonstrated via an application for estimating GHSZ. This study is preliminary and however significant towards a regional risk analysis of MH production in SCS.

Acknowledgement. The work is supported by the Chinese National Natural Science Foundation (with grant No. 41572267, 51639008 and 51239010), the Fundamental Research Funds for the Central Universities, and Guangxi Key Laboratory of Geomechanics and Geotechnical Engineering (12-KF-02).

References

1. Kvenvolden, K.A., Lorenson, T.D.: The Global Occurrence of Natural Gas Hydrate. American Geophysical Union, Washington, DC (2001)
2. Shi, D., Zheng, J.: The status and prospects of research and exploitation of natural gas hydrate in the world. Adv. Earth Sci. **14**(4), 330–339 (1999). (in Chinese)
3. Zhang, H., Zhang, H., Zhu, Y.: Gas hydrate investigation and research in China: present status and progress. Geol. China **34**(6), 953–961 (2007). (in Chinese)
4. Behrmann, J.H., Meissl, S.: Submarine mass movements and their consequences. Adv. Nat. Technol. Hazards Res. **41**(3), 1–12 (2010)
5. Li, G., Li, X.S., Chen, Q., Chen, C.: Numerical simulation of gas production from gas hydrate zone in Shenhu area, South China Sea. Acta Chim. Sin. **68**(11), 1083–1092 (2010). (in Chinese)
6. Su, Z., Chen, Z., Wu, N., He, Y.: Gas hydrate system and numerical modeling on accumulations at Shenhu area of northern South China Sea. In: National Conference on Porous Flow (2011, in Chinese)
7. Guan, J., Liang, D., Wan, L., Gu, R.: Analysis on dynamic methane hydrate accumulation simulation in Shenhu area of the northern South China Sea. J. Eng. Geol. **22**(5), 997–1002 (2014). (in Chinese)
8. Wang, X., Wu, S., Liu, X., Guo, Y., Lu, J., Yang, S., Liang, J.: Estimation of gas hydrate saturation based on resistivity logging and analysis of estimation error. Geoscience **24**(5), 993–999 (2010). (in Chinese)
9. Zhang, G., Liang, J., Lu, J., Yang, S., Zhang, M., Su, X., Xu, H., Fu, S., Kuang, Z.: Characteristics of natural gas hydrate reservoirs on the eastern slope of the South China Sea. Nat. Gas. Ind. **34**(11), 1–10 (2014). (in Chinese)
10. Zhuang, C., Chen, F., Cheng, S., Lu, H., Wu, C., Cao, J., Duan, X.: Light carbon isotope events of foraminifera attributed to methane release from gas hydrates on the continental slope, northeastern South China Sea. Sci. China Earth Sci. **46**(10), 1334 (2016). (in Chinese)
11. Yang, S., Liang, J., Lu, J., Qu, C., Liu, B.: New understandings on the characteristics and controlling factors of gas hydrate reservoirs in the Shenhu area on the northern slope of the South China Sea. Geosci. Front. **24**(4), 1–14 (2017). (in Chinese)
12. Geng, J., Shao, L., Wu, S., Wu, N., Zhang, G.: Further discussion on gas hydrate of ODP1144, South China Sea. Annual Meeting of Chinese Geoscience Union (2003, in Chinese)
13. Su, X.: Introduction of expedition 184 of ocean drilling program. Geoscience **2**, 175–176 (1999)
14. Gmbh, E.G.: ArcGIS server administrator and developer guide (2004)
15. National Oceanic and Atmospheric Administration (NOAA). http://www.noaa.gov/. Accessed 23 Oct 2017

16. National Center for Environmental Information (NCEI): ETOPO1 Global Relief. https://www.ngdc.noaa.gov/mgg/global/global.html. Accessed 23 Oct 2017
17. Wang, S., Song, H., Yan, W., Shi, X., Yan, P., Fan, S.: Stable zone thickness and resource estimation of gas hydrate in southern south china sea. Nat. Gas. Ind. **25**(8), 24–27 (2005)
18. Wang, Y., Hao, F., Zhao, Y.: Geothermal investigation of the thickness of gas hydrate stability zone in the north continental margin of the South China Sea. Acta Oceanol. Sin. **36**(4), 72–79 (2017)
19. Mckoy, V., Sinanoğlu, O.: Theory of dissociation pressures of some gas hydrates. J. Chem. Phys. **38**(12), 2946–2956 (1963)
20. Kamath, V.A., Holder, G.D.: Dissociation heat transfer characteristics of methane hydrates. AIChE J. **33**(2), 347–350 (1987)
21. Lundgaard, L., Mollerup, J.: Calculation of phase diagrams of gas-hydrates. Fluid Phase Equilib. **76**, 141–149 (1992)
22. Bejan, A., Rocha, L.A.O., Cherry, R.S.: Methane hydrates in porous layers: gas formation and convection - transport phenomena in porous media II - 14. In: Transport Phenomena in Porous Media II, pp. 365–396 (2002)
23. Wang, S., Yan, W., Song, H.: Mapping the thickness of the gas hydrate stability zone in the South China Sea. Terr. Atmos. Oceanic Sci. **17**(4), 815–828 (2006)

Optimization-Based Design of Stabilizing Piles

Wenping Gong[1(✉)], Huiming Tang[1], C. Hsein Juang[2],
James R. Martin[2], and Liangqing Wang[1]

[1] Faculty of Engineering, China University of Geosciences,
Wuhan 430074, Hubei, China
wenpinggong@cug.edu.cn
[2] Glenn Department of Civil Engineering, Clemson University,
Clemson, SC 29634-0911, USA

Abstract. Piles are widely adopted for stabilizing the unstable slopes or active landslides. The subject of stabilizing piles has been extensively studied in literature; however, the optimization of stabilizing piles is rarely reported. This paper presents an optimization-based design framework for landslide stabilizing piles, within which both effectiveness of the stabilizing piles (as reinforcement in the landslide) and cost efficiency could be explicitly considered and optimized. The design parameters of the stabilizing piles considered in this paper include pile diameter, spacing, length and position. The design objective is to simultaneously optimize the reinforcement (i.e., stabilizing piles) effectiveness and cost efficiency. In that the desire to maximize the reinforcement effectiveness and that to maximize the cost efficiency are two conflicting objectives, the outcome of this bi-objective optimization yields a Pareto front which depicts a trade-off between these two design objectives. With the obtained Pareto front, an informed decision regarding the design of landslide stabilizing piles can be reached. The effectiveness and the significance of this optimization-based design framework for landslide stabilizing piles are demonstrated through an illustrative example.

Keywords: Landslide · Stabilizing piles · Factor of safety · Cost
Design · Optimization

1 Introduction

In the Three Gorges Reservoir Area, China, the stabilizing piles have been extensively constructed [1]. With the aid of stabilizing piles, part of the lateral earth pressure from the upper unstable layer could be transferred to the lower stable layer [2, 3]; as such, the stability of the reinforced landslide or slope can be improved. The pile-slope or pile-landslide interaction, which further affects the reinforcement effectiveness and cost efficiency of the stabilizing piles, is influenced by various factors, including the surrounding soils and pile parameters. In a conventional design of stabilizing piles, the following procedures can be adopted: evaluating the lateral resisting force that is required to achieve the target factor of safety (FS), and selecting a stabilizing pile design that is capable of offering the required lateral resisting force [2, 4]. Here, the required lateral resisting force from the stabilizing piles can be estimated using either the coupled interaction analysis or the uncoupled stability analysis [5, 6].

© Springer Nature Singapore Pte Ltd. 2018
A. Farid and H. Chen (Eds.): GSIC 2018, *Proceedings of GeoShanghai 2018*
International Conference: Geoenvironment and Geohazard, pp. 45–53, 2018.
https://doi.org/10.1007/978-981-13-0128-5_6

According to Wang and Kulhawy [7], the essentials of good engineering designs hinge on the requirement of economics. As such, the design of stabilizing piles should be treated as a multi-objective optimization problem, in which the requirements from both reinforcement effectiveness and cost efficiency should be considered and optimized. This is similar to the designs of other geotechnical structures such as braced excavations [8] and shield tunnels [9]. However, most discussions on the design of stabilizing piles only deal with the reinforcement effectiveness, the discussions on the design of stabilizing piles which consider both reinforcement effectiveness and cost efficiency are only at the conceptual level. Thus, this paper will present a bi-objective optimization-based design framework for stabilizing piles, in which both reinforcement effectiveness and cost efficiency are explicitly considered and optimized.

2 Optimization-Based Design of Stabilizing Piles

2.1 Bi-Objective Optimization-Based Design Framework

In reference to the algorithm of the bi-objective optimization-based design of stabilizing piles shown in Eq. (1), the essence of the proposed optimization-based design of stabilizing piles is to seek an optimal stabilizing pile design (represented by a set of pile parameters $\mathbf{d}$) in the design space $\mathbf{DS}$ such that the target stability $\mathbf{TS}$ is satisfied, meanwhile both reinforcement effectiveness $\mathbf{R}$ and cost efficiency $\mathbf{E}$ are simultaneously optimized.

$$
\begin{aligned}
&\text{Find:} && \text{Pile parameters } \mathbf{d} \\
&\text{Subject to:} && \text{Design space } \mathbf{DS} \\
& && \text{Target stability } \mathbf{TS} \\
&\text{Objectives:} && \text{Maximizing reinforcement effectivenss } \mathbf{R} \\
& && \text{Maximizing cost efficiency } \mathbf{E}
\end{aligned}
\tag{1}
$$

In this study, all the geometric parameters of pile diameter (D), spacing (S/D), length (L) and position (X; see Fig. 2) are taken as the design parameters $\mathbf{d}$, $\mathbf{d} = \{D, S/D, L, X\}$, in which S represents the center-to-center spacing of the stabilizing piles.

2.2 Informed Design of Stabilizing Piles Based upon Optimization Results

In that the desire to maximize the reinforcement effectiveness $\mathbf{R}$ and that to maximize the cost efficiency $\mathbf{E}$ are two conflicting design objectives, this bi-objective optimization cannot lead to a utopia solution which is optimal with respect to both objectives. Instead, a set of non-dominated solutions could be identified, which are superior to all others in the design space; but among these non-dominated solutions, none of them are superior or inferior to others [9, 10]. As shown in Fig. 1, although non-dominated design $\mathbf{d}_1$ costs less and thusly implies higher cost efficiency $\mathbf{E}$, non-dominated design $\mathbf{d}_5$ yields higher reinforcement effectiveness $\mathbf{R}$; the utopia design $\mathbf{d}_0$ which is optimal with respect to both objectives could not be attainable in the design space $\mathbf{DS}$. These non-dominated

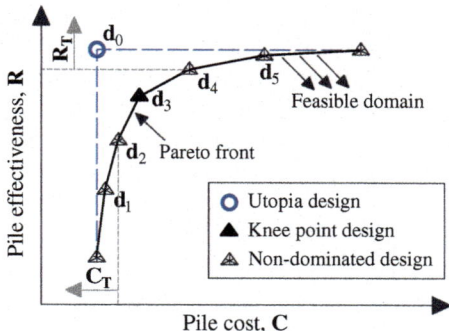

Fig. 1. Optimization results of the bi-objective optimization of stabilizing piles.

solutions collectively form a Pareto front, and this Pareto front could be obtained with either Non-dominated Sorting Genetic Algorithm version II, NSGA-II [11] or weighted sum-based approach [12, 13].

The obtained Pareto front depicts a trade-off between these two conflicting objectives, which can help render an informed design decision. For example, either the most cost-efficient design that is above a pre-specified level of reinforcement effectiveness $\mathbf{R_T}$ (see design $\mathbf{d_4}$ in Fig. 1) or the most effective design that is below a pre-specified level of cost $\mathbf{C_T}$ (see design $\mathbf{d_2}$ in Fig. 1) can be identified as the most preferred design in the design space **DS**. When a strong preference is not pre-specified by the owner or client, the knee point on the Pareto front (see design $\mathbf{d_3}$ in Fig. 1), which yields the best compromise between these two conflicting objectives, can be taken as the most preferred design in the design space **DS** [14, 15]. The knee point on the Pareto front may be identified with marginal utility function approach [14, 16], normal boundary intersection approach [8, 15], reflex angle approach [15] or minimum distance-based approach [18].

3   Illustrative Example

3.1   Geotechnical Characterization of the Illustrative Example

To demonstrate the effectiveness and significance of this optimization-based design of stabilizing piles, one illustrative example shown in Fig. 2, in terms of the design of stabilizing piles in a one-layer earth slope, is studied. The width (W) and height (H_1) of this earth slope are 20.0 m and 14.0 m, respectively. The depth of the underlying rock layer (H_2) is taken as an infinitely large value. The soil parameters of this example slope are listed in Table 1. In this study, the 2-D explicit finite difference program FLAC version 7.0 [17] is adopted as the solution model for evaluating the slope stability, the plane-strain condition is assumed in this stability analysis. The left-side boundary is set at 30.0 m away from the slope toe, the right-side boundary is set at 30.0 m away from the slope crest, and the bottom boundary is set at 30.0 m below the slope toe. In the discretization of the geometrical domain of this slope, the size of soil elements is set at 1.0 m. The bottom boundary is restrained vertically, and the left and right-side boundaries are restrained horizontally. The soil is modelled with the Mohr-Coulomb model, the stabilizing piles are modelled with beam elements, and the soil-pile interfaces are modelled with interface elements. The input parameters of the stabilizing piles and those of the soil-pile interfaces are listed in Table 1.

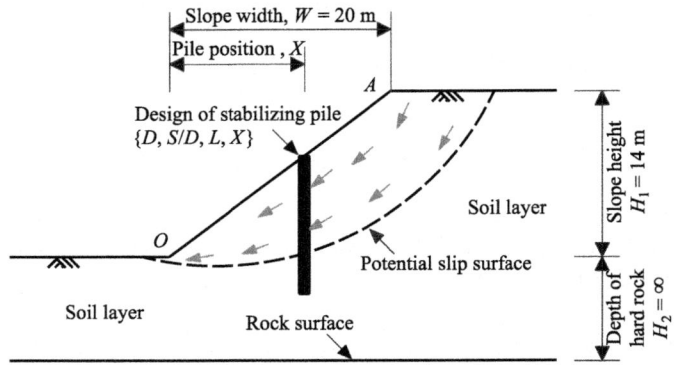

Fig. 2. Schematic diagram of the illustrative example.

Table 1. Input parameters for the illustrative example.

Category	Parameter	Value
Soil	Density (kg/m^3)	1,700
	Effective cohesion (kPa)	5.0
	Effective friction angle (°)	30.0
	Bulk modulus (MPa)	330
	Shear modulus (MPa)	150
Stabilizing piles	Density (kg/m^3)	2,500
	Young's modulus (GPa)	35.0
	Steel reinforcement ratio (%)	1.0
	Thickness of concrete protective cover (m)	0.05
	Yielding strength of steel bar (kPa)	345×10^3
	Compression strength of concrete (kPa)	39×10^3
Soil-pile interfaces	Normal stiffness (Mpa/m)	550
	Shear stiffness (Mpa/m)	550
	Cohesion (kPa)	5.0
	Friction angle (°)	30.0

With the input parameters shown in Table 1, the stability of this slope is analyzed, and the resulting FS of the unreinforced slope, denoted as FS_1, is 1.18. This FS_1 value is relatively low, thus this slope needs to be reinforced. For ease of construction, a discrete design space **DS**, formulated in Table 2, is adopted. The maximum pile spacing is set as $S/D = 4$ such that an effective soil arching could be formed between piles [4]. For a given slope, the longitudinal length is oftentimes fixed, thus the cost efficiency could be captured by the construction cost **C** of the stabilizing piles per longitudinal length. The construction cost **C**, in this study, is approximated by the volume of the stabilizing piles. A larger **C** value means that a larger budget is required on the slope reinforcement, thus implies lower cost efficiency **E**. The reinforcement effectiveness **R** is measured herein by the increase in the FS of the reinforced slope, expressed as $\Delta_{FS} = FS_2 - FS_1$, where FS_2 and FS_1 are the FS of the unreinforced slope and that of the reinforced slope, respectively. A larger Δ_{FS} value means that a larger improvement of the slope stability, thus implies higher reinforcement effectiveness **R**.

Table 2. Design space **DS** adopted for the illustrative example.

Design parameters **d**	Design pool (i.e., potential design values of the pile parameters)
Pile diameter, D (m)	{0.2 m, 0.4 m, 0.6 m, 0.8 m, 1.0 m}
Pile spacing, S/D	{1, 2, 3, 4}
Pile length, L (m)	{6 m, 8 m, 10 m, 12 m, 14 m, 16 m, 18 m, 20 m}
Pile position, X (m)	{1 m, 3 m, 5 m, 7 m, 9 m, 11 m, 13 m, 15 m, 17 m, 19 m}

3.2 Results of Bi-Objective Optimization-Based Design of Stabilizing Piles

With the bi-objective optimization algorithm of stabilizing piles shown in Eq. (1), the design of stabilizing piles for this example slope is readily conducted. For each of the candidate designs in the design space **DS** (see Table 2), the reinforcement effectiveness **R** and the cost efficiency **E**, in terms of the construction cost **C** and the increase in FS (Δ_{FS}), respectively, are evaluated, and the results are plotted in Fig. 3(a). Due to the different combinations of the pile parameters, the candidate pile designs of the same construction cost **C** can yield different reinforcement effectiveness **R**, the candidate pile designs of different construction cost **C** could yield similar reinforcement effectiveness **R**, as illustrated in Fig. 3(a). Thus, the reinforcement effectiveness **R** and cost efficiency **E** of the stabilizing piles might be simultaneously optimized through a careful

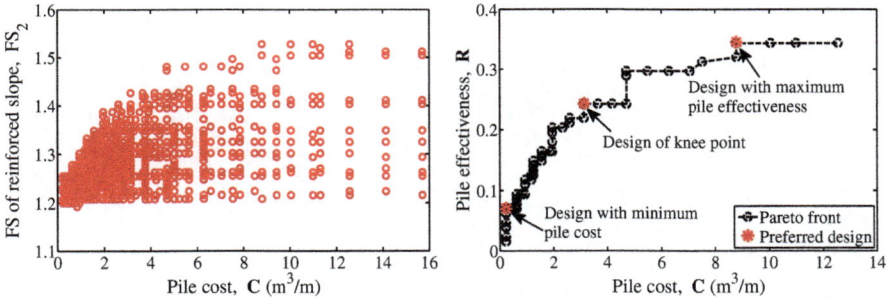

(a) Design objectives of all candidate designs   (b) Optimization results without target FS

Fig. 3. The results derived from the multi-objective optimization of stabilizing piles.

adjustment of the pile parameters. This is the theoretical basis for the presented bi-objective optimization-based design.

For illustration purposes, the target stability **TS** in this example is specified as 1.0, thus all the candidate pile designs depicted in Fig. 3(a) are classified as feasible designs. Then, a Pareto front that consists of 99 nondominated pile designs is established using the sorting algorithm of NSGA-II [11], as shown in Fig. 3(b). With the obtained Pareto front, an informed design decision can be made. For example, the knee point on the Pareto front is identified with the minimum distance-based approach [18], the result is shown in Fig. 3(b). As expected, on the left side of this knee point, a slight reduction in the cost C drastically decreases the reinforcement effectiveness R, this is not desirable; on the right side of this knee point, a slight increase in the reinforcement effectiveness R yields a huge increase in the cost C, this is not desirable, either. Thus, when the target level of the reinforcement effectiveness R or that of the cost C is not specified, the knee point on the Pareto front can be identified as the most preferred design in the design space DS.

Table 3. Optimization results for illustrative example.

Preferred design on Pareto front	Pile design parameters, **d**				Construction cost, **C** (m^3/m)	Reinforcement effectiveness, **R** = Δ_{FS}
	Diameter, D (m)	Spacing, S/D	Length, L (m)	Position, X (m)		
Minimum **C**	0.2	4	6	17	0.24	0.07
Knee point	1.0	3	12	9, 11, 13	3.14	0.24
Maximum **R**	0.8	1	14	11	8.80	0.34

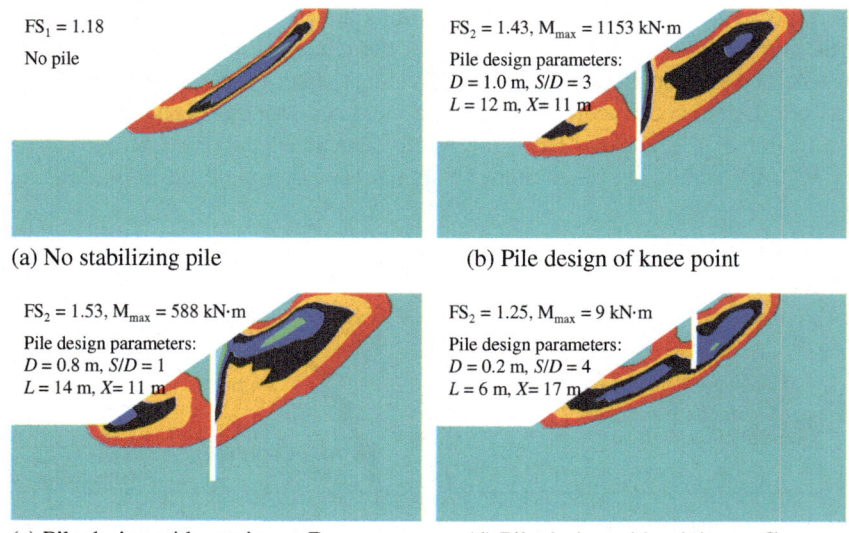

(a) No stabilizing pile

(b) Pile design of knee point

(c) Pile design with maximum **R**

(d) Pile design with minimum **C**

Fig. 4. Failure surfaces of the preferred designs on the Pareto front.

Next, two other designs on the Pareto front, one represents the design with the minimum construction cost **C** and the other the design with the maximum reinforcement effectiveness **R**, are identified and shown in Fig. 3(b). The pile parameters **d**, construction cost **C** and reinforcement effectiveness **R** of these three designs are tabulated in Table 3. The pile position of the knee point and that of the pile design with the maximum **R** are both located in the range of 9 m to 13 m, thus the optimal position of the stabilizing piles may be located around the middle part of the unstable slope. The pile length of the knee point and that of the pile design with the maximum **R** are 12 m and 14 m, respectively, which are larger than that of the design with the minimum **C** (i.e., $L = 6$ m), thus the pile length should be large. The data on the pile diameter and pile spacing indicate that large pile diameters and small pile spacings should be preferred.

3.3 Discussion on Optimization Results

Plotted in Fig. 4 are the potential failure surface of the unreinforced slope and those of the slopes reinforced with the three preferred designs on the Pareto front (see Fig. 3b). In Fig. 4(a), the depth of the failure surface of the unreinforced slope is small and which does not pass through the toe of the slope; whereas, the failure surfaces of the reinforced slopes extend to greater depths, and wider horizontal ranges can be affected by the failure of the slopes, as shown in Fig. 4(b), (c) and (d). In Fig. 4(b) and (c), the stabilizing piles are designed with the optimal position, sufficient length, large diameter and suitable spacing, the outputs are: the initial failure surface (of the unreinforced slope) is blocked by the stabilizing piles, the new failure surfaces (on both sides of the stabilizing piles) extend to greater depths, significant part of the lateral earth pressure from the upper unstable layer could be transferred to the lower stable layer and which can be indicated by the maximum bending moment of the stabilizing piles (i.e., $M_{max} = 1153$ kN m in Fig. 4b and $M_{max} = 588$ kN m in Fig. 4c), and thus the stability of the slope is improved (i.e., $FS_2 = 1.43$ in Fig. 4b and $FS_2 = 1.53$ in Fig. 4c). Whereas, in Fig. 4(d), the failure surface of the reinforced slope could not be blocked by the stabilizing piles, the lateral earth pressure from the upper unstable layer could not be effectively transferred to the lower stable layer (i.e., $M_{max} = 9$ kN m in Fig. 4d), and the improvement of the slope stability is limited (i.e., $FS_2 = 1.25$ in Fig. 4d), which can be attributed to the insufficient length of the stabilizing piles.

4 Concluding Remarks

This paper presents a bi-objective optimization-based design framework for stabilizing piles; in which, both reinforcement effectiveness and cost efficiency are explicitly considered and optimized, in addition to the requirement of target stability. The design objective of reinforcement effectiveness and that of cost efficiency are conflicting with each other. The output of this optimization is a Pareto front which depicts a trade-off between these objectives. This Pareto front helps render an informed design decision, in terms of the knee point on the Pareto front. The effectiveness of this design framework is demonstrated through an illustrative example: the design of stabilizing piles in a one-layer earth slope.

Note that the optimization of stabilizing piles is a challenging problem. Apart from the factors investigated in this paper, many other factors such as the groundwater, 3-D behavior of the reinforced slope, soil arching effect between the single row of stabilizing piles, interaction between the multiple rows of stabilizing piles, and target stability of the reinforced slope or landslide should be considered.

Acknowledgements. The financial support provided by the National Natural Science Foundation of China (No. 41702294) is acknowledged.

References

1. Tang, H., Hu, X., Xu, C., Li, C., Yong, R., Wang, L.: A novel approach for determining landslide pushing force based on landslide-pile interactions. Eng. Geol. **182**, 15–24 (2014)
2. Poulos, H.G.: Design of reinforcing piles to increase slope stability. Can. Geotech. J. **32**(5), 808–818 (1995)
3. Ashour, M., Ardalan, H.: Analysis of pile stabilized slopes based on soil–pile interaction. Comput. Geotech. **39**, 85–97 (2012)
4. Kourkoulis, R., Gelagoti, F., Anastasopoulos, I., Gazetas, G.: Slope stabilizing piles and pile-groups: parametric study and design insights. J. Geotech. Geoenvironmental Engineering **137**(7), 663–677 (2010)
5. Bishop, A.W.: The use of the slip circle in the stability analysis of slopes. Géotechnique **5**(1), 7–17 (1955)
6. Spencer, E.: A method of analysis of the stability of embankments assuming parallel interslice forces. Géotechnique **17**(1), 11–26 (1967)
7. Wang, Y., Kulhawy, F.H.: Economic design optimization of foundations. J. Geotech. Geoenvironmental Eng. **134**(8), 1097–1105 (2008)
8. Juang, C.H., Wang, L., Hsieh, H.S., Atamturktur, S.: Robust geotechnical design of braced excavations in clays. Struct. Saf. **49**, 37–44 (2014)
9. Gong, W., Wang, L., Juang, C.H., Zhang, J., Huang, H.: Robust geotechnical design of shield-driven tunnels. Comput. Geotech. **56**, 191–201 (2014)
10. Gong, W., Tien, Y.M., Juang, C.H., Martin, J.R., Zhang, J.: Calibration of empirical models considering model fidelity and model robustness - Focusing on predictions of liquefaction-induced settlements. Eng. Geol. **203**, 168–177 (2016)
11. Deb, K., Pratap, A., Agarwal, S., Meyarivan, T.: A fast and elitist multi-objective genetic algorithm: NSGA-II. IEEE Trans. Evol. Comput. **6**(2), 182–197 (2002)
12. Hajela, P., Lin, C.Y.: Genetic search strategies in multicriterion optimal design. Struct. Optim. **4**(2), 99–107 (1992)
13. Chen, W., Wiecek, M.M., Zhang, J.: Quality utility - a compromise programming approach to robust design. J. Mech. Des. **121**(2), 179–187 (1999)
14. Branke, J., Deb, K., Dierolf, H., Osswald, M.: Finding knees in multi-objective optimization. In: Parallel Problem Solving from Nature-PPSN VIII, pp. 722–731. Springer, Heidelberg (2004)
15. Deb, K., Gupta, S.: Understanding knee points in bicriteria problems and their implications as preferred solution principles. Eng. Optim. **43**(11), 1175–1204 (2011)

16. Gong, W., Wang, L., Khoshnevisan, S., Juang, C.H., Huang, H., Zhang, J.: Robust geotechnical design of earth slopes using fuzzy sets. J. Geotech. Geoenvironmental Eng. **141** (1), 04014084 (2014)
17. FLAC version 7.0.: Fast Lagrangian Analysis of Continua. Itasca Consulting Group Inc., Minneapolis (2011)
18. Gong, W., Huang, H., Juang, C.H., Wang, L.: Simplified-robust geotechnical design of soldier pile-anchor tieback shoring system for deep excavation. Mar. Georesour. Geotechnol. **35**(2), 157–169 (2017)

Vulnerability of Seafloor at Shenhu Area, South China Sea Subjected to Hydrate Dissociation

Xin Ju[1,2], Fang Liu[1,2(✉)], and Pengcheng Fu[3]

[1] State Key Laboratory of Disaster Reduction in Civil Engineering, Tongji University, Shanghai 200092, China
liufang@tongji.edu.cn
[2] Key Laboratory of Geotechnical and Underground Engineering, Tongji University, Ministry of Education, Shanghai 200092, China
[3] Lawrence Livermore National Laboratory, Livermore, CA, USA

Abstract. This study develops a hybrid model for investigating seafloor instability subjected to hydrate dissociation by combining the limit equilibrium slope stability analysis and numerical modeling of fluid flow and heat transport. This model is employed to study a slope configured according to the geological settings of hydrate reservoirs located at Shenhu area, South China Sea, under mild and sustaining warming scenarios. The vulnerable settings and the controlling factors are quantitatively identified through a parametric study. The results indicate that the slope is stable under the mild seafloor warming scenario in a short timescale. Under the sustaining warming scenario, the slope fails in the order of hundreds of years. The predicted slip onset is located at the top of the hydrate reservoir. The slope is particularly vulnerable to the presence of low permeable and tight sediments, and highly-concentrated hydrates. The permeability of the overburden layer plays a dominant role in determining the vulnerability of the slope, while the impact of the permeability, hydrate saturation, and porosity of the hydrate reservoir is secondary.

Keywords: Methane hydrate · Seafloor instability · THC coupled process South China Sea

1 Introduction

Gas hydrate is an ice-like form of water that contains gas molecules (mostly methane in nature) in its molecular cavities. Methane from gas hydrates may constitute a substantial future source of natural gas. Many countries, including China, have launched major gas hydrate research and development programs. Shenhu area at the continental slope of northern South China Sea (SCS) is viewed as a vast reservoir of MHs and a promising site for hydrate extraction [1–4]. The recent success of hydrate production trial at Shenhu area encourages an accelerating development in China towards mass hydrate production and also invokes a demand for thorough and careful environmental impact assessments, among which the identification of seafloor regions vulnerable to hydrate dissociation (due to either natural or anthropological perturbations) is a key issue.

© Springer Nature Singapore Pte Ltd. 2018
A. Farid and H. Chen (Eds.): GSIC 2018, *Proceedings of GeoShanghai 2018*
International Conference: Geoenvironment and Geohazard, pp. 54–62, 2018.
https://doi.org/10.1007/978-981-13-0128-5_7

Substantial evidence has suggested that hydrates dissociation is one of the main causes of seafloor instability [8]. The hydrate stability zone extends into the seafloor sediments down to a depth where temperature exceeds an equilibrium value required for hydrate stability. However, MHs in the marine sediments are prone to dissociate at a sufficient temperature rise (e.g. through climate change [5, 6]) or pressure drop (e.g. through falling sea level [7]). Hydrate dissociation causes loss of cementation and excess pore pressure, both of which are factors contributing to reduction in shear resistance of marine sediments. The connection between hydrates dissociation with seafloor instability has been investigated via theoretical [9, 10] and numerical [8, 11] models in the past decades. However, accurately evaluating the conditions for and risks of hydrates dissociation remains a challenging task, because stability of the slope with melting MHs is governed by interplays of thermodynamic, hydrodynamic, chemical and mechanical processes, each working at a specific timescale and affected by a large set of parameters. Additionally, quantitative analyses of seafloor stability for the specific geological settings (e.g. water depth, overburden thickness, seafloor terrain) of Shenhu area are still rare.

This study intends to develop a simple and robust method to quantify seafloor instability subjected to MH dissociation with a full consideration of thermo-hydro-chemical coupled processes that incorporates realistic parameters of Shenhu area. To identify the most vulnerable setting and the most important factor, we perform a parametric study for typical geological settings of Shenhu Area. The present study concentrates on how hydrate reservoirs react to hypothetical climatic conditions, and the model can be extended to other types of perturbations which will be addressed in our future work.

2   A Hybrid Approach for Stability Analysis of a Slope Subjected to Hydrate Dissociation

Figure 1 illustrates a submarine slope with a hydrate-bearing layer (hydrate reservoir) sandwiched by overburden and underlying layers. Note that each of the single layer can be refined with multiple layers to incorporate more realistic conditions. The top of the reservoir is prone to dissociate if the seafloor temperature rises due to sustaining heat input (e.g. resulting from climate change). If the slope is gentle and the lateral extents of the unstable mass are much larger compared with its thickness, it is reasonable to idealize the slope into an infinite slope where one-dimensional (1D) analysis is applicable. To quantify the stability of the infinite slope, we propose a hybrid model that combines the limit equilibrium analysis with a numerical simulator for fluid and heat flow analysis.

In general, according to the concept of the limit equilibrium analysis, the stability of an infinite slope is evaluated through a safety factor [9] computed from:

$$F_s = \frac{c}{\gamma' H \sin\beta \cos\beta} + \frac{(\gamma H + \gamma_w z)\cos\beta \tan\varphi}{(\gamma' H)\sin\beta} - u\frac{\cos\beta \tan\varphi}{(\gamma' H)\sin\beta} \tag{1}$$

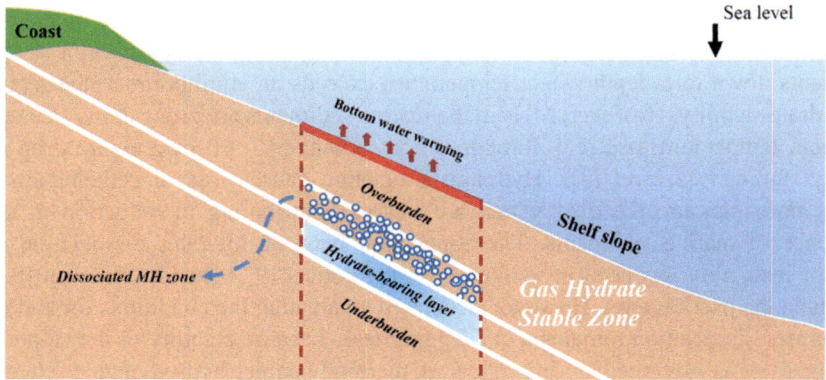

Fig. 1. A schematic sketch of a submarine slope with a hydrate reservoir subjected to seafloor warming

where β is the slope angle; H is the depth of a possible failure surface below the seafloor; z is the water depth; γ and γ' are the saturated and buoyant unit weight of the soil, respectively; c and φ are the cohesion and internal friction angle of the soil at the possible failure surface, respectively; u is the total pore pressure at the possible failure surface, which could even liquefy soils if its magnitude exceed that of total stress of soils.

To extend Eq. (1) to scenarios involving hydrate dissociation, two aspects need particular consideration. First, temporal and spatial variation in the strength parameters of the sediments during hydrate melting needs to be taken into account. For simplicity, we assume that cohesion c instantly vanishes when the dissociation front arrives, noting that a gradually-declining cohesion correlated to hydrate saturation would be more reasonable if available. The internal angle φ is nearly unaffected by dissociation according to published data [12]. Second, the spatial and temporal evolutions of pore pressure u must be quantified by considering the interplay between hydrate dissociation, heat diffusion, and fluid flow. In this proposed scheme, the evolving u is obtained from a vertical fluid flow and heat transport analysis using TOUGH + Hydrate, a simulator for multi-component, multi-phase fluid and heat flow in hydrate-bearing geologic systems. The results are then imported into Eq. (1) to yield F_s at different depths and time periods.

3   Model Set-up

The proposed hybrid approach is used to analyze a hydrate reservoir system conditioned to the geological settings of Shenhu area as listed in Table 1. The parameters adopted in the baseline simulation (presented in Sect. 4.1) are listed in Tables 1 and 2. In the parametric study (presented in Sect. 4.2), we fix the geometry (i.e. water depth, stratigraphy, and slope angle) and thermal properties of the slope, and concentrate the parametric analysis on the effect of intrinsic permeability of the overburden and underlying material $k_{ou}(0.0001 \sim 100 \text{ mD})$ and that of the hydrate reservoir $k_h(0.0001 \sim 100 \text{ mD})$, porosity of the sediments $n(10 \sim 60\%)$, and initial hydrate saturation S_{h0}(5 ~ 45%). These parameters either vary in wide ranges or significantly affect the variation of pore pressure.

The 1D numerical model for vertical fluid and heat flow analysis is 820 m high, composed of a 60-m thick overburden layer, a 60-m thick hydrate-bearing layer, and a 700-m thick underlying layer. Prescribed pressure and temperature are imposed to the top and bottom boundaries. The cells are 1 m tall in the overburden and underlying layers and 0.5 m in the hydrate-bearing layer. The initial pressure is assumed hydrostatic, and the initial thermal field complies with a constant geothermal gradient (given in Table 1).

We assume that seafloor instability is triggered by a seafloor temperature rising at a rate of 0.03°C/yr due to climate warming. Note that this assumption can be replaced by other triggering mechanisms. Here we consider two scenarios: (1) a mild warming scenario lasting 100 years, for which the baseline simulation is compared with published results; (2) a sustaining warming scenario for a sensitivity analysis seeking the time to slope failure under various conditions.

Table 1. Parameters of hydrate reservoirs at Shenhu, northern SCS.

Parameter	Typical range [13, 14]	Value in baseline simulation
Water depth	1108–1245 m	1176.5 m
Overburden thickness	20–200 m	60 m
Thickness of hydrate reservoir	20–100 m	60 m
Slope angle	0.1–71.4°	3.0°
Initial seafloor temperature	1–5 °C	3 °C
Geothermal gradient	43.0–67.7 °C/km	55.35 °C/km
Hydrate saturation	20–70%	45%

Table 2. Parameters and models used in the baseline simulation.

Parameter	Value
Initial salinity	35
Sediment density ρ	2.65 g/cm^3
Internal friction angle of sediment φ	30°
Dry thermal conductivity $k_{\theta d}$	3.3 W/m/K
Wet thermal conductivity $k_{\theta e}$	1.0 W/m/K
Intrinsic permeability of over- and under-burden k_{ou}	0.001 mD
Intrinsic permeability of hydrate-bearing layer k_h	10.0 mD
Porosity n (all formation)	40%
Initial hydrate saturation S_{h0}	45%
Relative permeability model [15]	$k_{rA} = \max\left\{0, \min\left\{[\frac{S_A - S_{irA}}{1 - S_{irA}}]^n, 1\right\}\right\}$ $k_{rG} = \max\left\{0, \min\left\{[\frac{S_G - S_{irG}}{1 - S_{irG}}]^{n_G}, 1\right\}\right\}$ $k_{irG} = 0.12, k_{irA} = 0.02, n = n_G = 4$
Capillary pressure model [16]	$P_{cap} = -P_0[(S^*)^{-1/\lambda} - 1]^{-1/\lambda}$ $S^* = (S_A - S_{irA})/(S_{mxA} - S_{irA})$ $S_{irA} = 0.11, P_0 = 12500$ Pa, $\lambda = 0.254$
Composite thermal conductivity model [17]	$k_\theta = k_{\theta d} + \sqrt{S_A}(k_{\theta w} - k_{\theta d}) + \phi S_I k_{\theta I}$

4 Results

4.1 A Mild Warming Scenario: The Baseline Simulation

During the 100-year warming, the profiles of hydrate and gas saturation obtained from the baseline simulation clearly indicate that the conditions for gas dissociation are not met under this temperature change condition. The pressure and temperature of the hydrate reservoir are still in the stable range. The safety factor of the slope F_s remains unchanged at approximately 11 because of minimal variation of pore pressure. This agrees well with the conclusion that deep oceanic case is insensitive to a mild seafloor warming at a short timescales [5].

4.2 The Parametric Study

If the temperature rising rate is maintained until hydrates ultimately dissociate, seafloor may finally become unstable ($F_s < 1$). Under this hypothetical scenario, this section presents the spatial-temporal evolution of F_s via a parametric analysis in order to identify the dominant parameter(s).

Figure 2 presents profiles of F_s, S_h, and u varying over time obtained from one simulation as an example ($S_{h0} = 20\%$, $\phi = 50\%$, $k_{ou} = 0.001$ mD, $k_h = 10$ mD). Given any time period, F_s reaches the minimum value ($F_{s,\min}$) at the interface between the hydrate reservoir and the overburden layer, where hydrate dissociation is initiated and pore pressure starts to build up (highlighted in pink in Fig. 2a). This indicates that the slide occurs the earliest at the top of the hydrate reservoir as a result of a sufficient temperature rise at the seafloor. Regardless of the depth, F_s decreases over time as hydrate dissociation front propagates downwards, and $F_{s,\min}$ reaches 1 at the top of the hydrate-bearing layer in 442 years (defined as the time of failure onset).

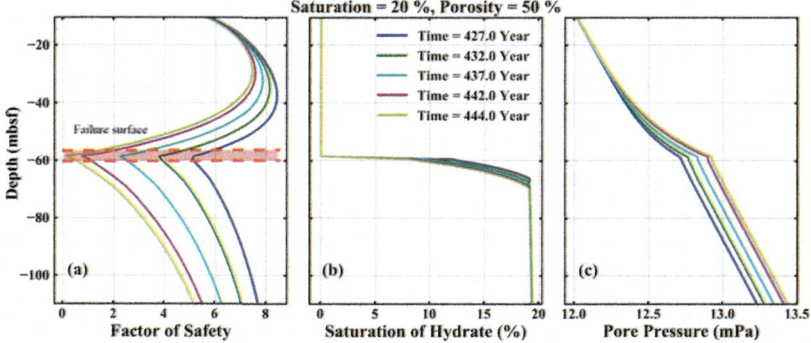

Fig. 2. The profile of (a) the safety factor, (b) hydrate saturation, and (c) total pore pressure at different time periods against depth in meter below seafloor (mbsf). The figures only plot the profiles from 427 to 444 years, during which the minimum safety factor $F_{s,\min}$ drops from 5 to 0 and slide occurs.

Each of the sub-figures in Fig. 3 plots $F_{s,min}$ over time when a single parameter varies while the other parameters remain at the baseline values. The curves of $F_{s,min}$ descend with increasing S_{h0}, and decreasing k_h, k_{ou} and n. This indicates that a submarine slope enclosing a hydrate reservoir with lower permeable sediments (in both the overburden and hydrate reservoir), higher hydrate concentration, and tighter geologic medium is more prone to failure under temperature rise. Particularly, as shown in Fig. 3 (b), the effect of k_{ou} is remarkable. If the overburden layer is highly permeable, seafloor instability is unlikely to occur because the pore pressure built up during hydrate dissociation could dissipate very rapidly. Particularly, if k_{ou} is greater than 1 mD, the failure timescale is at least in the order of thousands of years. In contrast, the effect of S_{h0} is fairly limited.

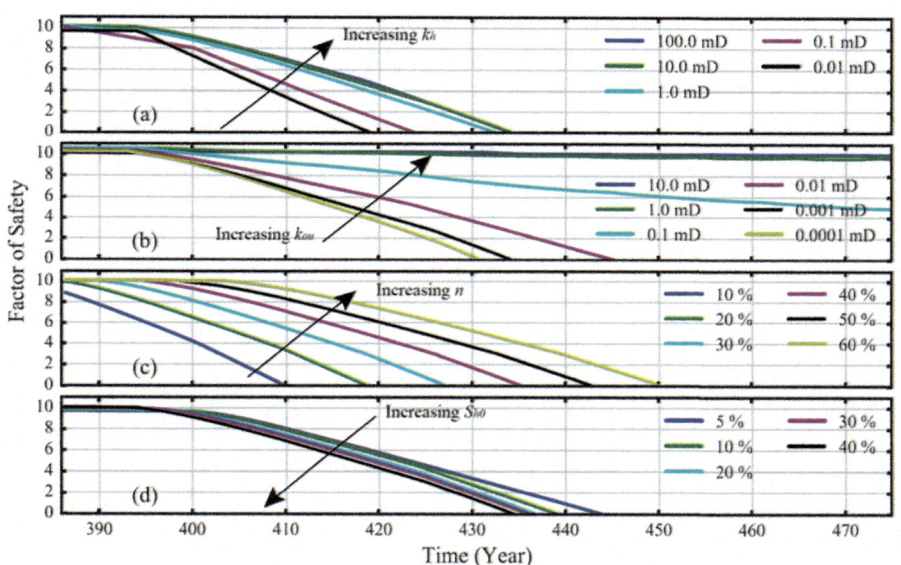

Fig. 3. Effect of (a) k_h, (b) k_{ou}, (c) n, and (d) S_{h0} on $F_{s,min}$ over time

We use the time of failure onset as a nominal descriptor for system vulnerability. An earlier onset indicates a more vulnerable slope system. Based on the result of a set of parametric simulations, each of the sub-figures of Fig. 4 plots the contours of the onset time in a space constituted by two parameters, resulting in a vulnerability map, to show the combined effect of the two parameters.

Figures 4a and b present the vulnerability maps in a k_{ou}-k_h space under high and low concentration hydrate reservoir, respectively. In both cases, the most vulnerable configuration appears at the bottom left corner with low k_{ou}-k_h; the failure onset is earlier with a lower S_h for different combinations of k_{ou} and k_h. This indicates that the hydrate-bearing slope is more vulnerable as S_{h0} increases. In both cases, k_{ou} plays a dominant role in affecting the onsets of slope failure, whereas k_h is less sensitive especially when k_h is in the range of 1 to 100 mD. However, when k_h drops below a critical value (approximately 1 mD) for both cases, the effect of k_h is enhanced. This is signaled by nearly-vertical dense contours at k_h equaling about 1mD.

Figure 4c presents the vulnerability maps in a k_{ou}-S_{h0} space. Similar to the results discussed above, k_{ou} plays a vital role in affecting the vulnerability of the slope, whereas the variation of S_{h0} has only marginal effect. This is signaled by flattened contours over a wide range of S_{h0} in the vulnerability map. The effect of S_{h0} is slightly enhanced when k_{ou} is less than 0.01 mD. A decrease of S_{h0} will delay failure onset because of the reduced volume of available hydrates in the porous media, i.e. reduced sources of pressure build-up.

Figure 4d presents the vulnerability maps in a n - S_{h0} space. The porosity of the sediments n has more significant impact on the failure onset than the initial hydrate saturation does, as the contours are nearly parallel to S_{h0}–axis. Failure delays with increasing n. This indicates that a slope with a tight reservoir with limited pore spaces is vulnerable during hydrate dissociation, because the released gas quickly fills the limited void spaces in the tight geological system and yields very rapid pressure buildup. These results show a good agreement with previous studies [18].

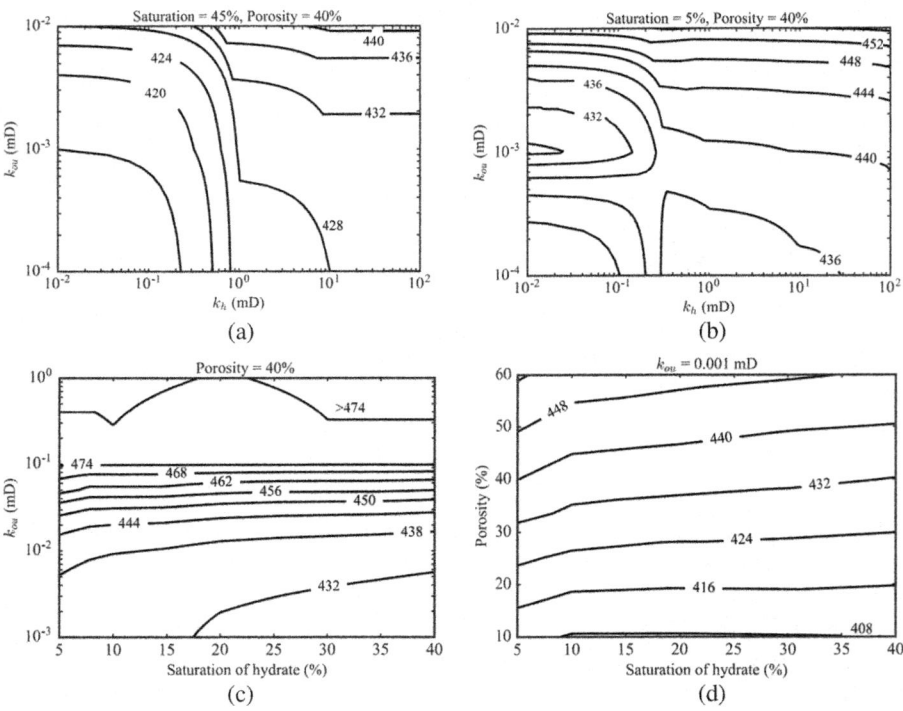

Fig. 4. Vulnerability maps in terms of slope failure onset time (in years) in (a) k_{ou}-k_h space under high hydrate concentration (S_{h0} = 45%), (b) k_{ou}-k_h space under low hydrate concentration (S_{h0} = 5%), (c) k_{ou}-S_{h0} space with porosity equaling 40%, and (d) n-S_{h0} space with permeability of the overburden equaling 0.001 mD

5 Conclusion

We developed a hybrid model to quantify the stability of seafloor subjected to hydrate dissociation. This model has a simple form within the conventional framework of limit equilibrium analysis of slope stability, and incorporates the complexity attributed to THC coupled processes in a hydrate-bearing geological system. This model is employed in this preliminary study of a slope configured according to geological conditions of hydrate reservoirs located at Shenhu area, SCS in order to identify the vulnerable settings and controlling factors. The results show the slope is stable under a mild seafloor warming scenario in a 100 years. Under the sustaining warming scenario, the slope may fail in the order of hundreds of years. The slope is particularly vulnerable to the presence of low-permeability and tight sediments overburden layer, and high hydrate concentration in the hydrate reservoir. The permeability of the overburden layer plays a dominant role in determining the vulnerability of the slope, while the permeability and hydrate concentration of the hydrate reservoir, and porosity of the geological system only have secondary effects.

The parameters examined here are limited to physical properties of the slope. Given that geometry settings of slopes have been proved to affect the instability [9], it should be valuable to conduct a more systematic and comprehensive parametric analysis incorporating the joint effects of both geometry conditions and physical properties of the slope. Additionally, the seafloor warming scenarios assumed in this study are rather rough and could be replaced by more accurate considerations of different perturbation at different rates and intensities. These simplifications used in this preliminary study will be resolved in our future work.

Acknowledgement. The work is supported by the Chinese National Natural Science Foundation (with grant No. 41572267, 51639008, and 51239010), and the Fundamental Research Funds for the Central Universities.

References

1. Mienert, J., Vanneste, M., Bünz, S., Andreassen, K., Haflidason, H., Sejrup, H.P.: Ocean warming and gas hydrate stability on the mid-norwegian margin at the storegga slide. Mar. Petrol. Geol. **22**(1), 233–244 (2005)
2. Song, Y., Yang, L., Zhao, J., Liu, W., Yang, M., Li, Y., et al.: The status of natural gas hydrate research in China: a review. Renew. Sustain. Energy Rev. **31**(2), 778–791 (2014)
3. Yu, X., Wang, J., Liang, J., Li, S., Zeng, X., Li, W.: Depositional characteristics and accumulation model of gas hydrates in northern South China Sea. Mar. Petrol. Geol. **56**(3), 74–86 (2014)
4. Matsumoto, R., Ryu, B.J., Lee, S.R., Lin, S., Wu, S., Sain, K., et al.: Occurrence and exploration of gas hydrate in the marginal seas and continental margin of the Asia and Oceania region. Mar. Petrol. Geol. **28**(10), 1751–1767 (2011)
5. Reagan, M.T., Moridis, G.J.: Dynamic response of oceanic hydrate deposits to ocean temperature change. J. Geophys. Res. Oceans **113**(12), 1–21 (2008)

6. Stranne, C., O'Regan, M., Dickens, G.R., Crill, P., Miller, C., Preto, P., Jakobsson, M.: Dynamic simulations of potential methane release from East Siberian continental slope sediments. Geochem. Geophys. Geosyst. **17**(3), 872–886 (2016)

7. Mestdagh, T., Poort, J., De Batist, M.: The sensitivity of gas hydrate reservoirs to climate change: perspectives from a new combined model for permafrost-related and marine settings. Earth Sci. Rev. **169**, 104–131 (2017)

8. Sultan, N., Cochonat, P., Foucher, J.P., Mienert, J.: Effect of gas hydrates melting on seafloor slope instability. Mar. Geol. **213**(1), 379–401 (2004)

9. Nixon, M.F., Grozic, J.L.: Submarine slope failure due to gas hydrate dissociation: a preliminary quantification. Can. Geotech. J. **44**(3), 314–325 (2007)

10. Xu, W., Germanovich, L.N.: Excess pore pressure resulting from methane hydrate dissociation in marine sediments: a theoretical approach. J. Geophys. Res. Solid Earth **111** (B1), B02104 (2006)

11. Kwon, T.H., Cho, G.C.: Submarine slope failure primed and triggered by bottom water warming in oceanic hydrate-bearing deposits. Energies **5**(8), 2849–2873 (2012)

12. Masui, A., Haneda, H., Ogata, Y., Aoki, K.: Effects of methane hydrate formation on shear strength of synthetic methane hydrate sediments. In: The Fifteenth International Offshore and Polar Engineering Conference. International Society of Offshore and Polar Engineers (2005)

13. Wang, L., Wu, S.G., Li, Q.P., Wang, D.W., Fu, S.Y.: Architecture and development of a multi-stage Baiyun submarine slide complex in the Pearl River Canyon, northern South China Sea. Geo-Mar. Lett. **34**(4), 327–343 (2014)

14. Li, G., Moridis, G.J., Zhang, K., Li, X.S.: The use of huff and puff method in a single horizontal well in gas production from marine gas hydrate deposits in the Shenhu area of South China Sea. J. Petrol. Sci. Eng. **77**(1), 49–68 (2011)

15. Stone, H.L.: Probability model for estimating three-phase relative permeability. J. Petrol. Technol. **22**(02), 214–218 (1970)

16. Van Genuchten, M.T.: A closed-form equation for predicting the hydraulic conductivity of unsaturated soils. Soil Sci. Soc. Am. J. **44**(5), 892–898 (1980)

17. Li, G., Li, X.S., Zhang, K., Li, B., Zhang, Y.: Effects of impermeable boundaries on gas production from hydrate accumulations in the Shenhu area of the South China Sea. Energies **6**(8), 4078–4096 (2013)

18. Thatcher, K.E., Westbrook, G.K., Sarkar, S., Minshull, T.A.: Methane release from warming-induced hydrate dissociation in the West Svalbard continental margin: timing, rates, and geological controls. J. Geophys. Res. Solid Earth **118**(1), 22–38 (2013)

Reliability Analysis of Sensitive Clay Slope with the Response Surface Method

Zhongqiang Liu[✉], Jung Chan Choi, Farrokh Nadim,
and Suzanne Lacasse

Norwegian Geotechnical Institute, 0806 Oslo, Norway
zhongqiang.liu@ngi.no

Abstract. The response surface method (RSM) is increasingly useful for reliability analysis of slopes, because it can integrate numerical slope stability evaluation and reliability analysis for the class of problems where each numerical simulation of slope stability is time-consuming and numerically intensive. The RSM approach is useful for constructing global approximations of the response of, for example, a soil slope under loading. A reliability analysis of a sensitive clay slope was carried out using the commercially available program PLAXIS with the RSM. Monte Carlo simulations (MCS) were done using the response surfaces established by the RSM to estimate the probability of slope failure. The results of the reliability analyses with response surface were compared with direct Monte Carlo simulations (MCS) and First Order Second Moment (FOSM) using the full PLAXIS model. The comparison indicate that the combined RSM-MCS reliability analysis leads to more accurate predictions of the slope stability than the FOSM approach and provides practically the same results as the more time-consuming MCS with the full PLAXIS model. The approach has the advantage of being computationally efficient (fast-running) when a large number of MCSs are needed.

Keywords: Slope stability · Response surface method · Sensitive clay
Reliability · Probabilistic methods

1 Introduction

Landslides in sensitive materials is a natural hazard that can be triggered by natural causes or by human actions. There are significant uncertainties in the way we calculate the factor of safety against failure. Because of regulation or tradition, the same value of factor of safety is applied to conditions that involve widely varying degrees of uncertainty. That is not logical. Risk and probabilistic analyses have reached a level of maturity that makes them effective to use in practice [1]. The reliability-based approach provides more insight than deterministic analyses alone. They provide a way to quantify the uncertainties and to handle them consistently. Reliability approaches, however, do not remove uncertainty nor do they alleviate the need for judgment. Reliability approaches also provide the basis for comparing alternatives, and focus on safety and cost-effectiveness. Site investigations, laboratory test programs, limit equilibrium and deformation analyses, instrumentation, monitoring and engineering judgment are necessary inputs to the reliability-based approach [1].

© Springer Nature Singapore Pte Ltd. 2018
A. Farid and H. Chen (Eds.): GSIC 2018, *Proceedings of GeoShanghai 2018*
International Conference: Geoenvironment and Geohazard, pp. 63–72, 2018.
https://doi.org/10.1007/978-981-13-0128-5_8

Currently, most geotechnical numerical programs do not have the probabilistic analysis function, thus the applications of reliability methods are still limited to problems with relatively simple limit state functions. Alternatively, response surface methods (RSM) can be adopted for complex geotechnical reliability problems. The reliability analysis is then realized through computationally efficient models that approximate the deterministic numerical solutions (e.g., [2, 3]).

The paper briefly describes the reliability approaches used (Sect. 2) before the concepts are applied to a case study of a slope in sensitive clay (Sect. 3). The paper introduces the method that can automate reliability analysis using the commercial geotechnical program PLAXIS and compares the mean factor of safety and nominal probability of slope failure obtained from the reliability approaches (Sect. 4).

2 Reliability Approaches Used

The objective of the reliability study was to make a probabilistic estimate of the factor of safety of the Finneidfjord landslide (1996) in mid-Norway. Two probabilistic approaches were used to obtain the mean of the factor of safety and a nominal probability of failure: the Monte Carlo simulation-based method (MCS) combined with the response surface method (RSM), the first order second moment method (FOSM). The results of the two approaches are compared in the paper. The RSM combined with the MCS approach was selected to provide an alternative method to apply reliability methods in geotechnical engineering.

2.1 Response Surface Method (RSM)

In the RSM, the probabilistic analysis is realised through computationally efficient models approximating the deterministic numerical solutions. A traditional second-order polynomial is commonly used as a surrogate model to approximate the unknown implicit performance function $g(x)$ (e.g. [2, 3]):

$$g(x) \approx a_0 + \sum_{i=1}^{k} a_k x_i + \sum_{i=1}^{k} a_{k+i} x_i^2 \tag{1}$$

where x_i is the ith element of x, k is the dimension of x, and a_i ($i = 0, 1, \ldots, 2k$) are unknown deterministic coefficients.

A commonly used central composite design proposed by [4] was employed to generate different possible combinations of independent random variables. The first scenario assigns the mean values for all random variables for a given design. Then, for each random variable, two sampling points are taken at the mean $\pm a \times$ one standard deviation of this random variable while assigning the mean to each of the other random variables. The a value can be determined from:

$$a = 2^{k/4} \tag{2}$$

2.2 Monte Carlo Simulation (MCS)

A Monte-Carlo simulation is a procedure which simulates stochastic processes by random selection of a value for each of the random variables in a model and obtains their joint probability density function. The Monte Carlo simulation approach is a powerful technique that is applicable to both linear and non-linear problems, but it can require a large number of simulations to provide a reliable probabilistic distribution of the response. To reduce the number of simulations required for a reliable evaluation, and to reduce the significant time demands for the calculations, several reduction or acceleration methods can be used. The Latin Hypercube Sampling (LHS) method (e.g. [5, 6]) is one such acceleration method well suited to a complex problem with many random variables. Python coding was used to realize the 100 Monte Carlo simulations enhanced with the LHS method. 100 MCSs are enough to obtain reliable results as the nominal probability of failure is extremely high in the study.

2.3 First Order Second Moment (FOSM)

The first order second moment method (FOSM) is one of the more common probabilistic approaches to solve probabilistically different geotechnical problems (e.g. [7, 8]). The FOSM approach provides analytical approximations for the mean and standard deviation of a parameter as a function of the mean and standard deviations of the various input random variables and their correlations. Details on FOSM approach can be found elsewhere (e.g. [9]).

3 Description and Triggering of the Finneidfjord Landslide

The catastrophic landslide at Finneidfjord in 1996 mobilized 1 million m^3 of sediments and caused four fatalities. The vicinity of the landslide, with contour lines illustrating the steepness of the embankment down the fjord was shown in Fig. 1. Geotechnical investigations before and after the landslide revealed large volumes of sensitive clay, particularly in the shore and foreshore regions, that were mobilized in the later stages of the landslide (e.g. [10, 11]).

The landslide occurred shortly after about 14 days of heavy rain. Reference [11] documented that, about several months before the landslide, the placement of 12 to 15,000 m^3 of fill, with a total height of 2.5 m, on the foreshore of the Fjord was the triggering mechanism for the landslide. Reference [12] found that a 30 cm weak layer about 3 m below seafloor was sandwiched between clay layers of very low permeability. This layer is thought to be the initial sliding plane.

4 Reliability Analysis of the Finneidfjord Slope

For the Finneidfjord slope, the probability of failure was investigated. The level of the fill height, from 0 to 2 m, was used as governing parameter. The thin weak layer was included in the numerical model used.

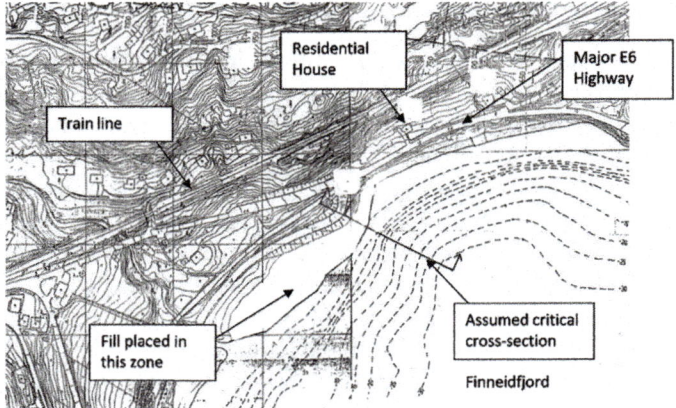

Fig. 1. Map of area prior to Finneidfjord landslide (after Gregersen [11]; Cassidy et al. [14]).

4.1   Numerical Model

The NGI-ADP material model [13] was used in the PLAXIS finite element analyses (www.plaxis.nl). The NGI-ADP model is an elasto-plastic model (therefore not accounting for strain-softening) that is able to account the anisotropy of the undrained shear strength. The finite element model adopted for the analyses is shown in Fig. 2. A mesh convergence study was done and showed that the mesh discretisation was sufficiently fine (scales in both directions are in meters) to give reliable numerical results.

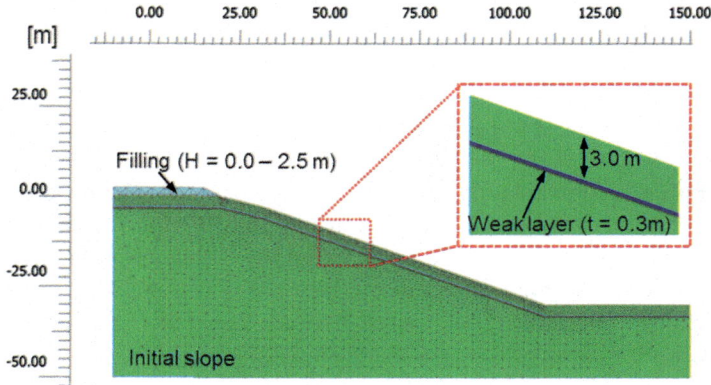

Fig. 2. PLAXIS finite element model for the Finneidfjord slope.

The probabilistic analysis was combined with the finite element method with an in-house scripting code. The analysis procedure controlled by Python scripting language with the Monte Carlo simulation is described in Fig. 3. The code generates random numbers from the statistical description of the input parameters. Latin hypercube sampling was used to generate near-random numbers efficiently. The code then fed the generated random numbers into the PLAXIS 2D model using the API functionality (www.plaxis.nl). The calculated results were stored for each iteration.

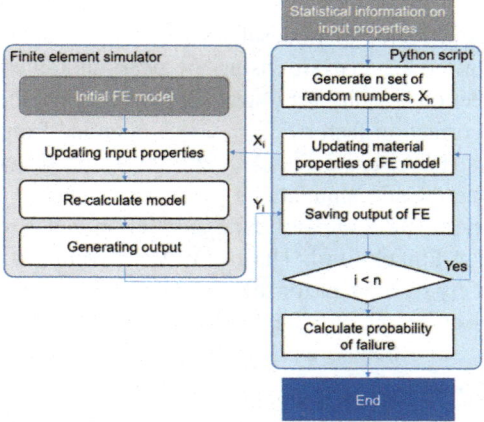

Fig. 3. Flow diagram with Python script of Monte Carlo simulations in finite element analysis.

4.2 Statistical Description of Soil Properties

Table 1 summarizes the soil strength profile used in the finite element analyses, including the mean, coefficient of variation and the probability density function. The values were based on the earlier studies of shear strength at the Finneidfjord location ([11, 12, 14, 15]). The anisotropy ratios s_{uDSS} / s_{uC} and s_{uE} / s_{uC} were taken as 0.7 and 0.4 respectively, where s_{uC} stands for the undrained shear strength in triaxial compression, s_{uDSS} stands for the undrained shear strength in simple shear and s_{uE} stands for the undrained shear strength in triaxial extension.

Table 1. Material properties used in assessment of probability of failure of Finneidfjord slope.

Parameter	Mean	Distribution	CoV
Shear strength in triaxial compression (s_{uC}) and increase in s_u with depth s_{u_inc}	10 kPa for depths $\leq$ 3m 0.33 p'_o for depths > 3m	Lognormal	0.15
Shear strength weak layer (s_{uC})	8 kPa	Lognormal	0.25
ϕ' of fill material	40°	Lognormal	0.10
Strain-softening correction factor	1.07, range 1.04 to 1.10	-	-

Results from both laboratory data and cone penetration tests showed that the sensitivity of the 30-cm weaker clay layer was greater than the rest of the clay, with values up to 9 [12]. The uncertainty in this thin clay layer was also significantly higher (CoV of 25%, whereas CoV is 15% in the rest of the clay).

As part of the large research NIFS project in Norway, [16, 17] proposed to account for the effect of strain-softening in elasto-plastic analyses with a correction factor on the factor of safety, $\gamma_M^{softening}$:

$$\gamma_M^{softening} = \gamma_M / F_{softening} \tag{3}$$

where γ_M is the material factor (partial safety factor) calculated by an elasto-plastic model (as in limit equilibrium analysis) using the peak undrained shear strength, and $F_{softening}$ is the correction required to account for the reduced capacity due to strain-softening. The correction factor $F_{softening}$ was obtained from a detailed series of finite element analyses (FEA) using PLAXIS and the strain-softening material model NGI-ADPSoft and Monte Carlo simulations [16].

In the present analyses, the strain-softening correction factor recommended in [16] was used for the clay at Finneidfjord. The factor was introduced as a quotient on the calculated factor of safety to account for the strain-softening that is not included in the elasto-plastic model used in the PLAXIS analyses. From the cited earlier studies, a mean value for $F_{softening}$ of 1.07 was used, with a range of 1.04 to 1.10. This parameter served as model uncertainty.

4.3   Results of Probabilistic Analyses of Finneidfjord Slope Failure

An example of the initial failure mechanism is illustrated in Fig. 4. The start of the failure was located in the thin weak clay layer at a depth of 3 m.

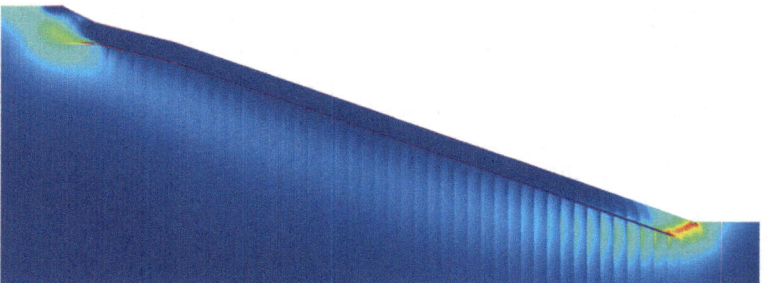

Fig. 4. Failure mechanism of the Finneidfjord slope.

The deterministic factor of safety obtained by PLAXIS analyses was 0.99 for no added fill, and 0.97 for a fill of 2 m. The possible reasons why the calculated factor of safety is not unity (1.0) at the time of the start of failure (fill height of 2.3 m) include:

– Uncertainty in the undrained shear strength of the weak clay layer. A change from 8 kPa to 7 or 9 kPa can be significant for the factor of safety, but difficult to measure in the laboratory.
– Uncertainty in the shear strength anisotropy, especially the shear strength in simple shear.

– Exact fill eight at initiation of failure.
– Time effects: the weak layer clay is not highly sensitive (Sensitivity of 9 only).
– Finite element modelling.

The mean probabilistic factor of safety and nominal probability of failure (at time of failure) for fill heights from 0 to 2 m are given in Figs. 5 and 6. The factors of safety and the probabilities of failure calculated by the RSM-MCS approach are comparable to the ones calculated by 100 MCSs. The FOSM results, however, differ from the RSM-MCS and MCS results. The assumed linearized limit state function around its mean point in the FOSM formulation is the explanation for the difference. There is however a different requirement of computer time to do the three types of probabilistic approaches. The Monte Carlo analyses combined with the PLAXIS analyses required about 5 days to run 100 simulations on a double 2.8 GHz Intel computer, whereas the RSM-MCS and the FOSM calculations could be done in a few hours.

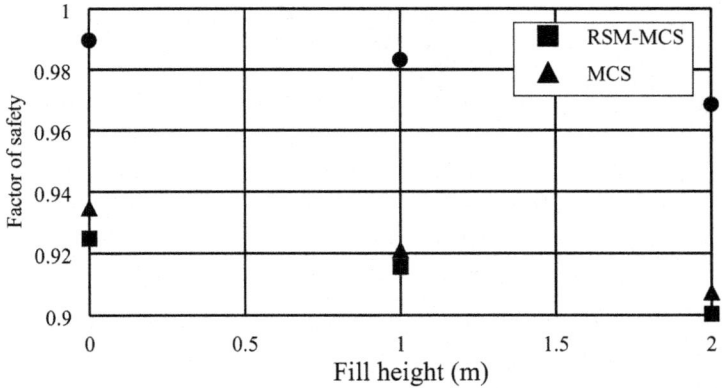

Fig. 5. Calculated mean factors of safety of Finneidfjord slope as a function of fill height.

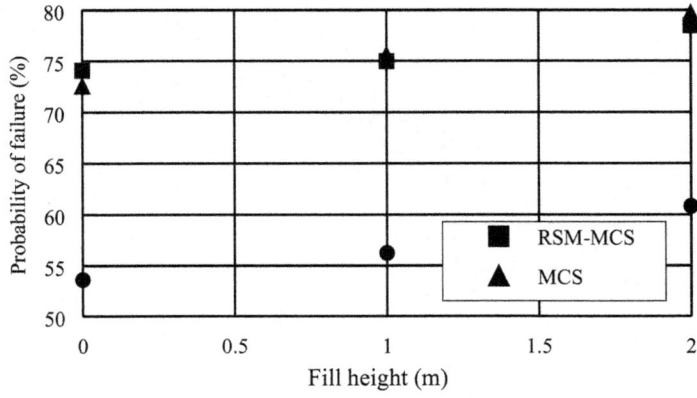

Fig. 6. Calculated nominal probability of failure of Finneidfjord slope as a function of fill height.

The three approaches suggested nominal probabilities of failure that are very high: 55% and above, with 80% for an added fill of 2 m. These values are considered as realistic. One would not allow a design when the estimated nominal probability of failure is 50% or more.

Figures 7, 8 and 9 compare the factors of safety calculated from 100 MCSs with the ones calculated by 100 RSM approach for no fill, 1 m fill and 2 m fill on the Finneidjord slope. The prediction of higher factor of safety using RSM approach appears to be more accurate than that of lower factor of safety, particularly for the factor of safety less than 1.

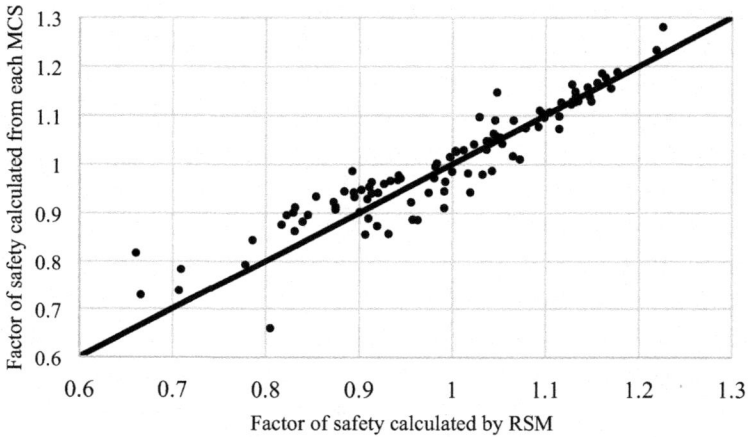

Fig. 7. Comparison of factor of safety from RSM and from PLAXIS for 100 MC simulations (Finneidfjord slope, no fill)

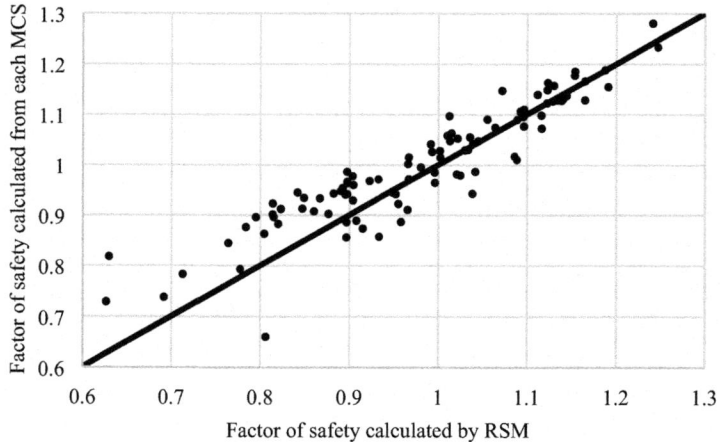

Fig. 8. Comparison of factor of safety of Finneidfjord slope with 1 m filling from RSM and from PLAXIS for 100 MC simulations (Finneidford slope, 1-m fill)

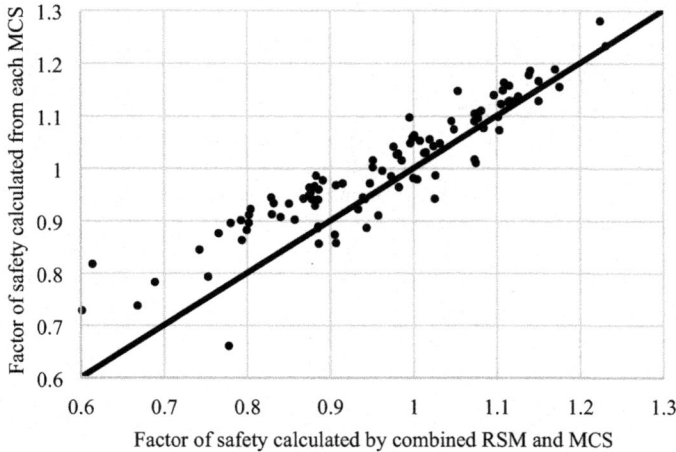

Fig. 9. Comparison of factor of safety of Finneidfjord slope with 2 m filling from RSM and from PLAXIS for 100 MC simulations (Finneidford slope, 2-m fill)

5 Conclusions

A combined RSM-MCS probabilistic approach is suggested to do the geotechnical reliability analysis, using stand-alone deterministic numerical programs. The results of the analyses in this paper suggest the following conclusions:

(1) The probabilities of failure calculated by RSM-MCS method are comparable to the ones calculated by MCS, which is more accurate than the ones calculated by FOSM for a nonlinear slope stability problem in the study;

(2) The values of mean safety factor of 0.92 to 0.90 calculated from the RSM-MCS approach are however less than unity. The difference may be due to the uncertainties in the shear strength parameters within the failure volume, and especially in the thin layer of weak clay, and a difference between the model and the actual in situ behaviour at failure;

(3) The prediction of higher factor of safety using the RSM approach appears to be more representative that the actual factor of safety.

(4) All three probabilistic approaches gave realistic nominal probability of failure, with the RMS-MCS and MCS analyses giving probability of failure closest to unity for the failed Finneidfjord slope.

The RSM-MCS approach applied herein is suitable to estimate the probability of slope failure and has the advantage of being computationally efficient (fast-running) when a large number of MCSs are needed.

References

1. Lacasse, S.: 55th Rankine Lecture. Hazard, risk and reliability in geotechnical practice. Paper submitted to Geotechnique (2017)
2. Xu, B., Low, B.K.: Probabilistic stability analyses of embankments based on finite-element method. J. Geotech. Geoenviron. Eng. **132**(11), 1444–1454 (2006)
3. Liu, Z.Q., Choi, J.C., Lacasse, S., Nadim, F.: Uncertainty analyses of time-dependent behaviour of Ballina test embankment. Comput. Geotech. **93**, 133–149 (2018)
4. Box, G.E., Wilson, K.G.: On the experimental attainment of optimum conditions. J. Roy. Stat. Soc. **13**, 1–45 (1951)
5. Iman, R.L., Helton, J.C.: An approach to sensitivity analysis of computer models, part 1. Introduction, input variable selection and preliminary variable assessment. J. Qual. Technol. **13**(3), 174–183 (1981)
6. Lacasse, S., Nadim, F.: Uncertainties in characterizing soil properties. In: Uncertainty in the Geologic Environment: From Theory to Practice. Proceedings of Uncertainty 1996, Geotechnical Special Publication, 58, pp. 49–75 (2016)
7. Kaynia, A.M., Papathoma-Köhle, M., Neuhäuser, B., Ratzinger, K., Wenzel, H., Medina-Cetina, Z.: Probabilistic assessment of vulnerability to landslide: application to the village of Lichtenstein, Baden-Württemberg. Ger. Eng. Geol. **101**, 33–48 (2008)
8. Liu, Z.Q., Lacasse, S., Nadim, F., Gilbert, R.: Reliability of API and ISO guidelines for bearing capacity of offshore shallow foundations. In: Schweckendiek, T. et al. (eds.) Proceedings of the Fifth International Symposium on Geotechnical Safety and Risk (ISGSR 2015), pp. 803–809. IOS Press (2015)
9. Alfredo, H.S., Tang, W.H.: Probability Concepts in Engineering. Emphasis on Applications to Civil and Environmental Engineering, 2nd edn. Wiley, Hoboken (2007)
10. Janbu, N.: Raset i Finneidfjord – 20 Juni 1996. Unpublished expert's report prepared for the County Sheriff of Nordland, Norway. Report no. 1, rev. 1 (in Norwegian) (1996)
11. Gregersen, O.: Kvikkleireskredet I Finneidfjord 20 juni 1996. NGI Report 980005-1. NGI, Oslo, Norway (in Norwegian) (1999)
12. Vanneste, M., Longva, O., L'Heureux, J.S., Steiner, A., Vardy, M.E., Morgan, E., Forsberg, C.F., Kvalstad, T.J., Strout, J.M., Brendryen, J., Haflidason, H., Lecomte, I., Kopf, A., Mörz, T., Kreiter, S.: Finneidfjord, a field laboratory for integrated submarine slope stability assessments and characterization of landslide-prone sediments: a review. OTC Offshore Technology Conference, OTC Paper P-686-OTC (2013)
13. Grimstad, G., Andresen, L., Jostad, H.P.: NGI ADP: anisotropic shear strength model for clay. Int. J. Numer. Anal. Meth. Geomech. **36**(4), 483–497 (2010)
14. Cassidy, M.J., Uzielli, M., Lacasse, S.: Probability risk assessment of landslides: a case study at Finneidfjord. Can. Geotech. J. **45**, 1250–1267 (2008)
15. Lacasse, S., Liu, Z., Kim, J., Choi, J.C., Nadim, F.: Reliability of slopes in sensitive clays. In: Thakur, V., L'Heureux, J.S., Locat, A. (eds.) Landslides in Sensitive Clays. ANTHR, vol. 46, pp. 511–537. Springer, Cham (2017). https://doi.org/10.1007/978-3-319-56487-6_45
16. NIFS: Natural Hazards – Infrastructure for flood and slides. Programplan 2012–2015 for etatsprogrammet (in Norwegian) (2012) ISSN 1501-2832
17. Fornes, P., Jostad, H.P.: Correction factors for undrained LE analyses of sensitive clays. In: Thakur, V., L'Heureux, J.S., Locat, A. (eds.) Landslides in Sensitive Clays. ANTHR, vol. 46, pp. 225–235. Springer, Cham (2017). https://doi.org/10.1007/978-3-319-56487-6_20

Effect of Particle Size Segregation in Debris Flow Deposition: A Preliminary Study

Lu Jing[1,2], Fiona C. Y. Kwok[1,2(✉)], Tao Zhao[2], and Jiawen Zhou[2]

[1] Department of Civil Engineering, The University of Hong Kong,
Pokfulam Road, Hong Kong, China
fiona.kwok@hku.hk
[2] State Key Laboratory of Hydraulics and Mountain River Engineering,
College of Water Resource and Hydropower,
Sichuan University, Chengdu 610065, Sichuan, China

Abstract. To understand the effect of size segregation in the depositional process of debris flows, both flume experiments at the laboratory scale and numerical simulations using the discrete element method (DEM) are performed. A variety of particle size distributions with coarse and fine particles are adopted. It is found that larger particles tend to reach the front of the final deposits, while small particles are accumulated at the tail of the flows. Quantitative agreement is achieved in the DEM simulations, where rolling resistance and geometric roughness at boundaries are adopted to account for the effect of particle shape. With the DEM results, the effect of segregation on the runout distance is studied from the perspective of energy dissipation. The progress of segregation is analyzed in detail, which revealed that segregation occur slowly while the flow is propagating rapidly over the slopes; it becomes significant during the deposition stage, where more large particles are found near the surface. The effect of segregation in debris flow deposition can help better predict the runout distance and impact pressure, which is crucial in the assessment and mitigation of debris flow-related natural hazards.

Keywords: Segregation · Debris flow · Flume experiment · DEM

1 Introduction

Particle size segregation can be found in debris flows where coarse and fine grains tend to separate themselves according to their different sizes [1]. In general, small particles settle towards the bottom while large particles drift towards the surface [2–4]. The occurrence of such segregation may enhance the runout distance and modify the deposit morphology, which are crucial in the assessment and mitigation of debris flow-related natural hazards [5–8].

To understand the effect of size segregation in the depositional process of debris flows, both flume experiments at the laboratory scale and numerical simulations using the discrete element method (DEM) are performed. Our focus is the effect of flow composition (i.e., particles size distributions) on flow dynamics, thus runout distance and deposit morphology. In the current preliminary work, we report a set of

© Springer Nature Singapore Pte Ltd. 2018
A. Farid and H. Chen (Eds.): GSIC 2018, *Proceedings of GeoShanghai 2018*
International Conference: Geoenvironment and Geohazard, pp. 73–80, 2018.
https://doi.org/10.1007/978-981-13-0128-5_9

experiments with different concentrations of large and small particles, and propose a DEM model which captures the main flow behaviors by using spherical particles, rolling friction, and a dissipative base with geometric roughness. By calibrating input parameters independently for each species (i.e., large and small grains), the segregating behavior of the mixture of large and small grains can be reproduced without further calibration. The effects of segregation on the flow dynamics and deposit morphology are discussed.

2 Experiments and Simulations

2.1 Experimental Setup

The experimental flume consists of a channel with two inclined segments and a horizontal part, as shown in Fig. 1. The upper and lower slopes have the length of 2 m and 1.5 m, and the angles of inclination of 38.3° and 17.5°, respectively. A tank (0.5 m long) is located at the top of the flume. The width of the whole channel is consistently 0.35 m. In this paper, we define the start of the horizontal channel as $x = 0$, x-axis points along the runout direction, and the z-axis points upward.

The materials we use are rock fragments taken from a site of natural debris avalanche in the Bayi Gully, Southwest China [9]. The size of "large" gains varies between 40 mm and 60 mm, while "small" grains are 10–20 mm in diameter. In this paper we report the results of four experiments, each has a different mass concentration of large grains, i.e., $\Phi_L = 0$, 33.3%, 66.7% and 100%. In each experiment, dry rock fragments with a total mass of 60 kg is well mixed and poured into the tank. The gate of the tank is removed quickly to initiate the flow, which accelerates on the upper slope and deposit on the lower regions. After the cease of each flow, we measure the weight of large and small grains every 0.2 m along the channel, producing accumulative percentage of mass for large, small, and all grains.

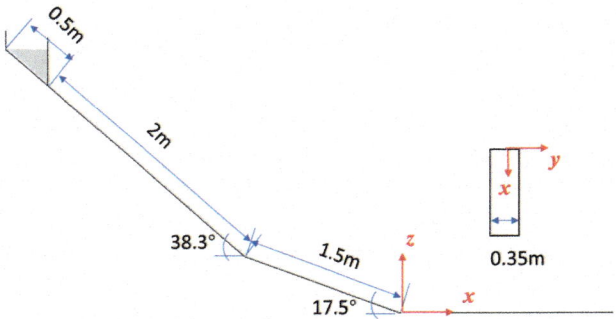

Fig. 1. Flume and coordinate system.

2.2 Numerical Model

Discrete element modeling of the experimental debris flows is performed using an opensource program, LIGGGHTS, which implements the Hertz model for the calculation of contact forces [10]. In this model, four material properties are used, namely, Young's modulus E, Poisson's ratio v, the coefficient of friction μ, and the coefficient of restitution e.

We use spherical particles with uniform size distribution to model the large (40–60 mm) and small (10–20 mm) grains. The initial mass of each test is reproduced, as shown in Fig. 2. Note that due to the effect of gravity, a slight initial separation of the two species can be observed for the 33.3% and 66.7% cases. However, such an initial segregation is negligible compared to the major segregation stages, as we shall see in the later analysis.

A layer of spherical beads is fixed as a rough bed on the surface of the entire channel [11]. Although in the experiments the bottom is made of flat plane, the roughened base is necessary in the DEM simulations due to the use of spherical particles. This treatment introduces several parameters, such as the size/distribution of base particles, and the contact parameters between the fixed and flowing particles. In this work, we arrange the base particles with a triangular close packing, and the particles size is set as 10 mm (see Fig. 2(e)). The choice makes sure that large grains can slide over the base layer, while small grains can accumulate at a lower slope, consistent with the experimental observations. The contact parameters between the flow and the boundary particles will be adjusted to match the experimental results, as presented later.

Another necessary treatment resulting from the use of spherical particles is to employ a rolling friction model in the flowing particles. A rolling friction parameter, μ_r, is used to account for the shape effect, which will be calibrated in the next section.

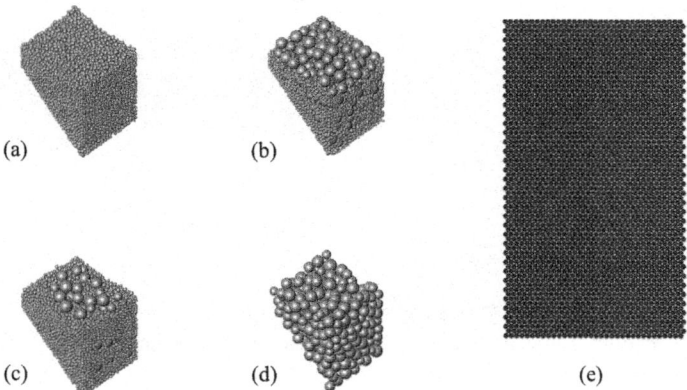

(a)   (b)

(c)   (d)   (e)

Fig. 2. Numerical models. (a–d) The initial mass of 60 kg with an overall mass concentration of large particles 0, 33.3%, 66.7% and 100%, respectively. (e) The triangular arrangement of base particles with a size of 0.01 m over the whole channel bottom.

2.3   Calibration and Validation

Our uniaxial compression tests on single grains suggest a stiffness of 50 kN/mm. In the DEM model we use $E = 10^9$ Pa and $v = 0.25$ for all particles, and we have verified that the results are not sensitive to these parameters. The tested angles of repose for large and small grains vary in a range of 31–36°, based on which we choose $\mu = 0.5$ and $\mu_r = 0.2$, which produce reasonable results in terms of the overall kinetics gained during each test. Since the collective behavior of a dense granular flow is not significantly affected by the coefficient of restitution, we choose a low value of $e = 0.2$ to avoid unphysical collisions for the highly-energetic frontal grains.

As the major parameters are chosen, we fine-tune the coefficient of friction between flowing particles and base particles, μ_b, to match the final deposit in the two cases with only small and large grains, respectively. Good agreement is achieved for both species with $\mu_b = 0.18$. The experimental and numerical deposits of the $\Phi_L = 0$ case (i.e., only small grains) are presented in Fig. 3, which show a similar depositional behavior at the toe of the lower slope. In Fig. 4(a) we plot the comparison of cumulative percentage of each species (i.e., large and small grains) when $\mu_b = 0.18$ is used. In general, the deposit morphology is captured. However, note that a large discrepancy appears at the front of the coarse-grain case. This is attributed to the fact that when large grains at the front leave the bulk, they can continue running for a longer distance than that in the experiment (where grains come to a halt easily due to shape effect). We mark these regimes of runny grains with the dashed lines in Fig. 4, which represent where only isolated grains present. The dash lines show that the discrepancy between numerical and experimental results mainly occur at these regimes.

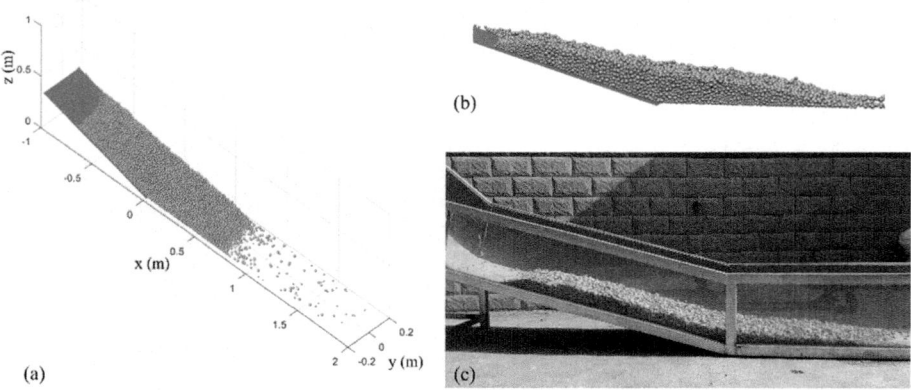

Fig. 3. Calibration of the $\Phi_L = 0$ case (purely small grains). (a) The final deposit in numerical simulation. (b) Side-view numerical deposit. (c) Side-view experiment deposit.

The value (i.e., $\mu_b = 0.18$) obtained through independent calibrations of the two species is then applied in the mixture cases where both large and small grains exist (i.e., $\Phi_L = 33.3\%$ and 66.7%). As expected, the overall depositional behaviors can be captured without further adjustment of parameters (Fig. 4(a)), and more importantly, the segregation behaviors are also well reproduced (Fig. 4(b)). By comparing the two panels in Fig. 4, it is clear that larger particles tend to reach the front of the final deposits, while small particles are accumulated at the tail of the flows.

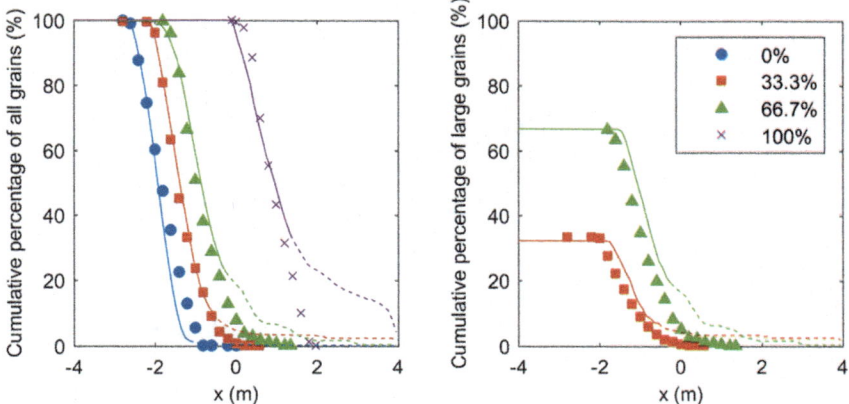

Fig. 4. Validation of numerical simulations by comparing the cumulative mass percentage of all (left panel) and large (right panel) grains along the channel. Symbols and lines are experimental and numerical results, respectively; colors and symbol types are explained in the legend. The dash lines represent frontal regions where grains remain loosely contacted (see text).

3   Results

3.1   Flow Dynamics

As seen in Fig. 4, despite the same initial mass (60 kg), thus the same potential energy, the runout distance is largely affected by the composition of debris flows. The case with purely large grains reaches a much longer distance after running out of the inclined channel. As the mass of small particles increases, the runout decreases.

Figure 5 shows the temporal evolution of the average kinetic energy, $\langle E_k \rangle$, for each case. Note that the vertical axis is in log scale. In the early stage ($t < 2$ s), the grains in the $\Phi_L = 100\%$ case gain an average kinetic energy which is greater than the other cases by a factor of 50–100. In other words, the energy dissipation increases significantly with the increasing number of small particles. We attribute this particle size dependency of energy dissipation to the number of particle–particle contacts [12]. Indeed, more contacts occur in a flow of more small particles, given the same total mass, thus a higher dissipation of energy through sliding and collisions. As we can observe from both the experiments and simulations, the $\Phi_L = 100\%$ case presents a loose state where grains remain poorly connected until they start to deposit, while the flow in other cases is generally dense due to the presence of a large number of small grains. To demonstrate this, in Fig. 6 we plot the coordination number, Z, for each flow, which measures the average number of contacts. In the initial state, all cases have

$Z > 4$. It drops dramatically towards zero since the materials start to flow. When the flow reaches a slower slope or the horizontal region, deposition occurs and the number of contacts start to increase. However, Z only increases to a value below 2 in the $\Phi_L = 100\%$, showing that the large grains therein are not flowing in a dense state.

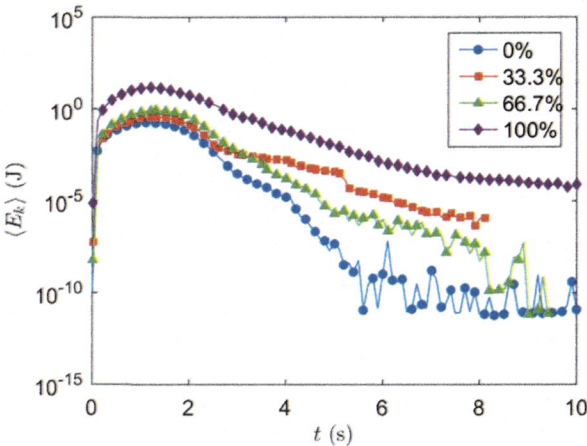

Fig. 5. Evolution of the average kinetic energy ($\langle E_k \rangle$) with time (t).

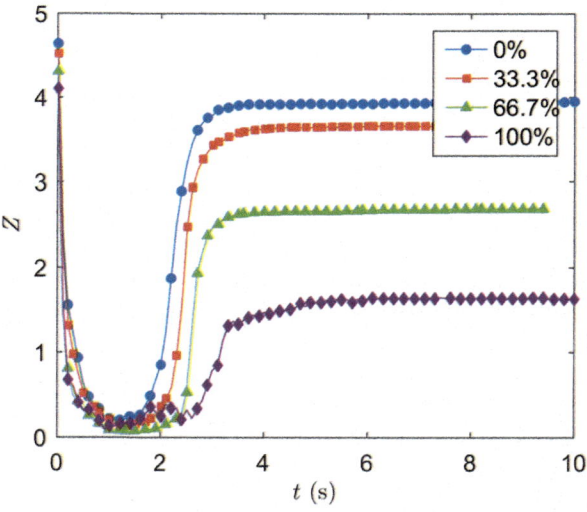

Fig. 6. Evolution of the average number of contacts (Z) with time (t).

3.2   Process of Segregation

Next, we analyze the process of segregation in the two mixture cases, i.e., $\Phi_L = 33.3\%$ and 66.7%. To monitor the occurrence of segregation, we divide the flow at any moment as vertical bins with a width of 0.5 m. In each bin (No. i), we calculate the normalized (by local flow thickness) centers of mass for large and small particles, which are denoted as C_L^i and C_s^i, respectively. The local degree of segregation is then defined as $\alpha^i = (C_L^i - C_s^i) + 0.5$. The addition of 0.5 is to define $\alpha_i = 0.5$ as the initial state when $C_L^i = C_s^i$. The overall degree of segregation, α, is calculated by averaging over all bins where both large and small particles exist. Standard deviation is also recorded as the averaging step is performed, which represents the deviation of the state of segregation at different locations.

Figure 7 shows the progress of segregation in the two mixture cases. As we noted in Fig. 2, a small amount of segregation already occurs in the initial state during the sample generation, which is however negligible compared to the later stage where α is close to 1. The standard deviation indicated by the length of error bars is the greatest before deposition occurs (around 2 s). When most of the flow is on the slopes, segregation takes place slowly. Segregation becomes more significant in the deposition stage, where nearly all large particles emerge on the surface.

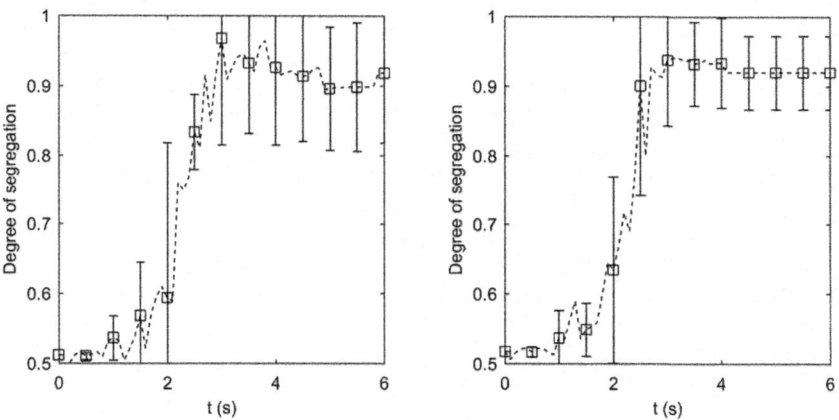

Fig. 7. Progress of segregation. Left: $\Phi_L = 33.3\%$. Right: $\Phi_L = 66.7\%$.

4   Concluding Remarks

In this paper, flume experiments and DEM simulations are performed to understand the effect of segregation in the deposition process of debris flow. The flow dynamics and the progress of segregation are analyzed. The work is a preliminary study towards a better understanding of segregation in debris flow-like natural hazards. Future work will focus on creating experimental debris flows that are more relevant to natural debris flows, and exploring the energy dissipations in relation to particle size distribution and the mechanical aspects of segregation with the aid of DEM modeling.

References

1. Johnson, C.G., Kokelaar, B.P., Iverson, R.M., Logan, M., LaHusen, R.G., Gray, J.M.N.T.: Grain-size segregation and levee formation in geophysical mass flows. J. Geophys. Res. **117** (F1), F01032 (2012)
2. Savage, S.B., Lun, C.K.K.: Particle size segregation in inclined chute flow of dry cohesionless granular solids. J. Fluid Mech. **189**, 311–335 (1988)
3. Gray, J.M.N.T., Thornton, A.R.: A theory for particle size segregation in shallow granular free-surface flows. Proc. Roy. Soc. A **461**(2057), 1447–1473 (2005)
4. Jing, L., Kwok, C.Y., Leung, Y.F.: Micromechanical origin of particle size segregation. Phys. Rev. Lett. **118**(11), 118001 (2017)
5. Kokelaar, B.P., Graham, R.L., Gray, J.M.N.T., Vallance, J.W.: Fine-grained linings of leveed channels facilitate runout of granular flows. Earth Planet. Sci. Lett. **385**, 172–180 (2014)
6. Zanuttigh, B., Lamberti, A.: Instability and surge development in debris flows. Rev. Geophys. **45**(3) (2007). https://doi.org/10.1029/2005RG000175
7. Zhou, G.G.D., Ng, C.W.W.: Numerical investigation of reverse segregation in debris flows by DEM. Granul. Matter **12**(5), 507–516 (2010)
8. Zhou, G.G.D., Wright, N.G., Sun, Q.C., Cai, Q.P.: Experimental study on the mobility of channelized granular mass flow. Acta Geol. Sin. **90**(3), 988–998 (2016)
9. Zhou, J., Huang, K., Shi, C., Hao, M., Guo, C.: Discrete element modeling of the mass movement and loose material supplying the gully process of a debris avalanche in the Bayi Gully, Southwest China. J. Asian Earth Sci. **99**, 95–111 (2015)
10. Goniva, C., Kloss, C., Deen, N.G., Kuipers, J.A.M., Pirker, S.: Influence of rolling friction on single spout fluidized bed simulation. Particuology **10**(5), 582–591 (2012)
11. Jing, L., Kwok, C.Y., Leung, Y.F., Sobral, Y.D.: Characterization of base roughness for granular chute flows. Phys. Rev. E **94**(5), 052901 (2016)
12. Utili, S., Zhao, T., Houlsby, G.T.: 3D DEM investigation of granular column collapse: evaluation of debris motion and its destructive power. Eng. Geol. **186**, 3–16 (2015)

Study on Slope Stability Effected by Creep Characteristics of Weak Interlayer in Clastic Rock Slope in Guangxi

Yingzi Xu[1(✉)], Fu Wei[2], Rikui Yan[2], and Jian Li[2]

[1] Guangxi Key Laboratory of Disaster Prevention and Engineering Safety, College of Civil and Architectural Engineering, Guangxi University, Nanning 530004, China
xuyingzi@gxu.edu.cn
[2] College of Civil and Architectural Engineering, Guangxi University, Nanning 530004, China
53803184@qq.com, 1609394677@qq.com, 1021894990@qq.com

Abstract. The decreased long-term shear strength of the weak interlayer is one of main reasons to occur landslide in the clastic rock region of Guangxi. The goal of this paper is to study the creep characteristics of weak interlayer of clastic rock slope in Guangxi and its influence to the long-term slope stability. Soil samples were collected from a typical clastic rock slope in Guangxi. The shear creep test results show that the long-term shear strength of the weak interlayer soil decreased to half of peak shear strength after long-term load. The modeling results show that the slope factor of safety decreased from 1.493 to 0.717 when the shear strength of weak interlayer soil decreased from peak to residual shear strength. The creep characteristics of the weak interlayer soil led to the significance loss of residual strength thus weakens the slope stability.

Keywords: Clastic rock · Slope · Weak interlayer · Creep characteristics
Residual shear strength · Slope stability

1 Introduction

Clastic rocks are widely distributed in Guangxi, China. Weak interlayers propagate and interbed inside the soft rock. When the rock slope contains weak intercalated layers, creep may cause slope stability issue. Xie and Sun [1] summarize the creep characteristics of several different saturated soft clay in Shanghai. Yan [2] studied the creep test at different consolidation pressures on the large-rocky landslide soil. In terms of the influence of the weak interlayer on the stability of the slope, Dai et al. [3] show that the mudded intercalation is the controlling factor of slope stability. Wang [4] analyzed the variation of the strength from the material composition, thickness, structure, production and load of the weak interlayer. By analyzing the failure model of the slope, the failure mechanism of the weak interlayer slope is revealed. Xu et al. [5] found that weak interlayer lead into a potential sliding surface under the influence of external factors.

The study on the soft interlayer of the clastic rock in Guangxi is rare in literature. This paper studied the weak interlayer of Guangxi clastic rock through triple shear creep

© Springer Nature Singapore Pte Ltd. 2018
A. Farid and H. Chen (Eds.): GSIC 2018, *Proceedings of GeoShanghai 2018*
International Conference: Geoenvironment and Geohazard, pp. 81–89, 2018.
https://doi.org/10.1007/978-981-13-0128-5_10

test. The creep characteristics and long-term strength problems were analyzed. The slope stability of rock slope under residual strength and peak strength were simulated.

2 Materials and Methods

2.1 Study Area

The clastic rock landslide in this study is located in Xilin County, Guangxi of China. The thickness of weak interlayer is from a few centimeters to several meters. The clastic rocks in Xilin are usually interlayer or interbedded with soft mudstone/shale, and hard sandstone/siltstone. The decreased strength of the weak interlayer affects the stability of the clastic rock slope.

2.2 Weak Interlayer Soil

The weak interlayer soil of the landslide is a detrital mudstone interlayer with red-brown a fully-weathered mudstone and oxidized mudstone debris. The soil characterization of the weak interlayer soil showed that the PI of the interlayer soil is 28.61 and the nonuniformity coefficient is 18. According to Code for Investigation of Geotechnical Engineering (GB50021-2001), the test soil is named as well-graded clay.

2.3 Consolidated Shear Creep Test

The consolidation shear creep test was used to obtain the residual strength parameters of the weak interlayer soil and creep characteristics. Test equipment is ZLB-1-type triple flow rheometer made by Nanjing Soil Instrument Inc. Graded loading method was used to conduct the creep loading test. When shear deformation of the specimen did not exceed 0.01 mm within 24 h, it was stable to load the next level of shear stress. The vertical pressures were 50 kPa, 100 kPa and 150 kPa in the creep tests. The peak strength of the samples was 65.26 kPa, 82.71 kPa and 100.24 kPa at 50 kPa, 100 kPa and 150 kPa vertical pressure respectively by direct shear test. After several exploratory tests of different loading stages, six loading levels were used. The horizontal shear displacement measurement test data were recorded until the horizontal shear displacement is up to the creep stability standard.

2.4 Numerical Model for Stability of Clastic Rock Slope with Weak Interlayer

The conceptual slope is a bedding soft clastic rock slope, which has 20–30 m high and 25–35 degree slope ratio. The surface of the slope is the Quaternary residual slope soil with the thickness of 1–2.5 m. The lower layers are the sandstones with different weathering degrees, which the shallow layer is about 3–6 m thick fully weathering sandstone filled with mud. The deeper layers are the thick strong weathered sandstone and the middle-weathered sandstone and weak weathered sandstone. There is a thin weathered mudstone layer between the fully weathered and the strong weathered sandstone. The property of the mudstone is much different from its surrounding sandstone, which can be considered a weak interlayer of this slope. The field data show that the groundwater level of the slope is about 2.45 m below the horizontal ground. Groundwater is mainly stored in sandstone and weathering fissures, where the permeability is strong. The weak interlayer is poor in permeability, where form a relative aquifer to gather water and soften the mudstone. The weak interlayer affects the slope stability. The simplified model of clastic rock slope with weak interlayer was shown in Fig. 1.

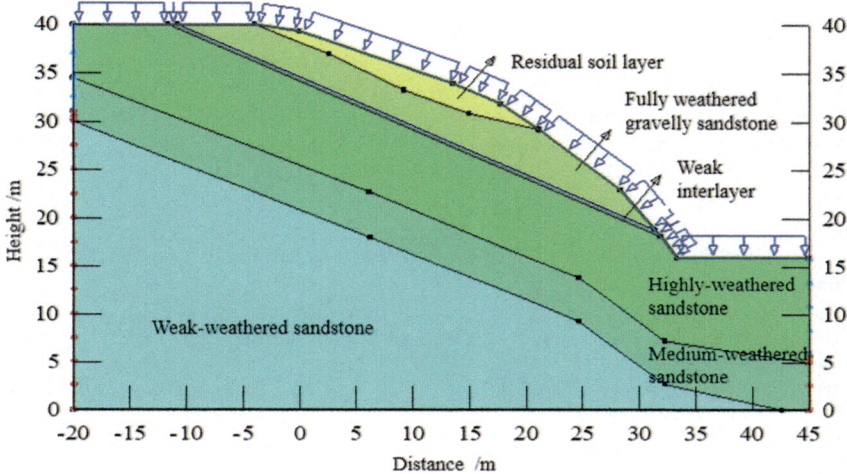

Fig. 1. Conceptual model of clastic rock slope

Hydraulic Boundary Conditions. Steady flow calculation was carried out by setting the head boundary and the flow boundary condition. The field data showed that the pore water was few in this area, and the fracture water was just exposed for small area. The bottom boundary was set as no flow condition. The total head of left and right boundaries were set as 31 m and 6 m. The slope stability of natural state was considered in this study. The rainfall intensity of the slope in natural state was set as 1114 mm/y as the flow boundary conditions, which was the average annual rainfall (1992–2013 about 22 years) in Xilin County.

Model Parameters. According to field data, the hydraulic conductivity and mechanical parameters of each rock and soil layer are shown in Table 1.

Stability Analysis Method. The limit equilibrium method is one of the widely used methods because of its simple mechanical model and convenient calculation. Based on Mohr - Coulomb theory of shear strength, this method pre-supposes a slip surface first, then divides this slip surface into lots of strips, and builds static equilibrium equations on these strips. The most dangerous slip surface and factors of safety could be obtained finally. Morgenstern-Price method, as the most rigid and accurate method of limit equilibrium method, is widely used to analyze slope stability. The Morgenstern-Price method meets the balance of force and the moment balance. The method has no

Table 1. Parameters of each rock and soil layer

	Unit weight γ/kN·m^{-3}	Saturated unit weight γ_{sat}/kN·m^{-3}	Cohesion c/kPa	Friction angle φ/°	Saturated hydraulic conductivity k_s/m·s^{-1}	Saturated volumetric water content θ_s/m^3·m^{-3}
Residual slope	18.70	19.90	20.48	18.59	60.20×10^{-6}	0.42
Fully weathered sandstone	20.38	21.26	48.80	20.70	1.462×10^{-6}	0.38
Strong weathered sandstone	22.60	23.30	93.30	24.50	0.951×10^{-6}	0.31
Weathering sandstone	25.30	25.60	845.20	33.70	0.520×10^{-6}	0.29
Weak weathering sandstone	26.50	26.70	1502.30	39.10	0.358×10^{-6}	0.28
Weak interlayer (peak)	19.60	20.30	47.76	19.28	0.056×10^{-6}	0.33
Weak interlayers (long-term)	19.60	20.30	20.33	10.20	0.056×10^{-6}	0.33

limitations of force among strips and slip surface. The Morgenstern-Price method was chosen for slope stability analysis in this paper.

SEEP module and SLOPE module are two modules of one software. SEEP module is a groundwater seepage analysis software and SLOPE module is a slope stability analysis software. For the slope stability analysis under the natural state, the steady seepage was analyzed by SEEP module, and then the seepage calculation results were imported to SLOPE module. Morgenstern-Price method was selected to calculate the factors of safety. The influence of weak interlayer soil strength parameters on slope stability is analyzed in this article. The strength parameters of weak interlayer include the peak strength from direct shear test and long-term strength from creep test.

3   Results and Discussion

3.1   Shear Creep for Weak Interlayer Soil

The laboratory test data were collated to obtain the creep curve of the weak interlayer soil specimen of Xilin County, Guangxi, under different vertical pressure (Fig. 2).

As shown in Fig. 2, the creep deformation of the clastic rocks weak interlayer soil is obvious, and there is instantaneous deformation in every shear stress level. At the same vertical pressure, the creep is mainly elastic deformation and deformation is small when the shear stress level is low. As the shear stress level increases, the creep rate increases in the stable creep stage. When the shear stress increases to a certain value,

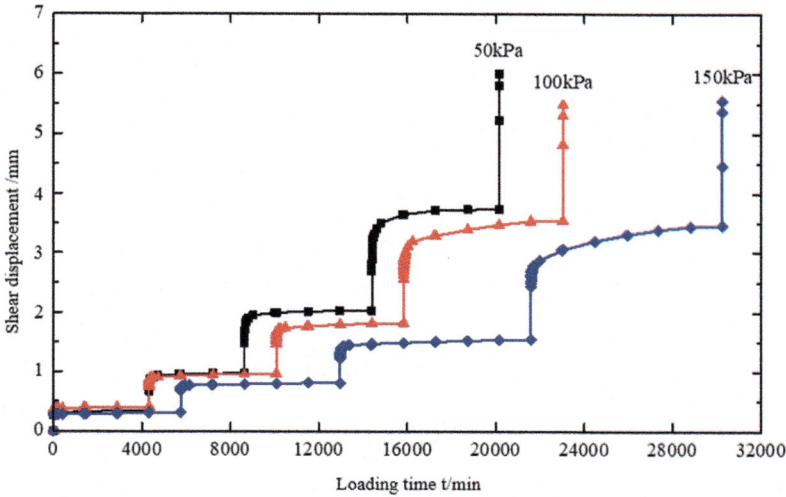

Fig. 2. Shear creep for weak interlayer soil

the deformation increases rapidly and enters the creep damage stage thus the specimen brittle damage occurs. At the same shear stress level, the greater the vertical pressure, the longer the time required for the sample to stabilize, that indicating the slope creep stabilization time has a direct relationship with the size of the overburden load.

3.2   Creep Curves of Weak Interlayer Soil by Staged Loading

The creep curves were processed according to the "Chen loading method" [6], and the "loading" creep curves of the samples under different vertical pressures are shown in Fig. 3.

Figure 3 show that the shear stress required for the destruction of the specimen increases with increasing vertical pressure. It is possible that the increase of the vertical pressure increases the consolidation degree of the soil and is favorable to the squeezing of the soil particles, so that the shear resistance increases with the shear stress of the specimen.

3.3   Isochronous Stress-Strain Curves of Weak Interlayer Soil

The stress-strain curves of the weak interlayer are shown in Fig. 4. It can be seen from the figure that the isochronal curves under different vertical pressures are composed of approximately linear and non-linear segments. With the increase of vertical pressure, the curve cluster is more and more biased towards the displacement axis, which fully reflects the weak interlayer creep characteristics.

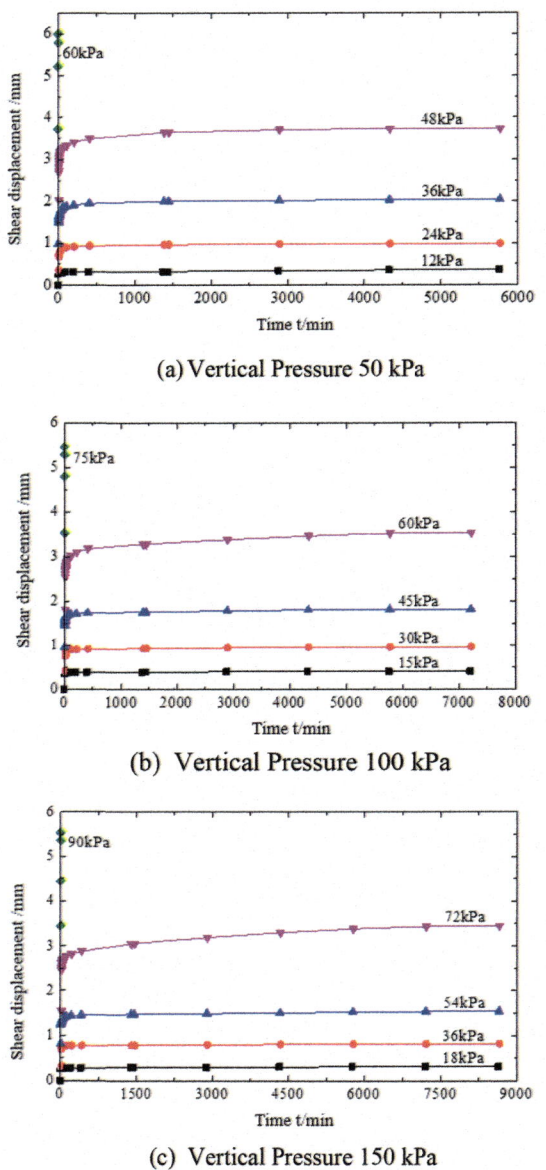

Fig. 3. The creep curves of weak interlayer soil by staged loading under different vertical pressures

3.4   Determination of Residual Strength for Weak Interlayer Soil

The residual shear strength values of the interlayer soil under different vertical pressures were obtained and compared with the peak shear strength. The results are shown in Table 2. The least squares method was used to linearly fit, and the fitting results are shown in Table 3.

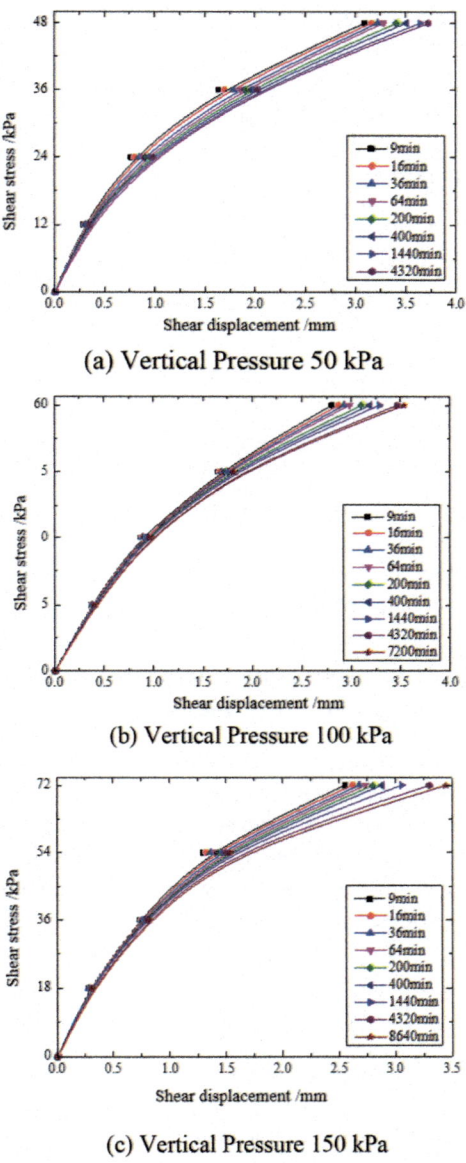

(a) Vertical Pressure 50 kPa

(b) Vertical Pressure 100 kPa

(c) Vertical Pressure 150 kPa

Fig. 4. The isochronous stress-strain curves of weak interlayer soil under different vertical pressures

It can be seen from the data in the table that the long-term strength of the weak interlayer soil is less than half of its peak strength value, and the strength index is correspondingly about half of the peak case, which indicates that the long-term strength of the weak interlayer soil is more unfavorable to Guangxi Xilin. The stability of the clastic rock slope in the design of the actual slope engineering, should consider the

Table 2. The peak strength and long-term strength of weak interlayer soil under different vertical pressures

Vertical pressure/kPa	Peak shear strength/kPa	Long-term shear strength/kPa	Decreasing amplitude/%
50	65.26	30	54.03
100	82.76	37	55.29
150	100.26	48	52.12

Table 3. The weak interlayer soil's shear strength parameters of peak strength and long-term strength

Strength index	Peak shear strength/kPa	Long-term shear strength/kPa	Decreasing amplitude/%
Cohesion/kPa	47.76	20.33	57.43
Internal friction angle/°	19.28	10.20	47.10

weak interlayer soil or the general strength of the residual strength of the soil layer to facilitate the long-term stability of the slope.

3.5 The Comparison of Slope Stability Under Peak Strength and Long-Term Strength of Weak Interlayer Soil

Through the numerical calculation, the slope stability coefficient corresponding to the peak intensity and the residual strength of the weak interlayer soil were divided into 1.493 and 0.717. The slope is stable when the weak interlayer is at the peak of the intensity, but the slope is unstable when considering its residual strength. The creep characteristics of the weak interlayer soil result in a large attenuation of the residual strength, which severely impacts the slope stability.

4 Conclusions

This paper used experimental tests and numerical simulations to study the creep characteristics of weak interlayer soil and its strength changes on clastic rock slope stability. A field site at Xilin County, Guangxi was used as conceptual model. The shear creep test of weak interlayer soil was conducted by using ZLB-1 type triple rheometer and two-dimensional slope seepage model based on fluid-solid coupling was developed. On the basis of analysis, the following conclusions were drawn from this study.

(1) The creep characteristics of clastic rocks in Xilin are obvious. The curves are developed in a nonlinear way. Low shear stress level, the elastic deformation of the main, a short period of time from the attenuation of creep transition to stable creep. At high shear stress levels, the creep rate of the stable creep stage increases and the time required for stabilization increases, and creep damage occurs when the shear stress level reaches a certain value.

(2) The greater the vertical pressure, the longer the creep stabilization takes longer, the greater the shear stress required for the failure of the specimen. In the actual slope engineering, the consolidation of the slope can be increased to enhance the slope stability.

(3) In the natural state, the long - term strength of the weak interlayer soil under different vertical pressures has a greater damage than the peak intensity, and the damage rate is 52.12%–55.29%, the peak strength index $c = 47.76$ kPa, $\varphi = 19.28°$. The long-term strength index $c = 20.33$ kPa, $\varphi = 10.20°$, the decrease is 57.43% and 47.10% respectively, which indicates that the long-term load will seriously reduce the shear strength of the weak interlayer soil.

(4) The creep characteristics of the weak interlayer will seriously weaken the stability of the slope, which often leads to the destabilization of the stability of the slope.

Acknowledgements. This research was supported by National Natural Science Foundation of China (NSFC) (No. 51369006) and the Project of Department of Land and Resources of Guangxi (No. GXZC2015-G3-3917-KLZB).

References

1. Xie, N., Sun, J.: Rheological properties of saturated soft clay in Shanghai. J. Tongji Univ. (Nat. Sci.) **03**, 233–237 (1996)
2. Yan, S., Xiang, W., Tang, H., et al.: Research on creep behavior of slip band soil of Dayantang landslide. Rock Soil Mech. **29**(01), 58–62 (2008)
3. Dai, G., Ling, Z., Shi, X., et al. Some engineering geological properties of soft interlayer and its mud layer in dam foundation of Gezhouba water control project. Acta Geol. Sin. **2**, 153–166, 170 (1979)
4. Wang, Z.: Stability Analysis of Weak Interlayer Rock Slope. Zhongnan University, Changsha (2003)
5. Xu, B., Qian, Q., Yan, C., et al.: Stability and strengthening analyses of slope rock mass containing multi-weak interlayers. Chin. J. Rock Mech. Eng. **28**(S2), 3959–3964 (2009)
6. Chen, Z.: Mechanical problems of long - term stability of underground roadway. Chin. J. Rock Mech. Eng. **1**(1), 1–20 (1982)

Stability and Deformation in Clayey Slopes with Varying Slope Density and Inclinations Subjected to Rainfall

Binod Tiwari[1]($\boxtimes$), Beena Ajmera[2], Mohammed Khalid[3], and Rosalie Chavez[3]

[1] California State University, Fullerton, 800 N. State College Blvd., E-419, Fullerton, CA 92831, USA
btiwari@fullerton.edu
[2] California State University, Fullerton, 800 N. State College Blvd., E-318, Fullerton, CA 92831, USA
bajmera@fullerton.edu
[3] California State University, Fullerton, 800 N. State College Blvd., Fullerton, CA 92831, USA
{mkhalid, rosaliechvz}@csu.fullerton.edu

Abstract. A reduction in the shear strength resulting from an increase in degree of saturation and pore pressures due to the infiltration of rainwater is often cited as a trigger for shallow slope failures. Models were prepared, in this study, to examine the influence of the density of the slope forming material and the inclination of the slope on the evolution of the seepage velocity, deformation mechanisms and stability of the clayey slopes. The relative compaction of the slope forming material ranged from 60% to 85% with slope inclinations ranging from 30 to 50°. The seepage velocity was found to decrease as the depth of the wetting front increased. Furthermore, the seepage velocity along the slope was greater than the seepage velocity at the head of the slope. The seepage velocity measured with tensiometers located throughout the slope matched well with the seepage velocity determined from the wetting fronts. When the relative compaction of the slope forming material was less than 75%, the slopes would tend to settle as the material became saturated with greater volumetric strains occurring in models with lower relative compactions. On the other hand, for slopes with relative compactions greater than 75%, the infiltration of rainwater would cause the slope to swell with greater negative volumetric strains at higher relative compactions.

Keywords: Rainfall-induced slope failures · Clayey slopes · Wetting front Deformation · Seepage velocity

1 Background

Causing billions of dollars of property damage and hundreds of human casualties each year, landslides are one of the worst natural disasters. Although there are several factors that can trigger landslides, the most common throughout the world is rainfall. Guzzetti et al. [1] identified rainfall intensity-duration thresholds required to trigger landslides

© Springer Nature Singapore Pte Ltd. 2018
A. Farid and H. Chen (Eds.): GSIC 2018, *Proceedings of GeoShanghai 2018 International Conference: Geoenvironment and Geohazard*, pp. 90–98, 2018.
https://doi.org/10.1007/978-981-13-0128-5_11

using statistical data analyses in different regions of the world. The thresholds will depend on a number of parameters including the soil type, slope gradient, and the relative compaction of slopes. The rate of infiltration of the rainwater into the ground, or the seepage velocity, will be one of the most important factors controlling the stability of the partially saturated slopes. Larger seepage velocities will result in a quicker movements of the wetting front, which will reduce suction and increase the unit weight of the slope material. Both of these changes will increase the likelihood of the slope instability. Factors including the density of the slope material, the inclination of the slope as well as the intensity and duration of the rainfall will impact the seepage velocity [2–7]. As part of this study, small scale experimental models were prepared in a Plexiglas container to examine the influence of the density of the slope material and duration of rainfall on the seepage velocity through the slope material and settlement experienced as a result of the rainfall.

2   Materials and Methods

2.1   Slope Material

For the preparation of the model slopes tested in this study, soil collected from a new construction project at the Titan Student Union at California State University, Fullerton was used. The properties of the slope material are presented in Table 1. The material was air-dried and sieved through the U.S. #4 sieve (sieve opening of 4.75 mm) prior to its use in the construction of any model.

Table 1. Geotechnical properties of slope material.

Property	Parameter value
Liquid limit	28
Plasticity index	8
Specific gravity	2.70
Clay	15%
Silt	10%
Sand	75%
USCS classification	CL
Max. dry unit weight	19.2 kN/m^3
Optimum moisture content	10.2%
Coefficient of permeability	2.1×10^{-5} cm/s

2.2   Preparation of Model Slopes

The experimental slopes were prepared in a Plexiglas container. The preparation of the slope model began with the installation of a 7 cm thick drainage layer. The drainage layer consisted of pea-sized gravel. It was overlain by geotextile that was 5 mm thick and acting as a filter between the slope material and the drainage layer. The slope was constructed on top of this geotextile in 5 cm thick lifts. The total dry weight of material required to obtain the desired relative compaction was determined and mixed with an

initial moisture content of 7.2% (3% dry of optimum). This material was, then, compacted to the desired lift thickness. All of the slopes were constructed to an inclination
of 40°. Nine different model slopes were prepared at different relative compaction
levels. The relative compactions used in each of these models are summarized in
Table 2. The slope models were constructed in a larger Plexiglas container with interior
dimensions of 1.2 m by 2.4 m and a height of 1.8 m. Some pictures from the model
preparation process in the Plexiglas container are presented in Fig. 1.

After the slope was constructed to the desired height and required slope inclination,
the slope was instrumented with eight miniature tensiometers and eight copper wires.

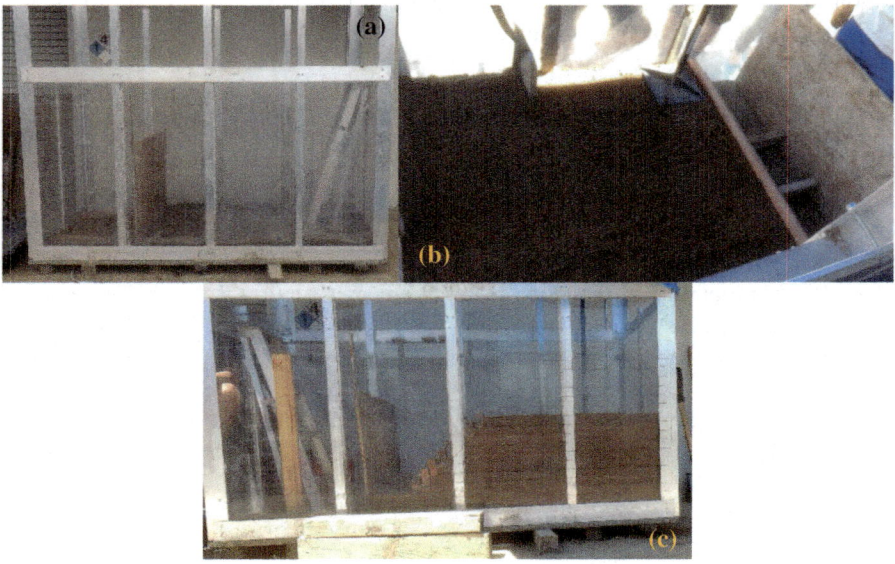

Fig. 1. Model preparation process: (a) Plexiglas container with gravel drainage layer and
geotextile, (b) compaction of soil lifts, and (c) model slope compacted to half height.

Table 2. Relative compaction in the models tested.

Model no.	Dry unit weight (kN/m³)	Relative compaction (RC, %)
1	13.1	68.4
2	11.6	60.4
3	14.2	73.8
4	12.5	65.0
5	15.4	80.0
6	12.1	63.0
7	13.4	70.0
9	13.4	70.0
10	14.4	75.0
11	12.5	65.0

The tensiometers had capacities to measure pore pressures from -180 kPa to 100 kPa. The copper wires were flexible and acted as inclinometers to determine the deformation in the slope. Figure 2 depicts the location of the instrumentation used in this study.

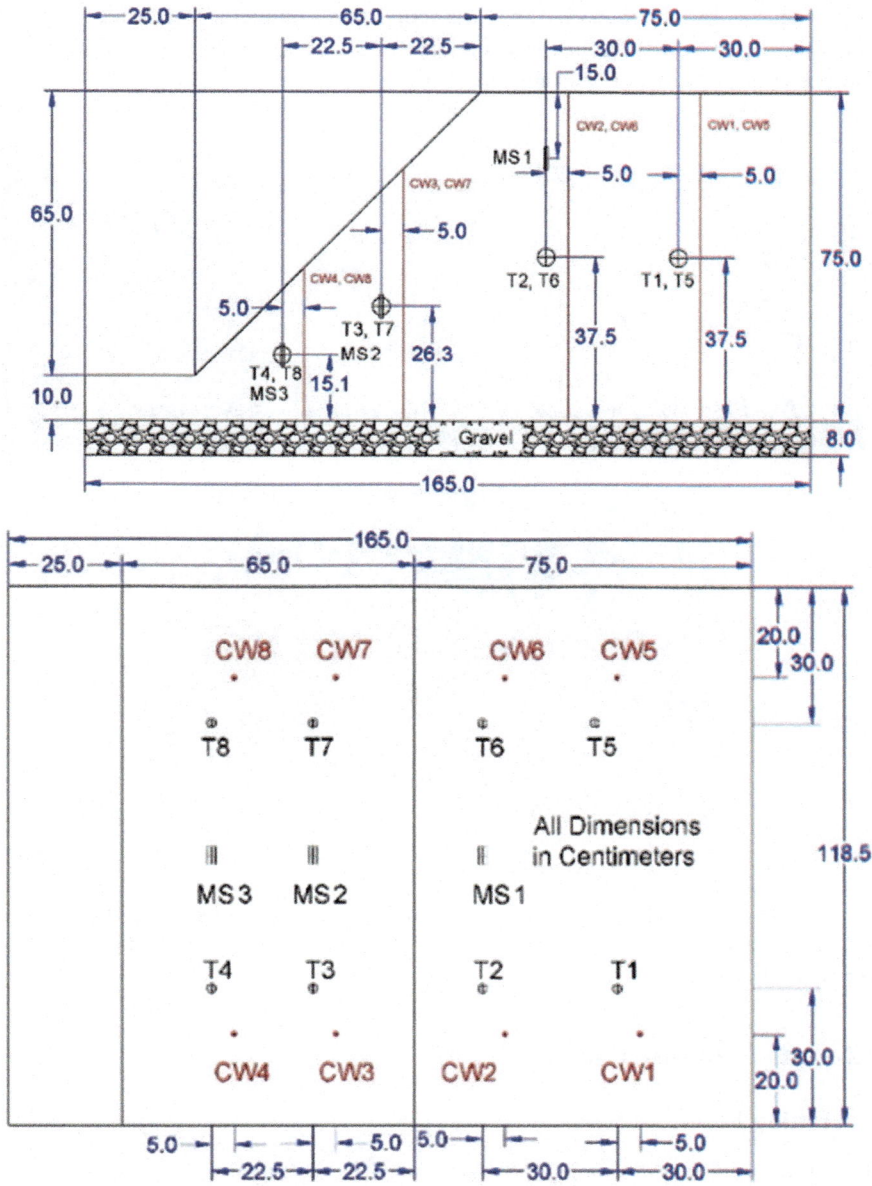

Fig. 2. Location of the instrumentation in the model slopes. Tensiometers are denoted with the letter T, while the copper wires are denoted with CW and soil moisture sensors are denoted with MS. All of the dimensions shown in centimeters.

The holes used to install the tensiometers were covered with a bentonite slurry in order to ensure that rainwater would not infiltrate rapidly through these holes. Three moisture sensors were also installed. The location of these moisture sensors are also included in Fig. 2. After the instrumentation was installed, the slopes were subjected to rainfall at an intensity of 30 mm/h using a rain simulator system. The location of the wetting front was recorded in 15 min intervals until the slope was completely saturated. Figure 3 contains some pictures of the instrumentation and the rain simulator system as well as the progression of the wetting front with time.

Fig. 3. (a) Instrumentation of the slope with tensiometers and copper wires, (b) completed slope model with installed rainfall simulator system, and (c) progression of wetting front with time during the application of rainfall.

3   Results and Discussion

3.1   Wetting Front Locations

Presented in Fig. 4 is an example of the location of the wetting front with time recorded during the application of rainfall in Model 7. The movement of the wetting fronts were recorded on both sides as well as the back of the Plexiglas container. The seepage velocities were calculated at the tensiometer locations using the wetting front data as well as with the tensiometer data for comparison. The recordings in the tensiometers installed in Model 7 are shown in Fig. 5. These recordings give an indication of the movement of the wetting fronts inside of the soil mass and will approach a value of zero suction when the wetting front reaches the tensiometers.

Figure 6 contains a comparison of the time required for the tensiometers to obtain suction values of 0 kPa and the time required for the wetting front to reach the location of the tensiometers. Also included in Fig. 6 is a line representing that both times were equal. As it can be seen, the differences between the two measurements are similar in some of the cases, while in other cases time required to cease suction was slightly more than the time required for the wetting front to reach the tensiometer location. Such differences were expected as the location of the tensiometer and the, location where the weting front advancement was measured were slightly different.

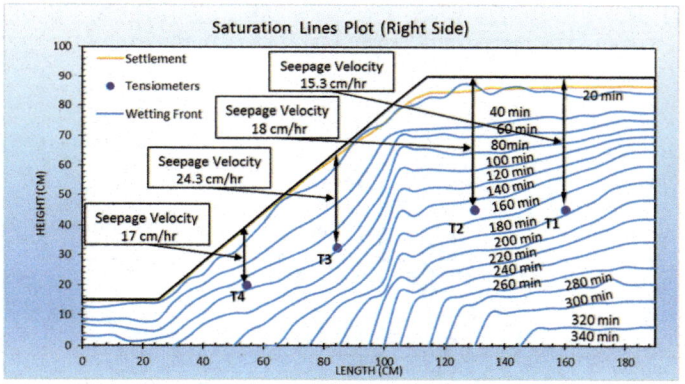

Fig. 4. Movement of wetting front with time in model 6.

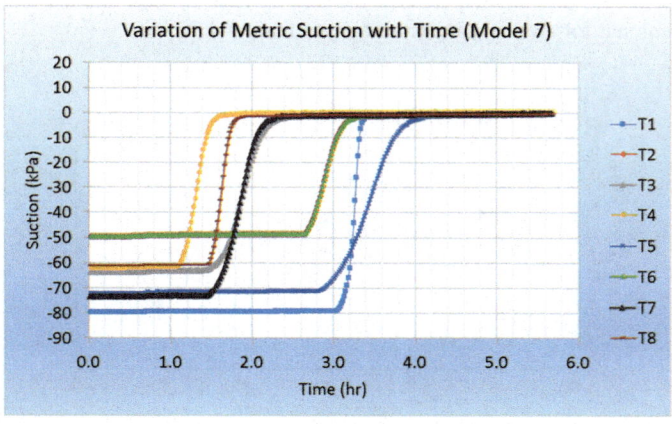

Fig. 5. Tensiometer readings with time in model 6.

The variation in the seepage velocity with relative compaction at the crest of the slopes is shown in Fig. 7. It can be seen from Fig. 7 that an increase in the relative compaction corresponded to a decrease in the seepage velocity when the relative compaction was greater than 63%. The seepage velocity in the model with a relative compaction of 60% appears to be lower than expected. However, this lower seepage velocity is attributed to the fact that the model slope experienced nearly 8% settlement in approximately the first hour after the application of rainfall. As a result, the slope material will densify impairing the infiltration of rainwater in the slope.

Figure 8 depicts the settlement and swelling experienced by the slopes as a result of the application of rainfall. In Fig. 8, the negative strain values represent settlement and positive strain values represent swelling of the slope. It can be seen that slopes with relative compactions initially less than 75% would experience settlement, while those with relative compactions initially greater than 75% would experience swelling with the application of rainfall.

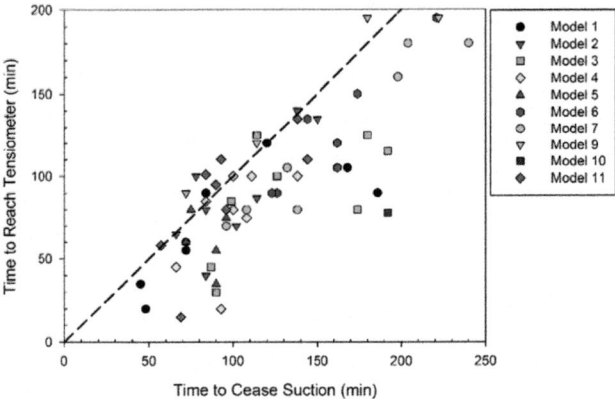

Fig. 6. Comparison of the time required for the wetting front to reach the tensiometer location and the time required for the tensiometer reading to reach 0 kPa.

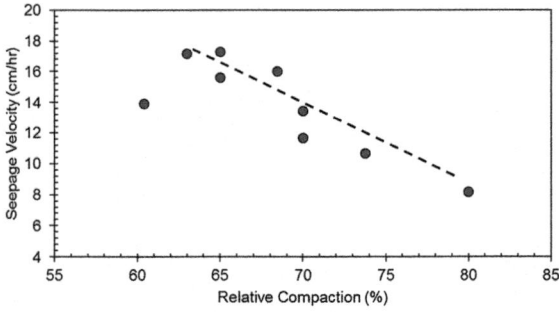

Fig. 7. Variation in the seepage velocity with relative compaction at the crest of the slopes.

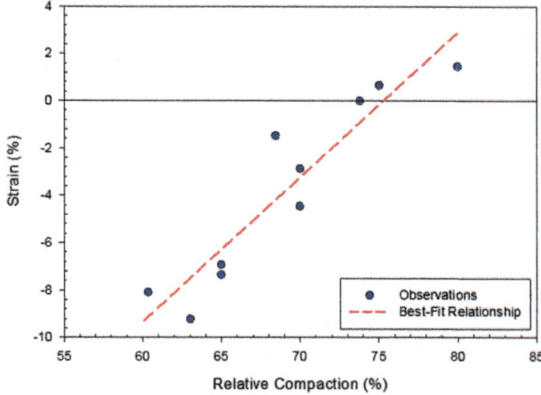

Fig. 8. Settlement (negative strain values) and swelling (positive strain values) experienced by the model slopes when subjected to rainfall until complete saturation was achieved.

4 Conclusions

Ten model slopes were prepared in Plexiglas container using soil obtained from a construction project at the Titan Student Union at California State University, Fullerton. The 40° inclined slopes were prepared to have different relative compactions ranging from 60% to 80% and were subjected to rainfall at an intensity of 30 mm/h. The location of the wetting front with duration of rainfall was recorded and used to determine the seepage velocity throughout the slope. The seepage velocity was found to decrease with an increase in the relative compaction. The model slopes were seen to experience settlement when the initial relative compaction was less than 75%, but would experience swelling when the initial relative compaction was greater than 75%.

Acknowledgements. The authors would like to thank California State University, Fullerton Instructionally Related Activities (IRA) Grant No. 3361 for the generous support provided to purchase the materials used in this study. Additionally, the efforts of the numerous students that assisted with the model preparation and data collection are greatly appreciated.

References

1. Guzzetti, F., Peruccacci, S., Rossi, M., Stark, C.P.: Rainfall thresholds for the initial of landslides in central and southern Europe. Meteorol. Atmos. Phys. **98**(3), 239–267 (2007)
2. Tiwari, B., Kawai, K., Caballero, S., Viradeth, P.: How rainfall and earthquake trigger shallow slides – experimental and numerical studies on laboratory prepared slopes. Int. J. Landslide Environ. **1**(1), 109–110 (2013)
3. Tiwari, B., Lewis, A., Ferrar, E.: Experimental simulations of rainfall and seismic effects to trigger slope failures. Geotech. Spec. Publ. **231**(1), 448–451 (2013)
4. Tiwari, B., Lewis, A.: Experimental modeling of rainfall and seismic activities as landslide triggers. Geotech. Spec. Publ. **225**, 471–478 (2012)

5. Xue, K., Tiwari, B., Ajmera, B., Hu, Y.: Effect of long duration rainstorm on stability of red clay slopes. Geoenviron. Eng. 3(12), 1–13 (2016)
6. Tiwari, B., Caballero, S.: Experimental model of rainfall induced slope failure in compacted clays. Geotech. Spec. Publ. 256, 1217–1226 (2015)
7. Tiwari, B., Tran, D., Ajmera, B., Carrillo, Y., Stapleton, J., Khan, M., Mohiuddin, S.: Effect of slope steepness, void ratio and intensity of rainfall on seepage velocity and stability of slopes. In: Proceedings of the Geotechnical and Structural Engineering Congress, vol. 1, pp. 584–590 (2016)

Exploring a Coupled Approach to Model the Geomechanical Processes of Sinkholes

Suraj Khadka[1], Zhong-Mei Wang[2], and Liang-Bo Hu[1(✉)]

[1] University of Toledo, Toledo, OH 43606, USA
Liangbo.Hu@utoledo.edu
[2] Guizhou University, Guiyang 50025, Guizhou Province, China

Abstract. Sinkholes may manifest in various forms of geological hazards or disasters, including gradual depression or rapid collapses in the underlying rock layer or overlay soil. In this paper a coupled approach is explored to study the chemo-mechanical processes involved in the dissolution dominated sinkhole processes, as well as the hydro-mechanical processes involved in the cover subsidence or collapse type of sinkholes. The first part of the analysis focuses on the dissolution kinetics and enhanced deformation processes. Specific solution rate of the constituent mineral (limestone or dolomite) and the surface area available for reaction are related via a chemo-mechanical coupling with the consideration of the damage-enhanced dissolution mechanism. The second part of the analysis explores the cover collapse type of sinkholes in which a critical mechanism is the growth and upward propagation of cavity. A strain-softening constitutive model is used to describe the strength evolution dependent on accumulated plastic deviatoric strain and erosion progression of soils around the cavity. The numerical results demonstrate the possibility of the presented coupling approach to better understand the time-dependent progressive processes of sinkhole formation and development.

Keywords: Sinkholes · Chemo-mechanical · Dissolution · Cover collapse

1 Introduction

Sinkholes over the past few decades have occurred in an alarming frequency, posing a serious threat to infrastructures and human safety. This has been a strong motivation to better understand the sinkhole mechanisms and develop effective mitigation strategies to this geohazard. Sinkholes, mostly a common phenomenon in karst geology, originate from the chemical dissolution of soluble carbonate bedrock that ultimately sets the way for sinkhole development. Sinkholes occur naturally but they also can be provoked by human activity, including intensive groundwater pumping for irrigation or residential taps, leakage of sewer system and water pipelines or by work above ground that destabilizes the karst rock below. The propagation mechanism in sinkhole formation is further ignited by the water level fluctuations, rapid drop-down in water level and rainfall of high intensity in short durations.

In general, the formation and development of sinkholes is strongly dominated by the geo-mechanical characteristics of soil and rock behavior complicated by the

© Springer Nature Singapore Pte Ltd. 2018
A. Farid and H. Chen (Eds.): GSIC 2018, *Proceedings of GeoShanghai 2018*
International Conference: Geoenvironment and Geohazard, pp. 99–107, 2018.
https://doi.org/10.1007/978-981-13-0128-5_12

chemical and hydraulic factors such as dissolution kinetics, porosity of minerals, hydraulic conductivity and flow rate. It has been widely accepted that sinkholes can be classified in two major categories, sinkholes formed in the karstic rock (limestone, gypsum or salt), and those formed in soils overlying the karstic rocks (Waltham et al. 2007). As shown in Fig. 1a and b, the dominant process behind sinkholes formed in rocks is the dissolution of soluble rocks. Dissolution process may be enhanced by the presence of caves or fissures that provide more specific surface for more dissolution to occur, these caves or fissures in turn continue to expand or grow as a result of dissolution. Collapse and caprock sinkholes are defined by fracturing, breakdown and collapse of bedrock slabs and arches as they gradually lose the support around dissolutional cavities (Fig. 1c and d). Cover type of sinkholes, i.e., sinkholes formed in soils are a more widespread geohazard. They are generally caused by the erosion, transport and failure of the soils that overlie cavernous rock. Because of its low strength compared to rock, which, if left over a cave can still be strong enough to stand for a long period of time, a soil arch over a void is inherently unstable and its collapse can occur rapidly. Their underlying mechanisms are extremely intricate and have traditionally received more attention from the geotechnical communities (e.g., Tharp 1999; Goodings and Abdulla 2002; Augarde et al. 2003; Rawal et al. 2017).

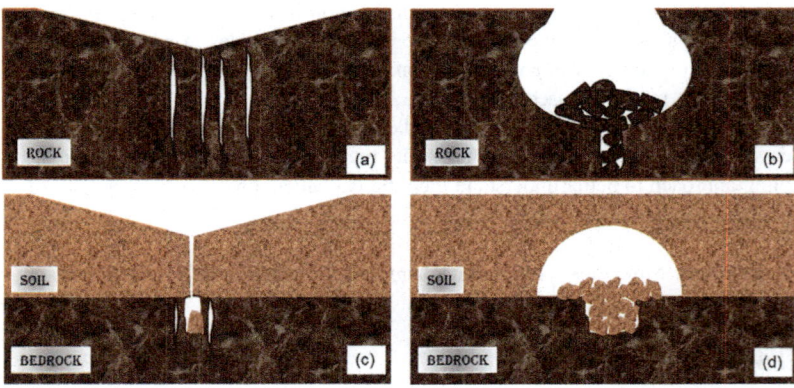

Fig. 1. A schematic representation of (a) subsidence in solution sinkhole; (b) collapse sinkhole; (c) cover subsidence sinkhole and (d) cover collapse sinkhole where a cavity grows in the soil cover.

The present study attempts to explore the coupled processes involved in the two different categories of sinkholes discussed above. First, the chemo-mechanical processes in the dissolution dominated sinkholes in karst rocks are examined, focusing on the kinetic rates of mineral dissolution and mechanical softening. Subsequently the hydro-mechanical processes involved in the cover-soil sinkholes are studied via a numerical simulation of various hydraulic scenarios and potential phenomena of suffusion and sloughing of soil around the cavity.

2 Mechanisms of Dissolution-Dominated Sinkholes

2.1 Dissolution Kinetics in Karst Rocks and Numerical Models

A general framework for the coupled processes of deformation and dissolution in karst rocks is briefly presented in this section, followed by numerical examples of the typical kinetic rate and dissolution enhanced deformation.

The dissolution enhanced damage can be addressed in constitutive formulations for karst rock minerals, considering the chemical changes and strength evolution. To better illustrate the possible approach, a chemo-plasticity model is presented here, assuming rigid plasticity. The yielding function can be generally defined as

$$f = f(\sigma_j, \, p_c) = 0 \tag{1}$$

$p_c = p_c\left(\varepsilon_q^{pl}, \, \xi\right)$, describes an isotropic size characteristic of yield locus (i.e., strength). It depends on a set of hardening and/or softening parameters that are either mechanical or chemical in nature. ε_q^{pl} is the deviatoric strain hardening parameter, defined as

$$\varepsilon_q^{pl} = \left(\frac{2}{3} \dot{e}_{ij}^{pl} \dot{e}_{ij}^{pl}\right)^{1/2}, \, \dot{e}_{ij}^{pl} = \dot{\varepsilon}_{ij}^{pl} - \frac{1}{3} \dot{\varepsilon}_{kk}^{pl} \delta_{ij} \tag{2}$$

Here the chemical parameter, ξ is chosen to be the mass removal of the material,

$$\xi = \Delta m / m_0 \tag{3}$$

It is the ratio of change of mass (or mol) to the original mass (or mol) of the mineral, thus it is confined to the range [0, 1] and can be directly used as a softening parameter. Its evolution is described by the dissolution kinetic rate which will be discussed in the next section.

This framework reflects two different and independent ways in which the material may become harder or softer. One is a classical deviatoric strain hardening and the other reflects the removal of mass in weakening of the material. The following derivation can be readily obtained based on the associated flow rule and Prager's consistency condition,

$$\dot{\varepsilon}_{ij}^{pl} = -\frac{1}{H} \frac{\partial f}{\partial \sigma_{ij}} \left(\frac{\partial f}{\partial \sigma_{kl}} \dot{\sigma}_{kl} + \frac{\partial f}{\partial \xi} \dot{\xi}\right), \text{ where } H = \frac{\partial f}{\partial \varepsilon_q^{pl}} \left[\frac{2}{3} \frac{\partial f}{\partial s_{ij}} \frac{\partial f}{\partial s_{ij}}\right]^{\frac{1}{2}} \tag{4}$$

$s_{ij} = \sigma_{ij} - \frac{1}{3} \sigma_{kk} \delta_{ij}$ is the deviatoric stress. It describes the dependence of plastic strain on stress (mechanical loading) and dissolution (chemical effect). The couplings summarized above can be used to model the deformation and dissolution of karstic rocks when complemented by the stress equilibrium and kinematic relationships for formulated boundary value problems.

The rate of dissolution of rocks is believed to generally depend on the solubility and specific solution rate constant of the constituent mineral, and the surface area available for reaction to occur, in an over-simplification of the potentially very complex reaction processes and kinetics. There are four typical karst rock minerals in nature: calcite, dolomite, gypsum and salt (halite). The first two are most common carbonate rocks in karst terrain and their rates are much slower than the latter two; gypsum has a moderately fast dissolution and salt dissolves very rapidly under normal environmental conditions. Although it is not the objective of the present study to compare the dissolution of different minerals, the following rate equation is adopted,

$$\dot{m} = kA\left(1 - \frac{C}{C^{\text{sat}}}\right) \tag{5}$$

The dissolution rate, $\dot{m}$ (mol/s), as also related to Eqs. (3) and (4), is dependent on the specific surface area, A (m^2/m^3). k is the rate constant (mol/m^2/s). C and C^{sat} (mol/m^3) are the concentration and the equilibrium (saturation) concentration of the mineral, respectively. Obviously C^{sat} can be related to the so-called solubility, S (kg/m^3), via $S = C^{\text{sat}} v_m$, v_m is the molar mass of the mineral.

2.2 Modeling Results of Dissolution Enhanced Deformation

In this section a numerical simulation is focused on a soluble cavity, assumed to constitute of calcite, as shown in Fig. 2a. Some results are briefly summarized here.

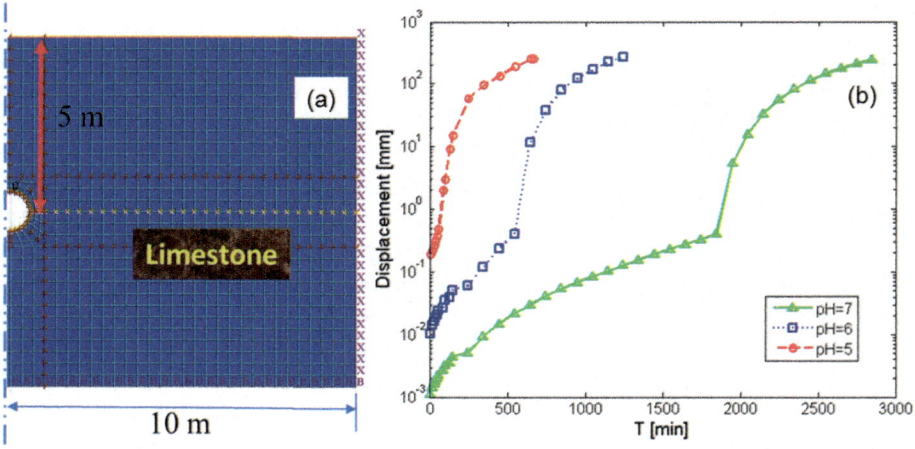

Fig. 2. (a) The simulated limestone formation around a cavity of 1-m in diameter, shown in half (due to symmetry). (b) Enhanced displacement at the top of the cavity under different levels of acidity.

The cavity has a diameter of 1 m and subject to chemical dissolution, under a constant surcharge from the top ground surface.

The chemical softening mechanism is introduced to the constitutive law for the simulated rock, a conventional Mohr-Coulomb failure function is modified with the shear strength parameter (e.g., cohesion) decreases as a result of chemical dissolution, and implemented in the computational software FLAC (Itasca 2011).

In the present simulation, the fact that the dissolution of limestone or dolomite in pure water is extremely low but can be accelerated considerably in acidic conditions (Sjoberg 1976) under which the dissolution rate of calcite could rise over several orders of magnitude, is also considered. A proportional relationship between the rate constant and [H^+] concentration (note that pH = log[H^+]) is used (Ciantia and Hueckel 2013).

Figure 2b shows the progressive development of the settlement at the top of the cavity, under different levels of acidity attacking the soluble calcite around the cavity. The dramatic rise in the enhanced deformation beyond approximately 0.1 mm is accompanied with the propagation in the plasticity zone around the cavity. It is evident that faster dissolution weakens the mineral more rapidly and induces more significant deformation. For simplicity, in the presented study a constant dissolution rate is used. As shown in Hu and Hueckel (2007), more intricate formulations can be also explored to consider the effect of volumetric strain (dilatancy) and micro-cracking, which can lead to increased surface area and thus enhanced dissolution: a two-way coupling may be necessary to fully address the complex interactions.

3 Mechanisms of Cover-Collapse or Cover-Subsidence Sinkholes

3.1 Potential Mechanisms Explored

Cover-subsidence or cover-collapse (dropout) sinkholes are most often caused by anthropogenic activities (USGS 1999). Cover-collapse sinkholes occur when covering sediment contains high amount of clay. As void enlarges, eventually the crown of the cavity thins to a point where soil can no longer support the overburden and collapses in a sudden manner (Fig. 1d). Such processes may be often triggered by water table changes associated with various hydraulic scenarios. The inner mechanisms are very intricate and thus far a variety of processes are considered to be capable of contributing to the cover types of sinkholes, such as loss of buoyant support, increased pore pressure, increased amplitude of water table variations, exposure to repeated saturation and drying (e.g., Brink 1984; Sowers 1996; Lei et al. 2001).

One particularly important process involved is the sloughing and suffusion of soil around the cavity (typically located near the bottom of the soil and/or the top of the underlying rock). The continual loss of soil around the cavity leads to the upward propagation of the cavity/void and eventually the collapse of the soil arch. There has been yet widely accepted theory established in a quantitative way that can adequately describe such a process. In the present study a potential erosion type of weakening of soil is explored. It is hypothesized that with the flow of water evolving around the cavity under a drawdown scenario, soil erosion process advances in a due course of time, and the soil strength is reduced due to the erosion, and eventually no longer able to support the overburden.

This concept is implemented in FLAC to model a two-layer soil/rock formation under a typical drawdown scenario. The erosion rate expressed as mass per unit time per unit area, $\dot{m}_e$ is typically assumed to be linear proportional to the hydraulic shear stress developed in excess of critical shear stress (τ_c), multiplied by an erosion coefficient, k_e

$$\dot{m}_e = k_e(\tau - \tau_c), \text{ when } \tau > \tau_c \tag{6}$$

The surface erosion rate constant, k_e is often expressed as $k_e = k_0\rho_d$, where ρ_d is the soil dry density and k_0 is an erosion coefficient that depends on sensitivity of material to hydraulic force.

A mass removal rate defined at any material point can be introduced as

$$\dot{\varsigma} = \dot{m}_e/m_0 = k_e(\tau - \tau_c)A/\rho V = k_0\rho_d(\tau - \tau_c)\tilde{A}/\rho \tag{7}$$

$\tilde{A}$ is the specific surface area, defined as A/V, the ratio of the surface area A to the total volume V. ρ_d is the bulk density of soil. This mass removal parameter, ς, is then introduced as a softening parameter in the constitutive law, describing the weakening of the soil due to erosion. In the present simulation a Mohr-Coulomb failure criterion is adopted, the effective cohesion (c') is related to the mass removal parameter (ς),

$$c' = c_0'(1 - \varsigma) \tag{8}$$

while the internal cohesion remains a constant. c_0' is the original effective cohesion. Obviously other forms of plasticity function can also be explored. In the present study a numerical simulation is conducted in FLAC (Itasca 2011) in which the above mentioned formulations are implemented with its programming language FISH.

3.2   A Numerical Simulation

A numerical simulation is performed using FLAC2D, focusing on the progression of the cavity deformation affected by soil sloughing/suffusion due to erosion and soil weakening. Figure 3a shows the cavity ($R = 1$ m) in a two-layered model consisting of a soil cover over a karst limestone.

A typical drawdown scenario is considered in this simulation. The water table is originally located at the top surface and subsequently lowered gradually. The erosion rate and associated strength change as described above is implemented in FLAC. Typical values for the erosion coefficient and critical shear stress are adopted from those reported previously in literature (e.g. Regazzoni et al. 2008; Wahl 2010), $k_0 = 1.86 \times 10^{-8}$ m^3/(N·s) and $\tau_c = 277.49$ Pa. The erosion is confined to a very small region around the cavity; in the present study it is imposed within the radius of 0.5 m around the cavity, and the shear stress in Eq. (6) is taken from the in-plane stress τ_{xy} in the stress field computed in FLAC. Whether such treatment is adequate is subject to

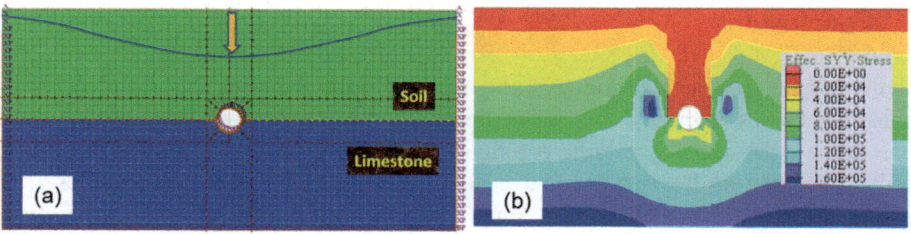

Fig. 3. (a) Geometry of a two-layer model simulated. (b) Effective stress contour after a drawdown of 0.26 cm/day for a time of 10^8 s with erosion coefficient $k_0 = 1.86 \times 10^{-8}$ m^3/(N·s).

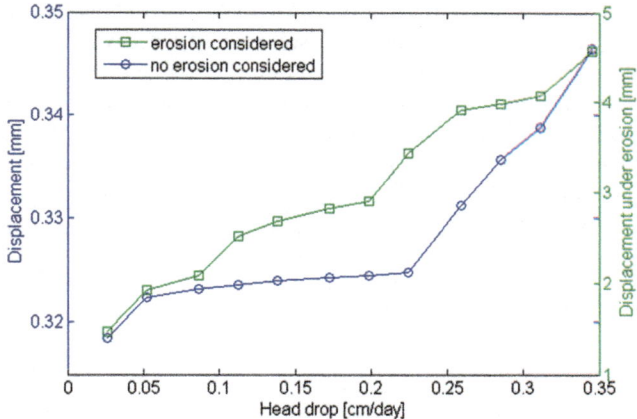

Fig. 4. Vertical deformation of the cavity during draw-down scenarios in a comparison between the results with erosion effects considered and those with erosion effects neglected.

further investigations, ideally a more reasonable approximation would be the shear stress along the flow direction.

Figure 3b shows a typical stress-contour map plotted for a rapid drawdown of 0.26 cm/day in erosion-incorporated constitutive model developed with FLAC. Figure 4 shows the deformation at the top of the cavity (after 10^8 s) versus the rate of head-drop, subjected to hydro-mechanical simulation with and without erosion being considered. It is noted that the plotting scales are intentionally established differently to highlight the comparison; while the effect of rapid or slow drawdown rate is not very significant when no erosion is considered, but quite substantial when erosion considered. It is also evident that the deformation increases considerably under the influence of erosion.

4 Conclusions

As the intricate interactive processes play a critical role in the formation and development of sinkholes, a coupled approach to the geomechanics of sinkholes is necessary and must address the interplay of different physical and/or chemical processes. The present study is focused on the two distinctive dominant mechanisms behind sinkholes. The dissolution-dominated sinkholes in karst rock are modeled in a chemo-mechanical framework. Kinetic rates of minerals are examined and subsequently used to demonstrate the dissolution enhanced damage at different levels of acidity. The presented formulation of sequential couplings of chemo-mechanical processes can be extended to address two-way, simultaneous couplings. Many intricacies about the complex evolution of fissure or fracture opening for enhanced dissolution remain an interesting subject for future studies.

The second part of the investigation is focused on the sinkholes in the cover soils. The simulated scenario of weakening of the soil around the cavity is attributed to the erosion. Numerical results show the enhanced deformation and damage during typical drawdown scenarios. While the coupled approach explored in the present study is successfully implemented in the numerical simulations, it should be noted that the underlying mechanisms of suffusion or sloughing of soils in sinkholes remain an intriguing subject for further investigations, a collective effort in the experimental, theoretical and numerical developments is still much needed to better understand the intricate processes involved.

References

Augarde, C.E., Lyamin, A.V., Sloan, S.W.: Prediction of undrained sinkhole collapse. J. Geotech. Geoenvironmental Eng. **129**(3), 197–205 (2003)

Brink, A.B.A.: A brief review of the South African sinkhole problem. In: Beck, B.F. (ed.) Proceedings of the First Multidisciplinary Conference on Sinkholes, pp. 123–127. Balkema, Rotterdam (1984)

Ciantia, M.O., Hueckel, T.: Weathering of submerged stressed calcarenites: chemo-mechanical coupling mechanisms. Geotechnique **63**(9), 768–785 (2013)

Goodings, D.J., Abdulla, W.A.: Stability charts for predicting sinkholes in weakly cemented sand over karst limestone. Eng. Geol. **65**(2), 179–184 (2002)

Hu, L.B., Hueckel, T.: Coupled chemo-mechanics of intergranular contact: toward a three-scale model. Comput. Geotech. **34**(4), 306–327 (2007)

Itasca: FLAC2D User's Guide, 5th edn. Itasca Consulting Group Inc., Minneapolis (2011)

Lei, M., Jiang, X., Yu, L.: New advances of karst collapse in China. In: Beck, B.F., Herring, J.G. (eds.) Proceedings of the Eighth Multidisciplinary Conference on Sinkholes and the Engineering and Environmental Impacts of Karst, pp. 145–151. Balkema, Lisse (2001)

Rawal, K., Wang, Z.M., Hu, L.B.: Numerical investigation of the geomechanics of sinkhole formation and subsidence. In: Brandon, T.L., Valentine, R.J. (eds.) ASCE Geotechnical Special Publication 277: Geo-Frontiers 2017: Transportation Facilities, Structures and Site Investigation, pp. 480–487 (2017)

Regazzoni, P., Marot, D., Courivaud, J., Hanson, G.J., Wahl, T.L.: Soil erodibility: a comparison between the Jet Erosion Test and the Hole Erosion Test. In: Proceedings of the Inaugural International Conference of the Engineering Mechanics Institute, Minneapolis, Minnesota, pp. 1–7 (2008)

Sjoberg, E.L.: A fundamental equation for calcite dissolution kinetics. Geochim. Cosmochim. Acta **40**(4), 441–447 (1976)

Sowers, G.F.: Building on Sinkholes: Design and Construction of Foundations in Karst Terrain. ASCE Press, New York (1996)

Tharp, T.M.: Mechanics of upward propagation of cover-collapse sinkholes. Eng. Geol. **52**(1), 23–33 (1999)

Wahl, T.L.: A comparison of the Hole Erosion Test and the Jet Erosion Test. In: Joint Federal Interagency Conference on Sedimentation and Hydrologic Modeling, Las Vegas, NV (2010)

Waltham, T., Bell, F.G., Culshaw, M.: Sinkholes and subsidence: karst and cavernous rocks in engineering and construction. Springer, Berlin (2007)

GIS Based Seismic Risk Analysis of Ahmedabad City, India

Tejas P. Thaker[1], Pankaj K. Savaliya[2(✉)], Mehul K. Patel[2], and Kundan A. Patel[2]

[1] Department of Civil Engineering, Pandit Deendayal Petroleum University, Gandhinagar 382007, India
tejas.thaker@sot.pdpu.ac.in
[2] Pandit Deendayal Petroleum University, Gandhinagar 382007, India
pankaj.scvl3@sot.pdpu.ac.in

Abstract. Earthquakes have been one of the most dangerous and calamitous hazards faced by mankind. Lots of researches have been focused on how to forecast, prevent and mitigate the effects caused by Earthquake or seismic hazards. This paper has been one such effort that utilized various properties pertaining to geotechnical, seismic and topographical aspects of Ahmedabad city in India to evaluate the vulnerability and susceptibility of the region to earthquakes. Factors such as population, water table variation, shear wave velocity, peak ground acceleration at rock and surface level, spectral acceleration and predominant frequency were mapped layer-by-layer using ArcGIS. Analytical hierarchy approach for multi criteria decision system was used to determine the relative importance of each parameter, thus giving us risk index for different areas of Ahmedabad city to seismic hazards. From the study it is found that all though most of the factors are having their higher values in south zone and/or in west zone, higher risk is to be seen in these zones, but population map indicates that south zone has higher population therefor giving higher risk at this zone.

Keywords: Seismic hazards · Arc GIS · Analytical hierarchy approach
Risk index

1 Introduction

Gujarat is one of the most seismic prone intercontinental regions of the world. It has experienced two large earthquakes of magnitude Mw 7.8 and 7.7 in 1819 and 2001 respectively and seven earthquakes of magnitude Mw 6.0, during the past two centuries. The intense aftershock activity of 2001 Bhuj earthquake is continuing. Through March 2008, 14 aftershocks with Magnitude 5.0–5.8, about 200 aftershocks with magnitude 4.0–4.9, about 1600 aftershocks of with magnitude 3.0–3.9, and several thousand aftershocks with magnitude <3 have been recorded. Regional seismicity has also increased with Mw 5.0 earthquakes and associated foreshocks and aftershock sequences [5]. Many authors have carried out various seismic studies in Ahmedabad and surrounding regions [9–12]. Thaker and Rao [6] carried out seismic hazard analysis

© Springer Nature Singapore Pte Ltd. 2018
A. Farid and H. Chen (Eds.): GSIC 2018, *Proceedings of GeoShanghai 2018*
International Conference: Geoenvironment and Geohazard, pp. 108–116, 2018.
https://doi.org/10.1007/978-981-13-0128-5_13

in Ahmedabad region. Mehta et al. [7] reviewed seismic hazards studies in Indian subcontinents. Rao et al. [8] carried out deterministic seismic hazard analysis of Ahmedabad region.

Ahmedabad, According to the Bureau of Indian Standards falls under seismic zone 3, in a scale of 2 to 5 (in order of increasing vulnerability to earthquakes). The Deccan basement, generally found in adjacent Cambay basin at 4 to 5 km depth is missing at Ahmedabad and the hard rock basement of granite rock occur at depth of 7 to 8 km. The upper 200–300 m below Ahmedabad city is mostly alluvium, formed by the meandering Sabarmati River in the past. Ahmedabad lies in two marginal faults and also on the intersection of two minor lineaments [6].

In past years, study has been done for Ahmedabad city using different methods [4, 6, 8] and by considering different factors, but they had limitations in terms of data collected, methods of collection for that data, resolution of satellite images etc. Moreover, no study has been done by consideration of combination of factors included in this study. In this GIS based analysis of risk index, technical factors and one socio-economic factor are considered. If study is only concerned for technical factors then risk index cannot be calculated, without taking population into account as without any considerable population the most vulnerable place due to seismic activities cannot be said contains high risk. Factors those considered in this analysis are: population, peak ground acceleration, predominant frequency, spectral acceleration, shear wave velocity, compressible wave velocity, standard proctor test values and ground water table.

This study contains layer by layer data analysis of factors mentioned above on the base map of Ahmedabad city in ArcMap (Arc GIS). Here, topo sheets (from survey department of India) of Ahmedabad city are used as base map in Arc GIS and data are added on topo sheets in various layers. AHP (analytical hierarchy process) is used to merge all the layers according to their weightage, which is calculated by making matrix and by prioritizing each factors. From integrated map made as a result of AHP, risk index is calculated for different areas of Ahmedabad city ranging from 1–5 (1 as less severe to 5 as high severe zones).

2 Factors Considered

2.1 PGA (Peak Ground Acceleration)

PGA is equal to the amplitude of the largest absolute acceleration recorded on an accelerogram at a site during a particular earthquake. In an earthquake, damage to buildings and infrastructure is related more closely to ground motion, rather than the magnitude of the earthquake itself. For moderate earthquakes, PGA is a reasonably good determinant of damage.

Here, PGA values are taken as average of values from different ground motion data carried out in the region in the separate analysis, further, standard deviation of the data set is added to the average value to keep it on higher side. PGA at rock level and surface level both for the Ahmedabad city are plotted through ArcGIS. Maps of PGA at surface level for depth 0 m, 1 m, 2 m, 3 m and 4 m are plotted. The map of PGA at surface level reveals higher values at south zone and north zone of the city 1 m (refer

Fig. 1), 2 m and 3 m depth where, central zone, west zone and east zone have the average lower values for PGA in comparison. PGA value for 4 m depth has average values concentrated over city area, lower values around the surroundings of the city and higher value of 0.915 g (g = gravitational acceleration) at the central zone near Shahpur. PGA value at surface level ranges from 0.062 g to 0.915 g for the city. PGA value at rock level has very slight change over the whole area of the city ranges from 0.056 g to 0.067 g (Fig. 1).

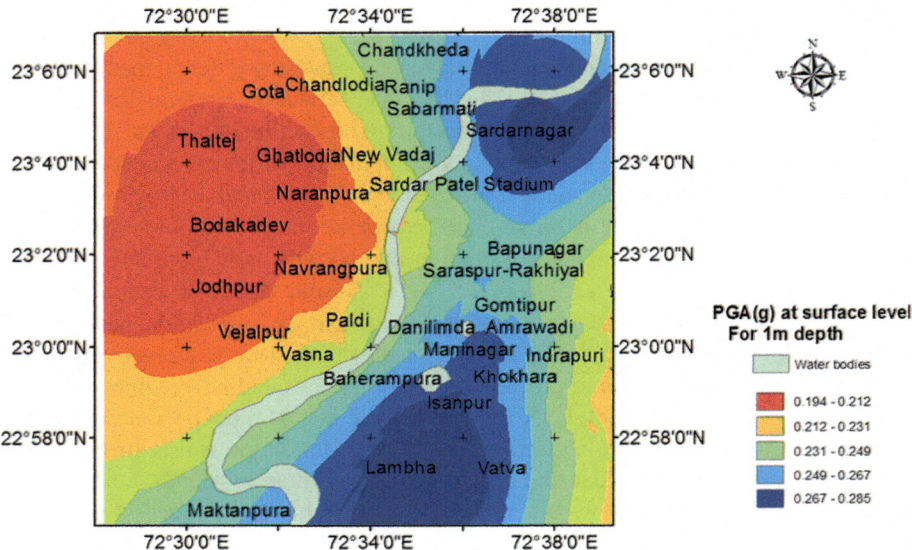

Fig. 1. PGA at surface level.

2.2 Predominant Frequency (PF)

Predominant frequency is the second prioritized factor obtained by carrying out ground response analysis. Map of predominant frequency show higher values at west zone and on some areas of north zone and central zone of the city. Predominant frequency value has normal average distribution over other areas ranges from 2.021 Hz to 4.355 Hz.

2.3 Population

Population is socio-economic factor and its importance is very much high. For example, if earthquake occurs in a region where there is less population than the damage will be less than highly populated regions.

Population is the major factor affecting to seismic risk of the city. The area and word wise population data have been collected from Ahmedabad Municipal Corporation and population density model for Ahmedabad city was prepared. This map reveals that the most of the population is accumulated over south zone, central zone and some areas of north zone and east zone. West side of the city has comparatively lower population density.

2.4   Spectral Acceleration (SA)

The spectral Acceleration is equally important parameters obtained from ground response modeling. The Maps for different spectral acceleration for different time (0.1 s, 0.2 s, 0.3 s, 0.5 s and 1 s) are plotted which conclude that spectral acceleration has the lowest value of 0.060 g for 1 s time and the highest value of 0.733 g for 0.2 s and 0.5 s. From maps of spectral acceleration, it can be said that south zone and north zone have average higher values of spectral acceleration.

2.5   Compressive Wave (P-Wave) Velocity (Vp) and Shear Wave (S-Wave) Velocity (Vs)

P-waves and S-waves (transverse waves) are the two main types of elastic body waves, called seismic waves in seismology. Here average velocity of the wave, over the particular depth is been taken. Average P-wave and S-wave velocity is determined by following formula.

$$Vavg(h) = \frac{h}{\sum ha/Vi}$$

Where, h is the particular depth.

Maps for depth of 10 m, 20 m and 30 m are plotted in ArcGIS by taking average P-wave velocity for particular depth. Distribution for the values of average velocities was discrete for all the depth, all over the city. P-wave velocities are ranging from 0.447 km/s to 0.743 km/s for different depth where S-wave velocities are ranging from 0.284 km/s to 0.530 km/s.

2.6   Ground Water Table (GWT)

The water table fluctuates both with the seasons and from year to year because it is affected by climatic variations and by the amount of precipitation used by vegetation.

For larger earthquakes, the presence of ground water tends to decrease the horizontal acceleration response on the surface of basin, while the opposite trend is observed for horizontal displacement response. For earthquake of smaller intensity, the seismic response of a basin is almost not affected by the presence of ground water. Thus, the presence of ground water will have larger effects on the seismic response of a basin for larger earthquakes than for small earthquakes [2].

Maps of depth for water table are plotted as before monsoon season and after monsoon season to compare the ground water depth change during the monsoon season. Average 3 m high is noted as change after monsoon over the city. Highest water table is seen at areas Gomptipur, Behrampur, Paldi and Vatva.

2.7   Standard Penetration Test (N-Value)

Extensive geotechnical data collected from various private and government organization and synthesized. The SPT N values for the city is taken from the and plotted for different depth (2 m, 4 m, 6 m, 8 m, 10 m and 12 m), which indicate that south zone and north-west zone have higher N values in general where other zones have average lower N values in comparison.

3 Multi Criteria Analysis (MCA)

MCA describes any structured approach used to determine overall preferences among alternative options, where the options accomplish several objectives. In MCA, desirable objectives are specified and corresponding attributes or indicators are identified.

In this study, **AHP** (analytical hierarchy process) is used to integrate all the maps created by Arc GIS. AHP was introduced by Saaty (1980) as a management tool for decision making in multi attribute environments. The fundamental approach of AHP is to break down a "big" problem into several "small" problems; while the solution of these small problems is relatively simple, it is conducted with a view to the overall solution of the big problem [1]. AHP involves building a hierarchy of decision elements (factors) and then making comparisons between possible pairs in a matrix to give a weight for each element and also a consistency ratio. It is based on three principles: decomposition, comparative judgment and synthesis of priorities [3]. It is a multiple criteria decision-making technique that allows subjective as well as objective factors to be considered in the decision-making process. AHP aims at quantifying relative weights for a given set of criteria on a ratio scale. Two features of AHP differentiate it from other decision-making approaches. One, it provides a comprehensive structure to combine the intuitive rational and irrational values during the decision-making process. The other is its ability to judge the consistency in the decision-making process.

Parameters are prioritized according to their importance and influence to the seismic activities and hazard. PGA was given the highest priority (9) and GWT was given the least priority (3). SPT-N value, Vs and Vp were given the same priority (5) (see Table 1).

Table 1. Priority of the parameters.

PGA (ground)	9
PF	8
Population	7
SA	6
Vs	5
Vp	5
SPT-N	5
PGA (rock)	4
GWT	3

If **pairwise comparison matrix** (see Table 2) A is n × n matrix, where n is the number of factors considered. Each component a_{jk} of the matrix represent the importance of the jth criterion relative to the kth criterion. If $a_{jk} > 1$, then the jth criterion is

Table 2. Pairwise comparison matrix.

	PGA at ground	PF	Population	SA	Vs	Vp	SPT-N	PGA at rock	GWT
PGA (ground)	1.00	1.12	1.28	1.50	1.80	1.80	1.80	2.25	3.00
PF	0.88	1.00	1.14	1.33	1.60	1.60	1.60	2.00	2.67
Population	0.78	0.87	1.00	1.67	1.40	1.40	1.40	1.75	2.33
SA	0.67	0.75	0.86	1.00	1.20	1.20	1.20	1.50	2.00
Vs	0.56	0.62	0.71	0.83	1.00	1.00	1.00	1.25	1.67
Vp	0.56	0.62	0.71	0.83	1.00	1.00	1.00	1.25	1.67
SPT-N	0.56	0.62	0.71	0.83	1.00	1.00	1.00	1.25	1.67
PGA (rock)	0.44	0.50	0.57	0.67	0.80	0.80	0.80	1.00	1.33
GWT	0.33	0.37	0.43	0.50	0.60	0.60	0.60	0.75	1.00
Sum	5.78	6.5	7.43	8.67	10.4	10.4	10.4	13.00	17.33

more important than the kth criterion, while if $a_{jk} < 1$, then the jth criterion is less important than the kth criterion. If two criteria have the same importance, then the entry a_{jk} is 1.

Normalized value (see Table 3) is determined by dividing each cell with the sum of the relative column. If sum of the column 1 is $\sum A$, then normalized value for cell 1 is $a_{11}/\sum A$ (a_{11} is the value of pairwise comparison in cell 1).

Table 3. Normalized value and weights of each parameter.

	Normalized value	C	Weights
PGA (ground)	0.17	1.558	0.173
PF	0.15	1.285	0.154
Population	0.13	1.212	0.135
SA	0.11	1.038	0.115
Vs	0.10	0.865	0.096
Vp	0.10	0.865	0.096
SPT-N	0.10	0.865	0.096
PGA (rock)	0.08	0.692	0.077
GWT	0.06	0.519	0.058

Weights (see Table 3) for each parameter are calculated by dividing sum of the row by the number of parameters considered in comparison. If the sum of one row is C, then weight of that factor should be C/9.

Then **weighted sum values** (see Table 4) for implementation in ArcGIS are calculated according to priorities of data. Sum of these values should be equal to the weight of that factor.

Table 4. Weighted sum values of each parameter

Factors	Weights	Value	Weighted sum value
PGA at surface level	0.173	0 m	0.0576
		1 m	0.0461
		2 m	0.0345
		3 m	0.023
		4 m	0.01153
Predominant frequency	0.154		0.154
Population	0.135		0.135
Spectral acceleration	0.115	0.1 s	0.0461
		0.2 s	0.0307
		0.3 s	0.023
		0.5 s	0.0153
		1 s	0.0077
S wave velocity (Vs)	0.096	10 m	0.048
		20 m	0.032
		30 m	0.016
P wave velocity (Vp)	0.096	10 m	0.048
		20 m	0.032
		30 m	0.016
SPT value	0.096	2 m	0.0271
		4 m	0.0226
		6 m	0.018
		8 m	0.0136
		10 m	0.0091
		12 m	0.0045
PGA at rock level	0.077		0.077
Ground water table	0.058	After monsoon	0.0387
		Before monsoon	0.0193

4   Result and Conclusion

Integration of maps (prepared from the data of different factors affecting) through AHP method gave the seismic risk index of the city. Integration of different factors are done layer by layer in ArcGIS using weighted sum tool. Risk index is divided in 1 to 5 as 1 is lowest risk and 5 is highest risk (see Fig. 2).

In general most of the GIS maps of factors such as PGA at surface level, SA, P wave and S wave velocities, ground water table and SPT-N are having their higher values in south zone and/or in west zone, higher risk is to be seen in these zones but population map indicates that south zone have higher population therefor giving higher risk at this zone. Almost all the factors has moderate values on both sides of the river, hence moderate risk is observed in these areas and due to less population in west zone it has lower risk index in the risk map.

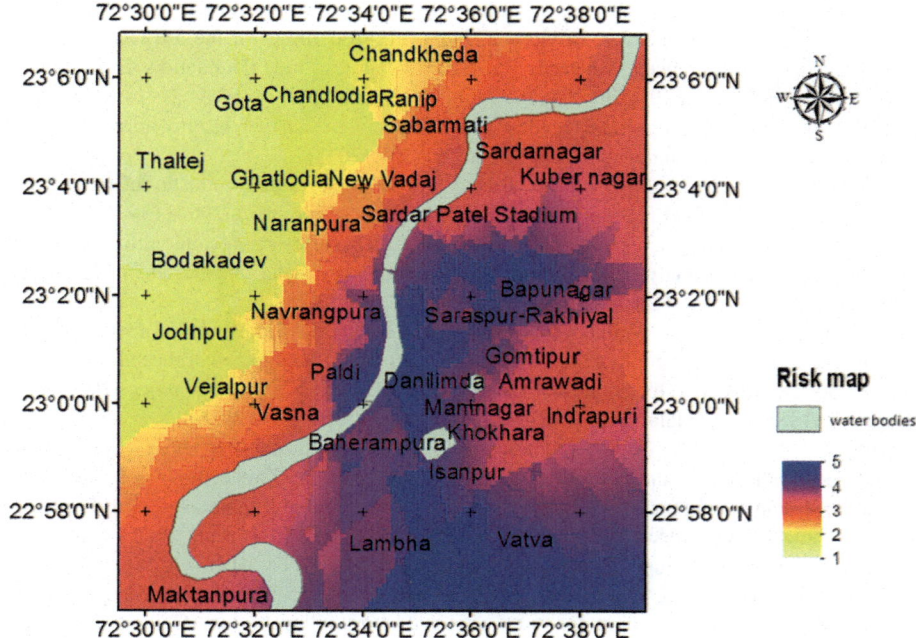

Fig. 2. Risk index map for Ahmedabad city.

This map can be used to estimate risk if any earthquake strikes in the region. Local Municipal Corporation can implement rules and regulations (for building heights, set-back distance, structural specifications) accordingly for construction of any residential or commercial buildings and bridges in high risk areas of the city. Prior to construction, structural designs can be improved accordingly to the risk factor accounted in the area.

References

1. Shapira, A., Simcha, M.: AHP-based weighting of factors affecting safety construction sites with tower cranes. J. Constr. Eng. Manag. **135**(4), 307–318 (2009)
2. Chen, H.-T., Ho, J.-C.: Effect of ground water on seismic response of basin. In: Proceedings of 12th World Conference on Earthquake Engineering, Auckland, New Zealand, 30th January–4th February (2000)
3. Malczewski, J.: GIS and Multicriteria Decision Analysis. Wiley, Hoboken (1999)
4. Thaker, T.P., Bhandari, T.S., Rao, K.S.: Seismic hazard analysis for Ahmedabad city. In: Proceedings of the Indian Geotechnical Conference, Roorkee, 22–24 December 2013 (2013)
5. Thaker, T.P., Rao, K.S., Gupta, K.K.: Ground response and site amplification studies for coastal soil, Kutch, Gujarat: a case study. Int. J. Earth Sci. Eng. **3**(4), 543–553 (2010)

6. Thaker, T.P., Rao, K.S.: Seismic hazard analysis for urban territories: a case study of Ahmedabad region in the state of Gujarat, India. In: Proceedings of the 3rd Geo-Shanghai International Conference, Advances in Soil Dynamics and Foundation Engineering, GSP240: ASCE, pp. 219–228 (2014)
7. Mehta, P., Thaker, T.P., Raghavendra, H.B.: Review of seismic hazard approaches for Indian subcontinents. Indian J. Tech. Educ. 200–205 (2017)
8. Rao, K.S., Thaker, T.P., Aggarwal, A., Bhandari, T., Kabra, S.: Deterministic seismic hazard analysis of Ahmedabad region, Gujarat. Int. J. Earth Sci. Eng. 5(2), 206–213 (2012)
9. Thaker, T.P., Rathod, G.W., Rao, K.S., Gupta, K.K.: Use of seismotectonic information for the seismic hazard analysis for Surat City, Gujarat, India: deterministic and probabilistic approach. Pure. Appl. Geophys. 169(1–2), 37–54 (2012)
10. Thaker, T.P., Rao, K.S.: Estimation of liquefaction hazard for Surat Urban Territory, South Gujarat, India Using geophysical and geotechnical investigations. In: Proceedings of the 19th Southeast Asian Geotechnical Conference and 2nd AGSSEA Conference (19SEAGC and 2AGSSEA), Kualalumpur, 31st May–3rd June, pp. 981–984 (2016)
11. Thaker, T.P., Rao, K.S.: Development of statistical correlations between shear wave velocity and penetration resistance using MASW technique. In: Proceedings of the 14th Pan-American Conference on Soil Mechanics and Geotechnical Engineering, 64th Canadian Geotechnical Conference and 5th Pan-American Conference on Teaching and Learning of Geotechnical Engineering, Ontario, Canada, 2–6 October 2011 (2011)
12. Thaker, T.P., Rao, K.S., Gupta, K.K.: Development of uniform hazard response spectra and shear wave velocity mapping for surat city and surrounding region, Gujarat, India. In: Proceedings of the 3rd International Conference on Geotechnical Engineering for Disaster Mitigation and Rehabilitation 2011 Combined with the 5th International Conference on Geotechnical and Highway Engineering (GEDMAR), Semrang, Central Java, Indonesia, May 2011, pp. 461–467 (2011)

Seismic Risk Assessment for Coimbatore Integrating Seismic Hazard and Land Use

E. Lalith Prakash, Sreevalsa Kolathayar[(⊠)], and R. Ramkrishnan

Department of Civil Engineering, Amrita School of Engineering, Amrita Vishwa
Vidyapeetham, Coimbatore, India
sreevalsakolathayar@gmail.com

Abstract. Indian cities are expanding not only in terms of built environment but
also in population. Multi-story buildings are rising rapidly to accommodate the
growing population, undesirably even in hazard prone areas. Such a scenario
calls for a proper disaster risk reduction program and plan to control the inevitable damage to lives and properties. Earthquakes are destructive only if the
factors that increase the damages prevail. Attention should be given to crowded
cities with people and infrastructure vulnerable to hazard. This study presents the
details of deterministic seismic hazard analysis (DSHA) done for Coimbatore
city of the state of Tamil Nadu, India using the latest available information on
seismicity of the region. The earthquake data was compiled from different
agencies and homogenized in a unified moment magnitude scale to create an
updated earthquake catalog. Seismotectonic map for the study area was prepared
by superimposing the earthquake events on the seismogenic sources. DSHA was
then performed by dividing the study area into grids of size $0.02° \times 0.02°$ (approximately 2 km $\times$ 2 km) using a MATLAB code, considering three different
attenuation relationships for the stable continental region. Land use (LU) map for
the region was developed from LANDSAT 8 data using various GIS platforms.
Hazard contour map prepared using ArcGIS, was then overlaid on the LU map to
comprehend the seismic risk of the region. It was observed that, though the wards
south-west of the city shows higher Peak Ground Acceleration (PGA) values, the
wards north-east of the city have larger and denser built-up areas, increasing its
vulnerability, in the event of an earthquake.

Keywords: Deterministic seismic hazard analysis · Peak ground acceleration
Land use · Remote sensing · Seismic risk

1 Introduction

1.1 General

Earthquakes have a devastating effect on human life and property when human
activities interfere with the occurrence of such a natural phenomenon. Populated human
settlements and congested built-up areas in earthquake prone regions amplify the risk to
both human life and property. Identification of such risk prone regions will be of
greater use for a comprehensive earthquake disaster mitigation plan. Presently, the
seismic zone map available for India as per the Bureau of Indian Standards

© Springer Nature Singapore Pte Ltd. 2018
A. Farid and H. Chen (Eds.): GSIC 2018, *Proceedings of GeoShanghai 2018*
International Conference: Geoenvironment and Geohazard, pp. 117–124, 2018.
https://doi.org/10.1007/978-981-13-0128-5_14

(IS1893-2002) broadly classifies India into four hazard zones based on past earthquake data. Previous studies indicate the limitations of the existing seismic zone map, as it projects unstructured regions of homogeneous hazard level as lumps of landmasses, lacking a scientific hazard assessment (Raghu Kanth and Iyengar 2006; Menon et al. 2010). In this study, a deterministic approach is adopted to analyze the seismic hazard for Coimbatore city. The deterministic hazard contour map thus generated was overlaid on a digitized ward wise map of Coimbatore city.

Indian cities follow a trend of rapid urbanization due to population concentration in urban centres and its suburbs, and also due to migration of rural population entailed by poverty. This rapid urbanization, also related to urban sprawl results in extremely high population intensity, which in turn leads to a gradual decline of urban services including disaster mitigation. An attempt is made here to identify and interpret the combined effect of seismic hazard and Land Use pattern for Coimbatore city.

1.2 Study Area of Coimbatore

The city of Coimbatore is located at the extreme west of the Indian state of Tamil Nadu (South India), spreading over an area of 105.5 sq. kms. As per the latest Census (2011), Coimbatore has a population of over a million (http://www.censusindia.gov.in/pca). Even though Coimbatore has experienced an earthquake of moment magnitude 6.3 in 1900 (the epicenter of the earthquake was located at 10°45′ North Latitude and at 76°45′ East Longitude), it was categorized in seismic zone Zero, in the first version of the IS 1893 (BIS 1962). Presently, Coimbatore city is categorized as Zone III as per the latest release of IS 1983 (BIS 2002), where Zone II corresponds to least hazard and Zone V corresponds to the highest hazard within the country. Coimbatore has not experienced major earthquakes after 1900. According to elastic rebound theory, the strain energy built up in the geologic faults for years are released during the event of earthquake; they are more likely to occur in regions where small or no seismic activity has been seen in recent past (Kramer 1996). Two earthquakes, each with a magnitude of around 5.0 on the Richter Scale, in Idukki and Coimbatore districts were observed in December 2000 and January 2001, respectively.

2 Methodology

2.1 Deterministic Seismic Hazard Analysis (DSHA)

An updated and homogenized earthquake data catalogue is the prime requirement for hazard estimation. Kolathayar et al. (2012) compiled a catalogue for the entire Indian subcontinent based on both historical and instrumental data. In the present study, the earthquake data catalogue for the study region was updated by including all the events till 2017 within a radius of 300 kms from political boundary of Coimbatore. All the earthquake events were homogenized in Moment Magnitude (M_w) scale using magnitude correlations specific to the Indian peninsular shield (Kolathayar et al. 2012). This earthquake data was declustered from aftershocks and foreshocks, rendering a Poisson distribution of earthquake events using the algorithm developed by Gardner and Knopoff (1974) modified by Uhrhammer (1986). Accurate Ground motion prediction equations (GMPE) for India are scarce due to lack of sufficient strong motion records in the region. For PGA estimation of the study region, the attenuation relationships developed by Raghu Kanth and Iyengar (2007), Atkinson and Boore (2006) and Campbel and Bozorginia (2003) were used. Raghu Kanth and Iyengar (2007) developed attenuation relations for Peninsular Indian region. Attenuation relations for Eastern North America (ENA) were developed by Campbell and Bozorgnia (2003) and Atkinson and Boore (2006). Similarities in the geological and tectonic settings of both the regions allows using the same equations for our study region. Cramer and Kumar (2003) studied the aftershocks of Bhuj earthquake and proved that the ground motion attenuation in Peninsular Indian shield and ENA are comparable. All three relationships were assigned equal weightage in a logic tree frame work for the estimation of PGA values.

DSHA was carried out using the earthquake sources and events data within a boundary of 300 kms from the outermost administrative boundaries of Coimbatore city in all four directions. The study area was divided into several grids of size $0.02° \times 0.02°$ (approximately 2 km $\times$ 2 km). Using the deterministic approach, the PGA values at the centre of each of these grids were estimated using a MATLAB code. All linear and point earthquake sources were considered for analysis. All earthquake events near the geologic faults in the area were considered and the event with maximum magnitude was taken as the governing event associated with that linear source. Controlling earthquake for the point of interest was then identified for all linear and point sources. PGA value at the central point of each grid corresponding to the controlling earthquake was then computed using the attenuation relationships as mentioned above, considering shortest source to site distance.

2.2 Land Use Mapping

Land Use map for the region was developed using open source GIS software platforms like QGIS and GRASS GIS, based on LANDSAT 8 remote sensing images of 30 m resolution (NIR, SWIR, visible); 100 m resolution (thermal); and 15 m resolution (panchromatic), obtained from USGS. The satellite images were geo-referenced, cropped and then extracted. Diverse patches in the landscape were identified on a False Colour Composite (FCC) generated from the georeferenced images (bands- green, red and NIR). Training polygons were selected on the FCC image by identifying and delineating various heterogeneous patches by overlaying it on a google earth satellite image. Pixels of deciduous and evergreen forest land, grasslands, farmlands, coconut, arecanut, mango and other plantations, built up areas including roads and highways, built structures, airports, waterbodies like rivers and lakes, and open areas were selected from different points on the FCC to form training polygons for LU mapping. Uniform distribution and accurate marking of these training polygons was ensured throughout the study area (Ramachandra et al. 2012). Built in Gaussian maximum likelihood algorithm was used to classify each pixel of the image, based on the training polygons previously selected on the FCC. Land Use was mapped based on six major categories like forest land, agricultural land, plantations, built up areas, water bodies and open areas.

2.3   Hazard Mapping and Overlaying

Seismic hazard contour map was developed using the PGA values corresponding to each grid points in QGIS. Land Use map generated was then overlaid on the hazard contour map to obtain a combined seismic risk map, which was then used to assess the risk pattern of the study area (Fig. 1).

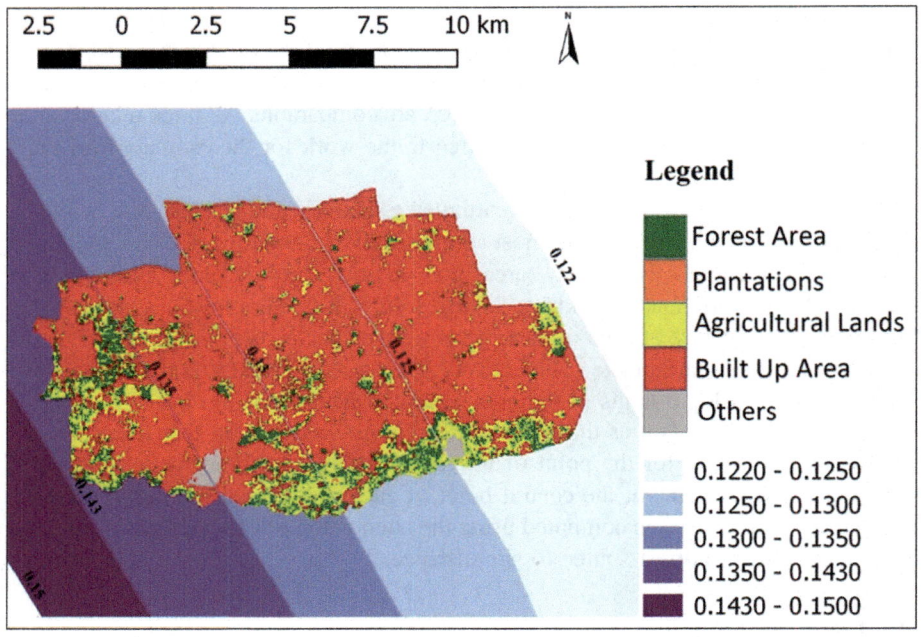

Fig. 1.   Land-use map overlaid on hazard contour map for Coimbatore city (All PGA values are expressed in terms of 'g')

A ward wise administrative base layer for Coimbatore city was developed using ArcGIS. The same was overlain on the combined hazard-Land Use Seismic risk map so as to enable accurate risk assessment of each ward of the city. (Fig. 2).

The hazard values obtained for the area were classified as high, moderate and low, based on relative comparison of PGA values, as compared to a maximum PGA of 0.143 g and a minimum 0.122 g. By visual interpretation, the urban areas was classified as high, moderate and low based on the concentration of built up area in the LU map.

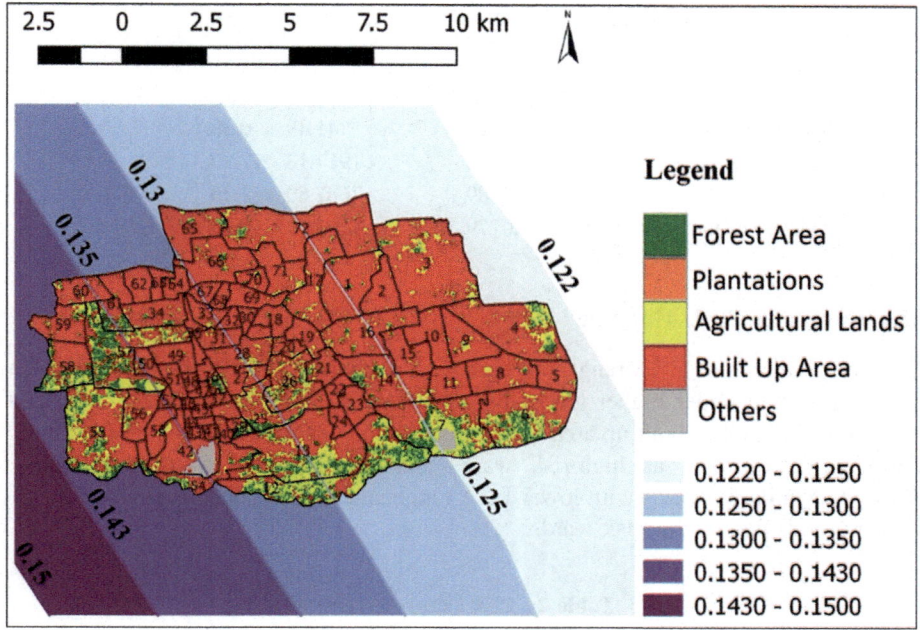

Fig. 2. Integrated ward wise seismic risk map (All values are expressed in terms of 'g')

3   Results and Discussions

3.1   Hazard and Land Use Map

Hazard contour mapping. Coimbatore falls within a PGA range of 0.122 g–0.143 g, with the southwestern zones of the city falling in the PGA range 0.135 g–0.143 g and northeastern zones in the range of 0.122 g–0.125 g (Fig. 1). The PGA distribution pattern was found to be in order with the ones estimated in previous similar studies (Anbazhagan et al. 2012), but with relatively lesser PGA values. The currently predicted PGA values for Coimbatore were found to be comparatively greater than the zone factors of 0.08 g specified in IS 1893 (BIS 2002).

Land Use classification. Coimbatore city spans over a total area of 106.56 sq.kms where only 29.67 sq.kms of the land is covered by forests, plantations, agricultural lands, water bodies and open areas altogether, the remaining being built up and populated areas. About 72% of the city is built up land (Table 1) indicating the presence of urban settlements in the form of all types of commercial buildings, public buildings, residential buildings, industrial buildings and other city infrastructure. Southern and eastern parts of the city has lesser built up area, indicating lesser vulnerability and risk of the region in the event of an earthquake (Fig. 1).

Table 1. Coimbatore city land use classification for the year 2016.

Description	Square kilometre	Hectares	Percentage cover
Forest area	9.0621	906.21	8.51
Plantations	0.4149	41.49	0.39
Agricultural lands	19.1718	1917.18	17.99
Built up area	76.8789	7687.89	72.16
Open areas and water bodies	1.0179	101.79	0.95

3.2   Integrated Risk Mapping

The wards with higher PGA range and higher concentration of built up areas indicate a larger population under higher risk, whereas the wards with higher PGA range and lesser concentration of built up areas indicate lesser population under higher risk. Such wards are categorized as high-risk wards and moderate-risk wards respectively (Table 2). Similarly, wards with lower PGA range and lower concentration of built up areas are classified as low-risk wards.

Table 2. Risk categorization.

Hazard*	Built up area	Risk
Moderate	Low	
Low	Low	Low
Low	Moderate	
High	Low	
Moderate	Moderate	Moderate
Low	High	
High	High	
High	Moderate	High
Moderate	High	

*High – 0.13 g to 0.143, Moderate - 0.125 g to 0.13 g, Low-0.122 to 0.125.

Hence, 35 wards out of a total of 72 wards of the city have been categorized as high-risk wards, 2 wards as low-risk wards and the rest as moderate-risk wards. Areas such as Selvapuram, PN pudur, VOC nagar, Agricultural university, Selvapuram south, RS Puram, Ukkadam, town hall, Saibaba colony and Tatabad are the major areas under high-risk. Gandhipuram, Sanganur, Lakshmi mills, Racecourse, PN palayam and Uppilpalayam are the major areas under moderate-risk, and Nanjundapuram, Singanallur, Keelakarai and Karumbu kadai are the major areas falling under low-risk according to the integrated seismic risk map (Fig. 2).

4   Conclusion

In the present study, an updated earthquake catalog was created for Coimbatore and surrounding regions considering historical and instrumental seismic events dataset from various agencies like Bhabha Atomic Research Centre (BARC), India Meteorological Department (IMD), Indira Gandhi Centre for Atomic Research (IGCAR), Kalpakkam, National Geophysical Research Institute (NGRI) Hyderabad, International Seismological Center (ISC) data file, Harvard seismology and USGS/NEIC catalogue etc., which was then homogenized and declustered to remove foreshocks and aftershocks. A state of the art Deterministic seismic hazard analysis was carried out considering three attenuation relationships (Raghu Kanth and Iyengar 2007; Atkinson and Boore 2006; Campbel and Bozorginia 2004) to estimate the peak ground acceleration at grid points selected in the study region. Spatial variation of seismic hazard in the region was presented as contour map using GIS tools.

Land Use map was developed for the city using LANDSAT-2016 images using GIS software tools. An administrative map for the Coimbatore city showing ward-wise boundaries was also developed using GIS software. An integrated risk map was developed for the region by overlaying all three maps. The south-western wards were identified to have maximum seismic risk compared to other wards. Even though Eastern and North-eastern wards have higher built-up land use area, risks in these wards are found to moderate to low, as these wards have low hazard values when compared to western part of the city. The findings from this study will be useful for town planners and engineers as directions for future expansion and infrastructure development of the city. These findings will also aid in seismic retrofitting of existing buildings and increasing the seismic resilience of city residence by inducing the earthquake risk awareness into public consciousness.

References

BIS-1893: Indian standard criteria for earthquake resistant design of structures, Part 1 – general provisions and buildings. Bureau of Indian Standards, New Delhi (2002)

BIS-1893: Recommendations for earthquake resistant design of structures. Bureau of Indian Standards, New Delhi (1962)

Raghu Kanth, S.T.G., Iyengar, R.N.: Seismic hazard estimation for Mumbai city. Curr. Sci. **91** (11), 1486–1494 (2006)

Menon, A., Ornthammarath, T., Corigliano, M., Lai, C.G.: Probabilistic seismic hazard macrozonation of Tamil Nadu in Southern India. Bull. Seismol. Soc. Am. **100**(3), 1320–1341 (2010)

Kramer, S.L.: Geotechnical Earthquake Engineering. Pearson Education Pvt. Ltd., Delhi (1996)

Anbazhagan, P., Gajawada, P., Parihar, A.: Seismic hazard map of Coimbatore using subsurface fault rupture. Nat. Hazards **60**(3), 1325–1345 (2012)

Kolathayar, S., Sitharam, T.G., Vipin, K.S.: Deterministic seismic hazard macrozonation of India. J. Earth Syst. Sci. **121**(5), 1351–1364 (2012)

Atkinson, G.M., Boore, D.M.: Earthquake ground motion prediction equations for Eastern North America. Bull. Seismol. Soc. Am. **96**(6), 2181–2205 (2006)

Campbell, K.W., Bozorgnia, Y.: Updated near-source ground motion (attenuation) relations for the horizontal and vertical components of peak ground acceleration and acceleration response spectra. Bull. Seismol. Soc. Am. **93**, 314–331 (2003)

Cramer, C.H., Kumar, A.: 2001 Bhuj, India, earthquake engineering seismoscope recordings and eastern North America ground motion attenuation relations. Bull. Seismol. Soc. Am. **93**, 1390–1394 (2003)

Raghu Kanth, S.T.G., Iyengar, R.N.: Estimation of seismic spectral acceleration in peninsular India. J. Earth Syst. Sci. **116**(3), 199–214 (2007)

Ramachandra, T.V., Aithal, B.H., Sanna, D.D.: Insights to urban dynamics through landscape spatial pattern analysis. Int. J. Appl. Earth Obs. Geoinf. **18**, 329–343 (2012)

Census India district wise population handbook search page. http://www.censusindia.gov.in/2011census/dchb/DCHB.html. Accessed 29 Mar 2017

Gardner, J.K., Knopoff, L.: Is the sequence of earthquakes in Southern California with aftershocks removed, Poissonian? Bull. Seismol. Soc. Am. **64**(5), 1363–1367 (1974)

Uhrhammer, R.A.: Characteristics of northern and central California seismicity. Earthq. Notes **1**, 21 (1986)

Assessment of Hydrodynamic and Deformation Characters of the Hongyanzi Landslide in the Pubugou Reservoir

Bing Han[1], Bin Tong[1(✉)], Jinkai Yan[1], and Jianhui Dong[2]

[1] China Institute of Geo-Environmental Monitoring, Beijing, China
tongbin1103@126.com
[2] School of Architecture and Civil Engineering,
Chengdu University, Chengdu, China

Abstract. This paper evaluated the hydrodynamic and deformation characters of the Hongyanzi landslide in Pubugou hydropower station reservoir area, Sichuan province, Southwest China. The fluctuations of slope groundwater level, slope surface displacement and deep seated displacement with the fluctuation of reservoir water level were monitored. The results showed that the fluctuation of groundwater level followed the reservoir water level with two month of time lag. Approximate 0.23 to 0.25 m/day of declination rate for reservoir water level could be benefit for maintaining the global stability of the bank slope when the elevation of reservoir water level varied from 820 to 847 m. A constant head difference between the groundwater level and reservoir water level is important to the stability of reservoir slope.

Keywords: Reservoir-induced landslide · Multi-parameter monitoring
Water level fluctuation · Slope deformation

1 Introduction

Reservoir landslides are very commonly seen geological hazards. Historical statistics show over 90% of reservoir landslides were caused by reservoir water level fluctuations, and more than 80% of them occurred during the first 3 to 5 years after the construction of the dam [1–6]. Accessing the temporal relations among reservoir water level fluctuation, the hydrodynamic and deformation characters of the slope are the key issue for advancing the understanding on failure mechanism. The periodic geological inspections for monitoring the hydrogeological condition and deformation characters of the slope during the reservoir water level fluctuation [7] provide the basis for advancing the analysis on reservoir landslides.

In this paper, a multi-parameter monitoring program was conducted to record the slope groundwater level fluctuation, slope surface displacement and deep seated displacement at various locations on the Hongyanzi landslide during the period of reservoir impoundment and drawdown from 2013 to 2014. Based on the results, the temporal relations among reservoir water level fluctuation, slope groundwater level, surface displacements, crack propagations and deep seated displacement could be

© Springer Nature Singapore Pte Ltd. 2018
A. Farid and H. Chen (Eds.): GSIC 2018, *Proceedings of GeoShanghai 2018*
International Conference: Geoenvironment and Geohazard, pp. 125–133, 2018.
https://doi.org/10.1007/978-981-13-0128-5_15

produced to provide guidance for reducing the reservoir landslide risk by considering the fluctuation of reservoir water level.

2 Overview of the Hongyanzi Landslide

The Hongyanzi landslide is located on the right bank of the Dadu River, a secondary tributary of the Yangtze River in Hanyuan County, Sichuan province, Southwest China. It is 23 km away from the Pubugou hydropower station in the upper stream region. Based on field inspections, the average slope of the Hongyanzi landslide is 27°. The slope of main scarp (40° to 50°) is steeper than the front of the sliding mass. The elevation of the slope toe and back-scarp is 810 m and 954 m, respectively. The length and width of the landslide is 600 and 580 m, respectively [7].

Prior to the installation of monitoring devices, the large deformations have occurred to landslide boundaries due to the fluctuation of reservoir water level, and the magnitudes reached up to 1.7 m and 1.5 m on the left and right boundary, respectively. The Hongyanzi landslide is an ancient accumulation landslide, and the sliding mass is mainly composed of the quaternary silty clay, cobble or gravel, and stones. The stratum of sliding bed is Jurassic reddish sandstone in the footwall of Hanyuan-Zhaojue fault, and Permian limestone in hanging wall of the fault [8, 9]. The Hongyanzi landslide faces the Dadu River in the north, and gullies in both east and west. The steep topography and large elevation differences from the scarp to the toe provide a significant potential for the slope groundwater to discharge into the Dadu River (Fig. 1).

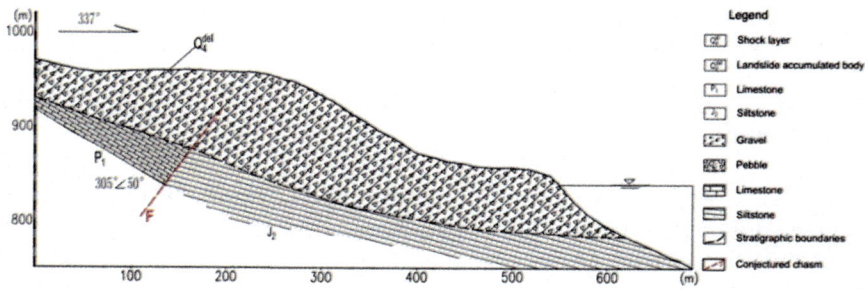

Fig. 1. Main geological profile of the Hongyanzi landslide

During the operation of the Pubugou hydropower station, the reservoir water level fluctuated from 790 m to 850 m, and the range of water-level-fluctuating zone was approximate 55 m. Based on historical records [9, 10], the Hongyanzi landslide remained stable before the construction of the hydropower station. After the impoundment, large deformation occurred to the slope during the declination of reservoir water level. The impoundment and power generation started on June and November 2010, respectively.

On February 2011, after 8 month of impoundment, large deformations started to occur and the accumulated downward deformation reached up to 1 m by the end of

2011. In 2012, simultaneous deformation continued to occur during the declination process of reservoir water level. From March to May 2013, the maximum deformation reached up to 2 m, and the accumulated deformation reached up to 3 m since the beginning of hydropower station operation. The majority of the deformations occurred during the declination of the reservoir water level, and no significant deformations were observed when the reservoir water level remained stable or increased. Due to the extensive deformation occurred, a multi-parameter monitoring program was conducted to access the hydrodynamic and deformation characters of the slope to advance the understanding on the major factors of inducing the deformations of the Hongyanzi landslide.

3 The Multi-parameter Monitoring Program

Since 2012, various monitoring equipment were employed to conduct a continuous monitoring on the surface deformation, deep-seated displacement, slope groundwater level, crack propagation on both left and right boundaries and atmospheric precipitations. Reservoir water level was measured by the water gauge. The distribution of the installed monitoring devices is shown in Fig. 2.

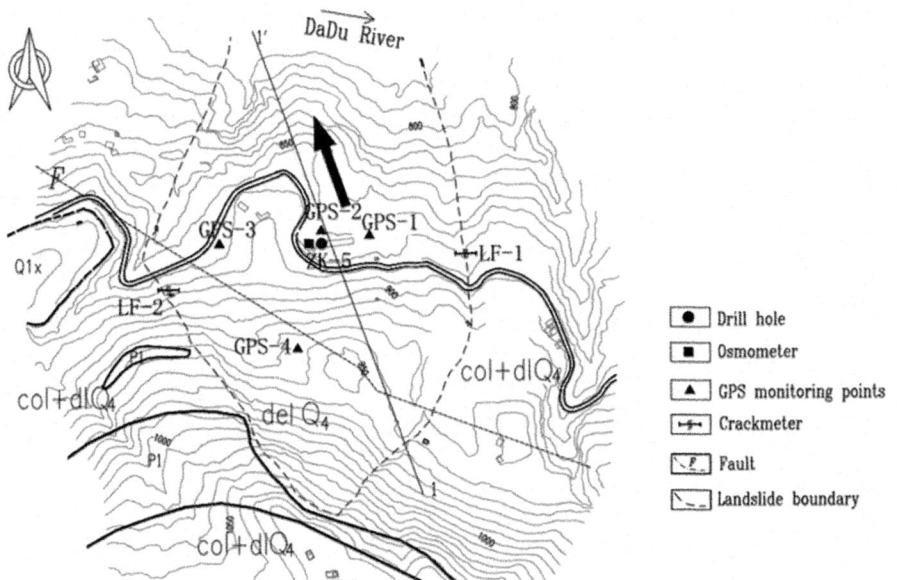

Fig. 2. The monitoring program scheme on the Hongyanzi landslide

To monitor the development of the slope deformation, totally five GPS stations were employed, of which four were installed within the landslide boundary as the monitoring stations and one was installed outside the boundary as the fixed monitoring station. A groundwater monitoring gauge was installed in the front of the landslide and its installation elevation was 825 m. One crack monitoring gauge was installed on both left and right boundaries of the landslide. Three deep seated displacement inclinometers were installed in the middle and front of the landslide. The monitoring program started in December 2012, and this study mainly focuses on the monitoring data received prior to December 2014.

4   Monitoring Data Interpretations and Discussions

4.1   Slope Groundwater Table

Figure 3 shows the fluctuation of slope groundwater level during the two cycles' of impoundment and drawdown in 2013 recorded by ZK-5-w. Because of the monitoring device technical errors, only the slope groundwater level data above 825 m were captured. As shown, a strong temporal correlation exists between the slope groundwater level and the reservoir water level. From January to August 2013, the impoundment started from May 2013 after four month of drawdown. From August to December 2013, the impoundment started from September 2013 after one month of drawdown. In both cycles, the fluctuation of slope groundwater level followed a similar pattern with the reservoir water level [11]. Based on the recordings from January 1st to March 15st 2013, the two curves are nearly parallel to each other, and that indicated a good consistency and constant head difference between slope groundwater level and reservoir water level. Since late March 2013, the increased declination speed of reservoir water level lead to an increased water head difference. Therefore, from April to July 2013, the head differences between slope groundwater level and reservoir water level were greatest, and caused a vital reduction on the slope stability as was confirmed by the slope deformation measurements presented in following section.

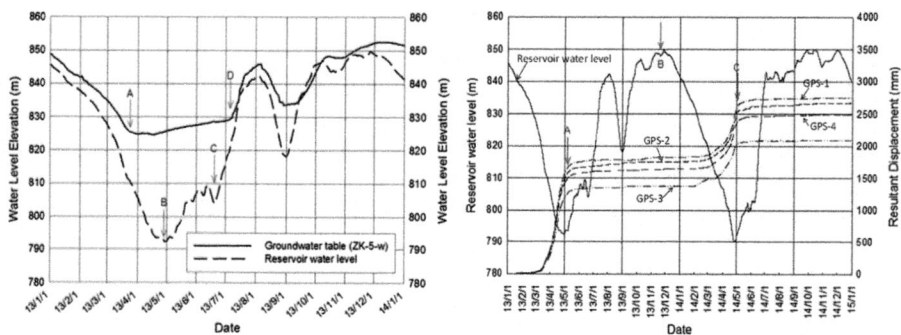

Fig. 3. Reservoir water level vs. slope groundwater level (ZK-5-w) vs. surface displacement

From May to August 2013, the reservoir started impoundment by increasing the water level from 792 m to 843 m, and the slope groundwater level also increased from 825 m to 845 m. Both levels reached the peak value at approximate the same time on August 2013. In general, in the second cycle, the fluctuations of reservoir water level and slope groundwater level followed the similar pattern as occurred in the first cycle, with only half of the time being used. From October 2012 to March 2013, the fluctuation rate of reservoir water level was almost identical to that of slope groundwater level based on ZK-5-w. The average declination rate for both curves is around 0.25 m/day by calculating the curve slope magnitude from January to mid of February 2013. Starting from February 15th to May 1st, the declination rate of reservoir water level increased to 0.57 m/day that almost doubled the previous value, which caused an increased water head elevation differences and seepage force [10, 11].

From January 2013 to late March 2013 (Prior to Point A in Fig. 3), the two curves are nearly parallel to each other because of the similar slope magnitude, which were calculated to be between 0.23 and 0.25 m/day as the water level declination rate. Meanwhile, during the period, no vital slope deformation measurements were captured as presented in following section. Therefore, above observations indicated that for the Hongyanzi landslide, a declination rate of reservoir water level about or lower than 0.23 to 0.25 m/day (the threshold value) could allow a consistent head difference between slope groundwater level and reservoir water level within the range of 810 to 845 m, and could be benefit to maintain the overall stability of the slope. Additional analysis and simulation is being conduct currently to quantify the influences of reservoir water fluctuation rate on the stability and hydro-geological condition of the landslide. Based on the measurements, a greater declination speed of reservoir water level would lead to a larger difference between slope groundwater level and reservoir water level, and longer time lag to maintain a stable head difference (Prior to Point A), so-called the longer restoration period of slope stability. The sudden drop of reservoir water level resulted in the lagging fall of slope groundwater level, and formed excess pore water pressure detrimental to the slope stability based on the recordings from August to September 2013.

4.2 Surface Displacement

Figure 3 shows the correlations between the fluctuations of reservoir water level and slope surface displacements recorded at different locations. As shown, a strong temporal correlation existed between surface displacement and reservoir water level. The recordings showed that large deformations occurred from February to April during the declination period. No large deformations were captured when the reservoir water level increased (during the impoundment period from May 2013 to March 2014, approximately). The rapid declination of reservoir water level induced an intense seepage force, and reduced the slope stability. The largest surface displacement of approximate 3 m was recorded by GPS-1, and followed by GPS-4, GPS-2, and GPS-3 based on the magnitudes of the recordings.

The deformations at the measuring positions were small at the beginning of March in 2013. From March 2013, significant deformations occurred and the maximum daily displacement reached 60 mm. The observations matched with the fluctuation of groundwater level due to the increased declination rate of reservoir water level, which doubled the value in previous time period. Hence, the water head differences inside and outside the slope increased rapidly. After May 2013, the deformation rate at all measuring positions decreased significantly, and gradually approached to zero after June. And that could indicate the recovery of the global slope stability.

Figure 4a, b and c show the displacements in the north, east, and vertical direction, respectively, to discover the deformation characters of the Hongyanzi landslide from January 2013 to January 2015. Overall, a good consistency between displacement and reservoir water level drawdown was observed. The sudden increases to the deformation occurred approximately from March 1st 2013 and February 1st 2014, corresponding to the beginnings of reservoir water level declination on January 1st 2013 and January 1st 2014, respectively. Therefore, approximate two month of time lag exists between the reservoir water level drawdown and slope deformation.

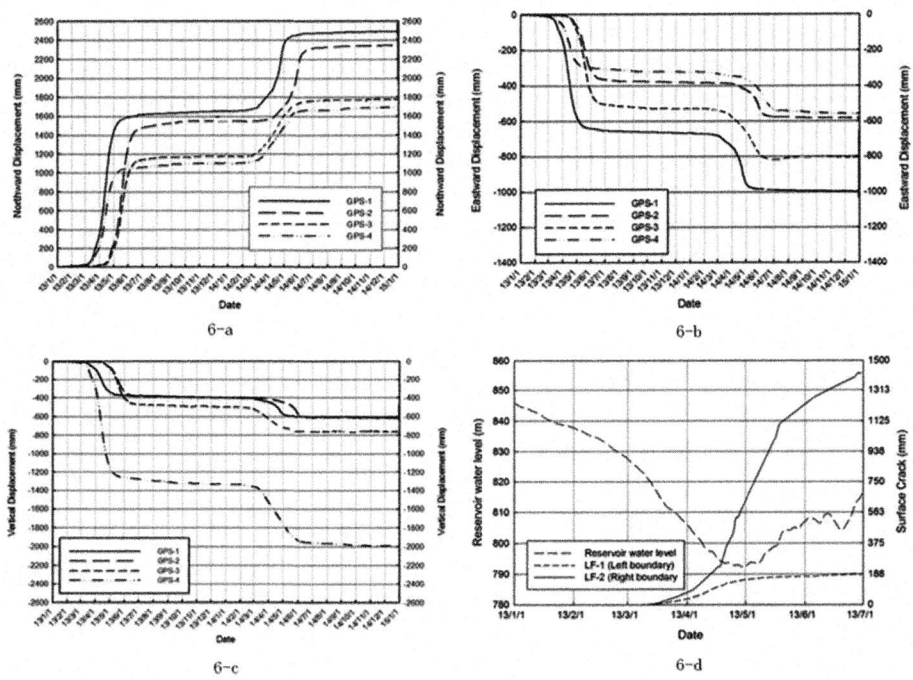

Fig. 4. Reservoir water level vs. surface displacements in different directions (a: northward displacements; b: eastern displacements; c: vertical displacements; d: displacements at right and left boundary)

In two years from January 2013, GPS-1 received the largest resultant displacement up to 2500 mm, and then followed by GPS-4, GPS-2, and GPS-3. The displacement vector to the north was larger than the magnitudes in the west. The displacement magnitude vector in the west direction was approximately same in the vertical direction. The direction of movements at all measuring positions were within NW 22 to 24 degrees, which can be regarded as the main sliding direction. From East to West, the displacement magnitude recorded at each measuring position gradually increased. This was caused by the impact of micro-topography and location of device installation relevant to the landslide boundary. Differing with the other measuring positions, GPS-4 was located near a scarp in the rear part of the landslide and the micro-topography at this location was steeper than the other measuring locations. The measurements provided a direct evidence of reservoir induced deformation magnitude and distribution characters under the influence of water level fluctuation. The recordings indicated deformation mechanism that the rapid declination of water level increased the hydraulic gradient of groundwater and seepage in the slope, which further changed the pore water pressure, decreased effective strength, and reduced the slope stability by reducing the buoyant force acting on the slope.

4.3   Deep Seated Displacement

The measuring depth for deep seated displacement at ZK-7-d is 18 m, 26 m, and 31 m below the slope surface. Figure 5 shows that small deformation less than 5 mm occurred at depth of 31 m and the deformations measured at the depths of 26 m and 18 m were much greater. The measurements provide a good support for determining the elevation of sliding surface and volume of sliding material within the depth range of 18 m to 26 m. Similar to the temporal correlation between surface displacement and slope groundwater level, a clear and consistent temporal correlation exists between the deep seated displacement and fluctuation of reservoir water level.

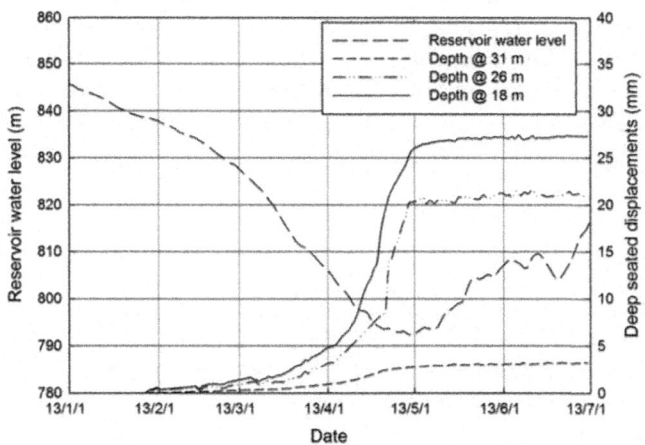

Fig. 5. Reservoir water level vs. deep seated resultant displacement (ZK-7-d)

4.4   Discussions

The SPSS software was utilized to conduct the correlation analysis among slope groundwater level, surface displacement (measured at GPS-2), reservoir water level, and precipitation. The results indicated that the difference between slope groundwater level and reservoir water level showed the most statistical importance to the slope surface displacement (correlation coefficient = 0.926), and followed by reservoir water table (coefficient = 0.267). The correlation between precipitation and slope surface displacement is comparatively lower. So the major triggering reason of the Hongyanzi landslide deformation is the rapid declination of reservoir water level; the gradually increased seepage pressure, which induced by the increased head difference between slope groundwater level and reservoir water level, caused the instability of the landslide when the equilibrium state of landslide was reached. In this case, the uplift pressure as generated during the impoundment process of the reservoir was relatively insignificant to reduce the stability of the landslide because nearly no obvious deformation was monitored during that period. The major reason is that the submerged area during reservoir impoundment is limited to the overall landslide area, and the generated buoyant force was insignificant to change the stability status of the slope.

5   Conclusions

The monitoring results of this study provided the in-situ data for advancing the understanding the internal correlations and working mechanism among reservoir water level fluctuation and bank slope stability. Following comments can be produced based on the monitoring data analysis: [1] the Hongyanzi landslide deformed significantly when the reservoir water level declined rapidly, and the slope deformation was mainly governed by the declination rate of the reservoir water level; [2] seepage pressure due to the difference between reservoir water level and slope groundwater level played a key role in controlling the slope deformation, and reducing the declination rate of reservoir water level can be benefit to slow down or cease the slope deformation; [3] based on the monitoring data on reservoir water level and slope groundwater level, the reservoir water level drawdown speed at approximate or lower than 0.23 to 0.25 m/day could be benefit to maintain global stability of the Hongyanzi landslide.

References

1. Tang, H.M., Li, C.D., Hu, X.L., Wang, L.Q., Criss, R., Su, A., Wu, Y.P.: Deformation responses of Huangtupo landslide to rainfall and the changing reservoirs of the Three Gorges Reservoirs. Bull. Eng. Geol. Environ. **74**(3), 933–942 (2016)
2. Yin, Y.P., Huang, B.L., Wang, W.P., Wei, Y.J., Ma, X.H., Ma, F.: Reservoir-induced landslides and risk control in Three Gorges Project on Yangtze River, China. J. Rock Mech. Geotech. Eng. **8**, 577–595 (2016)
3. Wei, J.B., Zheng, H.C.: Deformation characteristics of a preexisting landslide in reservoir area during reservoir filling and operation. Adv. Mater. Res. **15**, 2847–2851 (2011)
4. Yu, Y.J., Huan, Q.Z., Tang, H.M., Zhang, X.L.: Simulating the process of reservoir impoundment-induced landslide using the extended DDA method. Eng. Geol. **182**, 37–48 (2014)
5. Dai, F.C., Deng, J.H.: A large landslide in Zigui County Three Gorges Reservoir area. Can. Geotech. **6**, 1233–1240 (2004)
6. Jiang, J.W., Ehert, D., Xiang, W., Rohn, J., Huang, L., Yan, S.J., Bi, R.N.: Numerical simulation of Qiaotou landslide deformation caused by drawdown of the Three Gorges Reservoir, China. Environ. Earth Sci. **62**(2), 411–419 (2011)

7. Han, B.: Research on landslide monitoring and early warning in Ya'an Area. Ph.d. Dissertation. China University of Geosciences, Beijing (2016)
8. Huang, B.L., Yin, Y.P., Du, C.L.: Risk management study on impulse waves generated by Hongyanzi landslide in Three Gorges Reservoir of China on June 24, 2015. Landslides **13** (3), 603–616 (2016)
9. Wang, F.W., Zhang, Y.W., Hou, Z.T., Peng, X.M., Kiminori, A., Wang, G.H.: Movement of the Shuping landslide in the first four years after the initial impoundment of the Three Gorges Dam Reservoir, China. Landslides **5**(3), 321–329 (2008)
10. Liu, X.X., Zhang, X.S., Guo, R.Q.: Effects of drawdown of reservoir water level on landslide stability. Chin. J. Rock Mech. Eng. **24**, 1439–1444 (2005)
11. Huang, B.L., Yin, Y.P., Du, C.L.: Risk management study on impulse waves generated by Hongyanzi landslide in Three Gorges Reservoir of China on June 24, 2015. Landslide **12**, 603–616 (2015)

Review on the New Methods of Landslide Hazards Monitoring Methods in the Twenty-First Century

Haofeng Xing and Liangliang Liu[✉]

Tongji University, Shanghai 200092, China
liuliangliang@tongji.edu.cn

Abstract. Over the past few decades, the monitoring of landslide hazard has taken a new turn in the diversity of monitoring methods. It is intended to understand and grasp the characteristics of disruptive processes, to formulate appropriate prevention measures for the mitigation of its effects and decrease the casualties and property losses. In order to detect the stability conditions of landslides in different geological and environmental contexts, many new methods for landslide monitoring are proposed by dint of the advanced monitoring apparatus. To overcome a lack of comprehensive analyses of landslide hazards monitoring methods and popularize the scientific knowledge of landslide monitoring method for researchers, this paper reviews the application of several existing methods since the twenty-first century, including GPS (global position system), InSAR (interferometry synthetic aperture radar), LIDAR (light detection and ranging), TDR (time domain reflectometry) and BOTDR (brillouin optic time domain reflectometry), and simultaneously introduces the basic components and workflow of automatic monitoring and early warning system which displays immense superiority and potential in comprehensive monitoring of landslides. Based on previous researches in recent years, each method and its working principle and applicability are summarized to present the direction in further research and development by indicating the limitations and existing problems, respectively.

Keywords: Landslide hazards · Monitoring method · Early warning

1 Introduction

Landslide, defined as "the movement of a mass of rock, debris or earth down a slope", is responsible for at least 17% from all fatalities of natural disasters around the world, which is regarded as one of the most serious natural disasters (Cruden 1991; Jibson 2006). Because of the effects of frequent natural disasters and human engineering activities, landslide hazards are rather severe in many places worldwide (Table 1). In addition to their high threat to human lives and property, the landslide monitoring has caused widely concern from engineers and researchers. And it is not only a requisite part of landslide investigation, research, and prevention and control project, but also an effective way to obtain the information of landslide hazard prediction. Intrinsic factors (topography, geomorphology, soil regolith, stratum lithology, geological structure and

© Springer Nature Singapore Pte Ltd. 2018
A. Farid and H. Chen (Eds.): GSIC 2018, *Proceedings of GeoShanghai 2018*
International Conference: Geoenvironment and Geohazard, pp. 134–141, 2018.
https://doi.org/10.1007/978-981-13-0128-5_16

engineering properties) and extrinsic factors (hydrogeological condition, meteorological condition, land cover, fire, glacier outbursts, seismicity, natural erosion, volcanoes, reservoir water level and human activities) play significant roles in stability and inducement of landslide. In view of this, quite a few methods have been proposed to evaluate and analyze the stability of landslides by monitoring diverse physical and chemical data.

Table 1. General information on some major landslides in the twenty-first century.

Date	Location	Inducements	Casualties
13 Jan 2001	El Salvador	Earthquake	500–1700
9 Nov 2001	Amboori, Kerala, India	Seasonal rainfall	39
26 Mar 2004	Mount Bawakaraeng, South Sulawesi, Indonesia	Collapse of caldera wall	32
10 Jan 2005	La Conchita, California, United States	Deposit of debris flow	10
17 Feb 2006	Southern Leyte, Philippines	Rock-debris avalanche	1126
12 May 2008	Wenchuan, Sichuan, China	Earthquake	>20000
9 Aug 2009	Siaolin Village, Kaohsiung, Taiwan	Typhoon Morakot	500
8 Aug 2010	Zhouqu, Gansu, China	Rainfall	>1700
12 Jan 2011	Rio de Janeiro, Brazil	Fast moving rainfall, tropical storm	1300
16 Jun 2013	Kedarnath, Uttarakhand, India	Rainfall, flood	6074
2 May 2014	Argo District, Badakhshan Province, Afghanistan	Rainfall, old landslides	2700
30 Jul 2014	Ambegaon taluka, Maharashtra, India	Heavy rainfall	160
20 Dec 2015	Shenzhen, Guangdong, China	Excavated soils, seeper in muck collecting field	77

At present, the techniques and methods for landslide monitoring have been developed to a new level where the manual monitoring in the past is gradually moving to instrument monitoring, expanding towards high precision automatic remote sensing system. Monitoring equipment is constantly updated and the sophisticated equipment is introduced to landslide monitoring with the continuous development of computer technology and measurement technology.

2 Landslide Monitoring Methods

2.1 Global Position System Technology

GPS as a kind of technical means of contemporary geodesy can realize three dimensional geodetic deformation surveying with simple and convenient operation. It has the advantages of observatory stations without intervisibility, measurement without restriction of climatic conditions, simultaneous measurement of three dimensional coordinates of observed point, easy automation of entire system and so on. Especially in landslide monitoring, the difference between coordinates of the monitoring points in two periods is mainly concerned, rather than the coordinates of observed points themselves. And the systematic error of different times only affects the coordinate value but does not influence the deformation. Consequently, GPS has been extensively used in the deformation monitoring of landslide, ground subsidence, earthquake, ground fissure and other geological disasters.

With the completion of GPS positioning system in the latter stage of the 20th century, GPS technology is introduced for landslide monitoring in different degrees. Compared with the traditional methods, GPS allows a larger coverage and productivity with similar accuracy. Furthermore, it can work in all kinds of weather conditions and a direct line of sight between stations is not required. Traditional monitoring methods are constrained by the terrain, weather and equipment, so it cannot achieve continuous monitoring of deformation. Then the emergence of GPS provides an answer. Malet et al. (2002) first applied GPS to the continuous measurements of three-dimensional surface displacement of the Super-Sauze earthflow, which realized the long-term dynamic monitoring of landslides. Subsequently, GPS is gradually applied to landslide monitoring.

However, the employ of GPS is limited to the environmental characteristics of the site. For example, the GPS satellite signal is vulnerable to external environmental disturbance, such as occlusions, multipath effect etc. It can make the satellite signal become poor or even interrupted, which cannot be completed positioning. Besides, the cost of the equipment and its maintenance are still major obstacles to widen its routine application at the present time. In recent years, it is common to combine GPS with other monitoring technologies (e.g. LIDAR, InSAR, SAA and CR-InSAR) to make up for its defects, which obtains the remarkable results. At the same time, as the developing technology of electronic receiving equipment continues to mature, low-cost and high-precision GPS receiver has achieved great development, and it provides more convenient conditions for the development of GPS in landslide monitoring. GPS has the advantages of continuous, real-time, high precision, all-weather measurement and automation, and it will be more and more widely used in crustal deformation and geological hazards monitoring.

2.2 Interferometry Synthetic Aperture Radar Technology

InSAR is a microwave remote sensing technology and has been developed rapidly since the 21st century, which uses two or more synthetic aperture radar images to gain terrain elevation data or surface deformation maps based on the phase difference of the echoes received by satellite or aircraft. D-InSAR, a further development of the technology, can detect the topographical change on the scale of millimeter. And all-weather and large-scale data acquisition capability with high spatial resolution make the technology have important research significance in volcano monitoring, surface subsidence, landslide and earthquake deformation monitoring.

At the end of last century, Gabriel et al. (1989) put forward for the first time to use D-InSAR technology for monitoring surface deformation and demonstrated the feasibility of SAR interferometry to detect ground deformation. Achache et al. (1996) obtained six pieces of interferometry fringe patterns through the paired combination of data and D-InSAR technology processing. In comparison with the actual monitoring data, it is proved that the accuracy of the technique for monitoring landslide deformation is consistent with the traditional methods of ground monitoring. Whereafter, this method is broadly used in landslide research. At present, InSAR is mainly used to landslide susceptibility mapping and generate a wide range of high precision digital elevation model (DEM) of landslide by calculating the phase difference between the images.

The application of InSAR technology in mapping and monitoring of landslide has good prospects and great potential owe to its unparalleled advantages. But due to the limitations of the technology itself, phase mismatch influenced by ground vegetation, humidity and atmospheric conditions, delay in time and space and so on, restricts its development correspondingly. There are some shortcomings of InSAR used in landslide monitoring: the variation of atmospheric parameters, the error of the satellite orbit parameters and the change of surface coverages easily affect the measurement result; there is a phenomenon of radar beam overlapping radar shadow in the high mountain area; the time resolution of the archived data cannot meet the requirements.

For better using InSAR technology, permanent scatter (PS) technology is proposed, which keeps a high coherence in long time interval even if the length of baseline distance excesses its critical value, and it is smaller than the pixel size. This technique can effectively reduce the influence of air and noise on interference pattern. In addition, the integration of GPS and InSAR is better for the unification of GPS high temporal resolution characteristic and InSAR high spatial resolution characteristic, to complement each other. Therefore, technology integration and collaborative innovation is a major development trend of InSAR in landslide monitoring.

2.3 Light Detection and Ranging Technology

LIDAR is one of the latest technologies in active modern earth observation technology. Its principle is that the detection signal is transmitted to the target, and then the received signal reflected from the target is compared with the transmitted signal. After proper processing, the relevant information of the target can be obtained. By means of laser ranging, differential GPS and other techniques, the position, distance, angle and other observed data are obtained directly, and then the point cloud data of 3D coordinates of the earth's surface can be available to generate accurate DEM. LIDAR technology has a strong advantage on the ability to detect ground surface with high spatial and temporal resolution, large range of dynamic detection, partially through the forest cover, direct access to the high accuracy of three-dimensional information of surface and so on. Hence it is a new method to rapidly acquire high precision terrain information.

LIDAR technology can be categorized into terrestrial LIDAR and airborne LIDAR, depending on the position of the sensor. The former possesses high observation accuracy and high density, and can do continuous dynamic monitoring for long time. However, due to the limitation of the monitoring distance and range, it is applicable only to a single landslide. Compared to the former, airborne LIDAR can monitor wide range of large area landslide with the limited observation accuracy.

The greatest advantage of LIDAR remote sensing technology lies in quickly, directly and accurately detecting the real surface and ground elevation information, as an effective mean to direct obtain the terrain surface model. There is little doubt that the application of LIDAR will be more and more extensive. But there is more development space in the following areas: the extraction of high precision DEM based on LIDAR data is limited by some factors frequently, such as the characteristics of laser point data, ground condition and the adaptability of filtering algorithm to non-ground points; it is usually hard to use the LIDAR point cloud data obtained to extract landslide model and do the calculations for relevant parameters automatically and quickly; the limitation of the measurement accuracy of LIDAR leads to the result that it is only applied to the measurement of large deformation at present.

2.4 Time Domain Reflectometry and Brillouin Optic Time Domain Reflectometry Technology

TDR technology is a remote sensing test technology, which was first used in the power and communication industry with the intent to determine the failure and fracture of communication cables and transmission lines. In the early 1980s, TDR technology began to be used in engineering geological exploration and monitoring, especially in coalfield geological exploration. Until the end of 1990s, the department of transportation, California USA, applied TDR to remote and real-time landslide monitoring (Kane and Beck 1996). When TDR is used in landslide monitoring, first of all, it is necessary to transmit pulse signal to coaxial cable buried in detection hole. Then the cable will produce the reflected pulse signal once deformed, and the degree and position of cable deformation can be determined by analyzing and processing the reflection signal, so as to achieve the purpose of landslide monitoring.

BOTDR is a fiber optic sensing technology first proposed by Tkach et al. (1986) based on the combination between spontaneous brillouin scattering effect and optical time domain reflectometry (OTDR) technology. Subsequently, Japanese scholars made plenty of research works for the application of BOTDR technology in landslide hazards. BOTDR utilizes the linear relationship between the frequency shift quantity of natural brillouin scattering light in optical fiber and the axial strain of optical fiber to obtain the axial strain of fiber, which has been introduced into geological disaster for the purpose of monitoring the deformation of different parts of rock and soil mass. Normally, the optical fiber is fixed in the rock and soil mass with potential sliding signs (generally fixed in the structure of deformation area). When the deformation of rock and soil mass occurs, it will force the optical fiber to deform. And the deformation condition of the rock and soil mass can be acquired by measuring the deformation of optical fiber.

Both TDR and BOTDR can be used for the deep deformation monitoring of landslide, while both of them have their respective benefits and drawbacks. TDR has the advantages of low price, short monitoring time, high security, and can quickly provide digital monitoring data and send it to the remote control terminal, realizing the intelligent monitoring. But the method is not suitable for the inclination monitoring and determination of moving direction of landslide, and it can only determine the shear surface. Therefore, the application of TDR requires a combination of other equipment, such as borehole inclinometer and displacement meter. BOTDR has the characteristics of distributed technique, long distance supervising and measuring, long operational lifetime, high sensitivity, resistance to electromagnetic interference, etc. It can be easily carried out to monitor the various parts of landslide with a reasonable layout. But the engineering application of BOTDR also encounters a lot of practical problems. (1) The fixed mode of optical fiber is a difficult point in the application of optical fiber, especially for the slope with greatly undulate terrain. (2) BOTDR technology is based on Brillouin frequency shift which is sensitive to temperature and strain, and the two variables cannot be measured with the same frequency shift. That has caused the variable cross-sensitivity problem. (3) The improvement of spatial resolution restricts the measurement distance, which makes measurement distance and precision become a contradiction. (4) Fatigue effect of optical fiber and the calculation of strain transformation to deformation also limit the application on deep deformation monitoring of landslide to some extent.

3 Automatic Monitoring and Early Warning System

Along with the development of wireless communication technology and computer technology, the automation monitoring and early warning system has been developed and applied in the projects of flood disaster, mining and landslide. For major landslides, the system can achieve real-time monitoring, and reflect the real condition of landslides timely, accurately and integrally. If the abnormal situation occurs, it can accurately determine the security risks, in a timely manner to issue a warning signal, which is conducive to improving the comprehensive emergency response capacity and reducing the damage caused by major landslide hazards (Yin et al. 2010).

First of all, monitoring instruments arranged in the appropriate location of landslide can be served to survey landslide state information, and monitoring information is automatically collected and recorded by an expert data acquisition instrument. Then the collected data is transmitted to wireless base station with the aid of data transmission unit in a wireless way. Through the data receiving software, the server at remote surveillance and control center can receive data from internet in real-time. Finally the data are analyzed and processed by the matched software for predicting the development trend of landslide deformation and the possible damage degree. If an exception occurs, the control center will release early-warning information to mobile phone client to take measures to cope with the possible effects. This is the whole process of data transmission in the entire system (Fig. 1).

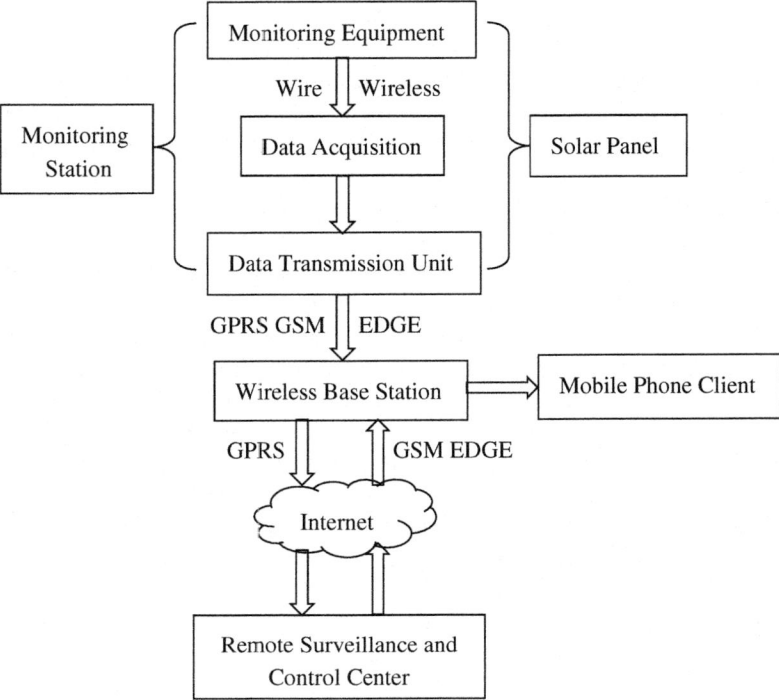

Fig. 1. Logic relation diagram of automatic monitoring and early warning system.

Generally, this system mainly includes three aspects: (1) data collection (2) data transmission (3) data analysis and early warning. The collection data of surface crack, deep displacement, slope deformation, underground water level and rainfall data of landslide hazard are fundamental, each of which is monitored by a variety of measuring instruments with reasonable accuracy and frequency. The transmission of data is connected to the entire system, which determines the timeliness of data and information. Single factor monitoring has been unable to meet the needs for accurate prediction of landslides. Normally the ground and deep displacement, precipitation and water level, pore water pressure and sliding force can be integrated together to establish real-time monitoring and early warning system. Data analysis and early warning is the key of whole system, the most pivotal of which is the selection of the criteria of landslide early warning. In most cases, the criteria of landslide early warning are established based on lots of cases of previously studied local landslides and are validated through the research object. Then the influence degree of different monitoring data on the state of landslide is considered comprehensively to establish different level early warning criteria of landslides (Yin et al. 2010). When one or more monitoring indexes reach the recommended criteria, the corresponding risk rank will raise to a higher level and the necessary measures need implementing. Because of its own instantaneity, comprehensiveness and automaticity, automatic monitoring and early warning system has experienced a rapid development in this century.

4 Conclusions

With the continuous progress of science and technology, more and more methods of landslide monitoring emerge and the technological development is getting more perfect. The trend of high-tech mean as a monitoring tool has become growing evident benefiting from remote sensing satellite technology. GPS, InSAR and LIDAR can monitor 3D surface displacement data, which is the most intuitive performance of landslide deformation instability. Howbeit there is a discontinuity between surface displacement and deep deformation, and it is necessary to consider the landslide stability stage demonstrated by deep deformation. As the new methods of deep displacement monitoring, TDR and BOTDR with timeliness and high sensitivity, can obtain further strengthened aided by the technological advances in optical fiber. The current landslide monitoring cannot be accomplished by a single monitoring method, which needs various monitoring means to integrate together. Each monitoring method has its own advantages and disadvantages, and several methods can be adopted to complement each other to better serve the landslide monitoring. Therefore, the automatic monitoring and early warning system with comprehensive monitoring capacity and early warning capacity is the development trend of landslide hazards monitoring in the twenty-first century.

Acknowledgements. The authors appreciatively acknowledge the financial support of the National Natural Science Foundation of China (Grant No. 41672273, 41272292).

References

Achache, J., Fruneau, B., Delacourt, C.: Applicability of SAR interferometry for operational monitoring of landslides. In: Proceedings of the 2nd ERS Applications Workshop, London, England, pp. 165–168 (1996)

Cruden, D.M.: A simple definition of a landslide. Bull. Int. Assoc. Eng. Geol. **43**(1), 27–29 (1991)

Gabriel, A.K., Goldstein, R.M., Zebker, H.A.: Mapping small elevation changes over large areas: differential radar interferometry. J. Geophys. Res. Atmos. **94**(B7), 9183–9191 (1989)

Kane, W.F., Beck, T.J.: Rapid slope monitoring. Civ. Eng. **66**(6), 56–58 (1996)

Jibson, R.W.: The 2005 La Conchita, California, landslide. Landslides **3**(1), 73–78 (2006)

Malet, J.P., Maquaire, O., Calais, E.: The use of global positioning system techniques for the continuous monitoring of landslides: application to the Super-Sauze earthflow (Alpes-de-Haute-Provence, France). Geomorphology **43**(1–2), 33–54 (2002)

Tkach, R.W., Chraplyvy, A.R., Derosier, R.M.: Spontaneous Brillouin scattering for single-mode optical-fiber characterization. Electron. Lett. **22**(19), 1011–1013 (1986)

Yin, Y.P., Wang, H.D., Gao, Y.L., Li, X.C.: Real-time monitoring and early warning of landslides at relocated Wushan Town, the Three Gorges Reservoir, China. Landslides **7**(3), 339–349 (2010)

Application of DEM to Investigate Mechanical Properties of Landslide Dam Materials

Zhen-Ming Shi[1], Hong-Chao Zheng[1], Song-Bo Yu[1($\boxtimes$)], and Tao Jiang[2]

[1] Key Laboratory of Geotechnical and Underground Engineering of Ministry of Education, Department of Geotechnical Engineering, Tongji University, Shanghai 200092, China
yusongbo@tongji.edu.cn
[2] Department of Underground Structures and Geotechnical Engineering, Shanghai Underground Space Engineering Design and Research Institute, Arcplus Group PLC, Shanghai 200002, China

Abstract. Landslide dams are formed when a river is blocked by a rock avalanche, landslide, or debris flow triggered by an earthquake, heavy rainfall, or some other factors. It is difficult to investigate the mechanical properties of landslide dams owing to their wide grain size range, short lifespan, and inaccessibility in mountain areas. The direct shear test is widely employed to obtain fundamental mechanical parameters of geotechnical materials quickly. However, it is difficult to reveal the failure mechanism in the absence of information regarding the internal stress and particle migration. Hence, this paper presents a discrete element method (DEM) to simulate direct shear tests on two (i.e., fine and coarse) kinds of typical dam-forming materials. The mechanical response of the dam materials was analyzed from the macro and micro perspectives. The stress-strain curve showed a prominent hardening behavior and volume contraction for the fine-grained soils but softening characteristics and volume dilatation for the coarse-grained soils. The principal stresses were gradually deflected, and the force chain was concentrated with increasing shear displacement. The horizontal and diagonal shear bands were developed for fine-grained and coarse-grained soils, respectively, in the stress deflection zone and concentrated region of particle rotation. The stress was distributed principally in skeleton particles for the coarse-grained soils but was uniform for the fine-grained soils.

Keywords: Landslide dam materials · Direct shear test · DEM
Macro and micro perspectives · Shear band

1 Introduction

Landslide dams are formed when a river is blocked by a large mass movement such as a rock avalanche, landslide, or debris flow triggered by an earthquake, heavy rainfall, groundwater, or some other similar factors. According to statistics, approximately 90% of landslide dams have a short longevity of less than one year and present overtopping failure due to the rise of the upstream water level [1]. An enormous breaching flood and

© Springer Nature Singapore Pte Ltd. 2018
A. Farid and H. Chen (Eds.): GSIC 2018, *Proceedings of GeoShanghai 2018*
International Conference: Geoenvironment and Geohazard, pp. 142–151, 2018.
https://doi.org/10.1007/978-981-13-0128-5_17

sediment pose disastrous risks to people's lives and property in the downstream areas. However, the wide grading, inconvenient transportation, and poor geological environment are extremely negative conditions for the onsite investigation of the mechanical properties of landslide dam materials and the further conduct of stability estimation and disposal measures.

Offering the benefits of simple operation and low costs, the direct shear test is widely employed to test the strength of geotechnical materials among various kinds of laboratory experiments. However, the direct shear test of landslide dam materials is hard to implement when constrained by the factors discussed above, and few cases have been reported in the literature. In addition, the conventional direct shear test has difficulty revealing the failure mechanism in the absence of information on the inner stress and particle migration. The DEM, based on the discontinuous media theory [2], has the outstanding advantages of simulating the interactions among the particles in the test and considering the granular composition, roughness, and compactness [3, 4]. Material behavior and failure mechanisms are the main research topics that use the numerical direct shear test. For material behavior, Masson [5] simulated a dense and loose 2D sample to research the change in the contact force orientations during the shear process. Liu [6] conducted a study on the friction between the internal surface of the shear box and the sample by means of a mixture of binary diameter cylinders with different compactness. Hartl [7] performed 90 numerical direct shear tests to study the influence of the particle shape, stress level, and packing density on the bulk friction at the ultimate shear stress. Ahad [8] investigated the macro and micro influence of particle size on the shear strength of coarse-grained soils. Regarding failure mechanisms, Sebastian [9] focused on the evolution of the crushing mechanism and found that the compressive behavior was counteracted by the dilating behavior due to particle rearrangement. Huang [10] conducted a series of biaxial compression tests on sand samples with different initial confining pressures and densities to analyze their critical state line and microstructure evolutions. Dong [11] performed numerical simulations of soil-rock mixtures to assess shear dilatation and shear contraction characteristics with shear displacement. Nevertheless, few numerical direct shear tests have been applied to geotechnical materials with significantly different gradations like landslide dam materials.

In this study, the DEM was used to simulate a series of direct shear tests of landslide dam materials for fine-grained and coarse-grained materials with wide granular compositions, to reveal their mechanical behavior from the macro and micro perspectives.

2 Model Setup and Numerical Implementation

2.1 Grain Composition of Model

There is large variation in the grain diameter of landslide dam materials, ranging from several micrometers for clay to decameters for boulders. Casagli [12] classified dam materials into two categories, namely, matrix-supported and grain-supported. The former involves the coarser particles that are scattered inside a prevailing fine matrix, despite not being in contact with each other, and the latter includes the coarser particles that are in contact with each other and have a fine matrix present at an interstitial level.

The M_w 7.9 Wenchuan earthquake triggered 828 landslide dams in the Long-menshan fault zone in China, of which the biggest was the Tangjiashan landslide dam with a height of 82 m and a reservoir capacity of 3.16×10^8 m^3 [13]. The grain composition of the specimens in the numerical direct shear test is based on the Tangjiashan landslide dam [14], as shown in Fig. 1. Two typical grading curves, shown in the thick black line type, were derived: fine-grained and coarse-grained soil, which correspond to matrix-supported and grain-supported dam materials, respectively. It should be noted that the two soil types were specifically defined based on the grading curves of a landslide dam, which are different from the conventional engineering classification of soil. To ensure calculation convergence, the particles with a grain diameter larger than 2 mm in the fine-grained soils were equivalent to those with a grain diameter of 2 mm, and the particles with a grain diameter smaller than 0.5 mm in the coarse-grained soils were equivalent to those with a grain diameter of 0.5 mm as current specification of soil test (Table 1) [15]. The uniformity coefficient, C_u, and the curvature coefficient, C_c, of the fine-grained soils were 4.3 and 1.5, respectively, and those of the coarse-grained soils were 23 and 2.8, respectively.

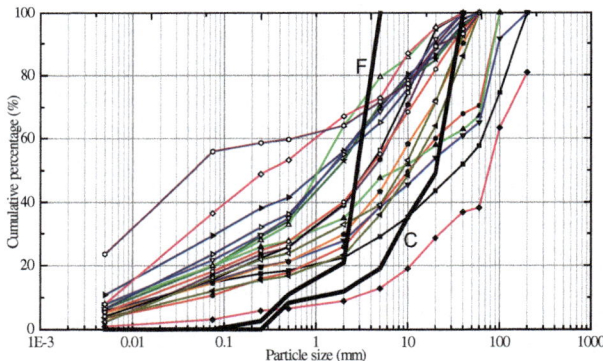

Fig. 1. Grading curve of Tangjiashan landslide dam

2.2 Numerical Implementation of Direct Shear Test

The numerical modeling was mainly composed of four steps: wall generation, graded particle generation, compaction and normal pressure exertion, and shear test implementation.

Wall Generation: Four orthogonal walls were generated, with an aspect ratio of 2, based on the standard for the soil test method. The wall width, B, in the model should be large enough to eliminate grain size effects, and the number of particles should also be optimized to save calculation time. The model dimensions were hence determined to be 40 mm × 20 mm for fine-grained soils and 300 mm × 150 mm for coarse-grained soils, as shown in Table 2.

Table 1. Grading curve of numerical modeling

Fine-grained soil	Diameter (mm)	2		0.5	0.25	0.075	
	Weight percentage (%)	79.17		9.88	8.27	2.68	
Coarse-grained soil	Diameter (mm)	40	20	10	5	2	0.5
	Weight percentage (%)	50.79	13.95	16.26	7.15	3.5	8.35

Table 2. Physical parameters of numerical model

Grain composition	Model	Model size (mm × mm)	Dry density (g/cm³)	Void ratio	Particle number	Relative compaction
Fine-grained soil	F1	40 × 20	1.82	0.48	5577	0.32
Coarse-grained soil	C1	300 × 150	1.82	0.48	18203	0.47

Graded Particle Generation: Spherical particles, shown in Fig. 2, were obtained using a Particle Flow Code [16] particle generator in three uniform layers as the given grain size distributions discussed previously. The initial positions of the particles in the box were determined randomly. In the particle settling simulation, the particles dropped from their original positions under gravity with interactional collisions. The input parameters of the numerical model were based on Liu [6] in Table 3.

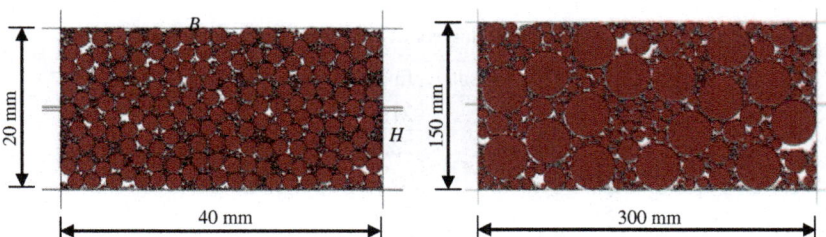

Fig. 2. Numerical modeling establishment: (a) fine-grained soil (b) coarse-grained soil

Compaction and Normal Pressure Exertion: The top wall moved downward at a constant speed of 0.5 mm/s until the desired compactness (i.e. 0.32 for F1 and 0.47 for C1) was achieved while the side and bottom walls remained static. Then, with the vertical movement of the top wall under the control of a numerical servo-mechanism, the required normal pressure between the top wall and the particle assembly was achieved.

Shear Test Implementation: After the applied normal pressure was set to a constant value, the shearing stage started with moving the upper and lower sections of the shear box at constant velocities of –0.75 and 0.75 mm/s, respectively, until the shear strain reached 5%. Meanwhile, the macro information such as shear stress, shear

Table 3. Input parameters for numerical simulation by DEM

Material parameters	Value
Grain density (g/cm^3)	2.7
Contact model	Linear elastic model
Interparticle friction coefficient	0.9
Interparticle normal stiffness (N/m)	1×10^8
Interparticle shear stiffness (N/m)	5×10^7
Wall friction coefficient	0.0
Wall normal stiffness (N/m)	1×10^{10}
Wall shear stiffness (N/m)	1×10^{10}
Time step (s)	4×10^{-6}

displacement, and vertical displacement, as well as the micro information such as contact force, principal stress, and particle rotation, were monitored and recorded.

2.3 Sensitivity Analysis of Coefficient of Friction

Considering that tangential motion is controlled by the Mohr–Coulomb theory, the coefficient of friction, among the intergranular micro parameters, has a significant influence on the macro mechanical behavior. Hence, a sensitivity analysis of the coefficient of friction to the mechanical parameters was investigated in the chapter (Table 4).

Table 4. Sensitivity analysis of the coefficient of friction

Model	Friction coefficient μ	Internal friction angle ϕ ($^\circ$)
F1	0.5	21.9
	0.7	23.3
	0.9	25.9
C1	0.5	24.4
	0.7	27.6
	0.9	30.3

The coefficient of friction was calibrated using the internal friction angles of the samples. Twenty-four direct shear tests were conducted under conditions of different coefficients of friction (0.5, 0.7, and 0.9) and normal stresses (100, 200, 300, and 400 kPa) for the fine-grained soils and coarse-grained soils. It was found that the internal friction angle gradually increased with the coefficient of friction for every model in Table 3. Based on the internal friction angle of the coarse-grained soils of the Tangjiashan landslide dams being in the range of 28° to 38° [17], the coefficient of friction was hence determined to be 0.9 so that the internal friction angles of samples F1 and C1 were 25.9° and 30.3°, respectively.

3 Macro Mechanical Behavior of Landslide Dam Materials

The macro mechanical behavior of the fine-grained soils was shown in Fig. 3. The stress-strain property of the fine-grained soils presented obvious strain hardening characteristics. The peak shear stress gradually increased with increasing normal stress, and the strain state basically remained as strain hardening. The shear contraction and shear dilatancy characteristics can be revealed by the shear dilatancy curve, wherein $y \sim x$. The volume variation generally showed a tendency of shear contraction but had a slight shear dilation in the final stage.

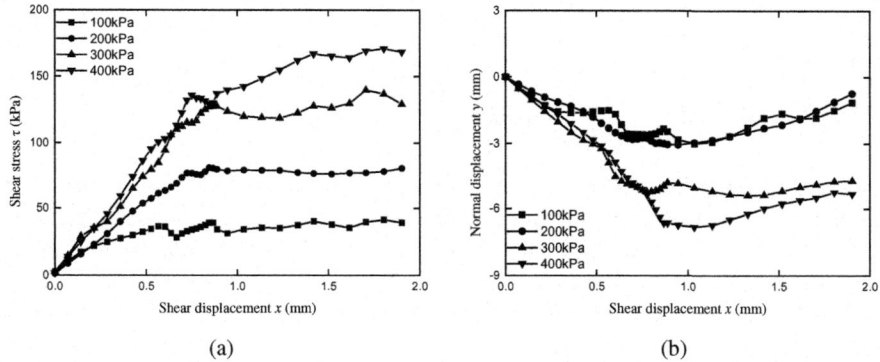

Fig. 3. Macro mechanical behavior of fine-grained soil: (a) stress strain curve; (b) shear dilatancy curve.

Compared with the fine-grained soils, the macro mechanical behavior of the coarse-grained soils shown in Fig. 4 was significantly different, presenting an obvious strain softening characteristic. An obvious peak exists in the stress-strain curve, and the shear stress gradually decreased in the post-peak stage. In addition, the peak shear stress of the coarse-grained soils was greater than that of the fine-grain soils under the same normal stress. Moreover, the volume variation uniformly presented a tendency of slight shear contraction at first and obvious shear dilation afterward, with increasing shear displacement. The contrasting behaviors of the fine-grained soils and coarse-grained soils are due to the differences in their grain compositions and relative compactions, shown in Table 2.

4 Micro Characteristics of Landslide Dam Materials

In the absence of an inner stress variable, the detailed shear failure mechanism is difficult to explore by means of a laboratory experiment. Nevertheless, the numerical direct shear test dominates for micro information. The force chain and principal stress fields were the primary mechanical information. It should be noted that each of the line segments and their thicknesses in the force chain diagram denotes the interaction force at the centroid of adjacent particles and the relative value of the contact force. The

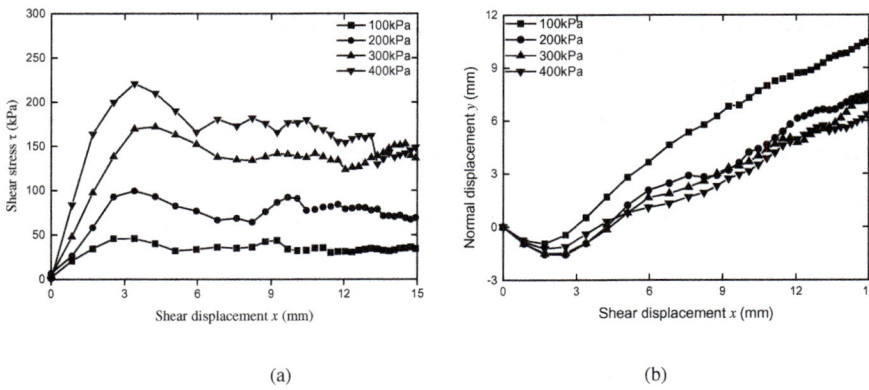

Fig. 4. Macro mechanical behavior of coarse-grained soil: (a) stress strain curve; (b) shear dilatancy curve.

macro-axis in the principal stress field represents the maximum principal stress, σ_1, while the minor axis is the minimum principal stress, σ_3. The compressive stress and tensile stress are labeled in black and red, respectively. Because of the lack of cement among the particles, there was purely compressive stress in the principal stress field.

At the initial stage of shearing, the force chain of model F1 (Table 5) distributed homogeneously, and the normal force prevailed without shear displacement. The horizontal contact force between the particles and the slide walls emerged due to the confining action of the side walls. The model was in biaxial compression from the principal stress field, where σ_1 and σ_3 were approximately vertical and horizontal, except in the neighboring region where there were horizontal contact forces.

Table 5. Shear failure process of F1 in normal stress 100 kPa

Time	Initial stage	Peak stage	Post-failure stage
Model			
Principal stress			
Force chain			

At the peak stage, the preponderant force chain of the model F1 transformed from normal to horizontal, which was basically parallel to the shear direction. The horizontal contact force was greater on the shear-imposed part of the wall but was significantly less on the remaining portion. In addition, the normal contact force was concentrated on the top and bottom walls but was sparse in the middle of the shear box. The σ_1 in the region neighboring the shear box changed prominently to be horizontal, but it was inclined to about 30° in the middle part of the box in the principal stress field. It was close to the angle ($45° - \theta_1/2$) between the failure plane and the minimum principal stress σ_3.

At the post-failure stage, a shear band developed in the stress deflection zone, between the two purple lines, where the shear resistance was supplied by the resistance of the particles.

At the initial stage of shearing, the force chain of model C1, shown in Table 6, distributed unevenly compared with model F1. The contact force of the coarse particles was greater than that of the fine particles because of their difference in soil fraction. A skeleton was developed by the coarse particles, resulting in the surrounding fine particles un-stressing. In addition, the normal force prevailed, which was similar to model F1. At the peak stage, the preponderant force chain of model C1 transformed from normal to horizontal but fastened largely on the coarse particles. Therefore, the shear resistance was sustained by the coarse particles, which was different from the fine-grained soils. Hence, the transfer patterns of the contact force were determined by the grain composition. At the post-failure stage, the shear band gradually diffused with increasing shear displacement. The particles in the shear band were staggered significantly, leading decreased occlusal force and shear stress (Fig. 4(a)).

Table 6. Shear failure process of C1 in normal stress 100 kPa

Time	Initial stage	Peak stage	Post-failure stage
Model			
Principal stress			
Force chain			

The particle migration state can be analyzed from the distribution of rotation shown in Fig. 5. The maximum angle of rotation and interval of F1 were 30.2 and 1 rad, respectively, and those of C1 were 42.3 and 2 rad, respectively. For the fine-grained soils, the particle movement was mainly concentrated in the horizontal shear band, which corresponds to the stress deflection zone of the force chain, shown in Table 5. Therefore, the macro shear contraction in Fig. 3 is attributed to the horizontal slip and the rolling of the particles subjected to the normal stress. For the coarse-grained soils, the skeleton particles were almost at rest while the surrounding fine particles slipped dramatically during the shear process, which appears concomitant with the maximum-minimum rotation and the diagonal shear band. Hence, the macro shear dilatancy in Fig. 4 arose from the dislocation of fine particles.

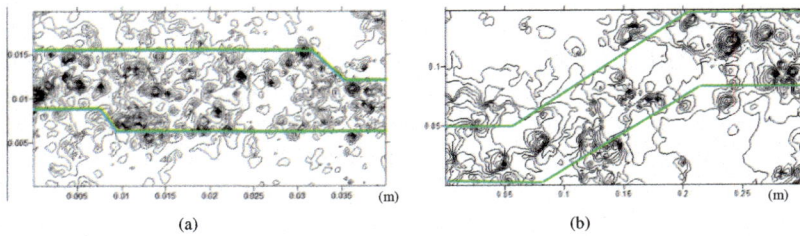

(a)   (b)

Fig. 5. Particle rotation contour map of (a) fine-grained soil F1 and (b) coarse-grained soil C1 at post-failure stage.

5   Conclusion and Discussion

A series of direct shear tests were simulated for fine-grained soils and coarse-grained soils of landslide dam materials based on the DEM, and their mechanical behaviors were revealed from the macro and micro perspectives. The main results are as follows:

(1) From the macro perspectives, the fine-grained soils with dry density 1.82 g/cm^3 presented obvious hardening characteristics. Nevertheless, it had remarkable softening characteristics for the coarse-grained soils with same dry density. The shear strength and internal friction angle of the coarse-grained soils were greater than those of the fine-grained soils with the same dry density and void ratio.

(2) Horizontal and diagonal shear bands were developed for fine-grained and coarse-grained soils, respectively, in the stress deflection zone and the concentration region of particle rotation. The force chain and principal stress fields were always changed at different shearing stages. The transfer patterns of the contact force were determined by the grain composition that distributed homogeneously in the fine-grained soils but concentrated on the coarse particles in the coarse-grained soil.

The rationality of the DEM numerical method applied to laboratory direct shear experiment of landslide dam materials was discussed. Further research will be focused on the particle irregularity; the particle shape was a regular sphere in this study.

Acknowledgements. The research reported in this paper was substantially supported by the Natural Science Foundation of China (No. 41502275 and No. 41372272), and the Shanghai Committee of Science and Technology (No. 14YF1403800).

References

1. Shi, Z., Ma, X., Peng, M., Zhang, L.: Statistical analysis and efficient dam burst modelling of landslide dams based on a large-scale database. Chin. J. Rock Mech. Eng. **33**(9), 1780–1790 (2014)
2. Cundall, P.A., Strack, O.D.L.: A discrete numerical mode for granular assemblies. Geotechnique **29**(1), 47–65 (1979)
3. Thornton, C., Zhang, L.: Numerical simulations of the direct shear test. Chem. Eng. Technol. **26**(2), 153–156 (2003)
4. Liu, S.H.: Simulating a direct shear box test by DEM. Can. Geotech. J. **43**(2), 155–168 (2011)
5. Masson, S., Martinez, J.: Micromechanical analysis of the shear behavior of a granular material. J. Eng. Mech. **127**(10), 1007–1016 (2001)
6. Liu, S.H., Sun, D., Matsuoka, H.: On the interface friction in direct shear test. Comput. Geotech. **32**(5), 317–325 (2005)
7. Hartl, J., Jin, Y.O.: Experiments and simulations of direct shear tests: porosity, contact friction and bulk friction. Granul. Matter **10**(4), 263–271 (2008)
8. Ahad, B.K., Ali, A.M.: Numerical and experimental direct shear tests for coarse-grained soils. Particuology **7**(1), 83–91 (2009)
9. Sebastian, L.G., Luis, E.V.: Discrete element method evaluation of granular crushing under direct shear test conditions. J. Geotech. Geoenviron. Eng. **131**(10), 1295–1300 (2005)
10. Huang, Z.Y., Yang, Z.X., Wang, Z.Y.: Discrete element modeling of sand behavior in a biaxial shear test. J. Zhejiang Univ. Sci. A **9**(9), 1176–1183 (2008)
11. Dong, Q.P., Lu, Z., Zhan, Y.X.: Particle flow modeling of soil-rock mixtures in-situ tests. J. Shanghai Jiaotong Univ. **47**(9), 1382–1389 (2013)
12. Casagli, N., Ermini, L., Rosati, G.: Determining grain size distribution of the material composing landslide dams in the Northern Apennine: sampling and processing methods. Eng. Geol. **69**, 83–97 (2003)
13. Fan, X., Westen, C.J.V., Xu, Q.: Analysis of landslide dams induced by the 2008 Wenchuan earthquake. J. Asian Earth Sci. **57**(6), 25–37 (2012)
14. Chang, D.S., Zhang, L.M., Xu, Y.: Field testing of erodibility of two landslide dams triggered by the 12 May Wenchuan earthquake. Landslides **8**(3), 321–332 (2011)
15. SL237-1999: Specification of soil test (1999). (in Chinese)
16. Itasca Consulting Group Inc.: PFC2D Particle Flow Code in 2 Dimensions Online Manual Table of Contents (2014)
17. Luo, G.: Analysis of blocking mechanism of Tangjiashan high-speed short-run landslide and dam-breaking mode of Tangjiashan barrier dam. Southwest Jiaotong University (2012)

Slope Stability Analysis Based on the Coupled Eulerian-Lagrangian Method

X. Y. Chen and L. L. Zhang[(✉)]

Shanghai Jiaotong University, 800 Dongchuan Road, Shanghai, China
lulu_zhang@sjtu.edu.cn

Abstract. Slope stability analysis based on the finite element method (FEM) is a powerful tool comparing with the traditional limit equilibrium methods. However, severe mesh distortion of the traditional Lagrangian FEM may result in convergence difficulty and the obtained factor of safety (FOS) is not reliable. In this study, a stability analysis based on the Coupled Eulerian-Lagrangian (CEL) method is presented. The method allows materials to flow between different grids regardless of mesh distortion. In an example problem, both shear strength reduction (SSR) technique and gravity increasing (GI) technique are utilized to study the capability of the proposed method. The result shows that the method can effectively simulate the entire process of failure to post failure state. The estimated factor of safety (FOS) rarely varies with the mesh density and agrees well with the previous studies based on the traditional Lagrangian FEM method. The FOS by the GI technique is slightly smaller than that by the SSR technique.

Keywords: Slope stability · Coupled Eulerian-Lagrangian Method
Large deformation

1 Introduction

Slope failure often results in catastrophic consequences. The slope stability analysis is widely used to evaluate the safety level of a slope in the geotechnical engineering field. Large quantities of stability analyses by the finite element method (FEM) have been developed [1–3]. Its advantage over the traditional limit equilibrium method (LEM) is that FEM simultaneously calculates the location of the critical slip surface, as well as the stress, strain and displacement. Nevertheless, it inevitably encounters significant mesh distortion [4]. This distortion may result in convergence difficulty and the analysis usually ends up at the onset of failure. Numerical errors or uncertainties may be introduced in evaluating the FOS value. Since slope failure is a problem of large deformation, a large deformation analysis method instead should be applied to the study of failure. Fortunately, several large deformation methods have been proposed in the last decades, including Coupled Eulerian-Lagrangian method (CEL), Smoothed Particle Hydrodynamic (SPH) and Material point method (MPM). Post failures of slope are usually studied by the SPH [5] and by the MPM [6]. Dey et al. [7] used the CEL method to well study the progressive failure of landsides with sensitive clay layers. However, very few studies on the stability analysis for getting reliable FOS values have

© Springer Nature Singapore Pte Ltd. 2018
A. Farid and H. Chen (Eds.): GSIC 2018, *Proceedings of GeoShanghai 2018*
International Conference: Geoenvironment and Geohazard, pp. 152–159, 2018.
https://doi.org/10.1007/978-981-13-0128-5_18

been made based on the large deformation methods listed above. The CEL method is chosen in this paper and the aim is to propose a slope stability analysis method under large deformation regardless of mesh distortion.

2 Theory of the CEL Method

The CEL method is based on the Eulerian formulation. The detained Eulerian formulation is available in previous studies [8, 9]. The mass, momentum and energy equations are written as Eqs. (1), (2) and (3) respectively.

$$\frac{\partial \rho}{\partial t} + \nabla \cdot (\rho \boldsymbol{v}) = 0 \tag{1}$$

$$\frac{\partial \rho \boldsymbol{v}}{\partial t} + \nabla \cdot (\rho \boldsymbol{v} \otimes \boldsymbol{v}) = \nabla \cdot \boldsymbol{\sigma} + \rho \boldsymbol{b} \tag{2}$$

$$\frac{\partial e}{\partial t} + \nabla \cdot (e \boldsymbol{v}) = \boldsymbol{\sigma} : \dot{\boldsymbol{\varepsilon}} \tag{3}$$

Where ρ is the density, $\boldsymbol{v}$ is the material velocity, $\boldsymbol{\sigma}$ is the Cauchy stress, $\boldsymbol{b}$ is the body force, e is the internal energy per unit volume and $\dot{\boldsymbol{\varepsilon}}$ is the strain rate respectively. Equations (1), (2) and (3) have a general form represented by Eq. (4).

$$\frac{\partial \phi}{\partial t} + \nabla \cdot \boldsymbol{\Phi} = \boldsymbol{S} \tag{4}$$

Where ϕ represents all the solution variables, $\boldsymbol{\Phi}$ is a flux function accounting for the convective effect and $\boldsymbol{S}$ represents a source term. Obviously, Eq. (4) is both temporally and spatially dependent. In Fig. 1, Eq. (4) is decomposed into two sequent steps as Eqs. (5) and (6) by the method of operator splitting [9].

$$\frac{\partial \phi}{\partial t} = \boldsymbol{S} \tag{5}$$

$$\frac{\partial \phi}{\partial t} + \nabla \cdot \boldsymbol{\Phi} = 0 \tag{6}$$

Fig. 1. The solving procedure of the CEL method.

During a time step, the Lagrangian step (Eq. (5)) is firstly solved with the material attached to the mesh; the Eulerian step (Eq. (6)) is followed and the deformed mesh recovers along with the remapping of all the solution variables onto the regular mesh through the transport algorithm [8]. It must be stated that the next analysis techniques are implemented in Abaqus/Explicit, where the CEL method has been incorporated.

3 Slope Stability Analysis Under Large Deformation

3.1 Constitutive Model

The elastic behavior of the soil is defined by the Young's modulus, E, and Poisson's ratio, υ. The Mohr-Coulomb criterion is chosen as the failure criterion:

$$\frac{(\sigma_1 - \sigma_3)_f}{2} = \frac{(\sigma_1 + \sigma_3)_f}{2} \sin \varphi + c \cos \varphi \tag{7}$$

Where, σ_1 and σ_3 are the principle stresses. The subscript f represents the failure state. The parameters of c and φ are cohesion and internal friction angle. These parameters can be described as effective stresses or total stresses depending on the actual measure method. Once the stress state at a point agrees with the failure criterion, plasticity flow occurs and the adjacent stresses are redistributed to attempt to meet a new balance until the entire slope ends up with a global failure.

3.2 Determination of the Factor of Safety (FOS)

The SSR technique adopts a series of strength reduction factors (SRF) by which the shear strength would be divided until to mobilize the slope failure. The slope fails once c_{SRF} and φ_{SRF} reach the mobilized c_f and φ_f when the FOS = SRF. This technique is widely used in all slope stability analyses including limit equilibrium [10], finite element [2, 3], and limit analysis approaches [11].

$$c_{SRF} = \frac{c}{SRF}, \varphi_{SRF} = \arctan\left(\frac{\tan \varphi}{SRF}\right) \tag{8}$$

Gravity increasing (GI) technique is also adopted and compared with the SSR technique. In Eq. (9), the GI technique multiplies the original gravitational acceleration g_0 by the gravity increasing factor (GIF) to get a greater g_{GIF} until bringing the slope to the failure point. Then the critical GIF is referred as the FOS value.

$$g_{GIF} = g_0 \cdot GIF \tag{9}$$

A successful obtaining of the FOS values also substantially relies on how the slope failure is defined. In the CEL method, when the slope initially fails can be determined by the abrupt change of the nodal velocity.

3.3   The Case Model

A typical homogeneous slope [3] is presented in Fig. 2(a). The Young's modulus is 10^5 Pa and the Poisson's ratio is 0.3. The soil is assumed as undrained clay with shear strength parameters as the Eq. (10) gives:

$$\varphi_u = 0°, \ c_u/\gamma H = 0.20 \tag{10}$$

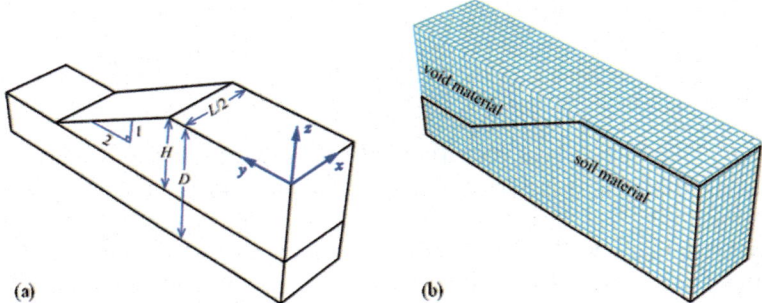

(a)   (b)

Fig. 2. The case model. (a) Geometric configuration of the slope, (b) Mesh strategy.

Figure 2(b) shows the mesh strategy. A regular spatial fixed mesh is divided into two regions by the slope geometry. The region filled with the slope is assigned with the soil material and the other is referred as the void elements which efficiently capture the slope boundaries by computing the material volume with the method of volume of fluid (VOF) [12]. Table 1 gives all the boundary conditions of the model.

Table 1. The boundary conditions of the case model.

Plane	Velocity components
$x = L/2, \ z = -D$	$V_x = V_y = V_z = 0$
$x = 0$	$V_x = 0$
$y = 0,$ $y = length$ in the y direction	$V_y = 0$
The remainder planes	Free boundary conditions

4   Results and Discussion

4.1   Evolution of the Slope Failure

The slope failure is monitored by the resultant velocities of characteristic material points on the slope surface. Figure 3(a) and (b) give the evolution of the resultant velocities with the SRF and the GIF. It is noted that velocities of all the material points on the slope surface have similar trends only different in magnitudes. The evolution of slope failure can be divided into three stages:

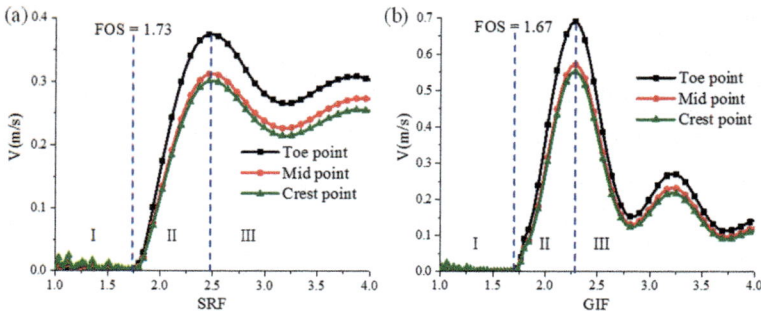

Fig. 3. The failure evolution ($L = 2H$): (a) the SSR technique, (b) the GIF technique.

In stage I, when SRF and GIF increase, the velocities keep zero approximately and the slope is stable. In stage II, velocities rise suddenly arising from the initial failure. Figure 4 shows the velocity field of the initial failure. The blue curves are streamlines along which the material resultant velocities are tangent and they represent current flow trends of the soil material. The material flows from the top to the toe along the slip surface and the SRF and GIF corresponding to the initial failure state are FOS values. They are 1.73 by the SSR technique (Fig. 3(a)) and 1.67 by the GIF technique (Fig. 3(b)). The rapid rising of velocities indicates an accelerated failure. In stage III, the slope has experienced a large plastic deformation and the geometry change enhances its ability to resist deformation. Therefore, the velocity in stage III begins to decline. Compared with the traditional Lagrangian FEM, here the slope deforms successively from initial failure to post failure without any mesh distortion. Figure 5(a) represents the initial failure (the connection point of stage I and stage II) and the shear band almost propagates to the top. Figure 5(b) is in post failure (stage II) and the slope with a completely developed shear band can not resist large deformation.

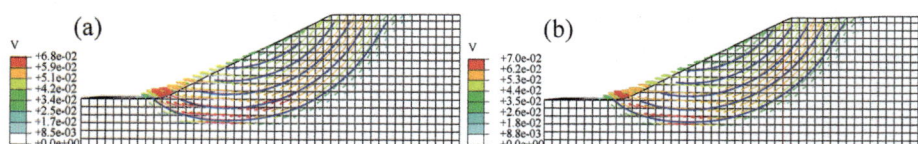

Fig. 4. Velocity field at initial failure. (a) By the SSR technique, (b) By the GIF technique.

4.2   Slope Length Effects on the Slope Failure

The FOS values for different slope lengths are compared with the study by Griffiths and Marquez [3].

In Fig. 6, When $L/H = 2$, the boundary effect is apparent and the slope has greater stiffness to resist deformation resulting in a maximum FOS value. When L/H = 8, the boundary effect is weakened and the slope has a smaller FOS value. If the value of the L/H ratio continues to increase, the boundary effect is almost vanished and the FOS

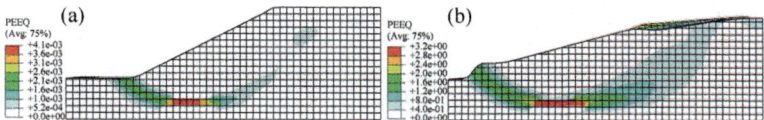

Fig. 5. Plastic strain contours under the SSR technique. (a) Initial failure, (b) Post failure.

values approach a constant value obtained from the plane strain state (PSS). These results are similar to the study by Li et al. [13] which found length effects lose significance when L/H ratio exceeds a certain value. Finally, FOS values obtained by the SSR technique are greater than those by the GI technique. The results by the 2D limit equilibrium are close to those of the plane strain state (PSS) and obviously conservative compared with 3D analyses especially when the L/H ratio is lower.

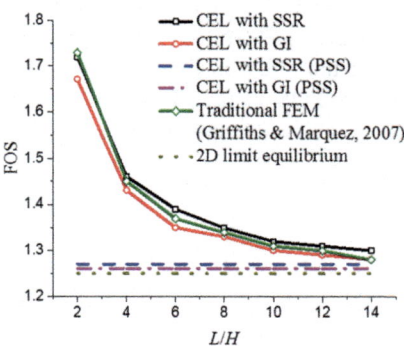

Fig. 6. FOS values obtained for different L/H ratios.

4.3   Sensitivity to Different Mesh Sizes

Sensitivity of the failure to mesh sizes is evaluated. Table 2 shows the FOS values corresponding to different element sizes with the SSR technique. The FOS values tend to be lower when a smaller mesh size is used, which is consistent with the result by traditional Lagrangian FEM in the study by Griffiths and Marquez (2007) [3]. Apparently, the FOS values corresponding to the element size of $l = 0.2H$ has larger errors of 4% compared with those below $l = 0.1H$. Below $l = 0.1H$, the results change little with a negligible errors below 2%. From the viewpoint of computation, the time costs will experience an exponential growth when the size varies from $l = 0.2H$ to $l = 0.05H$. As a suggestion, an element size of $l = 0.1H$ is a better choice by which exacter results and lower time costs can be well balanced.

Table 2. FOS values for different element sizes with the SSR technique.

L/H	2			8			14		
l/H	0.05	0.1	0.2	0.05	0.1	0.2	0.05	0.1	0.2
FOS	1.71	1.73	1.80	1.46	1.46	1.51	1.29	1.30	1.33
CPU time (s)	350.1	42.3	8.1	1239.6	179.2	33.5	2088.8	279.3	59.3

5 Concluding Remarks

Slope stability analysis by the CEL method, a large deformation FEM, is proposed in this study. The following conclusions can be made:

1. The process from initial failure to post failure is simulated without mesh distortion. The CEL method shows significant potentials in large deformation.
2. The new method is validated by the good agreement with the previous studies. The FOS values by GI technique tend to be slightly lower than those by SSR technique owing to the inertial effect in dynamic analysis.
3. The mesh size rarely affects FOS values. The element size of $l = 0.1H$ is a better choice by which exacter results and lower time costs can be well balanced.

References

1. Zienkiewicz, O.C., Humpheson, C., Lewis, R.W.: Associated and non-associated visco plasticity and plasticity in soil mechanics. Geotechnique **25**(4), 671–689 (1975)
2. Griffiths, D.V., Lane, P.A.: Slope stability analysis by finite elements. Geotechnique **49**(3), 387–403 (1999)
3. Griffiths, D.V., Marquez, R.M.: Three-dimensional slope stability analysis by elasto-plastic finite elements. Geotechnique **57**(6), 537–546 (2007)
4. Mohammadi, S., Taiebat, H.A.: A large deformation analysis for the assessment of failure induced deformations of slopes in strain softening materials. Comput. Geotech. **49**, 279–288 (2013)
5. Huang, Y., Zhang, W., Xu, Q., Xie, P., Hao, L.: Run-out analysis of flow like landslides triggered by the Ms 8.0 2008 Wenchuan earthquake using smoothed particle hydrodynamics. Landslides **9**, 275–283 (2012)
6. Bandara, S., Soga, K.: Coupling of soil deformation and pore fluid flow using material point method. Comput. Geotech. **63**, 199–214 (2015)
7. Dey, R., Hawlader, B., Phillips, R., Soga, K.: Progressive failure of slopes with sensitive clay layers. In: Proceedings of the 18th International Conference on Soil Mechanics and Geotechnical Engineering, pp. 2177–2180. Paris (2013)
8. Benson, D.J.: Computational methods in Lagrangian and Eulerian hydrocodes. Comput. Methods Appl. Mech. Eng. **99**, 235–394 (1992)
9. Benson, D.J., Okazawa, S.: Contact in a multi-material Eulerian finite element formulation. Comput. Methods Appl. Mech. Eng. **193**, 4277–4298 (2004)
10. Bishop, A.W.: The use of the slip circle in the stability analysis of slopes. Geotechnique **5**(1), 7–17 (1955)

11. Chen, J., Yin, J.H., Lee, C.F.: A three-dimensional upper-bound approach to slope stability analysis based on RFEM. Geotechnique **55**(7), 549–556 (2005)
12. Hirt, C.W., Nichols, B.D.: Volume of fuid (VOF) method for the dynamics of free boundaries. J. Comput. Phys. **39**, 201–225 (1981)
13. Li, A.J., Merifield, R.S., Lyamin, A.V.: Three-dimensional stability charts for slopes based on limit analysis methods. Can. Geotech. J. **47**, 1316–1334 (2010)

Study on Displacement Back Analysis of Spatial Variability of Soil Slope

Yixuan Sun and Lulu Zhang[✉]

Shanghai Jiaotong University, 800 Dongchuan Road, Shanghai, China
lulu_zhang@sjtu.edu.cn

Abstract. The in situ soil properties exhibit natural spatial variability due to various factors and are difficult to be estimated based on limited test data. In this paper, a probabilistic back analysis method is adopted to characterize the spatial variability of soil parameters based on the responses of displacement. The Karhunen-Loève (K-L) expansion method is used to discretize the random field of soil spatial variability. A finite number of basic random variables from the truncated K-L expansion are the random variables to be back estimated using the Markov Chain Monte Carlo (MCMC) method. To improve computational efficiency, a surrogate model based on the polynomial chaos expansion (PCE) is established to substitute the finite element model of slope stability analysis. A hypothetical example with a slope subject to surcharge load is presented to illustrate the accuracy and efficiency of the method. The spatially varied elastic moduli are estimated using the responses of displacement. It is found that the back analysis based on horizontal displacements is more accurate than that based on vertical displacements. When the correlation length decreases, the accuracy of the estimated field is reduced. The vertical correlation length has a greater effect on the estimation of soil spatial variability than the horizontal correlation length. When the coefficient of variation increases, the accuracy of the estimated field is reduced.

Keywords: Spatial variability · Back analysis · Karhunen-Loève

1 Introduction

In geotechnical engineering, accurate estimation of the soil parameters is required for accurate prediction of slope stability. The in situ soil properties exhibit natural spatial variability due to various factors [1, 2] and are difficult to be estimated using traditional methods. In recent years, the back analysis methods based on field responses are widely adopted for characterizing the spatial variability of soil parameters [3, 4]. However, few studies have been conducted to back estimate the spatial variability of soil parameters based on responses of displacement in a slope due to the large computational cost.

In this paper, the probabilistic back analysis method proposed in Yang et al. [5] is adopted to characterize the spatial variability of soil parameters based on the responses of displacement in a slope. The Karhunen–Loève (K-L) expansion method is used to reduce the dimensionality of the random vector to be back estimated using the Markov Chain Monte Carlo (MCMC) method. To reduce computational cost, a surrogate model

A. Farid and H. Chen (Eds.): GSIC 2018, *Proceedings of GeoShanghai 2018*
International Conference: Geoenvironment and Geohazard, pp. 160–168, 2018.
https://doi.org/10.1007/978-981-13-0128-5_19

based on polynomial chaos expansion (PCE) is established to substitute the finite element model of slope stability analysis. An example is presented to illustrate the accuracy and efficiency of the method. The effects of the correlation length and coefficient of variation (COV) of E on the back analysis of the spatially varied elastic moduli are discussed.

2 Displacement Back Analysis for Spatial Variability of Soil Parameters

2.1 Karhunen–Loève Expansion Method

In this paper, the spatial variability of soil parameters is characterized by random field and the K-L expansion method [6] is used to discretize the random field. The spatially varied soil properties (e.g., elastic modulus E) is characterized by a random field $E(\mathbf{x})$ which can be expressed as:

$$E(\mathbf{x}) = \mu(\mathbf{x}) + \sum_{i=1}^{\infty} \sqrt{\lambda_i}\theta_i\varphi_i(\mathbf{x}) \tag{1}$$

where $\mathbf{x}$ is the spatial location in the spatial domain $\mathbf{D} \subset \mathbf{R}^2$, $\mu(\mathbf{x})$ is mean of $E(\mathbf{x})$, θ_i is the ith independent standard normal random variable, $\varphi_i(\mathbf{x})$ and λ_i are the eigenfunctions and eigenvalues of the covariance function of $E(\mathbf{x})$, respectively. The K-L expansion is truncated to improve computational efficiency. The approximation of $E(\mathbf{x})$ can be expressed as:

$$\hat{E}(\mathbf{x}) = \mu(\mathbf{x}) + \sum_{i=1}^{n} \sqrt{\lambda_i}\theta_i\varphi_i(\mathbf{x}) \tag{2}$$

where n is the truncation level. Commonly, 95% of the total variability is preserved with n [7]. The random variables in Eq. (2) are to be estimated using MCMC method. The number of the random variables to be estimated is significantly reduced based on the K-L expansion method.

2.2 Surrogate Model Based on Polynomial Chaos Expansion

To improve the computational efficiency, a surrogate model based on PCE is established to substitute the finite element model of slope stability analysis. The input parameters of the model are a n-dimensional random vector $\boldsymbol{\theta} = \{\theta_1, \theta_2,..., \theta_n\}$, where θ_i is from Eq. (2). The approximate response of the model can be expressed as [8]:

$$P(\boldsymbol{\theta}) = \sum_{j=0}^{m-1} c_j\Psi_j(\boldsymbol{\theta}) \tag{3}$$

where m is the number of multi-dimensional orthogonal polynomials $\Psi_j(\boldsymbol{\theta})$, c_j are the unknown coefficients. As each θ_i is Gaussian, Hermite polynomials are adopted. The coefficient c_i can be calculated using the projection method:

$$c_i = \frac{E[\Psi_i(\boldsymbol{\theta})P(\boldsymbol{\theta})]}{E[\Psi_i^2(\boldsymbol{\theta})]} = \frac{\int \Psi_i(\boldsymbol{\theta})P(\boldsymbol{\theta})W(\boldsymbol{\theta})d\boldsymbol{\theta}}{E[\Psi_i^2(\boldsymbol{\theta})]} \tag{4}$$

where $E[.]$ denotes the expectation, $W(\boldsymbol{\theta})$ is the weight function of the polynomials. The numerical integration in Eq. (4) can be solved using the sparse grid collocation method based on the Smolyak algorithm [9].

2.3 Parameter Estimation Based on MCMC Method

The field measurements and responses of the surrogate model at n monitoring points are represented by the vectors $\hat{\mathbf{P}} = \{\hat{P}_1, \ldots, \hat{P}_n\}$ and $\mathbf{P}(\boldsymbol{\theta}) = \{P_1(\boldsymbol{\theta}), \ldots, P_n(\boldsymbol{\theta})\}$, respectively. The residual error between $\hat{\mathbf{P}}$ and $\mathbf{P}(\boldsymbol{\theta})$ is $\boldsymbol{\varepsilon} = \{\varepsilon_1, \ldots, \varepsilon_n\}$ which is given by:

$$\varepsilon_i(\boldsymbol{\theta}|\hat{\mathbf{P}}) = P_i(\boldsymbol{\theta}) - \hat{P}_i \tag{5}$$

The residual errors are assumed to be mutually independent and Gaussian-distributed with a mean of zero and a constant variance σ_ε^2, thus the likelihood function can be expressed as [10]:

$$l(\boldsymbol{\theta}|\hat{\mathbf{P}}) = \prod_{i=1}^{n} \frac{1}{\sqrt{2\pi\sigma_\varepsilon^2}} \exp\left(-\frac{(P_i(\boldsymbol{\theta}) - \hat{P}_i)^2}{2\sigma_\varepsilon^2}\right) \tag{6}$$

Based on Bayesian theory, the posterior distribution of $\boldsymbol{\theta}$ can be expressed as:

$$h(\boldsymbol{\theta}|\hat{\mathbf{P}}) = K \cdot l(\boldsymbol{\theta}|\hat{\mathbf{P}}) \cdot h(\boldsymbol{\theta}) \tag{7}$$

where K is a normalizing constant, $h(\boldsymbol{\theta})$ is the prior distribution of $\boldsymbol{\theta}$. As Eq. (7) is difficult to solve using analytical methods, an adaptive MCMC method entitled Differential Evolution Adaptive Metropolis (DREAM) algorithm [11] is adopted to generate random samples from the posterior distribution. The convergence of the algorithm can be calculated using the R criterion of Gelman and Rubin [12]. The convergence diagnostic R_{stat} value is required to be less than 1.2 to declare convergence to a stationary distribution. Details of the algorithm can be found in Vrugt et al. [11].

3 Illustrative Example

3.1 Slope Finite Element Model and Soil Parameters

The numerical model of a soil slope subject to surcharge load is established based on the finite element software ABAQUS. The slope is inclined at 40° and 20 m high.

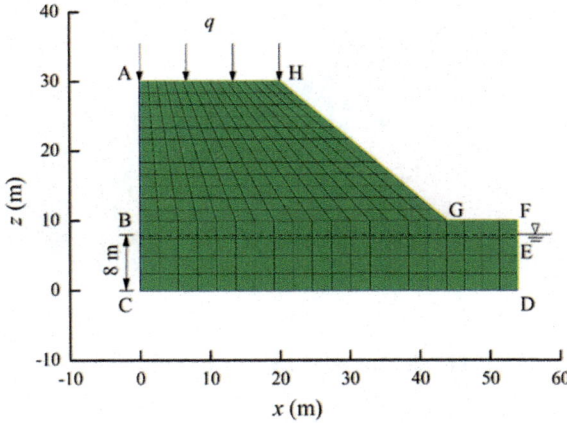

Fig. 1. Illustration of geometry and finite element mesh of the soil slope.

The horizontal displacements of vertical boundaries (AC, DF) are constrained. The bottom boundary (CD) is constrained in both directions. A surcharge load, $q = 10$ kPa is applied to top (AH) of the slope. The geometry and finite element mesh of the slope are illustrated in Fig. 1.

The soil in the slope is assumed to be sandy and the density of soil is 1800 kg/m^3. E is assumed to be spatially varied in the example. The mean and standard deviation of E are 10 MPa and 1 MPa, respectively. The Poisson's ratio λ is constant and equal to 0.3. The input parameters of the example are summarized in Table 1.

Table 1. Parameters of the illustrative example

Parameter	Example	Value	Unit
ρ	Density	1800	kg/m^3
μ_E	Mean of E	10	MPa
σ_E	Standard deviation of E	1	MPa
λ	Poisson's ratio	0.3	
q	Surcharge load	10	kPa
l_h	Correlation length in horizontal direction	50	m
l_v	Correlation length in vertical direction	10	m

In the example, E is assumed to follow a log-normal distribution. The truncation level is taken as 5 when l_h and l_v are 50 m and 10 m, respectively. A hypothetical true field of E is generated using a preselected set of variables and the spatially varied E are mapped into the mesh of finite element model. The spatial variability of E in the soil slope is illustrated in Fig. 2.

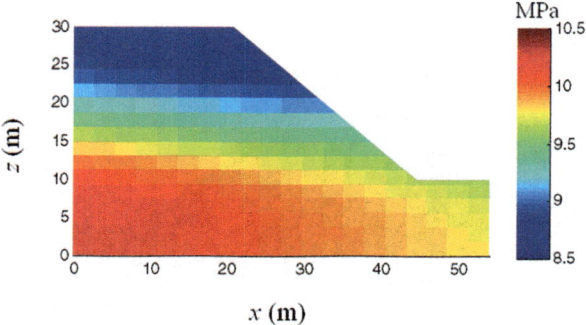

Fig. 2. Hypothetical true field of E

3.2   Surrogate Model and Artificial Data

In the example, the degree of PCE is chosen to be 2 so that the computational cost is small and the accuracy of PCE is acceptable. 81 sets of collocation points are selected using sparse grid collocation method. For each set of collocation points, a random field of E is generated using the K-L expansion method. The displacements of the 81 random fields are obtained using the finite element model. The unknown coefficients are calculated using the projection method.

A monitoring system with 12 monitoring points is set in the slope. Usually, horizontal displacements are measured using inclinometers or rod extensometers. Vertical deformations are normally measured by settlement cells, Borros anchor settlement points, extensometers or horizontal inclinometers. To simply the problem, the 12 monitoring points are selected from the FEM nodes. To consider the effect of measurement errors and model errors, the responses of displacement obtained from the finite element model are corrupted with 2% Gaussian noise to generate artificial data. The spatial variability of E is back estimated based on the artificial data.

4   Results and Discussions

10 Markov chains with 100000 samples are used in MCMC simulation to estimate the five parameters. In the example, 5000 samples are enough to declare convergence of each parameter to a stationary posterior distribution. The last 25% samples of the 10 Markov Chains are used as stationary samples from the posterior distributions of the parameters.

The root mean square error (RMSE) is used to measure the error between the estimated values and true value of E at each location and can be expressed as:

$$RMSE = \sqrt{\frac{\sum_{i=1}^{n}(E_i - \hat{E})^2}{n}} \tag{8}$$

where E_i is the ith estimated value of E at one location in the slope, n is the number of estimated values of E at the location, $\hat{E}$ is the true value of E at the location.

Figure 3(a) and (b) show the RMSE of the estimated fields with the stationary samples based on horizontal and vertical displacements, respectively. In general, the RMSEs at the locations close to the monitoring points are smaller than the RMSEs at the locations away from the monitoring points. The maximum RMSEs based on horizontal and vertical displacements are less than 2.5 kPa and 5 kPa, respectively. As the RMSEs of E are almost negligible compared to the values of E, the accuracy of the estimation is high when the monitoring points distribute over main body of the slope. As the maximum RMSE of the estimated fields based on horizontal displacements is smaller than that based on vertical displacements, the horizontal displacements are preferred for displacement back analysis of the spatial variability of E. Hence, inclinometers are preferred to set in the field in practical application.

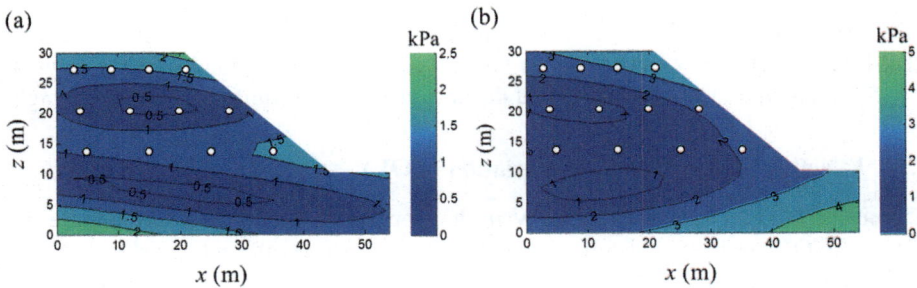

Fig. 3. RMSEs of the estimated fields with the stationary samples based on (a) horizontal and (b) vertical displacements with $l_h = 50$ m, $l_v = 10$ m.

5 Effects of Spatial Variability on Displacement Back Analysis

The stronger the spatial variability of the slope is, the more difficult it is to estimate the soil parameters in the slope. The spatial variability of E is significantly affected by the correlation length and coefficient of variation (COV) of E. Hence, the effects of the correlation length and COV on back estimation of spatially varied elastic moduli need to be analyzed.

5.1 Effect of the Correlation Length

To investigate the effect of l_h, l_h decreases from 50 m to 25 m. To investigate the effect of l_v, l_v decreases from 10 m to 5 m. For the two conditions, the random variables are back estimated based on the horizontal displacements of monitoring points.

Figure 4(a) and (b) show the RMSEs of the estimated fields based on the stationary samples with $l_h = 25$ m, $l_v = 10$ m and $l_h = 50$ m, $l_v = 5$ m, respectively. Comparing Fig. 3 with Fig. 4(a) and (b), the maximum RMSE increases by about 4 times and 80 times when l_h and l_v decreases by half, respectively. Hence, the accuracy of the estimation of E is reduced when the correlation length decreases. l_v has a much greater

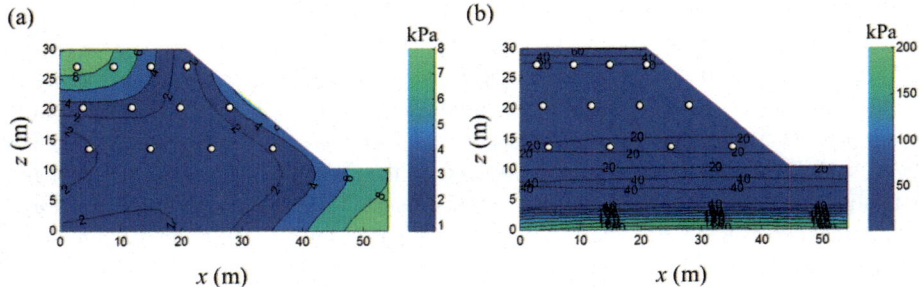

Fig. 4. RMSEs of the estimated fields based on the stationary samples with (a) $l_h = 25$ m, $l_v = 10$ m and (b) $l_h = 50$ m, $l_v = 5$ m.

effect on accuracy of the estimation than l_h. It should be noted that the maximum RMSE is 200 kPa which corresponds to a percentage error of about 2% for E when $l_h = 50$ m and $l_v = 5$ m, indicating that the accuracy of the estimation is still very high.

5.2   Effect of the Coefficient of Variation (COV)

The elastic moduli with different COV in the slope are back estimated to analyze the effect of COV on back estimation of E. Figure 5(a) shows the relationship between COV of E and the maximum COV of the estimated fields based on the stationary samples. The maximum COV of the estimated fields increases when COV of E increases, indicating that the stronger the spatial variability of the field is, the larger the uncertainty of the estimated fields is. Figure 5(b) shows the relationship between COV of E and the maximum RMSE of the estimated fields based on the stationary samples. The maximum RMSE can be negligible when COV of E is only 0.1. The maximum

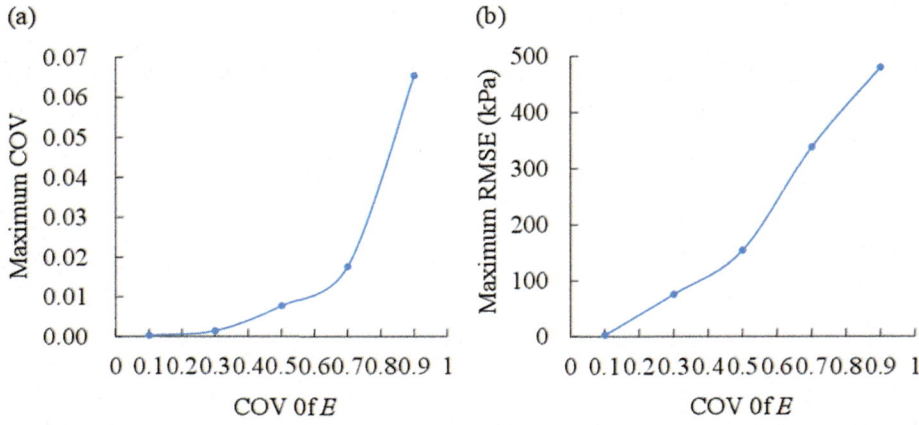

Fig. 5.   (a) Relationship between COV of E and the maximum COV of the estimated fields based on the stationary samples, (b) Relationship between COV of E and the maximum RMSE of the estimated fields based on the stationary samples.

RMSE increases when COV of E increases. Hence, the accuracy of the estimation of E is reduced with the increase of COV of E. It should be noted that the maximum RMSE is close to 500 kPa which corresponds to a percentage error of about 5% for E when COV of E is up to 0.9. Hence, the spatially varied elastic moduli can still be back estimated with high accuracy when the spatial variability in the slope is strong.

6 Conclusions

In this paper, a probabilistic back analysis method is adopted for displacement back analysis of spatially varied elastic moduli in the slope. The method is illustrated by a hypothetical example. The major conclusions are summarized below:

1. The spatial variability of E in the slope can be estimated efficiently using the method adopted when the monitoring points distribute over main body of the slope. The horizontal displacement is preferred for displacement back analysis compared to the vertical displacement.
2. The accuracy of the estimated field of E is reduced when the correlation length decreases. l_v has a much larger effect on accuracy of the estimation than l_h.
3. The accuracy of the estimated field of E is reduced with the increase of COV of E. In general, the spatial variability of E can still be back estimated with high accuracy when the spatial variability in the slope is strong.

References

1. El-Ramly, H., Morgenstern, N.R., Cruden, D.M.: Probabilistic slope stability analysis for practice. Can. Geotech. J. **39**(3), 665–683 (2002)
2. Lacasse, S., Nadim, F.: Uncertainties in characterising soil properties, vol. 201, pp. 49–75. Publikasjon-Norges Geotekniske Institutt (1997)
3. Trandafir, A.C., Sidle, R.C., Gomi, T., Kamai, T.: Monitored and simulated variations in matric suction during rainfall in a residual soil slope. Environ. Geol. **55**(5), 951–961 (2008)
4. Deng, J.H., Lee, C.F.: Displacement back analysis for a steep slope at the Three Gorges Project site. Int. J. Rock Mech. Min. Sci. **38**(2), 259–268 (2001)
5. Yang, H.Q., Zhang, L.L.: Efficient estimation of spatial varied soil properties based on field pore water pressure responses in a slope. In: Geo-Risk 2017, pp. 32–41 (2017)
6. Karhunen, K.: Über lineare Methoden in der Wahrscheinlichkeitsrechnung. Suomalainen Tiedeakatemia, Helsinki (1947)
7. Jiang, S.H., Li, D.Q., Cao, Z.J., Zhou, C.B., Phoon, K.K.: Efficient system reliability analysis of slope stability in spatially variable soils using Monte Carlo simulation. J. Geotech. Geoenviron. Eng. **141**(2), 1–13 (2015)
8. Ghanem, R.G., Spanos, P.D.: Spectral stochastic finite-element formulation for reliability analysis. J. Eng. Mech. **117**(10), 2351–2372 (1991)
9. Xiu, D.: Efficient collocational approach for parametric uncertainty analysis. Commun. Comput. Phys. **2**(2), 293–309 (2007)
10. Box, G.E.P., Tiao, G.C.: Bayesian Inference in Statistical Analysis. Wiley, New York (2011)

11. Vrugt, J.A., ter Braak, C.J.F., Clark, M.P., Hyman, J.M., Robinson, B.A.: Treatment of input uncertainty in hydrologic modeling: doing hydrology backward with Markov chain monte carlo simulation. Water Resour. Res. **44**(12), 5121–5127 (2008)
12. Gelman, A., Rubin, D.B.: Inference from iterative simulation using multiple sequences. Stat. Sci. **7**(4), 457–472 (1992)

Strategy for Improving Precision of Soil Liquefaction Potential

Sao-Jeng Chao, Hui-Mi Hsu, An Cheng, and Chien-Wei Pan[✉]

National Ilan University, Yilan, Taiwan
tk755620@gmail.com

Abstract. Ilan area is located between the east and the northeastern earthquake regions of Taiwan, which the earthquake occurs quite frequent. Langyang Plain is alluvial plain coupling with fine particles of loose granular soil and high groundwater level, the possibility of soil liquefaction in Ilan area is very high once the earthquake occurs. Therefore, soil liquefaction potential needs to be carefully investigated. In Ilan area the primary soil liquefaction areas was formed with precision is 1/25000 scale. February 6, 2016 Meinung earthquake in the southern Taiwan area brought a lot of liquefaction damages. Therefore, soil liquefaction once again gains people attention. The precision of soil liquefaction potential map is requested to improve to 1/2500 scale. In order to improve the precision of geospatial information effectively, the geological drilling density of 4 holes/km^2 are demanded. The drilling method employs the SPT method in general. The soil physical test is taken every 1.5 m or the soil layer change. This study adds about 213 boreholes of geological drilling data in the densely populated areas, low floor houses, old buildings and complex geological conditions of Ilan area. It is believed that we can use the drilling data to know the soil liquefaction potential more accurately. Finally, this study will use SPT-N HBF analysis method to calculate soil liquefaction potential value. The method is a simple evaluation technique to determine the occurrence of soil liquefaction in Taiwan with good accuracy and reliability. The proposed analysis in this study can be used to accomplish more precise soil liquefaction potential with different size earthquake occurs in the region.

Keywords: Soil liquefaction · Drilling · Borehole · Standard penetration test
HBF analysis method

1 Introduction

Taiwan situates in Circum-Pacific Seismic Belt, where earthquakes occur frequently. How to protect against and reduce disasters is an important strategy of government officials. On 6 February 2016, a strong earthquake of Richter local magnitude 6.6 occurred in the Tainan area of Taiwan, resulting in many serious disasters. Soil liquefaction is then becomes one of the government's concern. Consequently, soil liquefaction issue is taken seriously in order to achieve safety, prevention, early warning, etc. The soil liquefaction potential map must be improved for the accuracy at this critical moment. Especially, in densely populated areas, low floor houses, old buildings, and complex geological condition areas, the destruction comes after soil

© Springer Nature Singapore Pte Ltd. 2018
A. Farid and H. Chen (Eds.): GSIC 2018, *Proceedings of GeoShanghai 2018*
International Conference: Geoenvironment and Geohazard, pp. 169–177, 2018.
https://doi.org/10.1007/978-981-13-0128-5_20

liquefaction severely. Geologically, Ilan area is deposited mostly with fine particles contenting high content of weak soil, coupling with groundwater level very close to the surface. Once the earthquake occurs, the possibility of soil liquefaction is quite high.

Therefore, we need to add drilling holes in order to draw a more representative soil liquefaction potential map of intermediate level. As a result, the Ilan County Government can be more prepared for disaster prevention and early warning operations. Accordingly, people in Ilan area can get a safe and comfortable living condition. It is reason that we are obligatory to explore the soil liquefaction related problems in Ilan area.

2   Story of Soil Liquefaction

Soil liquefaction is the process of changing the soil from solid to liquid just as the name. For the period of liquefaction, the soil behavior is similar to the liquid. Due to the liquid cannot resist shear force and maintain its own shape, there is mobility and buoyancy, thus the structure will be subsidence, tilt or damage. The reason of resulting in this phenomenon is mainly caused by vibration, which may come from construction or earthquake, while most of the soil liquefaction is caused by the earthquake.

The study of liquefaction of granular soils using the concept of critical void ratio was first introduced by Casagrande in 1938. When a natural deposit of saturated sand that has a void greater than the critical void ratio is subjected to a sudden shearing stress (due to an earthquake, for example), the sand will undergo a decrease in volume and lead to the condition of soil liquefaction.

After that, many researchers kept on working in this topic. For example, Japanese scientists Mogami and Kubo (1953) suggested that the soil will be deformed as "liquefied" when cohesionless soil is disturbed. Following the 2 earthquakes of Alaska in US and Niigata in Japan in 1964, an intensive research has been carried out.

Seed (1979) mentioned that high pore water pressure would reduce the effective confining pressure of soil. At this point in time, the soil residual shear strength becomes lower and the soil mass is likely to produce deformation. As soon as the earthquakes occur, it is possible that the soil cannot afford the original stress all of a sudden. Consequently, the soil pore water increases and shear stress intensity becomes smaller. This condition provides the opportunity to produce soil liquefaction, when the excess pore water pressure continues to rise to the original effective stress (when $\sigma = \mu$). That is to say, when the earthquake induces the local average shear stress is greater than the local soil shear strength for resisting liquefaction, the soil is liquefied.

3   Liquefaction Influence Factors

The terrain of Langyang Plain is deposited by the soil particles carried down from the Langyang River, so that the soil is quite loose and soft, coupling with the conditions of groundwater close to the surface and located in the earthquake zone. Since earthquakes occur frequently, the seismic disaster caused by earthquake should not be ignored at all. Above mentioned conditions make the soil liquefaction problem needs to be assessed objectively and carefully in Ilan area. The factors of soil liquefaction are divided into three portions here: soil strength, groundwater level, and earthquake scale. These factors can be used to determine the occurrence of soil liquefaction in the region.

3.1 Resistance of Soil

Granular soils can commonly be judged by their relative density or standard penetration test. The soil particles of relatively tight condition are less prone to relative displacement, and their resistance to soil liquefaction is better. The process of soil liquefaction is that the soil particles changing the arrangement between the particles by repeated earthquake movements. Of course, after the moment of liquefaction, it will produce a more stable soil structure than the initial soil particle. Similarly, when the soil particles are stacked more closely with each other, the chance of soil liquefaction resulting in damage will be reduced greatly. Therefore, due to the getting closer relationship between the particles, easily liquefied soil particles can also have a good liquefaction resistance after the liquefaction process.

3.2 Groundwater Level

The effect of water in the soil has always been a matter of concern to civil engineers. Whether it is slope stability, excavation construction, and soil liquefaction, once the soil pores filled with water would make the soil to be in a relatively unstable state. In other words, when the groundwater level is closer to the surface, which means the more soil is saturated. When the earthquakes occur, its safety factor will become lower. On the other way, we know that the weak soil is more close to the surface, the impact to the liquefaction potential becomes more serious. When the soil elevation deep enough, one can make sure that the soil has good liquefaction resistance, thus the impact of liquefaction potential is smaller.

3.3 Earthquake Magnitude

Seismic factors are the most important influencing factors in three portions of soil liquefaction. When the earthquake occurs, it will release the energy from the epicenter to the periphery. These energies will transmit to the surrounding rock or soil through seismic waves. In the process of energies passing the soil, the shear stress inside the underground soil is generated. If the site of the soil liquefaction resistance is poor, it may generate soil liquefaction phenomenon. When the problem of soil liquefaction is discussed, the magnitude can be considered as the energy obtained for the soil. At the same time, the magnitude of the earthquake becomes greater, the shaking on the surface becomes more intense and the duration last longer. It is possible to continuous increase in pore water pressure, which increases the chance of soil liquefaction.

4   Geological Drilling Plan

Based on the geological drilling report, the soil liquefaction situation of the site can then be calculated and the potential map of regional soil liquefaction can be drawn. If the density of drilling holes can be increased, the soil liquefaction potential map will be better for accuracy, and the precise soil liquefaction situation in this area can be reflected.

In view of the February 6, 2016 Meinung earthquake happened in the Tainan area caused serious disaster, the soil liquefaction potential map needs to be refined. Especially in the Ilan region, considering the geological weakness and the groundwater level close to the surface, as well as the most critical part of locating in multiple seismic zones, soil liquefaction situation could be much more sensitive.

4.1   Existing Drilling Data Collection and Comparison

Prior to geological drilling, the first step must be done is to collect high-quality existing drilling hole data, such as from public constructions, large buildings, and other high credibility of the drilling report. The second step to compare them after evaluation in order to screen out the available drilling. Ilan County, for example, the first phase of soil liquefaction map to be reconstructed is Ilan city, Luodong Town, and Wujie Township. This study has collected about 350 borehole data as shown in Fig. 1 for their distribution.

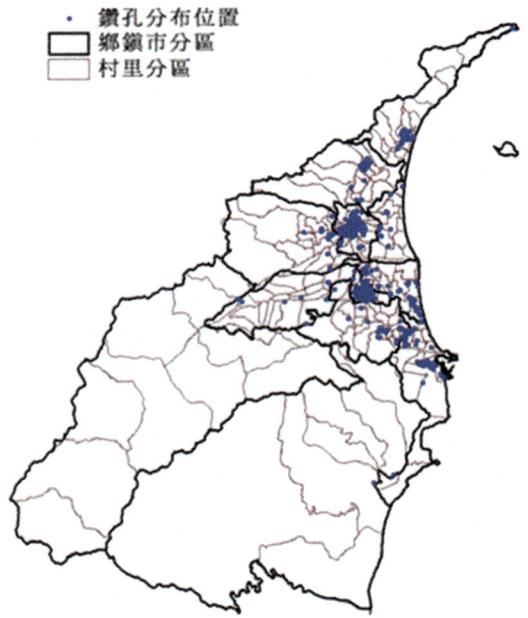

Fig. 1. Existing collected borehole data distribution map.

It can be clearly found that in the central parts of Ilan City and Luodong Town demonstrate more dense distribution of boreholes. The reason for these two places show higher degree density are the regional center of educational district for Ilan City and business district for Luodong Town. Besides, some other boreholes are located along the National Highway No. 5. because the design standards for highway construction are much strict than the country road, resulting in geological conditions of the highway must be considered more seriously.

In general, boreholes are mainly selected on the Langyang Plain while rare in the remote mountain area. As a matter of fact, the borehole distribution is quite uneven for the whole Ilan area. As a result, the existing collected borehole data cannot accurately accomplish the needs to establish a higher level soil liquefaction potential map for Ilan area. As a consequence, it is required to obtain more accurate information by means of supplementary drilling operation in Ilan area.

4.2 Drilling Holes Layout and Density

In order to make sure the new constructed potential maps of soil liquefaction to achieve intermediate precision, the key consideration is the drilling holes layout and density. The drilling density must be more than 4 holes/km^2 under the government specification. It is carefully arranged to increase the number of boreholes for areas of medium and high liquefaction potential, densely populated areas, low floors houses, old buildings, or those areas where the density is insufficient. In addition to the low potential areas or population living sparse and simple geological conditions, we have to reduce the drilling density so that the geological data obtained from the drilling holes can achieve maximum efficiency.

As the main factors affecting soil liquefaction are weak granular soil, high groundwater, etc. The drilling program of borehole distribution can be accomplish by considering following aspects: (1) The existing boreholes distribution map not only can be analyzed at the current drilling layout of the precision and accuracy, but also determine the accuracy of the liquefaction potential and decision to add the drilling holes; (2) The terrain and groundwater data over the years can be used to determine the distribution of weak soil layer and find out the areas where liquefaction is prone. It can also be used to correct drilling holes location and soil liquefaction potential value; (3) Data related to geology can comprehend the geological distribution, determine the correction of the drilling report and adjust the placement of the drilling holes position. Through the above mentioned steps, screening and adjustment boreholes location and other pre-operation can help to improve the accuracy of soil liquefaction potential map.

4.3 Drilling Technique and Supervision

Fully understanding the type of underground soil, distribution, and density is an important basis for judging soil liquefaction potential. This study takes the most widely used and most economical means of drilling technique to carry out drilling operations. The method for the standard penetration test, referred to as "SPT", is shown in Fig. 2, which is the site photo illustrating the field work condition. This test was carried out by a 63.5 kg (140 lb) hammer, which was free to fall into the drill pipe at a height of

76.2 cm (30 in.) so that punches the sampler into the soil, and records the number of blow at the same time. The blow count number is recorded by three sections of 15 cm (6 in.) for each penetration depth. The first section of the 15 cm (6 in.) penetration depth is used to determine the position of the sampler, and the sum of the second and the third segments of the 15 cm (6 in.) penetration depth is the N value. The test must reach the 45 cm (18 in.) or the N value reaches 100 blows.

Fig. 2. The field drilling photo in this study.

SPT drilling method is executed every 1.5 m while in the soil layer change should be carried out at once. All the field works are supposed to be judged by the site construction personnel according to experience to increase the basic information for soil liquefaction potential and verify the correctness of SPT drilling results. The samplers and test procedures used shall comply with ASTM D1586. It is worthwhile to mention here that the supervisors go to the sites all day long to watch the field work with very high expectation.

5   The Improvement of Soil Liquefaction Potential Map

Langyang plain is liquefied sensitive area without a doubt. The issue of life and property safety of inhabitants is surely a big matter of concern when a more violent or closer earthquake happens around the Ilan area. The improvement of liquefaction potential map accuracy is helpful for the assessments as well as the prevention of soil liquefaction disaster. However, how to decide the soil liquefaction occurs in the Ilan area, we can use the drilling data to understand the details for distribution of underground soil and groundwater level. With drilling data, we can calculate and draw the soil liquefaction potential map with the earthquake scale. Therefore, the accuracy of the soil liquefaction potential map is the important issue we must consider and try our best to improve the precision of the map.

5.1   Primary Map Precision

Ilan region soil liquefaction primary drawing precision is of 1/25000 scale, as shown in Fig. 3, by the Central Geological Survey. From the primary announced map, it can be seen that Ilan area is prone to soil liquefaction. The soil liquefaction primary in the figure is illustrated with three colors to distinguish them: ▮ represents the low potential area (Liquefaction potential index PL < 5); ▮ represents the medium potential area (Liquefaction potential index $5 \leq PL \leq 15$); ▮ represents the high potential area (Liquefaction potential index PL > 15).

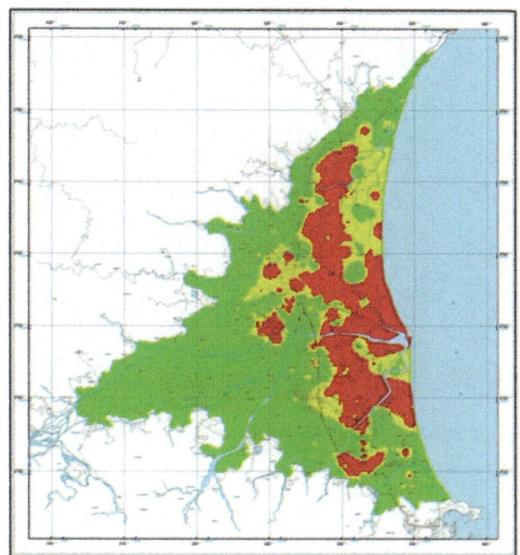

Fig. 3. Ilan area primary liquefaction potential map (courtesy of Central Geological Survey).

From the primary soil liquefaction map, Ilan city, Luodong Town, and Wujie Township contains the medium and high potential region. Additionally, the primary results of this assessment can be found to have a greater relevance between liquefaction and geological conditions. Alluvial plains area and the mountain area have totally different liquefaction potential situation. Liquefaction may occur in the area with the deposition of loose granular soil. The direction of the coastal of the Pacific oceans is likely to have a tendency to high liquefaction development trend.

5.2   Intermediate Precision

Ilan soil liquefaction intermediate precision is set to be 1/2500 scale, so that the density of geological drilling needs 4 holes/km^2. In the future study, the proposed supplementary drilling holes will be executed at the three townships in Langyang plain, which is shown in Fig. 4. The gray point in the figure is the existing collected drilling holes in the three townships. The red point in the figure is the supplementary drilling point will be added. Supplementary drilling holes have tried the best to evenly distribute within the three townships.

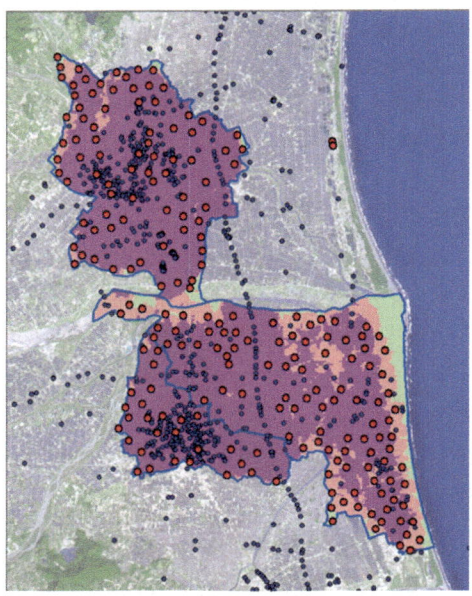

Fig. 4. Supplementary drilling holes distribution map (courtesy of CECI).

Moreover, the boreholes density in the figure can be used to view the supplementary drilling plan. The low density of drilling holes will result in the interpolation less supreme and can lead the predicted result inadequate to reflect the soil liquefaction situation. Therefore, there is expected accuracy when drawing liquefaction potential maps. This study adds about 213 boreholes of geological drilling data in the densely populated areas, low floor houses, old buildings and complex geological conditions of Ilan area. The drilling density in the figure is illustrated with three colors to distinguish them: ■ represents the boreholes density of 4 holes/km^2 or more; ■ represents the boreholes density is above 2 holes/km^2 but less than 4 holes/km^2; ■ represents the boreholes density is less than 2 holes/km^2.

The three townships (Ilan City, Luodong Town, and Wujie County) within the drilling density of 4 holes/km^2 and above covers the total area of about 95%, which can be anticipated to improve the accuracy of soil liquefaction potential map. As a final point, this study will employ the SPT-N HBF analysis method to calculate soil liquefaction potential, which is a quite simple evaluation method to determine the occurrence of soil liquefaction in Taiwan, which has been validated with good accuracy and reliability. This method can be fine-tuned for different earthquake scale to evaluate the soil liquefaction potential.

6   Conclusion

Ilan area is located in such an unique environment when the earthquake occurs would suffer quite high possibility of soil liquefaction. Although the earthquake in recent years in Ilan region did not carry serious liquefaction disasters, one cannot be ignored that Ilan area is undoubtedly highly liquefied sensitive. In view of this, it is necessary to improve the precision of soil liquefaction potential map, so that the assessment and improvement of soil liquefaction disaster can be carried out promptly.

In the current point, the improvement of soil liquefaction potential map project is mainly executed for Ilan city, Luodong Town, and Wujie Township. The initial arrangement for distribution of drilling holes has been roughly completed with the required density of 4 holes/km^2. The drilling method employs SPT standard penetration test method for the sampling every 1.5 m or soil layer change and perform the soil physical properties general test. Follow-up progresses such as data documentation, analysis operation, and related internet system construction for their benefit need to wait for the completion of the supplementary drilling operation.

References

Mogami, T., Kubo, K.: The behaviour of sand during vibration. In: Proceedings of 3rd International Conference on Soil Mechanics and Foundation (1953)

Seed, H.B.: Soil liquefaction and cyclic mobility evaluation for level ground during earthquake. J. Geotech. Eng. Div. ASCE **105**(GT2), 201–255 (1979)

Research of Formation Mechanism for the III# Landslide on a Hydropower Station Reservoir in Yellow River

Wang Yunnan[1(✉)], Liu Zhenghong[1], Cao Jie[1], Tang Guoyi[1],
Ren Guangming[2], and Qiu Jun[3]

[1] China JK Institute of Engineering Investigation and Design,
Xi'an 710043, China
605725462@qq.com
[2] State Key Laboratory of Geohazard Prevention and Geoenvironment
Protection, Chengdu University of Technology, Chengdu 610059, China
[3] Guizhou Transportation Planning Survey & Design Academe Co., Ltd.,
Guiyang 550001, China

Abstract. Toppling-deformation tends to occur in the anti-dip bedded slope, but recently we find that the steep bedding rock slope also exists this mode. The study of the formation mechanism has an important significance to the evaluation of the slope stability. This paper takes the III# landslide which is in front of a dam in Yellow River as an example. Based on the engineering geological survey, the failure process of the landslide is studied by qualitative analysis and numerical modeling. The results indicate that the failure process of the landslide includes rock mass unloading, toppling deformation and sliding. The results of using finite element match to that of discrete element calculation, which proves the rationality of the conclusion.

Keywords: Landslide · Formation mechanism · Toppling deformation
Numerical modeling

1 Introduction

The toppling-deformation is an instability mode which is prevalent in the anti-dip layered slope. However, with the development of a large number of engineering projects, this type of deformation failure mode is found in the bedding rock slope [1, 2]. According to the field investigation, the left bank of jinchuan hydropower station has the toppling deformation. The deformation and landslides in qingchuan county studied by Wang Zhetao were formed by the dumping [3]. Sun [4], Zhongkai [5] in the investigation also found that there is toppling deformation in the consequent slope; Ren Guangming has studied the deformation and failure characteristics of this slope and reconstructed its evolution by numerical simulation [6]. There are instances of the toppling deformation of the slope in the foreign countries, such as the deformable landslide of the Canadian Rockies [7, 8].

© Springer Nature Singapore Pte Ltd. 2018
A. Farid and H. Chen (Eds.): GSIC 2018, *Proceedings of GeoShanghai 2018*
International Conference: Geoenvironment and Geohazard, pp. 178–189, 2018.
https://doi.org/10.1007/978-981-13-0128-5_21

III# landslide is located on the right bank in the front of the dam, and the formation mechanism research of the landslide plays an important role in the stability evaluation and the deformation trend judgment.

The field survey find that the landslide is located in the steep bedding slope. According to the geological environment conditions and characteristics, the deformation and failure modes of landslide belong to the toppling deformation and sliding-tension. On the basis of geological survey, the formation and evolution of this landslide is discussed by means of discrete element, finite element and qualitative analysis.

2 Geological Conditions

III# landslide is located on the right bank of the proposed dam, with the upstream boundary about 800 m away from the dam axis, and the downstream boundary about 600 m away from the dam axis. The leading river is basically straight, the river flows to NE25°, the horizontal water level is about 2760 m, the river width is about 40 – 60 m. Landslide bedrock exposed surface, slope gradient change between 35° – 50°. The trailing edge of the landslide is about 130 m in width, the leading edge is about 200 m, the slope length is about 270 m, and the relative height difference of the landslide is about 160 m, as shown in Fig. 1.

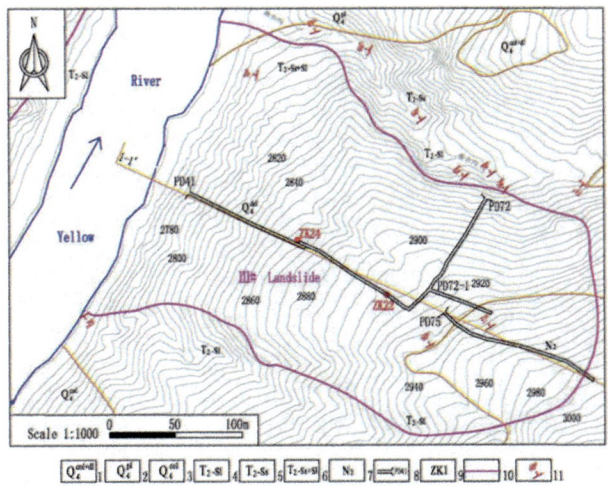

Fig. 1. Plan of the landslide. (1- Jinpingzi; 2- Diluvial deposit; 3- Gravelly soil; 4-Killas; 5-Sandstone; 6- Sandstone splint rock; 7-Clay rock; 8-Footrill; 9-drill; 10-Landslide boundary; 11-Attitude of rock;)

The landslide is mainly composed of Triassic slabs and the mixing layers of slate and sandstone, and the rock masses are thin laminates. Landslide bedrock strata tendency is between $300° – 350°$, with the dip of $68° – 82°$ and landslide main sliding direction of NW310°. The landslide main slip direction for NW310°, which is perpendicular to the river flow. According to the 《Design specification for slope of hydropower and water conservancy project》 (DLT5353), the slope of the III# landslide is steep bedding rock slope.

There is no large-scale fault development in the landslide area. Landslide fissure in both groups, within the scope of the main development respectively NE20° NW < 20° and NW285° SW < 25°. NE20°NW $\angle 20°$ fissure and interlayer displacement zone can be combined into a landslide slipping surface.

3 Characteristics

The thin-layer rock mass in the middle of the slope is close to be upright. The strata attitude is NE55° SE $\angle 80°$ (see Fig. 2). The attitude of leading edge rocks is NE78° SE $\angle 62°$, where the rock mass is broken with the signs of rock formation(see Fig. 3). Landslide back-up 3000 m elevation location there is about 1 m high scarp (see Fig. 4), The strike is parallel to the river. The top of the downstream boundary and the adjacent slopes form a wedge-shaped groove (see Fig. 5), showing signs of collapsing, initially estimating the collapse volume at 5,000 m³.

Fig. 2. Dumping rock.

There are three galleries in the landslide, PD41, PD72 and PD75. According to the results of the elastic velocity test of the wall in PD41, the pre-30 m wave velocity is below 1000 m/s; Within the range of 30 m – 47 m, the wave velocity is between 1000 m/s – 2000 m/s. After 47 m, the wave velocity is above 2400 m/s. According to the results of the wave velocity test in the galleries, it is assumed that 30 m is the slide belt location, 30 – 50 m is the influence zone of the slip zone, and the position greater

Fig. 3. Anti-dip layered rock mass of the slope toe.

Fig. 4. Scarp of the back edge

than 50 m belongs to the slope bedrock part. A fracture zone was also found at 17 m in pd72-1 hole, and the upper and lower layers of the broken band formed a set of bow fracture bands, presumably as a slip belt. There are obvious curved tortuous sections at 80 m in the back edge of the gallery. According to the above distortion and failure characters, the position of slide belt is obtained. According to the deformation and failure characteristics of the above slope table and the cave, the position of the sliding belt is obtained. The height of the trailing edge of the landslide is about 3000 m, and the shear opening is below 18 m of PD41, as shown in Fig. 6.

Fig. 5. Downstream boundary.

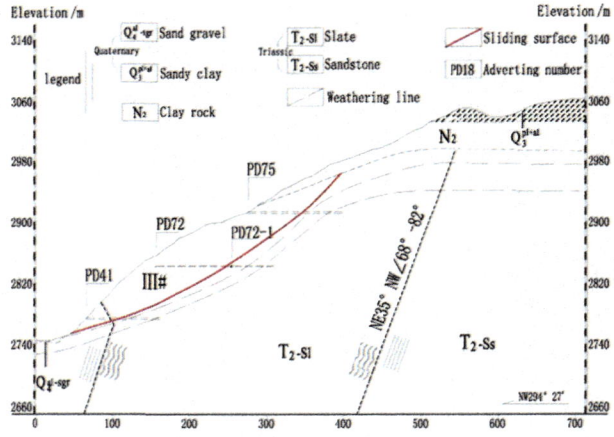

Fig. 6. Profile map of the landslide (1-1′).

4 Landslide Formation Mechanism

4.1 Landslide Formation Factors

Strata Lithology. The bedrock in the landslide is mainly composed of Triassic slabs and the mixing layers of slate and sandstone. The strata appear thin layer, poor mechanical properties, weak weathering ability. Rock levels and weathering fissure provide channel for surface water infiltration.

Geological Structure. The slope of the III# landslide is the steep bedding rock slope. Attitude of rocks is NE30° – 80°/NW ∠68° – 82°. There are two groups of fissures in the bedrock: NE20°NW ∠20° and NW285°SW ∠25°, which provide boundary conditions for landslide.

Neotectonics. The landslide is in the northeast of the Qinghai-Tibet plateau. Along with the lower cutting of the river, the front edge of the slope forms a steep slope, which provides the blank conditions for the formation of the landslide.

Hydro-geological Conditions. The area is in a semi-arid and high-cold climate zone with low rainfall. Groundwater is mainly pore water and bedrock fissure water. The long-term occurrence of fissure water will reduce the physical and mechanical parameters of the rock mass and induce the occurrence of landslide. In conclusion, the physical and mechanical properties of the strata and the steep slope are the internal factors of the deformation of the slope. The joint fissure of the bedrock development provides the boundary condition for the landslide. On the basis of neotectonic movement and groundwater softening rock formation, the landslide is finally formed.

4.2   Geological Analysis of Landslide Formation Mechanism

According to the above analysis of the topography, the material composition and the basic characteristics of the landslide, it is concluded that the landslide is induced by the toppling deformation of the steeply inclined interbedded sand slate, forming the phenomenon of "nodding" [9]. The specific formation of the landslide can be divided into the following three stages:

The first stage is the initial deformation. As the river cuts, the river valley deepens and the shoreslope expands, the stress distribution of slope rock mass is different from the unloading of slope: the slope is the stress distribution area and the toe of the slope area is the shear stress concentration zone. The closer the main stress traces near the slope, the gravitational field and the tectonic stress field are closer to the surface. Under this stress, the steeply stepped rock mass in the slope foot will initially deform to the temporary surface [6].

The second stage is the toppling-deformation. As the leading edge deformation intensifies, it provides the space for the deformation of the middle slope. The trailing edge of the slope produces the unloading rebound, and the influence of the stress concentration on the pull is formed along the pull crack in the steep rock formation. The trailing edge of the slope produces an unloading spring back, coupled with the effect of the tensile stress concentration, forming a crack along the steeply sloping layer. Under the action of initial disturbance and pore water pressure [10], the middle layered rock mass slips along the structural plane. But due to the lower rock block and bending deformation occurs, and under the action of the maximum principal stress bending effect, turn to the temporary direction of the dumping. When the rock formation is bent to a certain extent, it can lead to the formation of a broken section at the root of the rock formation. The fracture of the trailing edge along the rock and fracture surface formation tracking development.

The third stage is the sliding-tension. The fractured deformation of the middle part of the slope extends to the trailing edge of the slope, and the fracture surface of the front and the trailing edge constitutes the fracture zone of the slope. At this point, the slope is transformed from the toppling-deformation to the sliding-tension. Under the action of self-weight, tectonic movement, groundwater and rainfall, the fracture zone eventually penetrates, and the overall slip occurs, forming a landslide. The position of the landslide shear outlet retains some deformation marks, which is reflected in the anti-warping of the rock (see Fig. 3). In the process of the landslide, the downstream rock mass is extruded and pulled, resulting in the collapse of the slope and the formation of wedge-shaped grooves to form the downstream boundary (see Fig. 4).

4.3   Numerical Simulation of Landslide Formation Process

In order to reproduce the whole process of the landslide, the two-dimensional discrete element software UDEC is used to simulate the deformation process under the condition of self-weight [11–16]. Then the stress and strain characteristics of the slope are analyzed by two-dimensional finite element software phase.

Calculation Model. Based on the deformation and failure characteristics of the landslide and the topographic features of the surrounding area, it is deduced that the original slope is steep and the depth of the valley is shallow. According to the morphology of adjacent slope, the original slope of the landslide is reversed, and the corresponding calculation profile is shown in the figure below (Fig. 7, Tables 1 and 2).

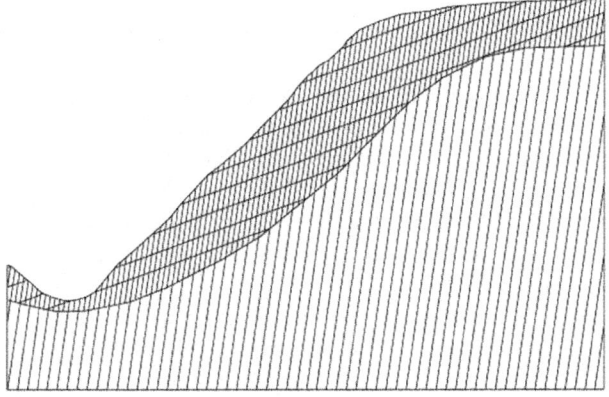

Fig. 7. UDEC model of slope.

Table 1. Parameters of rock mass.

Zone	Natural density/ (Kg/m3)	Cohesion/MPa	FrictionAngle/°	Bulk modulus/GPa	Shear modulus/GPa
Weathered	2500	0.58	34.4	14.68	9.8
Bedrock	2650	1.50	45.0	21.20	15.3

Table 2. Parameters of structural planes.

Structural	Tensile strength/MPa	Cohesion/MPa	Internal friction angle/°	Normal stiffness/ (GPa/m)	Shear stiffness/ (GPa/m)
Cracks	0.02	0.18	32.5	6.3	5.2
Level	0.35	0.5	40	8.5	7.4

Calculate Parameters. According to the combination of field test, engineering analogy and experience, the physical and mechanical parameters of each medium in the model are determined, as shown in the following table.

Calculation Results. The process of numerical simulation has a total of 250,000 steps, which directly reflected the whole process of the development of the landslide. According to the simulation results, the concrete formation process of landslide can be divided into the following stages.

In the lower river valley, the front edge of the slope becomes steeper, and the lower part of the rock is deformed in the direction of the air under the condition of gravity. The deformation of the leading edge slope provides space for the deformation of the middle slope. The middle strata are bent and deformed in the direction of the air. The trailing edge of the slope forms a unloading fracture and produces a crack along the steeply inclined plane, as shown in Fig. 8.

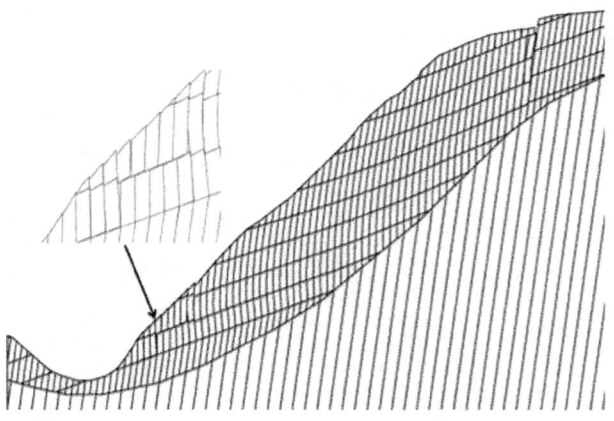

Fig. 8. Slope deformation characteristics (80000 steps).

Calculated to 0.11 million steps, the model finds that with the river valley continuing to cut and weathering, the front edge deformation slope extends to the slope and the surface rock begins to deflect in the direction of the air. The central rock mass has a tendency to dump deformation in the direction of the air, and the rock mass with large deformation has a local shear slip along the inclined plane. The trailing edge cleft faces the internal development of the slope, the rock in the vicinity of the crack is relatively broken, and the slope has a certain degree of subsidence, forming a stepping stone, as shown in Fig. 9.

Calculated to 14,000 steps, the model finds that on the basis of the above deformation, the front dumping rock mass is slippage along the inclined plane. The deformation of the central rock mass is aggravated, and the fracture zone is formed with the leading edge sliding surface and the trailing edge (see Fig. 10). Calculated to 0.25 million steps, the fracture zone is completely separated from the bedrock, and the whole slope is sliding down and blocking the valley. It can be seen from Fig. 11 that

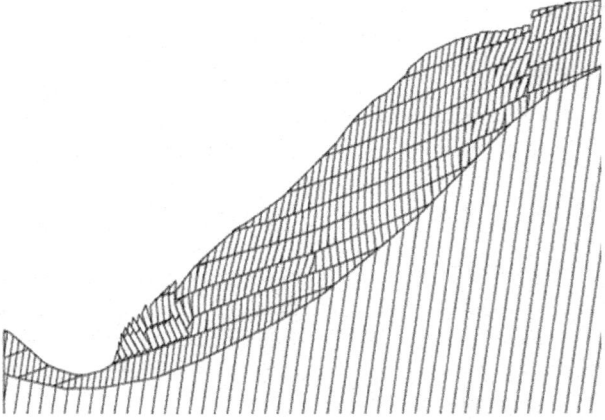

Fig. 9. Slope deformation characteristics (110000 steps).

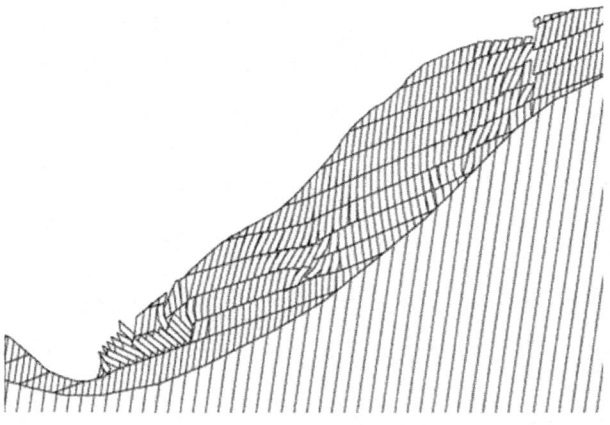

Fig. 10. Slope deformation characteristics (140000 steps).

the displacement of the front edge of the slope is the largest, close to 100 m, and the overall displacement of the slope is above 50 m. Thus, slope deformation terminates.

The stress and strain of the deformed slope are analyzed by finite element numerical simulation. The results show that the stress distribution of the slope is obviously controlled by the gravitational field, and the tensile stress of the slope is about 250 kPa. The minimum principal stress is zero when the slope table is unloaded under the valley. There is a certain degree of stress concentration near the foot and middle crushing zone, and there is a significant stress discontinuity distribution at the location of the fracture zone (see Fig. 12). From the shear strain distribution of the slope (see Fig. 13), it can be concluded that the shear strain mainly occurs in the front and middle crushing zones of the slope, especially in the central crushing zone, which is significantly larger than the surrounding area. And consistent with the results of discrete element analysis.

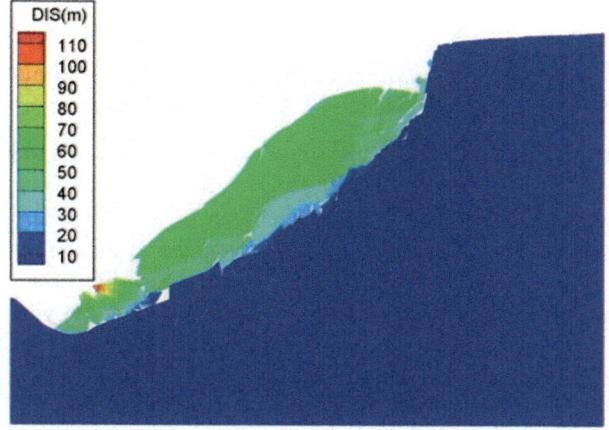

Fig. 11. Slope displacement (250000 steps).

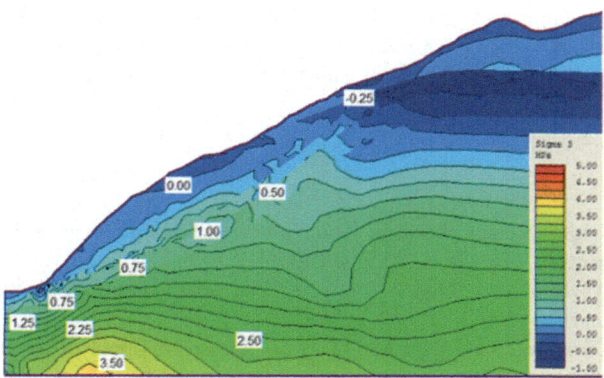

Fig. 12. Counter of σ3.

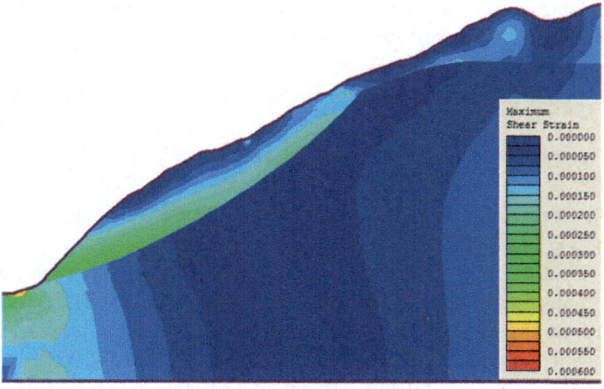

Fig. 13. Distribution of shear stress region.

5 Conclusion

The following conclusions are drawn based on the geological environment conditions, the basic characteristics of the landslide, and the results of numerical simulation.

(1) The toppling-deformation can occur in the steep bedding rock slope. The development of strata inclination and joint group is the main influencing factor.
(2) The formation mechanism of the landslide can be summarized as three stages, namely the slope unloading, the toppling dumping and the sliding-tension.
(3) The formation mechanism of landslide shows the composite failure mode of slope changes from toppling-deformation to slip-tension. It is shown that under the condition of certain slope structures, the slope has a compound instability mode, and different instability modes are distributed in different stages of slope development.

Funding. The authors would like to thank financial support provided by the Research on construction conditions of the countries in "The Belt and Road", the National Key Technology Support Program of China (No. 2013BAJ06B00), the National Key Research and Development Program of China (No. 2017YFD0800501), the Science and Technology Coordination and Innovation Project of Shaanxi Province in China (No. 2016KTZDSF03-02).

References

1. Runqiu, H.: Large scale landslides and their sliding mechanisms in China since 20th century. Chin. J. Rock Mech. Eng. **26**(3), 433–454 (2007)
2. Shang, J.: Analysis on Slope Stability of Jin-chuan Hydropower Station. Lanzhou University, Lanzhou (2007)
3. Zhetao, W.: Unstable Slopes in the Main City of Qing-chuan County. Chengdu University of Technology, Chengdu (2010)
4. Sun, Y., Wang, Y.: The Characteristics and genetic mechanism analysis of landslide of one hydroelectric power station of its water transfer way on Southwest. Res. Soil Water Conserv. **14**(4), 305–308 (2007)
5. Zhongkai, D., Qingyu, Z., Hai, L., et al.: Evolution mechanism and stability prediction of a reservoir bank landslide. Gansu Water Resour. Hydropower Technol. **46**(11), 10–11 (2010)
6. Guangming, R., Min, X., Guo, L., et al.: Study on toppling deformation and failure characteristics of steep bedding rock slope. Chin. J. Rock Mech. Eng. **28**(S1), 3193–3200 (2009)
7. Osaka, O.: A toppled structure with sliding in the Siwalik Hills. Nepal. Eng. Geol. **64**(4), 339–350 (2002)
8. Mcaffee, R.P., Cruden, D.M.: Landslides at Rock Glacier Site, Highwood Pass, Alberta. Can. Geotech. **33**, 685–695 (1996)
9. Li., Y., Dong, Z., Zhou, H.: Analysis of the cause of East Tang Jiawan ancient landslide and its modern reviving mechanism. Chin. J. Undergr. Space Eng. **8**(3), 652–658 (2012)
10. Guangming, R., Dexin, N., Gao, L.: Studies on deformation and failure properties of anti-dip rock mass slope. Chin. J. Rock Mech. Eng. **22**(S2), 2707–2710 (2003)
11. Baocheng, Z., Congxin, C., Xiaowei, L., et al.: Modeling experiment study on failure mechanism of counter-tilt rock slope. Chin. J. Rock Mech. Eng. **24**(19), 3505–3511 (2005)

12. Rujiao, T., Xuchao, Y., Ruilin, H., et al.: Numerical simulation of deformation process and failure mechanism of reservoir bank slope with large-scale anti-dip rock mass. J. Eng. Geol. **17**(4), 476–480 (2009)
13. Xiaoping, Z., Yusheng, L., Xiaobing, C., et al.: Numerical simulation for the problem of project slope toppling deformation on the right of a hydropower station dam abutment in Lancang river. J. Eng. Geol. **16**(3), 298–303 (2008)
14. Genlan, Y., Runqiu, H., Ming, Y., et al.: Engineering geological study on a large scale topping deformation at Xiaowan hydropower station. J. Eng. Geol. **14**(2), 165–171 (2006)
15. Beichuan, H., Sijing, W.: Mechanism for toppling deformation of slope and analysis of influencing factors on it. J. Eng. Geol. **7**(3), 213–217 (1999)
16. Boren, W., Kunjun, Y., Guolin, P., et al.: Failure mechanism of Yangtai landslide in mountain area of South Anhui province. J. Eng. Geol. **21**(2), 304–310 (2013)

Tectonics Dominated Hypogene Karst Geo-Hazard Mechanism Models in China's Tunnel Construction

Jianxiu Wang[1,2,3(✉)], Linbo Wu[1], Wuji Liu[1], Tao Cui[1], Yu Zhao[1], Yao Yin[1], and Xiaotian Liu[1]

[1] College of Civil Engineering, Tongji University, Shanghai 200092, China
wang_jianxiu@163.com
[2] Key Laboratory of Geotechnical and Underground Engineering of Ministry of Education, Tongji University, Shanghai 200092, China
[3] CCCC Key Laboratory of Environment Protection and Safety in Foundation Engineering of Transportation, Guangzhou 510230, China

Abstract. Hypogene karst was believed to develop only in small scale in China. However, in construction of China's deep tunnels, large-scale hypogene karst and corresponding geo-hazards were encountered frequently. The hypogene karst uncovered by deep tunnel construction probably induce water inrushes, mud bursts, karst collapses and even geo-hazards chains. Understanding the potential geo-hazard mechanism induced by hypogene karst is significant for both karst tunnel construction and hypogene karst evolution research. Typical cases of hypogene karst geo-hazards encountered during construction of China's deep tunnels such as Zhongliangshan tunnel group, Yuanliangshan tunnel, Yesanguan tunnel and Dayaoshan tunnel, were presented. Large-scale fold controlling model, large-scale water-rich fault model and complex tectonic deformation model based on tectonic deformation patterns and intensity were proposed and analyzed for the mechanisms of geo-hazards of hypogene karst in the tunnel construction. It can act as a model reference for hypogene karst geo-hazards prevention and control in karst tunnel construction.

Keywords: Hypogene karst · Deep tunnel · Geo-hazard chain
Mechanism mode

1 Introduction

Hypogene karst is generally considered to be formed at depth where dissolution power of the water is produced independent of the base level of erosion [1, 2]. In China, the phenomenon has been studied since 1970s, it was once believed that the scale of hypogene karst cannot be very large [3], the common base level of karst erosion should be at the depth of 350 m, which means that the karst below 350 m belongs to hypogene karst. However, the deep karst conduits discovered in Dayaoshan tunnel in Guangdong Province [4] indicate that the scale of hypogene karst can be large, and the depth of the base level of karst erosion is different in different areas.

© Springer Nature Singapore Pte Ltd. 2018
A. Farid and H. Chen (Eds.): GSIC 2018, *Proceedings of GeoShanghai 2018*
International Conference: Geoenvironment and Geohazard, pp. 190–199, 2018.
https://doi.org/10.1007/978-981-13-0128-5_22

Hypogene karst is acquiring more and more concerns. The geo-hazards of hypogene karst in tunnel constructions are reported frequently recently. Intense water inrushes and collapses out-broke, followed by subsidence on the ground surface, during the construction of the Sol-an tunnel, the longest railroad tunnel in South Korea [5]. In China, geo-hazards of hypogene karst in tunnel, such as Qiyueshan tunnel in Sichuan province [6], Zhongliangshan tunnel groups in Chongqing Municipality [7], Yesanguan tunnel [8] in Hubei Province, have been reported. These geo-hazards usually present as excessive water inrushes, surrounding rocks collapses, loss and even exhaustion of water resource in the engineering area and its adjacent areas, subsidence on the ground surface etc.

2 Typical Cases of Hypogene Karst Geo-Hazards in Tunnel

2.1 Zhongliangshan Tunnel Group

Zhongliangshan-Geleshan area in Chongqing Municipality holds low mountain geomorphology, and the main tectonic feature in this area is the Guanyinshan anticline (Fig. 1a). Tunnels, such as the Geleshan tunnel in Chongqing-Huaihua Railway, the Zhongliangshan tunnel in Chengdu-Chongqing Railway, the Zhongliangshan tunnel in Chognqing-Suichuan Highway, the Zhongliangshan tunnel in Xiangyang-Chongqing Railway, the Zhongliangshan tunnel in Chongqing Rail Transit Line 6, have been constructed through the Guanyinshan anticline, they are called the Zhongliangshan tunnel group. The Zhongliangshan-Geleshan area has been highly karstified, soluble carbonate rocks in both lower Triassic Jialingjiang Formation and middle Triassic Leikoupo Formation constitute the principal karst zone and groundwater enrichment zone, which means well connected karst conduits and caves probably exist inside these soluble carbonate rocks. The elevations of the tunnels are between 250 m and 350 m. They are far lower than the groundwater level (Fig. 1a). However, lots of geo-hazards induced by hypogene karst, such as tunnel water inrushes (Fig. 1b), mud bursts, collapses, exhaustion of water resource (Fig. 1c), subsidence on the ground surface (Fig. 1c) were reported during tunnel constructions in the area [9, 12].

Forty eight points of water inrushes broke out in the Zhongliangshan tunnel of Xiangyang-Chongqing Railway after heavy rainstorm from 31 July to 1 August 1980 [7]. These water inrushes indicated that the groundwater system in hypogene zone links with the surface water system tightly in this area, and this close connection enables the surface water system to recharge the groundwater system. Therefore, insufficient drainage during tunnel construction can evoke intense water inrushes, while excessive drainage probably cause exhaustion of surface water resource.

During construction of the Zhongliangshan tunnel of Xiangyang-Chongqing Railway, serious subsidence on the ground surface in the engineering area and its adjacent areas occurred as well. 29 subsidence points with a total area of 2319 m^2 were reported. Among the collapse sinkholes, depth of the deepest one reached 13 m, the area of the largest one reached 400 m^2 [7]. These phenomena were generated by large dewatering cones formed after plenty of surface water and groundwater getting drained. Since groundwater level in Zhonglaingshan-Geleshan area usually is very shallow,

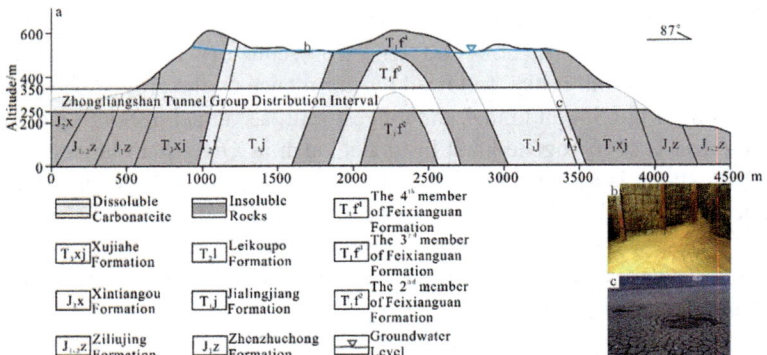

Fig. 1. Hypogene karst geohazards during the construction of Zhongliangshan tunnel group. (a) Engineering cross-section of the Guanyinxia anticline (modified after [9]). (b) Water inrush occurred during the construction of Zhongliangshan tunnel in Chongqing Rail Transit Line 6 [10]. (C) Water exhaustion and subsidence on the ground surface at Liujiawan Reservoir during the construction of the Zhongliagnshan tunnel group [11].

resulting in many springs developing around the surface, formation of those dewatering cones leads to intensive increase in effective stress inside solum and base rocks, and this means that instability both of the solum and karst rocks becomes exacerbated. Consequently, surface collapses of large scale get evoked.

2.2   Yuanliangshan Tunnel

The Yuanlaingshan tunnel is the longest tunnel in the Chongqing-Huaihua Railway. Its total length is 11.07 km long with 7.10 km through soluble limestone strata. The tunnel is divided into the Maoba syncline section, the Tongmaling anticline and the Lengshuihe shallow buried section based on physiognomy along the tunnel. The landform is controlled by tectonic deformations. During construction, five large-scale karst caves with high-pressure fillings were encountered, and many intense water inrushes, mud bursts resulted in great loss [13].

Among the five discovered large-scale karst caves, three are located in the Maoba syncline section (Fig. 2). The 1# and the 3# high-pressure water filled cave are developed inside the NNE limb and the SSW limb respectively, and the 2# cave presents near the syncline axis. The 2# cave is the largest one, and contains many especially large-size limestone blocks, which implies that large-scale karst collapse once occurred near the syncline axis, and that the stability of rock mass near the syncline axis is especially low. The reason for these karst phenomena and geo-hazards inside the syncline section can be analyzed from three primary aspects. Firstly, the Maoba syncline shows high pressure and rich in groundwater, underground rivers and spring groups can be found in the altitude of 850–900 m, constituting the base level of drainage. In the tunnel, the hydrostatic pressure of confined groundwater can be as high as 4.42–4.60 MPa [15]. Secondly, both fissures produced by bedding decollement and extensional fractures develop inside the strata by the folding process are conducive to migration of groundwater. Finally, a series of high-dip angle faults with good per-meability, water head difference between these faults and aforementioned extensional fractures and fissures work together, resulting in inverted siphon process [13], and this process enables groundwater to reach deep strata, and accelerates circulation of groundwater in hypogene karst zone. Long-term karstification can lead to significant increase in permeability of carbonate rocks [16]. Consequently, large-scale hypogene karst phenomena can develop in Maoba syncline (Fig. 2). And if tunnel constructions need to cross these karst-caves or disturb them intensively, excessive water inrushes, mud bursts and even karst collapses can be provoked.

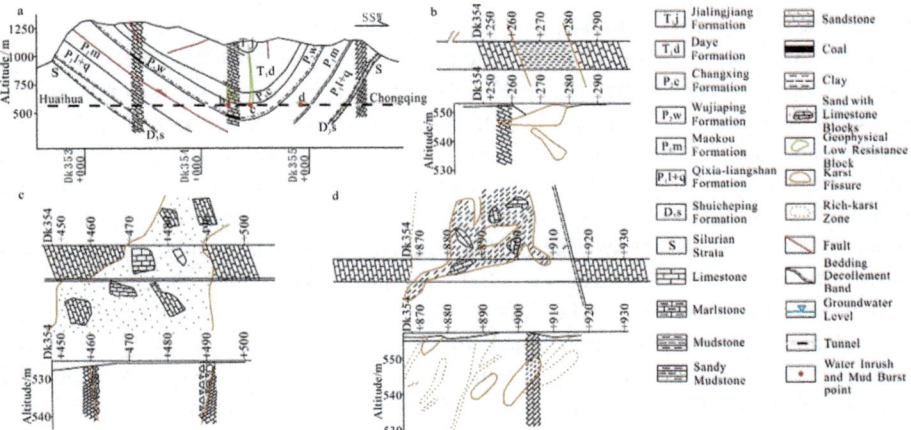

Fig. 2. Hypogene karst geohazards during the construction of the Maoba syncline section of Yuanliangshan tunnel (modified after [14]). (a) Engineering cross-section of the Maoba syncline. (b) Plane graph and cross-section of the #1 filled karst cave. (c) Plane graph and cross-section of the #2 filled karst cave. (d) Plane graph and cross-section of the #3 filled karst cave.

2.3 Yesanguan Tunnel

The Yesanguan tunnel is 13.1 km long, and it is the longest one in the Yichang-Wanzhou Railway. Serious water inrush, mud burst and collapse occurred during construction on 5 August, 2007 at the Kutaoxi-Taoziping section of the tunnel (Fig. 3).

Strata in the Kutaoxi-Taoziping section are carbonate rocks of varying solubility, and the elevation of groundwater level is 200 m higher than the tunnel in this section (Fig. 3a). Tracing tests of karst water in the Kutaoxi-Taoziping section [8] show that karst water in the Shuidongping karst depression is the primary source for the water inrush on 5 August, 2007. Sediments of the water inrush and mud burst from the interior toward the exterior, ordinally turn out to be big limestone blocks (larger than 0.25 m^3) of lower Permian Qixia Formation, muddy or carbonaceous semi-solificated breccia blocks, sands and gravels formed in underground karst river, and limestone blocks of lower Permian Qixia Formation again. Around the geo-hazard points, regional transpressional faults F_{17}, F_{18} both cross with the tunnel at large angles

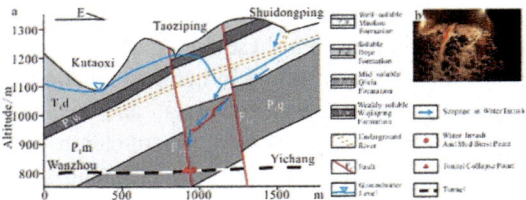

Fig. 3. Hypogene karst geohazards during the construction of the Kutaoxi–Taoziping section of Yesanguan tunnel. a Engineering cross-section of the Kutaoxi–Taoziping section (modified after [8]). b Water inrush that occurred during the construction of the Yesanguan tunnel [17].

(Fig. 3a), resulting in rock-masses between them fracture seriously, and fault F_{18} turns out to be strongly permeable [18]. Additionally, another regional Wangbei fault (F_{13}) also develops near the tunnel in the Kutaoxi-Taoziping section, but extends almost parallel to the tunnel and inclines to it, effectively promoting connections among these fractures and faults. Thus karst water can migrate smoothly in those fractures and faults. These materials imply that karst water from Shuidongping karst depression, after entering the Zhoujiabao underground river, can migrate to fault F_{17}, F_{18} along the boundary zone between the well permeable lower Permian Maojiaping Formation and the medium-permeable lower Permian Qixia Formation. Then when the tunnel crosses the fault F_{18}, water inrushes can be evoked easily, since this water-rich fault can import tons of karst water from aforementioned sources into the tunnel. Meanwhile, rich fine fillings produced by karstification and faulting can also burst into the tunnel along with karst water. Consequently, instability of the country rocks increases intensively, resulting in large-scale collapse.

2.4   Dayaoshan Tunnel

The 14.295 km long Dayaoshan tunnel in the Beijing-Guangzhou Railway crosses the NNE extending Banguao karst basin (Fig. 4). During its construction, many times of damaging water inrushes, mud bursts, collapses, subsidence on the ground surface have taken great loss of life and property [21, 22].

The Banguao basin, actually, is a large-scale syncline, strata in its hinge zone are mainly carbonate rocks, and those of the two limbs are mainly clasolites such as quartz sandstone, conglomerate. Its cymbiform landform helps to collect surface water, and surface water resource is very rich in this area with an average annual rainfall of 1500 mm [19]. As a result, level of groundwater is very shallow, and numbers of springs can be found on surface of the hinge zone (Fig. 4b).

Tectonic deformations in Banguao basin are very complex, the three high-dip angle thrust fault F_5, F_8, F_9 are the principal regional faults. And fault F_5, F_9 act as the boundaries between carbonate rocks in the hinge zone and other rocks in the two limbs. Layers between fault F_8 and F_9 show intensive folding deformations, accompanying with dense joints. Thus fractures and fissures in rock masses between the faults, especially those between fault F_8 and fault F_9 develop well, resulting in strong permeability [23] and rich groundwater inside these fault zones. Groundwater tests in the tunnel show that groundwater between fault F_8 and F_9 is karst cave water, and that

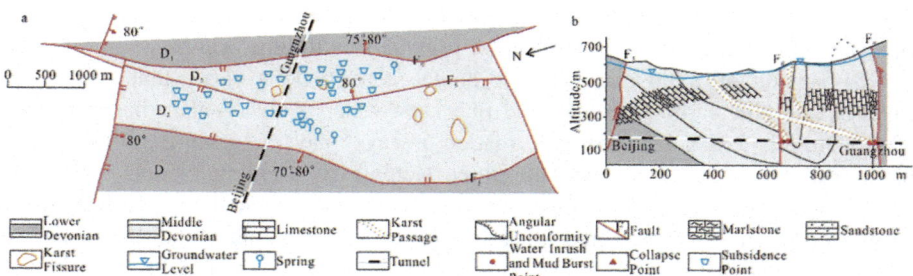

Fig. 4. Hypogene karst geohazards during the construction of the Banguao syncline section of Dayaoshan tunnel. (a) Engineering plane-section of the Banguao syncline (modified after [19]). (b) Engineering cross-section of the Banguao syncline (modified after [20]).

groundwater of the other sides of the faults is fracture water and fault water [20]. Hydrogen isotope tests show that about 75% groundwater stored inside the carbonatite rocks in this area is formed during recent 51 years [19]. Salt trace tests further indicate that karst conduits in the hypogene karst zone connect well with subsidence points on the surface through karst passages such as karst funnels, sinkholes (Fig. 4b) [20]. Thus the surface water system can replenish the deep groundwater system rapidly in this section, which means that water inrushes and mud bursts can be evoked easily when the tunnel crosses the Banguao basin.

During tunnel construction in the Banguao basin, large-scale precipitation inside the soil and fractured rock masses generated by drainage and water inrushes, can lead to serious instability of the solum and rock masses due to intensive decrease in effective stress. Besides, strong migration of groundwater inside the faults and fractures can wash away amounts of cements inside the rock masses and soil. These processes result in large-scale subsidence points on the surface and serious collapse in the tunnel, especially in the zone between fault F_8 and fault F_9 (Fig. 4a).

3   Hypogene Karst Modes Generating Tunnel Geo-Hazards

Through analyzing the aforementioned typical geo-hazard cases of hypogene karst in tunnel constructions [24, 25], water inrushes, mud bursts, tunnel collapses, loss of surface water resources and subsidence on the surface present as the typical geo-hazards during tunnel constructions in hypogene karst zones. These geo-hazards always lead to project delay, many casualties, property losses, and even deterioration of ecosystem, which may damage society stability seriously. During the construction of the Zhongliangshan tunnel group, excessive loss of both surface water and groundwater resources, and large-scale subsidence on the ground surface were generated, consequently, drinking water problems and housing dangers forced residents in the area to rally at the engineering zones and to protest [11]. And for excessive loss of karst water was caused by drainage during construction of the Huayingshan tunnel in Guangan-Chongqing Railway, poor-quality water in the coal layers intruded into ground water aquifers, resulting in quality of groundwater deteriorating intensively

[26]. Therefore, to take measures to deal with these geo-hazards effectively and eco-nomically, it is necessary to determine the primary controlling factors.

Geo-hazards of hypogene karst in tunnel constructions show that coexist of large-scale soluble carbonate rocks and rich karst water is always their basic precon-dition. On this basis, tectonic deformation always plays a leading role in these geo-hazard formation processes. Thus three tectonics controlling models were sug-gested to characterize formation mechanisms of typical hypogene karst geo-hazards in tunnel constructions. The three models are large-scale fold controlling model, large-scale water-rich controlling model and complex tectonic deformation controlling model respectively (Fig. 5a–c).

In large-scale fold controlling model of hypogene karst (Fig. 5a), a tunnel crosses deep karst zone, and general tectonic deformations of the surrounding rocks are dominated by large-scale folds. Amounts of decollement fractures develop well during folding process, and these fractures can act as good passages for groundwater migra-tion, making large-area karstification below the base level of erosion possible. As a result, water inrushes and mud bursts in hypogene karst zone can be evoked when tunnel constructions disturb the surrounding rocks intensively (Fig. 5d). Large-scale water inrushes and excessive drainage in deep karst tunnel means serious water loss of both upper groundwater and surface water resource, and this can result in intensive instability of solum and karst rocks when the initial level of groundwater is very shallow, causing subsidence on the surface (Fig. 5f). Geo-hazards of hypogene karst occurring during tunnels construction in large-scale anticline area, e.g. those in the Zhongliang-shan tunnel group, respond just like so. In large-scale syncline area, backsiphonage would arise when high water-head difference between water levels of the two limbs exists, which can accelerate migration of groundwater, resulting in large-scale deep karst phenomena. Besides, extensional joints usually develop well in hinge zone of synclines, which can not only result in groundwater migration, but also lower stability of rock masses in the hinge zone. Hence, water inrushes, mud bursts and karst collapses all occur frequently in fold hinge zone of hypogene karst during tunnel constructions (Fig. 5e). Actually, many regional folds, e.g. the Jura-type folds in Hubei, Sichuan Province and Chongqing Municipality, formed together with sets of high-dip angle faults (Fig. 5a, F1 and F2). And these faults, cooperating with decollement fractures, can constitute backsiphonage model when they hold high per-meability [27], facilitating hypogene karstification inside synclines. Typically, the Maoba syncline section of Yuanliangshan tunnel in Chongqing-Huaihua Railway develops large-scale hypo-gene karst by this kind of backsiphonage process [14].

In large-scale water-rich fault controlling model of hypogene karst (Fig. 5b), high-permeable faults connected with surface (Fig. 5b, F3 and F4) import karst water from surface or underground-river into deep karst zone. Then water inrushes and mud bursts can happen if the rock masses are disturbed intensively by tunnel construction. And tunnel collapses may break out on the condition that hypogene karstification develops well and the rock masses are fractured intensively by faulting process (Fig. 5g), just as the case of Yesanguan tunnel (Fig. 3). Large-scale water replenish-ment into deep karst zone through faults results in excessive loss of surface water resource, and subsidence on the ground surface where the initial groundwater level locates near the surface.

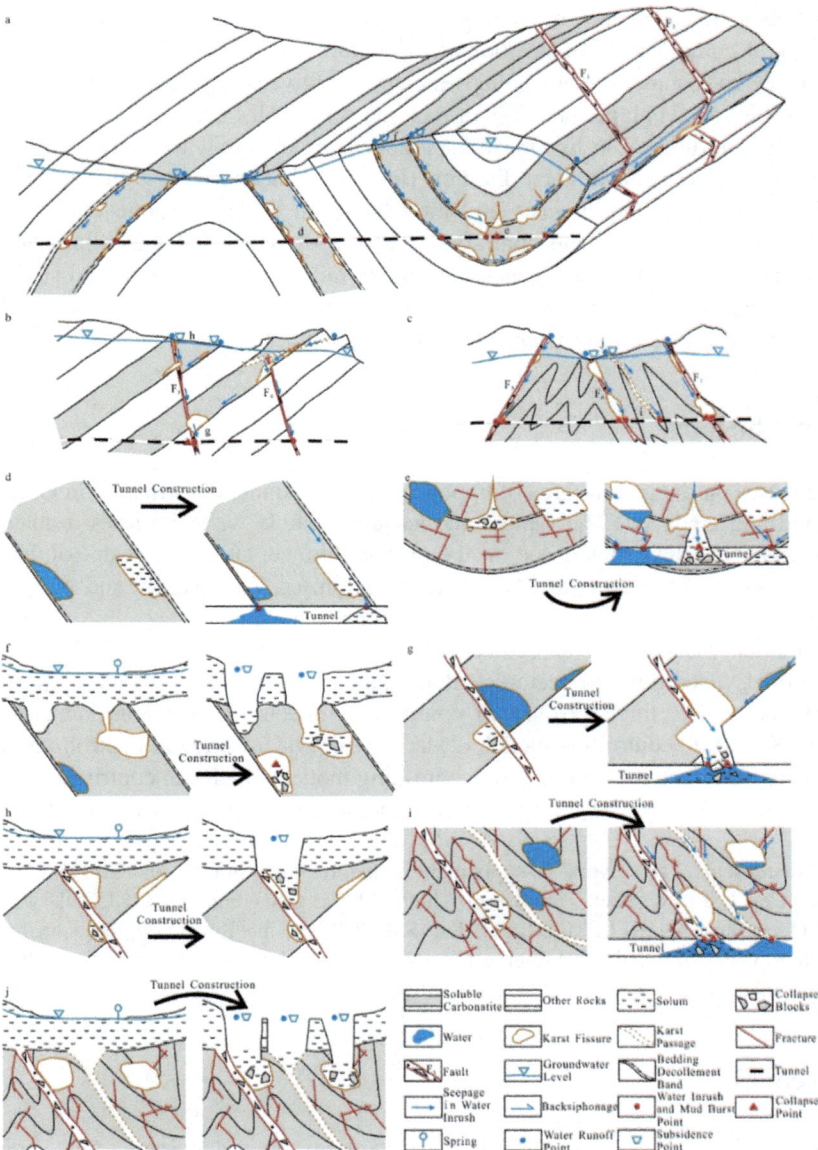

Fig. 5. Formation mechanism models of typical geo-hazards in hypogene karst zones during tunnel construction in China. (a) Large-scale fold-controlling model. (b) Large-scale water-rich fault-controlling model. (c) Complex tectonic deformation-controlling model. (d) Geo-hazards occurred in the limb of fold. (e) Geo-hazards occurred in the hinge of fold. (f) Geo-hazards occurred on the ground surface of folding area. (g) Geo-hazards occurred inside tunnel. (h) Geo-hazards occurred on the ground surface. (i) Geo-hazards occurred inside tunnel. (j) Geo-hazards occurred on the ground surface.

In hypogene karst area of complex tectonics, the surrounding rocks of tunnel usually hold complicated compositions and fracture intensively, resulting in low stability of the rock masses. Thus collapses can be motivated easily during tunnel construction (Fig. 5i). High permeability of rock masses enables that karst fissures develop abundantly, and tunnels in hypogene karst zone can connect with upper groundwater through faults, joints and karst fissures (Fig. 5c). Therefore, insufficient drainage probably leads to water inrushes and mud bursts in hypogene zone during tunnel construction (Fig. 5i). On the contrary, excessive drainage and water inrushes can result in conspicuous loss of upper groundwater and subsidence on ground surface in water-rich area (Fig. 5i), just like geo-hazards occurring during construction of Dayaoshan tunnel in the Beijing-Guangzhou Railway (Fig. 4a).

4   Conclusions

Typical cases on the geo-hazards of hypogene karst in tunnel construction in China are presented and analyzed. The typical geo-hazards include water inrushes, tunnel collapses, water-loss and subsidence on the ground surface, and large-scale soluble carbonate rocks and abundant karst water are the premises. The features and intensity of tectonic deformation always act as the controlling factors.

Formation mechanisms of geo-hazards in hypogene karst zone are divided into three models according to tectonics deformation patterns and intensity of tunnel country rocks. The three geo-hazard controlling models of hypogene karst include (1) large-scale fold controlling model, (2) large-scale water-rich fault controlling model and (3) complex tectonic deformation controlling model, which can contribute much to prevention and control for geo-hazards of hypogene karst in tunnel construction.

Acknowledgments. This work is supported by the research grant (2014CB046900) from National Key Basic Research Program of China, GDUE Open Funding (SKLGDUEK1417), LSMP Open Funding (KLLSMP201403, KLLSMP201404), the Key Discipline Construction Program of Shanghai (Geological Engineering, No. B308) and the Foundation of China Railway No. 2 Engineering Group Co., Ltd. (No. 201218).

References

1. Xu, M., Mao, B.Y., Zhang, Q.: Modern deep karst progress and prospects. Adv. Earth Sci. **05**, 495–500 (2008)
2. Dublyansky, Y.V., Klimchouk, A.B., Spötl, C., Timokhina, E.I., Amelichev, G.N.: Isotope wall rock alteration associated with hypopene karst of the Crimean Piedmont Ukraine. Chem. Geol. **377**, 31–44 (2014)
3. Karst Research group in Institute of Geology, Chinese Academy of Sciences (KGIGCAC). China Karst Research. Science Press, Beijing (1979)
4. Xu, J.C., Huang, S.X.: Mechanism of burst mud and spring water of the Dayaoshan tunnel. J. Railw. Eng. Soc. **02**, 83–89 (1996)
5. Song, K.I., Cho, G.C., Chang, S.B.: Identification, remediation, and analysis of karst sinkholes in the longest railroad tunnel in South Korea. Eng. Geol. **135–136**, 92–95 (2012)

6. Wang, W., Miao, D.H.: Characteristic and engineering countermeasures of F11 high pressure rich water fault in Qiyueshan tunnel in Wanzhou railway. Railw. Stand. Des. **08**, 81–86 (2010)
7. Fu, K.L.: Analysis of karst ground collapse and water inflow in Zhongliangshan tunnel in Yusui highway. Hydrol. Eng. Geol. **02**, 107–110 (2005)
8. Wu, L., Wan, J.W., Chen, G., Zhao, L.: Causes of Yesanguan Wanzhou railway tunnel "8.5" water burst accident. Carsl. Sin. **2**, 212–218 (2009)
9. Gong, R.: Study on the impact of tunnel construction on the groundwater environment of the partition style karst water-rich anticline-a case study of Guanyin Gorge anticline. Chengdu University of Technology, Chengdu (2010)
10. Liu, C.M.: The study of water and sand inrush disaster and prevention method in tunnel engineering. Chongqing Jiaotong University, Chongqing (2013)
11. Something about Geleshan tunnel construction tianya BBS.cn. http://bbs.tianya.cn/post-no04-2290437-1.shtml. Accessed 12 Sept 2017
12. Zhao, J.F.: Tunnel construction inflow impact on the surrounding groundwater system and environmental effects. Chengdu University of Technology, Chengdu (2004)
13. Zhang, M.Q., Liu, Z.W.: The analysis on the features of karst water burst in the Yuanliangshan tunnel. Chin. J. Geotech. Eng. **04**, 422–426 (2005)
14. Zhang, N.Y., Jiang, L.W.: A typical case history of infilled karstic caves formed by reverse siphonic circulation in deep phreatic zone—preliminary analysis of the formation mechanism of infilled huge caves exposed by Yuanliangshan tunneling in deep phreatic zone of Maoba syncline. J. Eng. Geol. **04**, 455–469 (2010)
15. Liu, Z.W., He, M.C., Wang, S.R.: Study on karst water-burst mechanism and prevention countermeasures in Yuanliangshan tunnel. Rock Soil. Mech. **2**, 228–232, 246 (2006)
16. Wang, J.X., Zhu, H.H., Yang, L.Z.: Influence of long-term draining groundwater in karst tunnel on permeability of its surrounding rock. Rock. Soil Mech. **05**, 715–718 (2004)
17. A large flooding accident suddenly occurred in Yichang-Wanzhou Railway Yesanguan tunnel xinhuanet.com. http://www.hb.xinhuanet.com/zhuanti/ysgts.htm. Accessed 16 Sept 2017
18. Xu, H.X.: The research on geological characteristic and water inrush prevention in construction of Yesanguan tunnels. China University of Geosciences, Beijing (2010)
19. Chen, C.Z., Mou, R.F.: Analysis of karst water inflow system in Dayaoshan tunnel. J. Eng. Geol. **01**, 36–46 (1993)
20. Zhang, K.C., Dou, P.S., Mou, R.F., Su, J., Shu, W.: Experimental study of karst gushing connectivity in Dyaoshan tunnel. Geotech Invest Sur 01:35–39, 34 (1992)
21. Shu, W., Wang, S.C.: Karst inflow and ground subsidence problem of Dayaoshan tunnel. J. Railw. Eng. Soc. **2**, 145–151 (1989)
22. Wang, J.X.: Theoretical and applied analysis on hydrochemical-hydraulic damage in tunnel surrounding rock. Southwest Jiaotong University, Chengdu (2002)
23. Liao, S.Y.: Remediation methods of Dayaoshan tunnel seepage. Railw. Eng. **02**, 32–33 (2004)
24. Wang, Y.C.: Mechanism and control measures of collapse of mountain tunnel. Zhejiang University, Hangzhou (2010)
25. Wang, G.B.: Study on catastrophe evolution mechanism of karst water inrush in Wuchiba tunnel of Lurong Expressway. China University of Geosciences, Wuhan (2012)
26. Liu, D., Yang, L.Z., Yu, S.J.: On ecological environment problems and effects caused by discharge from Huayingshan tunnel. J. Southwest Jiaotong Univ. **03**, 308–313 (2001)
27. Han, J.L., Wu, S.R., Wang, H.B.: Geological disasters chain. Earth Sci. Front. **06**, 11–23 (2007)

Vulnerability Analysis of Shanghai Metro Network Under Water Level Rise

Yanjie Zhang[1(✉)], H. W. Huang[1], D. M. Zhang[1], and B. M. Ayyub[2]

[1] Tongji University, Shanghai 200092, China
7zhangyj@tongji.edu.cn
[2] University of Maryland, College Park, MD 20742, USA

Abstract. Shanghai, as an estuarine and coastal metropolitan city, is physically and socio-economically vulnerable to typhoon induced flooding. Water level rise is proposed as the summation of sea level rise due to global warming, land subsidence and storm surges. Water level rise in Shanghai region will threaten the safety of its underground space and impact the efficient operation of its metro system. This study utilized the ArcGIS flood model to assess the potential impact of water level rise on the metro system of Shanghai by generating inundation maps under projected 1 m, 2 m, 3 m, 4 m and 5 m WLR scenarios. On this basis, quantitative vulnerability analyses were conducted according to the change of topological structure of Shanghai metro network and corresponding network global efficiency with the increase of water level rise. The result shows that when water level rises over 4 m, the integrity and connectivity of Shanghai metro network would be significantly disrupted. This study has identified vulnerable areas, metro stations and metro lines, therefore, could provide references for future prevention and mitigation of flooding damages.

Keywords: Water level rise · Metro network · Vulnerability analysis
Shanghai

1 Introduction

Shanghai is a typical estuarine and coastal city, located at Yangtze Estuary and at the edge of the East China Sea. Its average altitude is 4 m according to Wusong Elevation System. With a land area of 6,340 km^2 and a population of around 24 million, Shanghai is physically and socio-economically vulnerable to flooding risks. To assess the vulnerability of flood disaster in Shanghai area, it is significant to focus on the relative sea level rise, which is mainly the combined result of eustatic sea level rise due to global warming and land subsidence caused by the groundwater extraction [1]. Shanghai is also frequently affected by flooding disasters due to extreme typhoon storm surges [2]. Storms are expected to become more violent in terms of rates and intensities due to rising sea level rise [3, 4]. Consequently, the degree of flooding disaster will be strengthened when storm surges coincide with high tides resulting from rising relative sea level. Therefore, Water Level Rise (hereafter referred to as simply WLR) is adopted herein to measure the combined impact of relative sea level rise superimposed storm surges.

© Springer Nature Singapore Pte Ltd. 2018
A. Farid and H. Chen (Eds.): GSIC 2018, *Proceedings of GeoShanghai 2018*
International Conference: Geoenvironment and Geohazard, pp. 200–207, 2018.
https://doi.org/10.1007/978-981-13-0128-5_23

Currently, the Shanghai Metro system is one of the world's largest rapid transit systems. There are 14 lines and 305 stations in operation, with a route length of more than 580 km by the end of 2016. According to the *Shanghai Comprehensive Traffic Operation Annual Report 2015*, Shanghai metro accounts for 46% of the total passenger volume of city public transportation, with an average daily flow of 8.41 million. Shanghai metro network has become a key component in the public transport system of the city. To guarantee efficient and safe operation of the Shanghai Metro, vulnerability analysis is significant to cope with flooding disasters and manage emergency.

This paper develops a model to quantify the vulnerability of Shanghai metro network against flooding risks due to water level rise. Section 2 provides insight into the study and prediction of water level rise in Shanghai area to reveal the future flooding risks. Section 3 demonstrates the impact of water level rise on Shanghai metro by generating inundation maps under several projected WLR scenarios. Quantitative vulnerability analyses for two hypothetical situations are presented in Sect. 4. Some conclusions are drawn at the end of the paper.

2 Water Level Rise in Shanghai City

Global warming and associated sea level rise have posed a major threat to the city of Shanghai. According to the report "*Mapping Choices: Carbon, Climate, and Rising Seas—Our Global Legacy*", 76 percent of the Shanghai region would eventually be underwater if the Earth warms by 4 °C by 2100. In a view of historical data, during the past 30 years (1978-2007), the mean sea level of the coastline in Shanghai area has risen 115 mm (approx. 3.8 mm/a) according to *China water level bulletin 2015*. It is predicted that Shanghai coastal sea level will rise 75–150 mm in the next 30 years.

Land subsidence directly endangers the safety of urban flood defence by decreasing the land altitude. Long-term monitoring results have revealed that the amount of land subsidence in Shanghai is 1.973 m from 1921 to 2007 [5]. Shanghai is still experiencing subsidence at a yearly average value of 10 mm, which is more than twice the rate of sea level rise.

Considering estuary sea level rise, land subsidence, and tectonic subsidence, water level rise in Shanghai area was projected to be 170 mm by 2030 and 390 mm by 2050 [6]. Extreme storm flood elevations were estimated to reach 7.17 m in 2030 and 7.39 m in 2050 by using a simplified algebraic summation of the water level rise, monitoring maximum storm surge and the maximum astronomical high tide level [6]. Therefore, it is necessary to assess the flooding risks for the city of shanghai.

3 Impact of WLR on Shanghai Metro

In this study, flooding vulnerability of the metro network in Shanghai is assessed by comparing the projected water level to the real elevation of a metro station's entrances and exits. For this purpose, the topography of Shanghai was defined based on the digital elevation data (including longitude, latitude and altitude) of Shanghai area provided by *International Scientific & Technical Data Mirror Site*. The digital elevation data set has a grid resolution of 90 m and an elevation resolution of 1 m. Then the location of each metro station was determined by the geographical coordinates captured from Google map. Subsequently the ground elevation of each station was derived based on the digital elevation data of Shanghai. According to the *Code for Design of Metro (GB 50157-2013)*, the station entrance-exit is generally designed to be 450 mm above its ground. Finally, the elevation of station entrance-exit was obtained by adding 450 mm to the corresponding ground elevation.

Inundation map of Shanghai was generated by ArcGIS 10.1 software. The flood model was built as follows. First, the areas below a projected elevation (i.e., water level rise) were identified. Then the areas connective to the Huangpu River and East China Sea were selected and determined to be the real inundation zone. Finally, the stations within the inundation zone were identified to be flooded.

Five different scenarios of water level rise were made: 1 m, 2 m, 3 m, 4 m and 5 m. As shown in Fig. 1, flooded zones and flooded metro stations are presented by characteristic colors layer by layer. Non-flooded metro stations are denoted as green dots. The scale of inundated area and the number of flooded metro stations increase with the increase of water level rise. According to the results of this flood model, there is one station flooded under 1 m and 2 m WLR, and three under 3 m WLR. Under the WLR of 4 m, nearly half of the Shanghai region would be inundated and the number of flooded stations jumps to 31. When WLR reaches 5 m, most of the city would be affected, except for the central urban area, with a total of 77 stations impacted by flooding. Since most stations are densely located in this area, they are safe from flooding even under 5 m WLR scenario. Table 1 shows the distribution of flooded stations on each metro line under five different WLR scenarios. It is indicated that all of the 14 lines would be affected to some degree under 5 m WLR. Metro Line 11 would have most flood-risk exposure stations, followed by Metro Line 2.

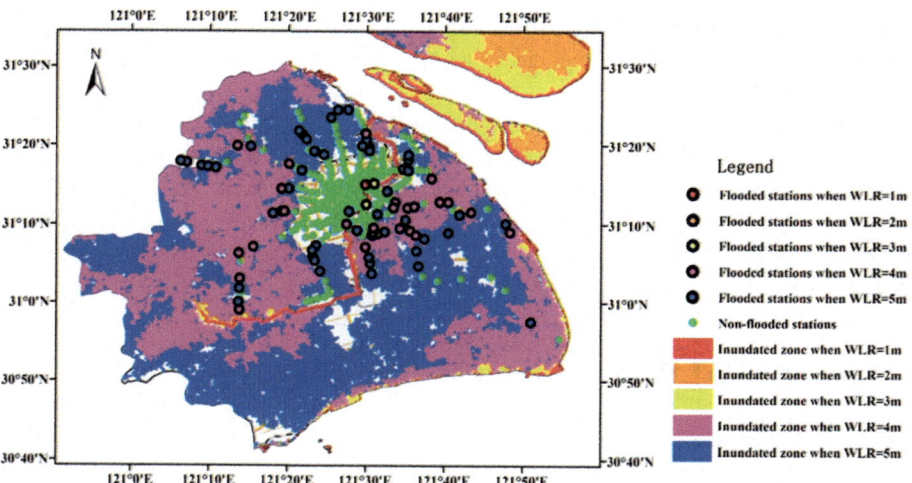

Fig. 1. Inundation map of Shanghai under projected WLR of 1 m, 2 m, 3 m, 4 m and 5 m.

Table 1. Number of flooded stations on each metro line.

Metro line	WLR = 1 m or 2 m	WLR = 3 m	WLR = 4 m	WLR = 5 m
Line 1				3
Line 2			8	12
Line 3			1	5
Line 4		2	2	3
Line 5				4
Line 6			2	6
Line 7			2	7
Line 8			1	5
Line 9			4	6
Line 10			2	4
Line 11			8	19
Line 12	1	1	2	5
Line 13			1	2
Line 16			2	6
Total	1	3	31	77

4 Vulnerability Analysis

4.1 Method

Based on the results of proposed flood model, further vulnerability analysis of Shanghai metro network was then carried out through the calculation of network efficiency. Shanghai metro system is a typical small-world and scale-free network under the L-space topological space [7]. Figure 2 shows the topological structure of Shanghai metro network, in which the black dots denote underground stations and the red ones stand for metro stations on viaduct. Among the 305 metro stations, 215 of them are underground and the other 90 are placed high above ground. The above-ground stations and their links are on urban viaducts. For an underground metro station, if the projected water level is higher than its entrances or exits, the metro tunnel and the station are supposed to be flooded. For a station above ground (i.e., station on viaduct), even though the entrances and exits get inundated, both the station and the metro line won't be affected by the high water level. In this context, two cases were studied in this paper:

- Case I: Suppose the station above ground will still remain in normal operation even though its entrance-exit parts get inundated.
- Case II: Suppose the station above ground will be suspended due to the inundation of its entrance-exit.

According to the theory of complex network, network global efficiency E_f was employed to measure network connectivity. This indicator provides an appropriate

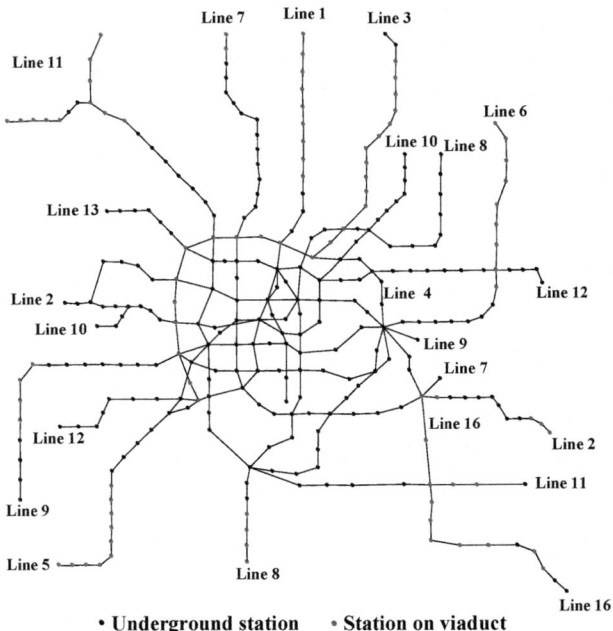

Fig. 2. Topological structure of Shanghai metro network.

basis for assessing network vulnerability [8]. The network global efficiency E_f is defined as follows:

$$E_f = \frac{1}{N(N-1)} \sum_{i \neq j} \frac{1}{d_{ij}} \tag{1}$$

where N represents the number of nodes in a topographical network and d_{ij} denotes the path length between node i and node j. The inverse of d_{ij} essentially reflects the connectivity between the two nodes. The network global efficiency E_f is equal to the mean connectivity over all pairs of nodes in a network.

With the identification of flooded stations under a given WLR, topological mapping of Shanghai metro network was achieved by removing the disrupted nodes and associated links. Further, the network global efficiency E_f of the affected metro network by flooding was calculated.

4.2   Results

On the basis of Case I and Case II proposed above, the disrupted topological structures of Shanghai metro network under different projected WLR scenarios were obtained, as shown in Fig. 3. The disrupted topological structures in Case I are the same as that in Case II when water level rise is no greater than 3 m. The metro network would be more severely impacted in Case II under water level rise of 4 m or 5 m. For both cases,

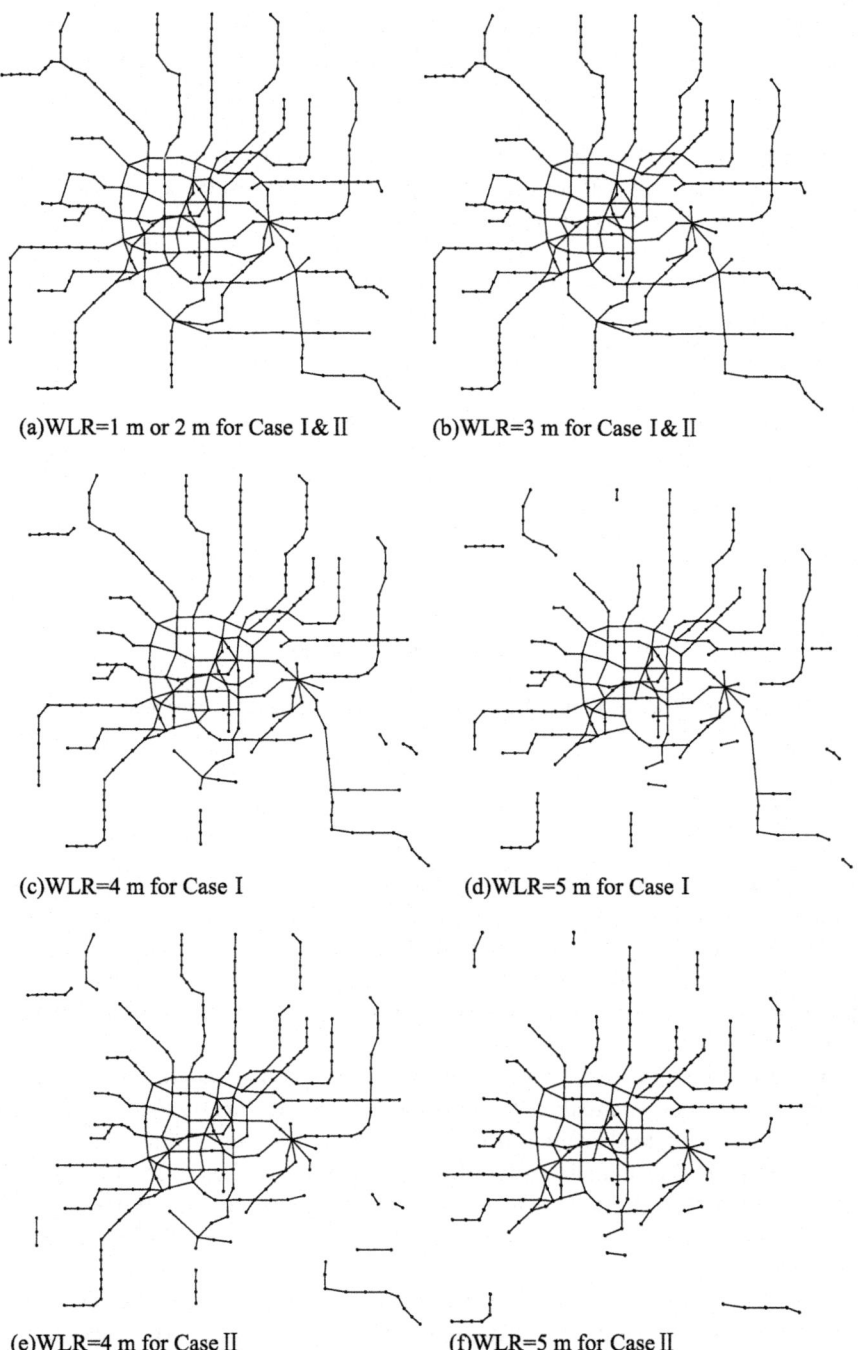

(a)WLR=1 m or 2 m for Case I & II   (b)WLR=3 m for Case I & II

(c)WLR=4 m for Case I   (d)WLR=5 m for Case I

(e)WLR=4 m for Case II   (f)WLR=5 m for Case II

Fig. 3. Topological structure of Shanghai metro network under 1 m, 2 m, 3 m, 4 m, 5 m WLR for Case I and Case II.

With WLR higher than 3 m, the network integrity and connectivity of would be clearly damaged as a result of many stations flooded. Moreover, network global efficiency E_f was computed under each WLR both for Case I and Case II. As seen in Fig. 4, the value of E_f sharply decreases when water level rises over 3 m, felling around 25% at 4 m WLR and around 40% at 5 m WLR for Case I, felling around 30% at 4 m WLR and around 50% at 5 m WLR for Case II, respectively. The result is consistent with the outcome of inundation projection of metro stations.

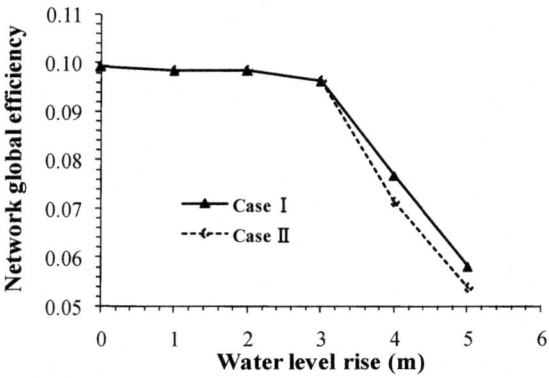

Fig. 4. The change of network global efficiency with the increase of water level rise for Case I and Case II.

5   Conclusion

Shanghai region is experiencing rapid sea level rise and land subsidence, which poses a direct threat to low-lying metro station entrances and subsequently influence the network efficiency of Shanghai metro system. This paper built an inundation model for vulnerability analysis of flooding risk due to water level rise. This model could identify the potential flooded metro stations under a projected WLR and quantify the associated network service efficiency.

Considering the difference in height between underground stations and the stations on viaduct, two response situations (Case I and Case II) were analyzed. Shanghai metro network would be more damaged in Case II under WLR higher than 3 m. For both cases, the network global efficiency would decrease sharply when water level rises 4 m due to the disruption of network integrity and connectivity. Each metro line in Shanghai metro system would be affected to some degree under 5 m WLR, and Line 11 suffers the most.

In addition to the direct impact of flooding, water level rise will also lead to corrosion and deterioration of structural materials through saltwater intrusion into aquifers and destruction of the seawall and drainage system [9, 10], which is beyond the scope of this paper. Feasible mitigation and adaptation strategies should be put forward to cope with the possible threats of land subsidence, sea level rise and associated storm surges. Efforts should be made to strengthen coastal defenses, to improve the drainage system, and to develop and promote the early warning system and emergency response system. Particularly, in order to enhance the resilience of metro system in Shanghai, local planners should account for the inundation situations when considering future development of the metro network.

It should be noted that the topological structure of Shanghai metro network was regarded as an undirected and unweighted network in this paper. Specific weights were not attached to any network node or link in the calculation of network global efficiency. Further study is needed to investigate the WLR vulnerability of metro system by considering weighting factors such as passenger volume.

Acknowledgements. This study is substantially supported by the Natural Science Foundation Committee Program (No. 51608380, 51538009), by Shanghai Rising-Star Program (17QC 1400300) and by Shanghai Science and Technology Committee Project (17DZ1204205). Hereby, the authors are grateful to these programs.

References

1. Cheng, H.Q., Wang, D.M., Chen, J.Y.: Study and prediction of the relative sea level rise in 2030 in Shanghai area. Progressus Inquisitiones de Mutatione Climatis **11**(4), 231–238 (2015)
2. Wang, J., Gao, W., Xu, S.Y., et al.: Evaluation of the combined risk of sea level rise, land subsidence, and storm surges on the coastal areas of Shanghai, China. Clim. Chang. **115**, 537–558 (2012)
3. Nicholls, R.J., Hoozemans, F.M.J., Marchand, M.: Increasing flood risk and wetland losses due to global sea-level rise: regional and global analyses. Glob. Environ. Chang.-Hum. Policy Dimens. **9**, 69–87 (1999)
4. Karim, M.F., Mimura, N.: Impacts of climate change and sea-level rise on cyclonic storm surge floods in Bangladesh. Glob. Environ. Chang.-Hum. Policy Dimens. **18**(3), 490–500 (2008)
5. Gong, S.L., Yang, S.L.: Effect of land subsidence on urban flood prevention engineering in Shanghai. Scientia Geographica Sinica **28**(4), 543–547 (2008)
6. Yin, J., Yin, Z.E., Yu, D.P., et al.: Hazard analysis of extreme storm flooding in the context of sea level rise: a case study of Huangpu river basin. Geogr. Res. **32**(12), 2215–2221 (2013)
7. Du, F., Huang, H.W., Zhang, D.M., et al.: Analysis of characteristics of complex network and robustness in Shanghai metro network. Eng. J. Wuhan Univ. **49**, 701–707 (2016)
8. Latora, V., Marchiori, M.: Is Boston subway a small-world network. Physical **A314**, 109–113 (2002)
9. Cai, F., Su, X.Z., Liu, J.H.: Coastal erosion in China under the condition of global climate change and measures for its prevention. Prog. Nat. Sci. **19**, 415–426 (2009)
10. Gu, Y.B., Zheng, X.Q., Xu, L.L., et al.: Risk survey research on seawater intrusion in Shanghai. J. Ocean Tech. **34**(6), 108–111 (2015)

Ground Response Analysis of Ahmedabad Region to Measure Earthquake Hazard Assessment

Tejas P. Thaker[1]([⊠]), Raviraj Dave[2], and Chirag Joshi[2]

[1] Pandit Deendayal Petroleum University, Gandhinagar, Gujarat, India
Tejas.Thaker@sot.pdpu.ac.in
[2] CEPT University, Ahmedabad, Gujarat, India

Abstract. Natural disasters are inevitable and unannounced; however, human-kind can develop early warning systems to take preventive measures. Few seconds of high intensity earthquake creates massive loss to property and human lives. The present paper proposes to suggest means to mitigate high impact earthquake with analysis of ground response in Ahmedabad region that can help defining design of earthquake resistant structure. The paper presents a study of complete ground response analysis to help determining propagation of seismic waves through the earth to the top of the bedrock beneath the particular site and then determining how the ground surface motion is influenced by the soil which lies above the bed rock. In the current study, equivalent linear analysis method for ground response study is used. The parameters suggested in the paper and their effects on site response in Ahmedabad to be considered during any construction activity as geotechnical parameter to avoid any disaster during earthquake can help mitigate future disasters in the region. The different types of geotechnical parameters are used for the ground response analysis for Ahmedabad region which is situated in Gujarat, India. The impacts of far field and near field earthquake motions are applied on the ground data of Ahmedabad. The result of Peak Spectral Acceleration vs. Time, Peak ground acceleration vs. depth and predominant frequency vs. amplitude ratio graphs are obtained. Ground response spectra for various regions in Ahmedabad are obtained and that are compared to the ground response spectra of Indian standard code to define variation.

Keywords: Peak shear wave velocity · Shear wave velocity
Ground response spectra

1 Introduction

Natural disasters are indispensable and it is not possible to get full control over them. Recognition of the hazard is one of the most important components of disaster management. The proposed study deals with hazard analysis and the first step towards the mitigation and prevention from the earthquake as the result will be useful to define building codes for the city as well as the land use planning. IS 1893:2002 for design of earthquake resistant structures, is developed on broad scale data. In addition, with each major earthquake, modification of code is required, which indicates some inadequacy of code.

© Springer Nature Singapore Pte Ltd. 2018
A. Farid and H. Chen (Eds.): GSIC 2018, *Proceedings of GeoShanghai 2018.*
International Conference: Geoenvironment and Geohazard, pp. 208–215, 2018.
https://doi.org/10.1007/978-981-13-0128-5_24

Serious problems are faced in geotechnical earthquake engineering when conducting ground response study. Ground response study is a process to evaluate the dynamic stress and strains for determining the earthquake forces which are capable of disturbing stability of earth or earth retaining structures by predicting ground surface motion for development of "Design response spectra". In the earthquake, the damage at a site is greatly influenced by the response of soil. This study of complete ground response analysis proposes to help determining propagation of seismic waves through the earth to the top of the bedrock beneath the particular site and then determining how the ground surface motion is influenced by the soil that lies above the bedrock.

Gujarat state has seen 24 incidents of earthquakes. From that, Ahmedabad has seen earthquakes of 1843 and 1864, which caused moderate to severe damage. Ahmedabad city is a leading commercial and industrial city of Gujarat in India. It is situated at 23.01° N latitude and 72.61° E longitude. Ahmedabad is situated on the banks of Sabarmati River. Soil in Ahmedabad region is younger alluvial deposits. 2001 Bhuj earthquake lasted for 2 min and the epicenter of this earthquake was about 9 km southwest from the village Chobari in Bhacau taluka of Kutch district. This earthquake measured at a magnitude of 7.7 on the Moment Magnitude Scale and had X intensity according to Modified Mercalli Scale. In Ahmedabad, fifty multi-storey buildings collapsed and several hundreds of people died because of it.

Ground response analysis helps geotechnical engineers to predict the natural period of soil deposit and liquefaction potential and determining motion amplification at a particular site and then providing data to structural engineers in the form of response spectra for determining safety in designing geotechnical structure in earthquake prone area. In this project, the analysis is proposed to be carried out on the different soil parameters, which would affect the amplification of the seismic waves by near field earthquake or far field earthquake on Ahmedabad region using software "Deep Soil V6.1". This project aims to determine the parameters and their effects on site response in Ahmedabad region which should be considered during any construction activity as geotechnical parameter to avoid any disaster during earthquake.

2 Literature Review

The ground response analysis has been carried out by many researchers but there is some limitations in their research which are discussed here.

For the calculation of shear wave velocity Govindaraju et al. (2004) used the formula suggested by Japan Road Association (1980) which is derived based on Japan soil data and its applicability to Indian soil condition is questionable and the effect of near field earthquake is not considered. Rao and Ramana (2009) considered the random earthquake history of magnitude 6 and 6.5 in moment magnitude scale which is hypothetical instead of considering past earthquake motion. Tests were conducted only for the high strains.

Govindaraju and Bhattacharya in their work considered seismic motion and synthetic ground motions having bedrock level acceleration in the limited range. However, the influence of ground motion having a wide range of variation in bedrock level acceleration is not considered. Chatterjee and Choudhury have used correlation for evaluating shear wave velocity, which is derived based on the soil data of Kolkata city. The applicability of that empirical relation in different location in India is questionable.

Ranjan in his work in Dehradun city has considered only near field earthquake effect but far field earthquake effect is not evaluated. Hwang and Lee (1991) in their work have used hypothetical reference cases for analysis which does not reflect real ground profile. Thaker et al. (2009) have carried out one dimensional ground response analysis of coastal soil near Naliya, Bhuj, Gujarat.

3 Parametric Study

Parametric study is a study of identification and analysis of different types of parameters, which are affecting the earthquake. This type of study is also known as sensitivity analysis. This study is carried out at the suitable range of parameter and the results will be helpful in ground response analysis. The parameters which are identified by this study are depth of bedrock and Shear wave velocity. Graph of Peak ground acceleration, Peak spectral acceleration and response spectra with respect to these parameters are obtained. The input motion for this analysis is Chichi earthquake motion.

3.1 PGA Variation with Bedrock Depth

Variations in PGA, for sand, are obtained with respect to the bedrock depth for particular value of shear wave velocity and unit weight of sand.

Parameters Considered:

Unit weight: 20 kN/m^3

Reference curve: Seed and Idriss (1970)

Bedrock condition: Rigid bedrock

Shear wave velocity: 250 m/s

Result:

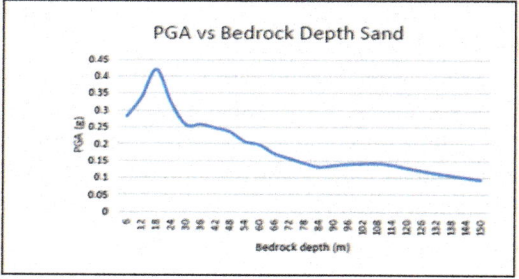

Fig. 1. PGA vs Bedrock depth for sand.

Variation in PGA is considerable up to 84 m bedrock depth. After this depth, variation in PGA is not significant (Fig. 1).

3.2   PSA Variation with Bedrock Depth

Variations in PSA, for sand, are obtained with respect to the bedrock depth for particular value of shear wave velocity and unit weight of sand.

Parameters Considered:

Unit weight: 20 kN/m^3

Reference curve: Seed and Idriss (1970)

Bedrock condition: Rigid bedrock

Shear wave velocity: 250 m/s

Result:

Peak spectral acceleration (PSA) is obtained at depth of 18 m (Fig. 2).

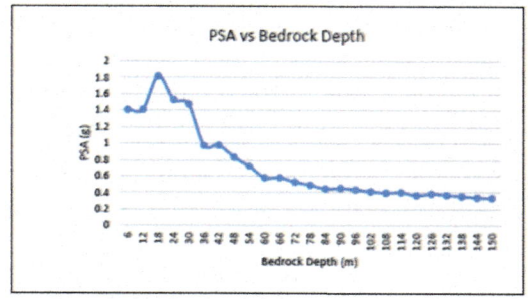

Fig. 2. PSA vs Bedrock depth.

3.3   PGA Variation with Shear Wave Velocity

Variation in PGA with respect to bedrock depth is obtained for different shear wave velocity values for sand.

Parameters Considered:

Unit weight: 20 kN/m^3

Reference curve: Seed and Idriss (1970)

Bedrock condition: Rigid bedrock

Results:

With increase in shear wave velocity, PGA value increases. As the shear wave velocity increases, PGA results at more bedrock depth (Fig. 3).

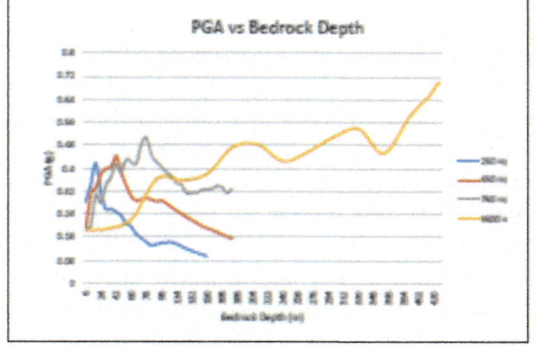

Fig. 3. PGA vs Bedrock depth.

3.4   Variation of Response Spectra with Bedrock Depth

Variation in spectral acceleration with respect to bedrock depth is obtained for different shear wave velocity values for sand.

Parameters Considered:

Unit weight: 20 kN/m^3

Reference curve: Seed and Idriss (1970)

Bedrock condition: Rigid bedrock

Results:

For particular shear wave velocity, as depth of bedrock increases, PSA shifts towards right i.e., occurs at more time period.

At larger bedrock depth, variation in response spectra is insignificant (Fig. 4).

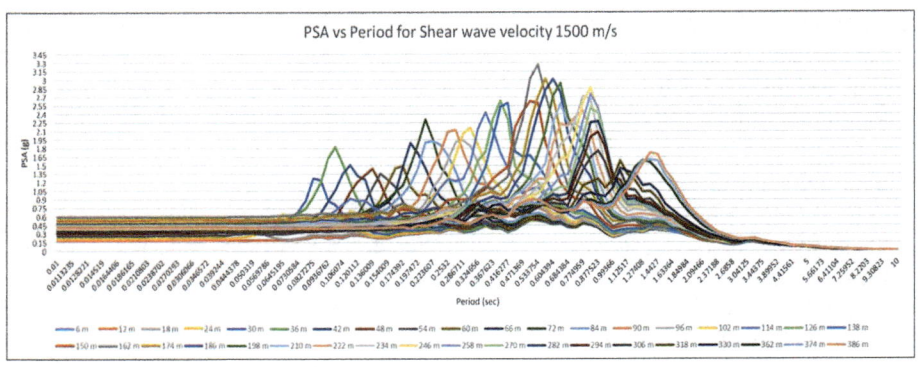

Fig. 4. PSA vs Period for shear wave velocity.

4   Ground Response Analysis

Ground response analysis is carried out at Ahmedabad city for earthquake hazard assessment. The soil data was collected from 600 borehole data for every region in Ahmedabad. The analysis is being carried out in "Deep Soil V6.1" software by using equivalent linear analysis and result of Spectral acceleration vs. time which is also known as response spectra and amplitude ratio vs. frequency which have derived the predominant frequency for every region. This analysis is carried out for near field and far field earthquake motion.

Near-field earthquakes have some characteristics that differ from far-field ones. These earthquakes have higher accelerations and restricted frequency content in higher frequencies than far-field ones. In addition, their records have pulses in beginning of record with high period and high domain.

Input motions have significant impact on ground response spectrum. Due to the uncertainties associated with the characteristics of rock motions and site parameters, it is felt that if only one site response analysis is performed for a site using single input motion then, calculated results may be unreliable. To overcome these issues, ten acceleration time histories simulated on firm rock condition compatible with the earthquake hazard in terms of probable magnitudes of 5.5 and 6.3 at distances of 10, 20, 30, 40 and 55 km from local sources were selected and for far field earthquake motion Bhuj earthquake motion was studied (Figs. 5, 6 and 7).

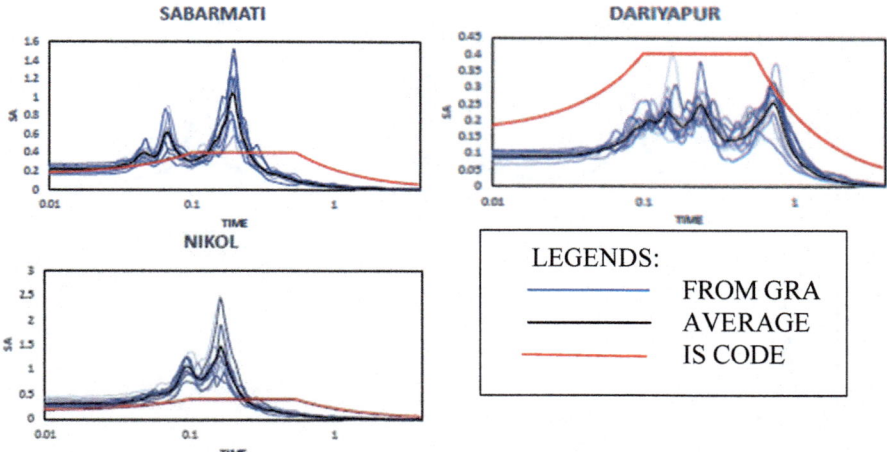

Fig. 5. Ground response spectra of various highly impacted location in Ahmedabad region for near field motion.

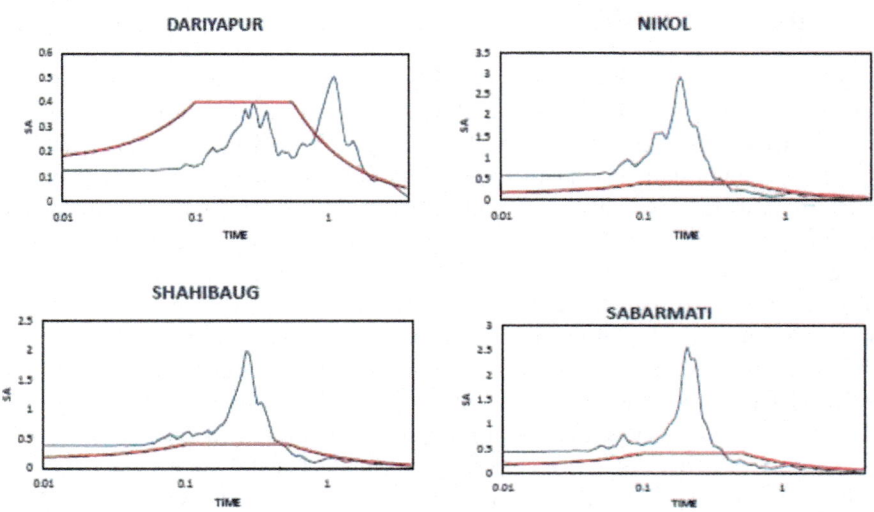

Fig. 6. Ground response spectra of various highly impacted location in Ahmedabad region for far field motion.

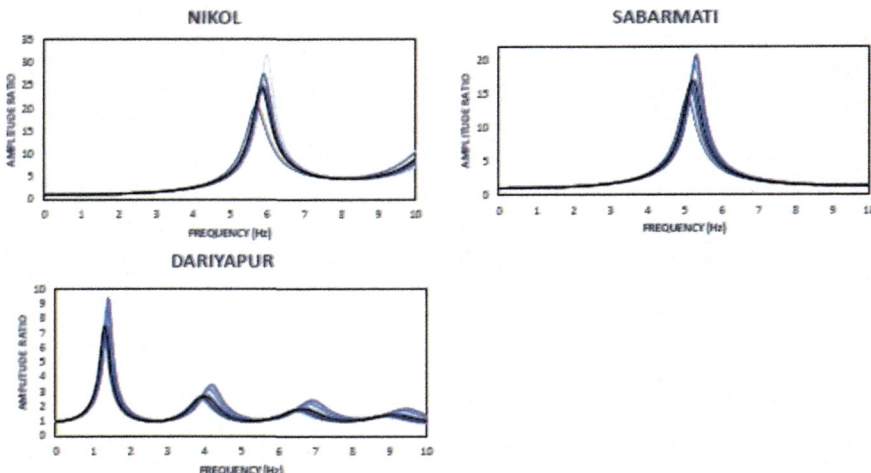

Fig. 7. Amplitude ratio vs Frequency graph for sever region for identification of predominant frequency.

5   Results and Conclusion

For near field earthquake motion the graph between Spectral acceleration and time for different locations are obtained but several regions have significant variation when it is compared with Indian standard code. This study will also help in modification in Response spectra in Indian standard code. This result will be helpful for analyzing the impact of massive earthquakes and mitigating the damages in best possible way. Various areas in Ahmedabad show variable ground response and the damage or loss in these areas may depend on the intensity of the quake and the ground response to it. The local civic body may benefit from such study as they can decide on approval for high-rise and low-rise buildings depending on the ground response. The study also helps identifying factors of variable loss and damage the earthquake of 2001 when differing damage was observed in different areas of Ahmedabad.

This type of graph is also obtained for far field earthquake and the location is noted down who have higher difference compared to Indian standard code.

Predominant frequency also obtained from the graph between Frequency and Amplitude ratio and from that sever region for predominant frequency are identified.

References

Chatterjee, K., Choudhury, D.: Variations in shear wave velocity and soil site class in Kolkata city using regression and sensitivity analysis. Nat. Hazards **69**(3), 2057–2082 (2013)

Choudhury, D., Phanikanth, V.S., Mhaske, S.Y., Phule, R.R., Chatterjee, K.: Seismic liquefaction hazard and site response for design of piles in Mumbai city. Indian Geotech. J. **45**(1), 62–78 (2015)

Govindaraju, L., Ramana, G.V., Hanumantarao, C., Sitharam, T.G.: Site specific ground response analysis. Curr. Sci. **87**(10), 1354–1362 (2004)

Hanumanthrao, C., Ramana, G.V.: Site-specific ground response analyses at Delhi, India. Electron. J. Geotech. Eng. **14(D)**, 1–16 (2009)

Hwang, H.M., Lee, C.S.: Parametric study of site response analysis. Soil Dyn. Earthq. Eng. **10** (6), 282–290 (1991)

Indian Standard, 1893: Criteria for Earthquake Resistant Design of Structures, Bureau of Indian Standards, Part 1, New Delhi (2002)

NISA II/DISPLAY III: Training Manual, Engineering Mechanics Research Corporation Version 8.0, Michigan, USA (1988)

Seed, H.B., Idriss, I.M.: Soil moduli and damping factors for dynamic response analyses. Report no. EERC-70/10, Earthquake Engineering Research Center, University of California, Berkeley (1970)

Thaker, T.P., Rao, K.S., Gupta, K.K.: One dimensional ground response analysis of coastal soil near Naliya, Kutch, Gujarat, geotechnics in infrastructure development (GEOTIDE). In: Proceedings of an Indian Geotechnical Conference-09, Guntur, Andhra Pradesh, pp. 249–253 (2009)

Thaker, T.P., Rao, K.S., Gupta, K.K.: Ground response and site amplification studies for coastal soil, Kutch, Gujarat: a case study. In: Proceedings of the International Conference on Advances in Concrete Structural and Geotechnical Engineering (ACSGE 2009), October 2009, BITS Pilani, Rajasthan, India (2009)

Investigation on Characteristics of Large-Scale Creep Landslides

Zhongqing Chen[1,2(✉)], Fei Zhang[3], Jinyuan Chang[1,2], and Yue Lv[2]

[1] Centre of Rock Mechanics and Geohazards, Shaoxing, China
Q_CHEN_YK@163.com
[2] Shaoxing University, Shaoxing, China
[3] Zhejiang Geological Exploration Bureau for Non-ferrous Metals,
Shaoxing, China

Abstract. Two large-scale creep landslides occurred at Taipingshan in Chenxi township of Shangyu District in July 11, 2015 and at Chengjiashan in Xinchang County in November 28, 2016, respectively. The field monitoring of cracks, surface displacements and deep horizontal displacements of the landslides were carried out. Such examinations were combined with site geological investigation, site engineering geological survey and permeability test. The characteristics, formation mechanism and failure mode of the landslides were investigated. The results show that: (1) the total volume of Taipingshan landslide is around 1.05×10^6 m^3 with 150 to 180 m wide and 75 to 105 m wide at front and trailing edges, respectively, and the landslide is mainly composed of Quaternary slope bed and fully weathered granite; (2) the total volume of Chengjiashan landslide is around 4.0×10^6 m^3 with 480 m wide and 100 m wide at front and trailing edges, respectively, and the landslide is a typical large-scale soil landslide in basalt platform; (3) heavy rainfall and human engineering activities are two major factors causing new deformations of large-scale creep landslides; (4) Large-scale creep landslide is usually tractive and it is in a state of creeping deformation with slow speed of sliding until obvious deformation will occur if there are heavy rainfall or human engineering activities to take place.

Keywords: Large-scale creep landslide · Deformation · Basalt platform
Weathering granite soil · Field monitoring

1 Introduction

Landslides are the main geological disaster in Zhejiang Province, which are dominated by medium and small soil landslides composed of weathered residual layers of bedrock [1–3]. A large number of studies have been carried out on landslide prevention and monitoring and early warning around the world. The study of soil landslides in Zhejiang province is also abundant. For example, Yin et al. (2003), Zhang et al. (2005) and Zhang (2006) studied the early-warning and prediction system of sudden geological disasters in Zhejiang Province and established an early-warning and forecasting system of geological disaster based on WEBGIS and real-time rainfall information [4–6]. Zhu and Ma (2011) and Yu et al. (2006) developed a warning and forecasting system for landslides and debris flows aiming at the geological, topographical and climatic conditions of

© Springer Nature Singapore Pte Ltd. 2018
A. Farid and H. Chen (Eds.): GSIC 2018, *Proceedings of GeoShanghai 2018*
International Conference: Geoenvironment and Geohazard, pp. 216–225, 2018.
https://doi.org/10.1007/978-981-13-0128-5_25

Zhejiang Province [7, 8]. But the deformation characteristics and deformation mechanism of large-scale soil landslide in Zhejiang province are still less developed currently.

Taipingshan landslide is located in southwest of Chenxi township of Shangyu District of Shaoxing (see Fig. 1). New landslide deformations such as local collapse of retaining wall and cracks of road occurred in July 11, 2015 after the first landslide took place in the year 2000. The total volume of the landslide was about 1,050,000 m^3. Chenjiashan landslide is located in northwest of Chengtan town of Xinchang County. In the early morning of November 28, 2016, new landslide deformations such as foundation breaking of houses and cracking of road occurred after continuous heavy rainfall after the first landslide occurred in 1980s. The total volume of the landslide was about 3,940,000 m^3.

Fig. 1. Sketch map of landslide location

Geological environment conditions of Taipingshan landslide and Chenjiashan landslide were investigated in this paper as well as surface monitoring of crack width and horizontal displacement of the landslides was carried out. Basic characteristics, formation mechanism and sliding failure mode of large-scale creep landslides were finally investigated.

2   Geological Environment Condition of Landslide Area

2.1   Topographic Features

According to the satellite and ground survey, The Taipingshan landslide is a hilly geomorphic unit. The slope of natural slope in landslide area is 15 to 30° with slow slope in the middle ridge and relatively steep slope in the valleys. The landslide is about 330 m long, 150 to 180 m wide at the front edge, 75 to 105 m wide at trailing edge and the elevation difference between the front and trailing edge is about 161.5 m (see Fig. 2a). The front edge of the landslide extends to Baixuanxian road. The slope of the front edge is steep with height of 3 to 13 m and slope of 50 to 60° due to irregular sand excavation and other artificial slope cutting activities.

Chenjiashan landslide belongs to the edge of basalt platform and has typical geomorphic features of East Zhejiang basalt platform. The landslide area tilts from west to east. The slope of natural slope is 5 to 15° with around 25° at the front edge. The landslide is about 1250 m long, 480 m wide at the front edge, 120 m wide at trailing edge and the elevation difference between the front and trailing edge is about 161.5 m (see Fig. 2b). The height of the front edge is around 5 to 15 m due to construction of industrial park.

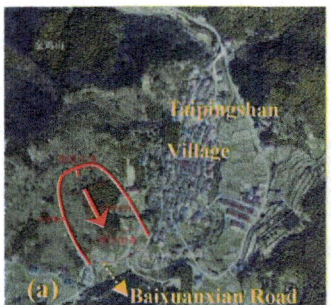 

Fig. 2. Topographic map of landslide area: (a) Taipingshan landslide (1:10000), (b) Chenjiashan landslide (1:2500)

2.2   Weather Conditions

The landslide belongs to the mid latitude subtropical monsoon climate zone. The average annual precipitation is 1424 to 1626.8 mm. The rainfall is mainly concentrated in the rainy season from May to June and the typhoon season from July to September.

2.3   Stratum Lithology

The bedrock strata of Taipingshan landslide are mainly late Yanshan monzonitic granite and there are fully weathered layer, intense weathering broken layer and middle weathered layer from top to bottom. The surface of natural slope is quaternary residual slope layer, which is mainly composed of silty clay containing gravel with low to medium compressibility. The valley and valley side are quaternary diluvial slope layers, which are mainly composed of silty clay containing gravel with medium compressibility. The quaternary diluvial slope layers are one of the main parts of the landslide body.

The strata from old to new of Chenjiashan landslide area includes layer of retaceous sandstone of Guantou formation, layer of tertiary Pliocene basalts of Shengxian formation containing lacustrine sedimentary facies, and quaternary unconsolidated deposits consisted of silty clay containing gravels (blocks). The sandstone layer distributes in the front of the whole basalt platform. The layer of Pliocene Shengxian formation mainly distributes in the west, southwest of the landslide area and the main lithology is olivine basalt with settled layer intercalated with cohesive soil. The Quaternary unconsolidated deposits are distributed in most areas of the exploration area and are also the areas where landslide occurs. The deposits are composed of silty clay containing gravels (blocks) with soft to hard plastic conditions and the content of the gravels (blocks) is about 15–30% with uneven distribution. The average thickness of the Quaternary unconsolidated deposits is about 11 m.

2.4 Hydrogeological Conditions

According to the properties, burial conditions and drainage modes of runoff of the aquifer in the landslide area, the groundwater can be divided into pore phreatic water of loose rock stratum and fissure water of bedrock.

The pore phreatic water in Taipingshan landslide occurs in the Quaternary deposits and the groundwater table is located at 1.5 m–18 m distant from the ground surface with lower table in the front the landslide. The horizontal and vertical permeability coefficients of the diluvial slope deposits are about 1.57×10^{-4} cm/s and 1.23×10^{-4} cm/s, respectively. Atmospheric precipitation is the main recharge source and the groundwater level and its seepage flow are greatly affected by the weather. The front edge of the landslide area is often in saturated state after heavy rainfall. The fissure water of bedrock in Taipingshan landslide occurs in moderately weathered and strongly weathered fissures of monzonitic granite.

The pore phreatic water in Chenjiashan landslide occurs in the layer of silty clay containing gravels. The groundwater table is usually located at 0.8 m–3.0 m distant from the ground surface and is located at 6 m–10 m distant from the ground surface near industrial park. The permeability coefficient of the deposit is about 4.63×10^{-4} cm/s. Atmospheric precipitation is also the main recharge source and the groundwater level varies greatly. The fissure water occurs in layers of retaceous sandstone and tertiary Pliocene basalts.

2.5 Seismicity Conditions

According to seismic ground motion parameter zonation map of China (GB18306-2001) [9], the seismic peak ground accelerations of both Taipingshan and Chenjiashan landslide areas are less than 0.05 g (g denotes gravitational acceleration) and the corresponding basic seismic intensity is determined to be VI degree.

3 Basic Characteristics of Landslide

3.1 Morphological Characteristics and Boundary of Landslide

The front edge of Taipingshan landslide is located in the toe of natural slope and area of Baixuanxian road (see Fig. 2a), which was highly affected by human engineering activities. The sides of the landslide are about 220 to 250 m long. Extended shear fissures were produced on these sides and obvious vertical displacement occurs on both sides of the fissures. There is no scattering of sill and bedrock exposure at the trailing edge.

The front edge of Chenjiashan landslide is located in the industrial park area (See Fig. 2b), which was highly affected by human engineering activities including constructions of Jiaochengxian road and industrial park. Obvious cracks and arching of concrete pavements were found in this area (see Fig. 3). The width of middle part of the landslide is about 430 m and the slope is about 6 to 23°. The road cracking was also found in this area (see Fig. 4). The boundary on both sides of the landslide is not obvious. The northeast side is about 1100 m long and the southwest side is about 1000 m long. There is no obvious deformation at trailing edge of the landslide and the slope of this area is about 12°.

Fig. 3. Arching of concrete pavement

Fig. 4. Cracking of concrete pavement

3.2 Deformation Characteristics and Monitoring Data Analysis of Landslide

Monitoring of Surface Cracks. Compared with the landslide deformation in the year 2000, around 20 new cracks were found on the front, trailing, middle and sides of the landslide and the original cracks also increased obviously. In order to obtain the landslide deformation characteristics of shallow soil mass, 29 monitoring points were set up and the width of the 20 cracks was monitored over a period of 120 h (From 14:30 July 14, 2015 to July 17, 2015). Figure 5 shows typical monitoring results of No.4 and No.13 cracks, which suggest that the crack width increased obviously in the first 60 h. It can be obtained that the landslide deformation was mainly affected by heavy rainfall and the instability deformation stage of Taipingshan landslide was mainly from the late July 11th to the July 14th.

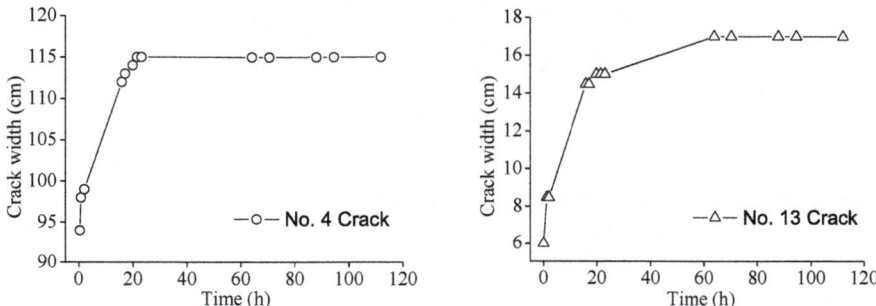

Fig. 5. Typical curves of crack width versus time

During the period from 2006 to 2008, due to constructions of industrial park obvious deformations such as house cracking, road cracking and heave of retaining wall in the middle part and the front edge of Chenjiashan landslide were found as well as groundwater overflows from front edge of the landslide. The monitoring results from October 2011 to March 2016 showed that the road heave was 15 to 50 cm and the maximum deflection of retaining wall was 1.6 m and the ground dislocation was up to 10 cm in front edge of the landslide (See Fig. 3, 6 and 7).

Fig. 6. The deflection of retaining wall

Fig. 7. Ground dislocation

Monitoring of Deep Horizontal Displacements. 3 monitoring points of deep horizontal displacement were installed in front edge, trailing edge and middle part of Taipingshan landslide, respectively, and the inclinometer of CX-3C type were used. Figure 8 shows typical monitoring results from July 21, 2015 to August 24, 2015, which suggests that the main horizontal displacement of the landslide occurred in shallow part and the maximum displacement was 4.4 mm. It can also be obtained that the Taipingshan landslide has been at a stable stage since July 21, 2015.

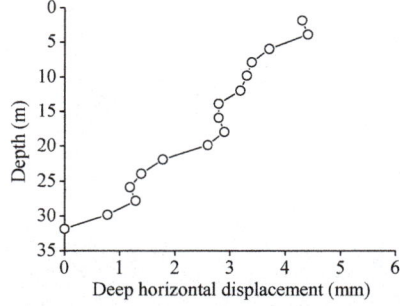

Fig. 8. Typical horizontal displacement curve

4   Formation Mechanism and Sliding Failure Mode of Large-Scale Creep Landslide

4.1   Formation Mechanism

Geographic and Geomorphic Conditions. The slope of natural slope in landslide area is slow and the average slope is not more than 30°. The relative elevation difference between the front and trailing edges of landslide can be generally more than 150 m. There is usually artificial steep slope formed by human engineering activities such as artificial slope cutting in the front of landslide.

Lithologic Conditions of Strata. There are large Quaternary sediments on the slope of landslide areas and the thickness can usually be up to 30 m, which provides a rich source of material for large-scale creep landslide. There are soft structural planes in the landslide such as weathered silty clay layer underlaid weathered sand layer of monzonitic granite, which softens easily while meeting water. Such lithologic conditions of strata provide necessary conditions for the occurrence and development of the landslide.

Conditions of Geological Structure. The conditions of geological structure are not necessary for deformation of large-scale creep landslide. Taipingshan landslide is located in the northeast of Changshan-Shaoxing deep fault zone and is far away from main ruptures, while Chenjiashan landslide is located in the west of Lishui-Yuyao deep fault zone. The faults and folds are not developed in these landslide areas. Combined with the seismicity conditions, the earth's crust in these areas can be determined to be stable.

Hydrogeological Conditions. The permeability of large-scale creep landslide is usually weak and the runoff of groundwater is poor. Therefore the groundwater level of landslide increases easily under condition of continuous heavy rainfall and then the weight of landslide increases simultaneously and the shear strength of soil near structural planes decreases obviously. As far as Taipingshan landslide is concerned, the permeability coefficient of fully weathered silty clay located at bottom of the landslide was 10^{-6} cm/s and the shear strength indices of silty clay decreased obviously after heavy rainfall, which were unfavorable to stability of overlying sliding mass.

Human Engineering Activities. Human engineering activities are important factors causing instability of large-scale creep landslide especially for tractive landslide. Due to human engineering activities such as slope cutting in the front of landslide, an empty surface will be formed and then the equilibrium state of landslide body will be destroyed, which will lead to local shear failure because of stress concentration. As far as Taipingshan landslide is concerned, the construction of Baixuanxian road and sand excavation at the front edge of the landslide caused cracks at trailing edge of the landslide and house cracking at front edge of the landslide. As far as Chenjiashan landslide is concerned, the construction of Jiaochengxian road and the construction of industrial park caused significant deformation of the landslide.

Meteorological Factors. The statistical results of rainfall data in July 2015 in Chenxi township of Shangyu District are shown in Fig. 9, which suggests that the cumulative rainfall was close to 250 mm due to the influence of the rainy season before July 11th and the daily rainfall of Chenxi Township reached more than 280 mm in July 11th due to the typhoon Chan-hom, which landed in Zhejiang province in July 11th. Based on comparison of meteorological rainfall data and occurrence time of landslide deformation in July 2015, it can be obtained that heavy rainfall was main factor causing instability of Taipingshan landslide.

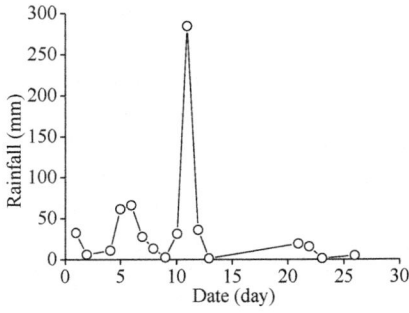

Fig. 9. Daily rainfall of Chenxi Township in July 2015

4.2 Sliding Failure Mode

According to the analysis of the deformation characteristics and the formation mechanism of large-scale creep landslide in regions of Shaoxing in Zhejiang province, it can be concluded that both Taipingshan landslide and Chenjiashan landslide belong to tractive landslide and that large-scale creep landslide is unlikely to be disposable overall sliding but be periodic creep deformation and failure under the influence of heavy rainfall and human engineering activities such as artificial slope cutting and so on.

5 Conclusions

1. Taipingshan landslide and Chenjiashan landslide located in Shaoxing area of Zhejiang province are two typical large-scale creep landslides, the volume of which are about 1,050,000 m^3 and 3,940,000 m^3, respectively. Both of these sliding masses are Quaternary sediments mainly composed of silty clay containing gravels overlaid weathered monzonitic granite or olivine basalt.
2. Heavy rainfall and human engineering activities are main factors to induce instability of large-scale creep landslide. The geomorphic factors, lithologic conditions of strata as well as hydrogeological factors are also important for the formation of large-scale creep landslides except geological structure factors.
3. Large-scale creep landslide is usually a type of tractive landslide, which is in a state of creeping deformation with slow speed of sliding. This type of large-scale landslide shows obvious deformation mainly under conditions of heavy rainfall and human engineering activities and generally remains stable.
4. Considering there are roads and an industrial park at the foot of these landslides, some stability procedures such as building a channels system to manage shallow water, covering the landslides slopes with geomat to avoid the effect of erosion caused by rainfall and building anti-slide piles should be adopted.

Acknowledgements. The authors would like to extend their appreciations to the support from Zhejiang public interest research project (No. 2016C33052), Shaoxing public interest research projects (No. 2015B70034; No. 2015B70035).

References

1. Wang, Z.P.: Current situation of geological disasters and the prevention measures in Zhejiang Province. J. Catastrophology **16**(4), 63–66 (2001)
2. Gian, S.G., Bao, Q.Y.: Present situation of geological hazard in Zhejiang province and preventing countermeasure. Geol. Zhejiang **14**(1), 90–96 (1998)
3. Yu, H.M., He, J.Y.: Study on the development law of landslide in Zhejiang. In: Engineering Geology Committee of Geological Society of China. The Eighth National Conference on Engineering Geology, vol. 1, pp. 217–220. Editorial Department of engineering geology Journal, Beijing (2008)
4. Yin, K.L., Zhang, G.R., Gong, R.X., Wang, K.Z.: A real time warning system design of geo-hazards supported by Web-GIS in Zhejiang Province, China. Hydrogeol. Eng. Geol. **3**, 19–23 (2003)
5. Zhang, G.R., Yin, K.L., Liu, L.L., Xie, J.M., Du, H.L.: A real-time regional geological hazard warning system in terms of WEBGIS and rainfall. Rock Soil Mech. **26**(8), 1312–1317 (2005)
6. Zhang, G.R.: Spatial prediction and real-time warning of landslides and it's risk management based on webGIS. China University of Geosciences, Wuhan (2006)
7. Zhu, X.S., Ma, T.H.: Review and Prospect of early warning and prediction of abrupt geological hazards in Zhejiang. Zhejiang Land Res. **7**, 22–24 (2011)
8. Yu, F.H., Ma, T.H., Zhang, Y.S., Xia, Y.Z.: Application and study on abrupt geological hazards for early-warning and prediction in Zhejiang Province. Chin. J. Geol. Hazard Control **17**(1), 36–39 (2006)
9. GB18306–2001: Seismic ground motion parameter zonation map of China. General Administration of Quality Supervision, Inspection and Quarantine of the People's Republic of China. Standards Press of China, Beijing (2001)

Experimental Studies on Landslide Dam Stability Under Surge Action

Peng Ming[⊠] and Jiang Mingzi

Department of Geotechnical Engineering,
Tongji University, Shanghai 200092, China
pengming@tongji.edu.cn

Abstract. The rapid rise and fall of the water level in the upstream reservoir area of a landslide dam may trigger slope sliding into the reservoir, leading to huge surge wave. The stability of landslide dams under the action of landslide surge is seldom studied and of the target in this paper. Landslide dam model tests with the height of 80 cm and width of 317 cm were performed in a water flume with a full length of 42 m. Three sine waves with the maximal height of 5, 10 and 20 cm were produced in the upstream reservoir with water depth of 55 cm. It is found from the tests that the surge wave would cause erosion and sliding on the upstream slope. A steady and flat slope would remain when the wave height is low. The erosion became larger and the finally steady slope became flatter with the increase of the wave height. When the surge wave surpasses the dam crest, significant erosion would occur in the downstream slope, leading to overtopping failure. The pore water pressure periodically varied with the wave. The area far from the upstream slope has a delayed and weak response to the pore pressure of the wave.

Keywords: Wave flow flume · Landslide dam · Landslide surge

1 Introduction

In some areas where geological structure is active and extreme climate disasters occur frequently, it is easy to induce dammed lake. Due to the rapid rise of water level in the upstream reservoir area, a large number of landslides may be induced, which can easily cause large surge, and will cause serious damage to the dam. When the landslide surge acts on the landslide dam, it will bring serious threat to the life and property safety of the downstream people.

At present, the most model test of landslide dam is to study overflow in the natural state. There is a lot of research in this area. For example Zhang et al. [1] carried out a flume test for the different internal slope and material. Wang et al. [2] considered the influence of the upstream peak flow on the dam breach. The research on the stability of landslide dam under the action of landslide surge mainly contains the following aspects: Risley et al. [3] calculated the overtopping wave volume of Usio landslide dam in Pakistan. Xu et al. [4] studied the effect of landslide surge on the dam under the conditions of different landslide water entry area, landslide height and distance from the water entry point to the dam site. Lin et al. [5] established an ISPH model to simulate the process of landslide surge climbing over the dam.

© Springer Nature Singapore Pte Ltd. 2018
A. Farid and H. Chen (Eds.): GSIC 2018, *Proceedings of GeoShanghai 2018*
International Conference: Geoenvironment and Geohazard, pp. 226–232, 2018.
https://doi.org/10.1007/978-981-13-0128-5_26

The existing research methods are mainly theoretical analysis and numerical analysis methods. Therefore, in order to better assess the impact of landslide surge on landslide dam stability, the corresponding model tests are needed. In this paper, the stability of landslide dams under the action of landslide surge is studied in wave flume model tests.

2   Experimental Design

2.1   Test Device

The experiment was carried out in the wave flume experiment system of Tongji University. The flume length is 42 m, width is 0.80 m, height is 1.25 m. The measurement system of pore water pressure consists of pore water pressure gauge and dynamic strain gauge. During the test, the deformation and displacement of the dam are monitored by high definition camera.

2.2   Dam Material

In this paper, the landslide dam model is formulated according to the particle size distribution curve of Donghekou landslide dam [6], as shown in Fig. 1. The particle size distribution of the dam material represents the particle distribution characteristics of the fine-grained landslide dam. The basic physical properties of dam material, as shown in Table 1. In the experiment, the method of stratified compaction is adopted and the thickness of each layer is 10 cm.

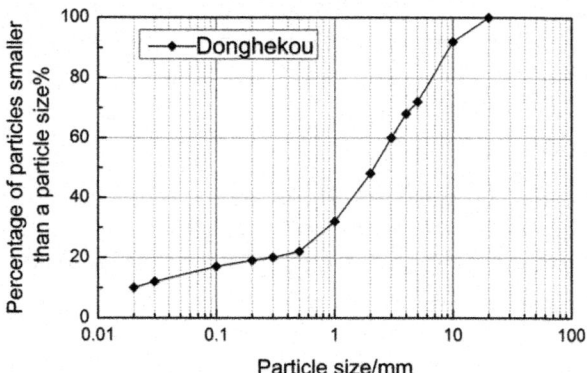

Fig. 1. Grain size distribution curve of Donghekou landslide dam

Table 1. Basic physical properties of dam material

Dry density	Minimum dry density	Maximum dry density	Compactness
1.78	1.44	2.05	0.642

2.3   Model Shape, Size and Layout of Test Equipment

In order to study the general law of stability of the landslide dam under the action of landslide surge, this paper does not choose a specific landslide dam as the prototype. According to the size and shape characteristics of the landslide dam, combined with the size limitation of the wave flow flume, the selected dam shape and size is shown in Fig. 2.

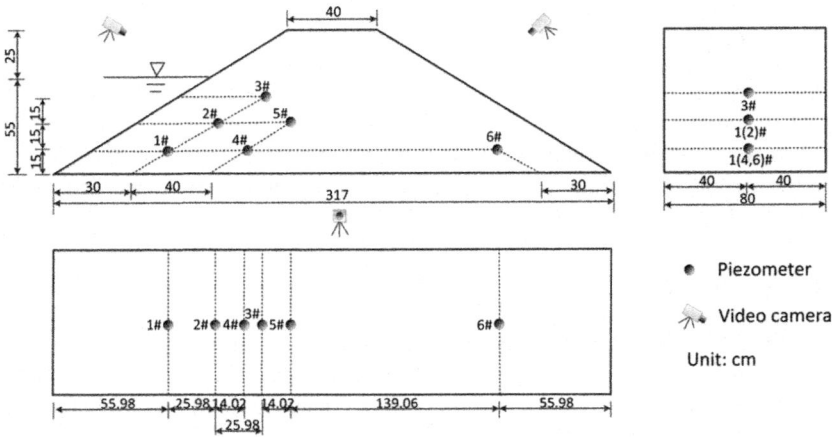

Fig. 2. Layout of instrument and equipment

In the test, 6 gauges are parallel to the upstream dam slope and are arranged in the central position of the dam along the width of the dam. In addition, in order to monitor the deformation and displacement of the dam during the test, three cameras are set up, as shown in Fig. 2.

2.4   Test Scenarios

Three scenarios were set up in the test. The upstream water level of the dam is 55 cm, and the wave period is set to 2 s. The wave heights are 5 cm, 10 cm and 20 cm respectively, as shown in Table 2. After each test scenario is completed, the dam is refilled and repeated the test steps.

3   Experimental Results and Analysis

3.1   Erosion Characteristics

In scenario 1, the wave height is 5 cm (Hw:5 cm). The failure process of the dam under this scenario is shown in Figs. 3(a) and 4. After the wave starts, the waves scour the upstream dam slope and gradually forms a scour datum plane, as shown in the dashed line in Fig. 3(a). The subsequent scour processes are developed on the basis of this

Table 2. Test scenarios

Scenario	Water depth (cm)	Wave height (cm)	Period (s)	Dam material	Wave form
1	55	5	2	Donghekou landslide dam	Regular wave
2		10			
3		20			

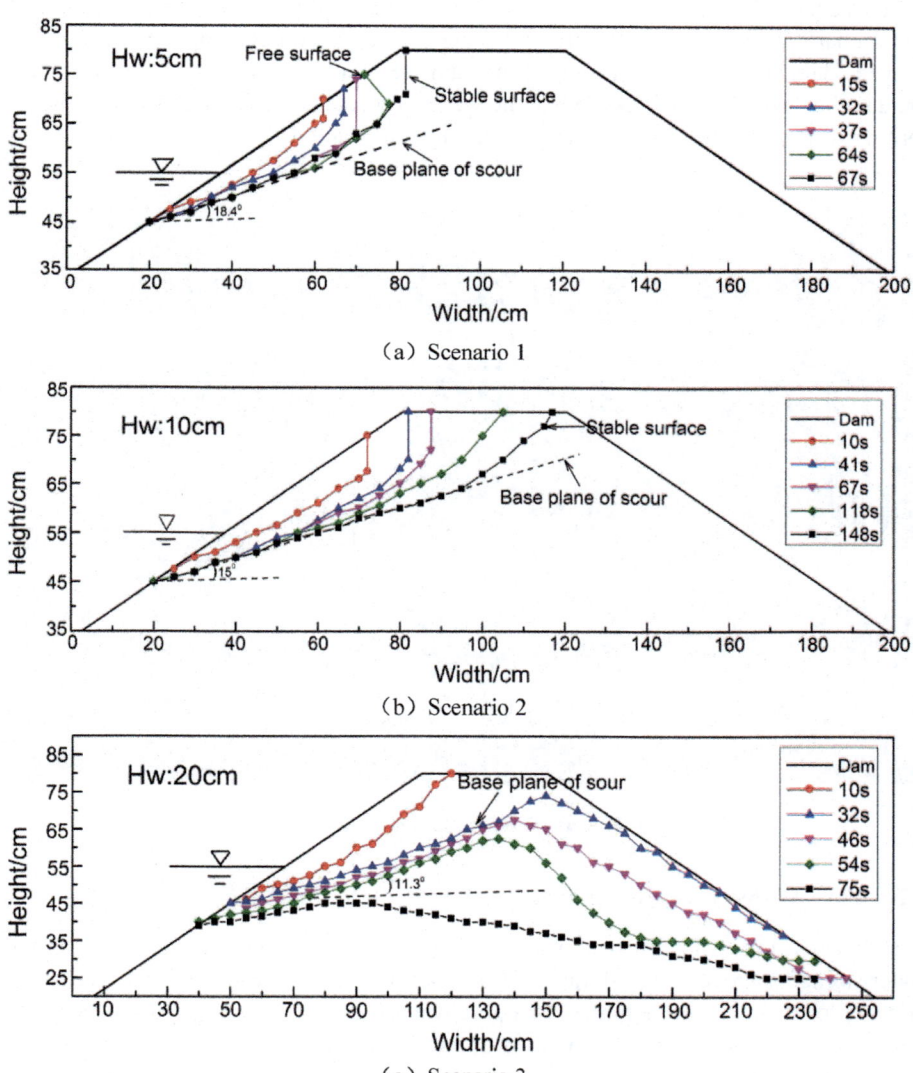

(a) Scenario 1

(b) Scenario 2

(c) Scenario 3

Fig. 3. Schematic diagram of dam failure

datum. In the process of wave scouring, the upstream dam slope will continue to suffer from local instability. Finally, as the wave climbing height is not enough to cause further scour of the dam, the upstream dam slope will form a stable scour surface and the process continues until wave making ends.

In scenario 2, the wave height is 10 cm (Hw:10 cm). The erosion process of wave to the dam is similar to the scenario 1, as shown in Figs. 3(b) and 4. In this scenario, the failure area of the dam is further increased, and finally a stable scour surface is also formed.

In scenario 3, the wave height is 20 cm (Hw:20 cm). Unlike the working scenario 1 and scenario 2, the waves not only will scour the upstream dam slope, but also climb over the dam crest, causing erosion of the dam crest and downstream dam slope. Then the dam began to rapidly burst. When the wave stop erosion, a stable erosion surface is formed or the dam breaks completely (see Figs. 3(b) and 4).

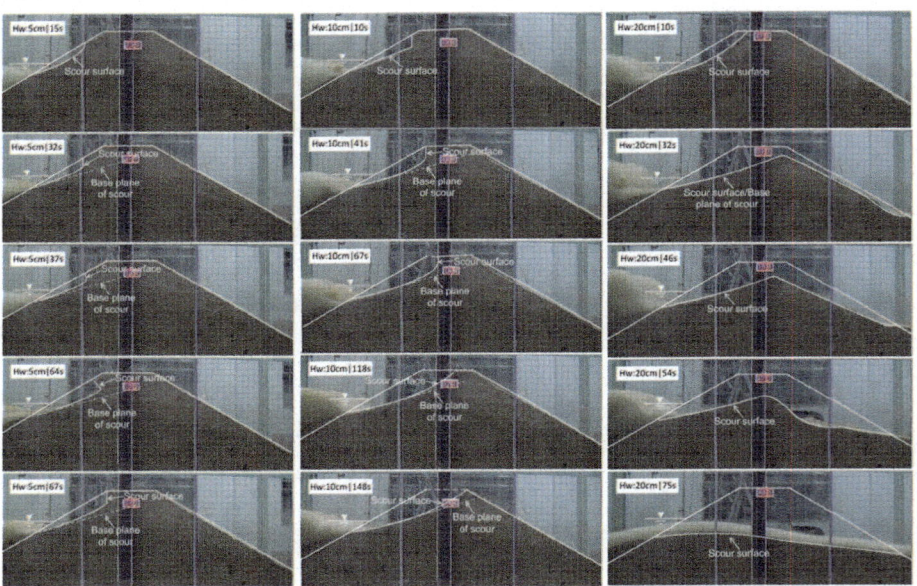

Fig. 4. The dam failure process

3.2 Pore Water Pressure Analysis

In scenario 1, the change of pore water pressure is shown in Figs. 5(a) and 6(a). In the water storage stage, the growth rate of 1-5# pore pressure is the same during the rising of water level and the growth rate of 6# pore water pressure is obviously smaller. In the static stage, the pore water pressure at each measuring point is relatively stable, and the pore pressure value is related to the location of each measuring point. By Fig. 6(a), it can be seen that in the wave making stage, the 1-3# pore pressure varies periodically with the wave, and the magnitude is basically the same as the height of wave. However, the pore pressure at 4-6# monitoring point has weak response to the wave. In the stage

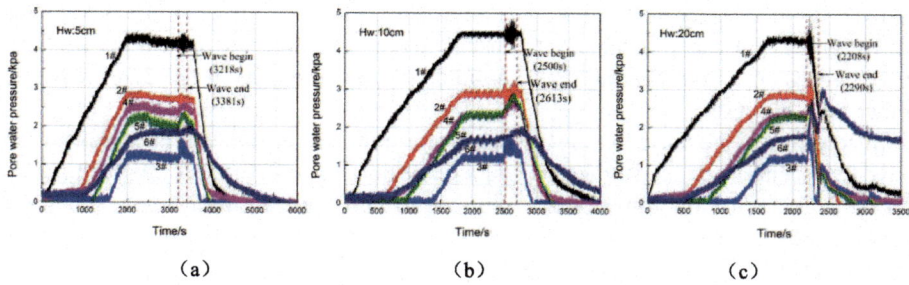

Fig. 5. Pore water pressure response in the dam

of water releasing, the pore pressure at each measuring point of the dam decreases gradually. The decrease rate of pore pressure of 1-5# is the same, but the rate of 6# is obviously smaller, as shown in Fig. 5(a).

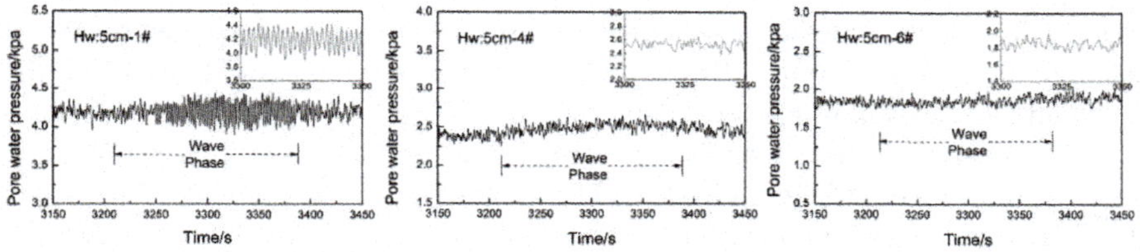

（a）Wave height:5cm

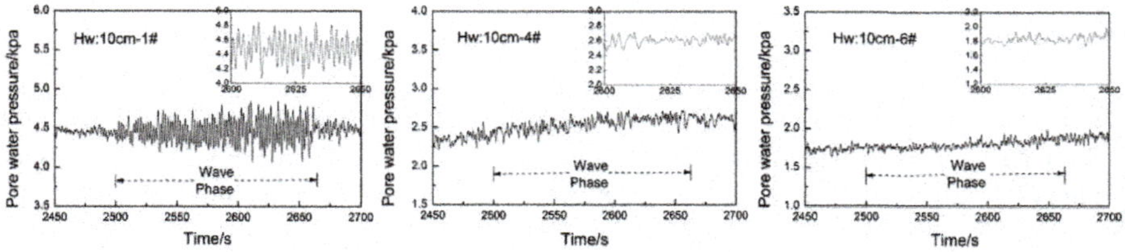

（b）Wave height:10cm

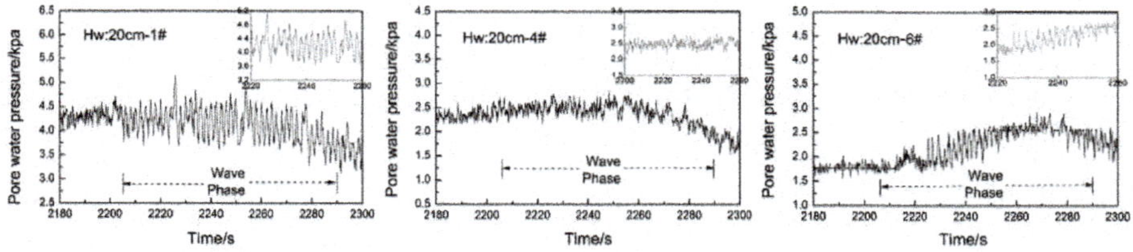

（c）Wave height:20cm

Fig. 6. Pore water pressure response in the wave making process

In scenario 2, the change of pore water pressure in dam is similar to that of scenario 1 in the process of water storage, static installation and drainage, as shown in Fig. 5(b). In the process of wave making, the amplitude of pore pressure in the dam increases further (see Fig. 6(b)).

In scenario 3, the variation law of pore water pressure in the dam is similar to the above two scenarios in the storage and static stage, as shown in Fig. 5(c). In the process of wave making, the amplitude of pore pressure in dam continues to increase. When the dam burst, the pore water pressure inside the dam would decline rapidly in the process of wave making (see Fig. 6(c)).

4 Conclusion

The stability of landslide dams under the action of landslide surge is studied in wave flume model tests with different wave heights. The following conclusions can be drawn.

The surge wave would cause erosion and sliding on the upstream slope. A steady and flat slope would remain when the wave height is low. The erosion became larger and the finally steady slope became flatter with the increase of the wave height. When the surge wave surpasses the dam crest, significant erosion would occur in the downstream slope, leading to overtopping failure.

The pore water pressure periodically varied with the wave and the magnitude is basically the same as the height of wave. The area far from the upstream slope has a delayed and weak response to the pore pressure of the wave. When the dam burst, the pore water pressure will decline rapidly.

References

1. Zhang, J., Cao, S.Y., Yang, F.G.: Experimental study on scouring of landslide dam. J. Sichuan Univ. **42**(5), 191–196 (2010)
2. Wang, Z.H., Chen, H., He, L.J.: Experimental study on effect of upstream burst peak on Dam breaching. Yellow River **37**(5), 38–41 (2015)
3. Risley, J.C., Walder, J.S., Denlinger, R.P.: Usoi Dam wave overtopping and flood routing in the Bartang and Panj Rivers, Tajikistan. Nat. Hazards **38**(3), 375–390 (2006)
4. Xu, F.G., Yang, X.G., Zhou, J.W.: Experimental study of the impact factors of natural dam failure introduced by a landslide surge. Environ. Earth Sci. **74**(5), 4075–4087 (2015)
5. Lin, P., Liu, X., Zhang, J.: The simulation of a landslide-induced surge wave and its overtopping of a dam using a coupled ISPH model. Eng. Appl. Comput. Fluid Mech. **9**(1), 432–444 (2015)

Stability Analysis of Rainfall-Triggered Landslides Considering the Previous Climatic Conditions

Ming Peng[1,2(✉)], Bingxin Li[1,2], and Zhenming Shi[1,2]

[1] Department of Geotechnical Engineering,
Tongji University, Shanghai 200092, China
pengming@tongji.edu.cn
[2] Ministry of Education Key Laboratory of Geotechnical and Underground
Engineering, Tongji University, Shanghai 200092, China

Abstract. In the past few decades, extreme rainstorms and dry weather have been occurring frequently. Despite of lots of studies on slope stability analysis, however, the study of rainfall-triggered landslides under the influence of the previous climatic conditions is less considered. This paper presents a case study on a rainfall induced landslide by taking account of three typical previous climatic conditions: normal climatic condition, extreme drought condition and extreme wet condition. Rainfall infiltration and slope stability analysis are conducted within the finite element software of Geostudio with different rainfall intensity. The analysis results demonstrate: under normal climatic condition, the rainfall intensity has no significant influence on the slope stability. Under extreme drought condition, due to the initial infiltration rate of soil is very small, long term and short intensity rainfall is easy to cause shallow landslides; short term and large intensity rainfall will result in large runoff flood. Under extreme wet condition, long term and short intensity rainfall will significantly reduce the stability of the slope, leading to deep landslides.

Keywords: Climate environment · Rainfall infiltration · Unsaturated soils
Slope stability

1 Introduction

Over the past century, the global climate change has frequently caused extreme weather events. According to the changes of temperature, rainfall and evaporation in Hong Kong area in recent decades, the trend of climate warming and frequent changes directly show the phenomena [1]. In recent years, the research on the mechanism of geological disasters caused by climate change has attracted the attention of scientists all over the world, and has become an important scientific problem in the field of geological disaster research. Wang [2] proposed the concept of extreme geological disasters and gave a grade to assess the risk of disasters. Zhang et al. [3] analyzed the effect of climate on the generation of possible geo-hazards. The previous researches are only through statistical methods to analyze the relationship between the occurrence of geological disasters and climate fluctuations. Griffiths et al. [4] analyzed the stability of

© Springer Nature Singapore Pte Ltd. 2018
A. Farid and H. Chen (Eds.): GSIC 2018, *Proceedings of GeoShanghai 2018*
International Conference: Geoenvironment and Geohazard, pp. 233–239, 2018.
https://doi.org/10.1007/978-981-13-0128-5_27

unsaturated slopes under steady seepage and evaporation conditions by the mean of finite element method. However, the existing researches only study the effect of rainfall and evaporation characteristics on the stability of unsaturated slope, rarely considering previous climatic conditions. Therefore, this paper studies the influence of rainfall on slope stability under three previous climatic conditions.

This paper presents a case study on a rainfall induced landslide by taking account of three typical previous climatic conditions. Rainfall infiltration and slope stability analysis are conducted within the finite element software of Geostudio with different rainfall intensity.

2   Numerical Analyses

2.1   Saturated-Unsaturated Slope Model

This paper takes the Sau Mau slope in Hong Kong as the research slope model. The height of the slope is 30 m, and the slope angle is 32° (Fig. 1a). The slope consists of two soil layers; the upper layer is silt, with a porosity ratio of 0.7 and a saturated permeability of 4.93×10^{-6} m/s. The lower layer is clay, with a porosity ratio of 0.8 and a saturated permeability of 8.74×10^{-9} m/s. The following Table 1 shows the typical physical and mechanical parameters of the two soils. The boundary conditions are shown in Fig. 1b. The ground surface is subjected to either evaporation or rainfall infiltration. When the rainfall intensity is less than the saturation permeability of the surface soil, it is set to the flow boundary, otherwise the head boundary. The two sides of the slope below the initial ground water table are set to the head boundary and above the initial ground water table are set to zero flow boundary. The bottom boundary is impervious. The model uses unstructured quadrilateral and triangular meshes. As the surface conditions vary with climatic conditions, in order to deal with such a sharp boundary changes, the grid within 1 m of the slope surface is encrypted, with a total of 2975 units.

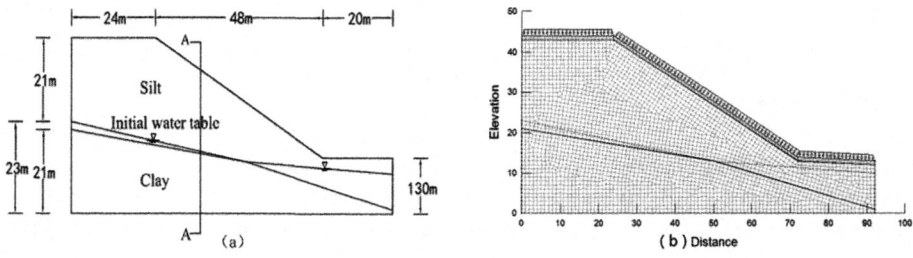

Fig. 1. Slope model: (a) Profile of the soil slope, (b) Boundary conditions and FE discretization

In the study, the soil-water characteristic curves and permeability functions for the two soils, shown in Fig. 2, are generated based on the residual saturation, porosity ratio and saturated permeability values following the methods developed by Fredlund and Xing [5, 6].

Table 1. Soil types and properties considered in slope stability analyses

Soil type	e	S_r (%)	k_s (m/s)	φ' (deg)	c' (kPa)	γ (kN/m^3)
Silt	0.7	10	4.93E-06	30	5	20
Clay	0.8	15	8.74E-09	20	10	20

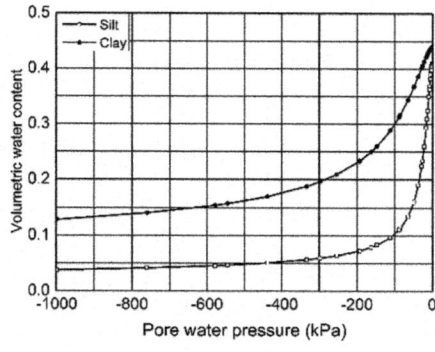 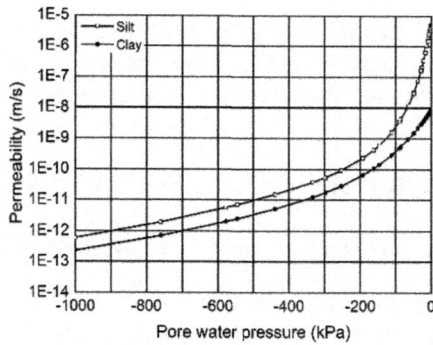

Fig. 2. The soil-water characteristic curves and permeability function for the slope soils

The constitutive equation is proposed by Fredlund and Morgenstorn [7] for unsaturated soils.

$$\varepsilon = C_t d(\sigma - \mu_a) + C_a d(\mu_a - \mu_w) \tag{1}$$

C_t and C_a are, respectively, compressibility of the soil structure with respect to changes in $(\sigma - \mu_w)$ and $(\mu_a - \mu_w)$, μ_a is the air pressure, and μ_w is the pore water pressure.

2.2   Simulate Previous Climatic Conditions

Three previous climatic conditions are normal climatic condition, extreme drought condition and extreme wet condition. The finite element analysis software Geostudio is employed to produce the distribution of initial pore water pressure to simulate previous climatic conditions by evaporation or rainfall infiltration, which is used to analyze rainfall infiltration and slope stability.

Normal Climatic Condition
The transient seepage analysis module is used to set initial ground water table and then automatic generate the distribution of initial pore water pressure to simulate normal climatic condition.

Extreme Drought Condition
The slope was continuously evaporated with the annual average rate of evaporation, 1402 mm/year, for 60 days on the basis of the normal climatic condition, using the transient seepage analysis module [8].

Extreme Wet Condition

The slope was subjected to a rainfall with the intensity of 17 mm/h for 7 days and then the water naturally infiltrate into the ground water table (i.e. in 13 days) on the basis of the normal climatic condition.

2.3 Rainfall Characteristics

The rainfall intensities respectively are 12, 24 and 36 mm/h. Because rainfall patterns may have potential effects on slope stability, it is assumed that rainfall intensity is constant. The total rainfall amount is set to 144 mm in the three cases.

3 Results and Discussion

3.1 Distributions of Pore Water Pressure

This section studies the influence of rainfall on the distribution of pore water pressure of slope under different previous climatic conditions. The following Fig. 3 shows the distributions of pore water pressure in different cases.

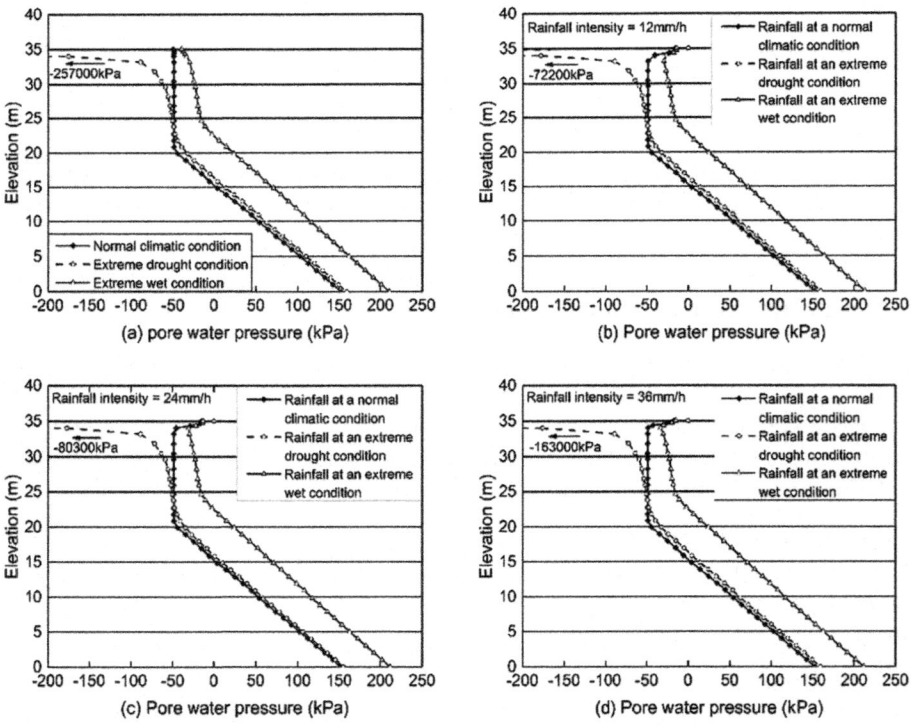

Fig. 3. Distributions of pore water pressure in the slope under different previous climatic conditions (a) 0 mm/h, (b) 12 mm/h, (c) 24 mm/h, (d) 36 mm/h

Given a rain event of a limited duration, the distribution of pore water pressure show that:

1. Under normal climatic condition, the pore water pressure of the soil on the surface increases during the rainfall process. With the increase of rainfall intensity, the depth of rainfall infiltration decreases because of the occurrence of runoff.
2. Under extreme drought condition, the pore water pressure of the soil on the shallow layer quickly becomes saturated during the rainfall process. The suction of the soil below the shallow layer is very large, which hinders further infiltration. Rainfall mainly affects the stability of slope surface.
3. Under extreme wet condition, the variation trend of pore water pressure on the slope surface soil is similar to that under normal climatic condition. The pore water pressure of slope soil is generally high.

3.2   Stability of Slope

In this section, the influence of different rainfall intensity on slope stability and slope failure form is discussed under different previous climatic conditions. Note the rainfall amount as a constant value of 144 mm (Fig. 4).

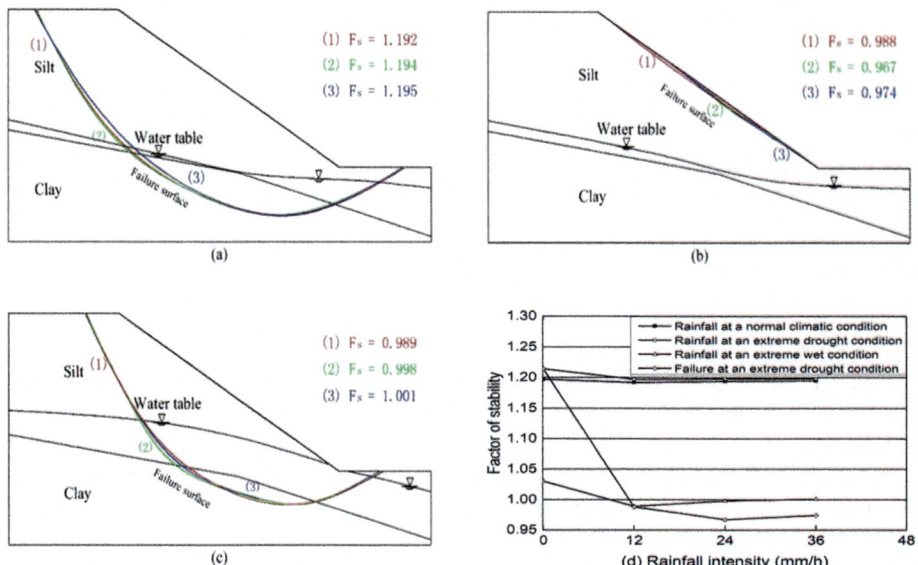

Fig. 4. Sliding surfaces of different rainfall intensity under different previous climatic conditions (a) Normal climatic condition, (b) Extreme drought condition, (c) Extreme wet condition, (d) Factor of stability

1. Under normal climatic condition, because the rainfall amount is limited to 144 mm, the slope stability is only a small drop during the rainfall process. The influence of different rainfall intensity on slope stability is very small (Fig. 4a).

2. Under extreme drought condition, because of the limited rainfall, the slope has remained stable during the rainfall process, but it is prone to cause debris flow. In order to study the failure forms of slope, different intensities of rainfall are applied until the slope is destroyed. (1) The rainfall intensity of 12 mm/h, when the rainfall reaches 288 mm, the slope is shallow sliding. (2) The rainfall intensity of 24 mm/h, when the rainfall reaches 384 mm, the slope is shallow sliding. (3) The rainfall intensity of 36 mm/h, when the rainfall reaches 360 mm, the slope is shallow sliding (Fig. 4b).
3. Under extreme wet condition, the stability of the slope is at the critical state. During the rainfall process, the slope is prone to deep sliding failure. The influence of rainfall intensity on slope stability is not significant with the same rainfall amount (Fig. 4c).

3.3 Analyses of Runoff

The pore water pressure is closely related to the permeability of unsaturated soil. Meanwhile, the rainfall intensity also determines the amount of rainfall infiltration. The variation of runoff rate with time in the A-A profile is as follows (Fig. 5).

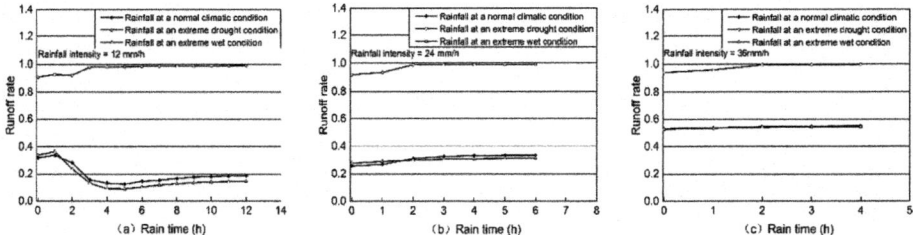

Fig. 5. Runoff rate and cumulative runoff unit area under different previous climatic conditions (a) 12 mm/h, (b) 24 mm/h, (c) 36 mm/h

1. Under normal climatic condition, with the development of rainfall infiltration, the permeability of soil in the slope surface is increased. When the rainfall intensity is 12 mm/h, most of the rainfall infiltrates into the slope, and a small part of rainfall flows in the form of runoff. With the increase of rainfall intensity to 36 mm/h, the runoff rate gradually increased to about 50%, and more than half of the rainfall infiltrates into the slope.
2. Under extreme drought condition, due to the low permeability of the surface soil, rainfall mainly flows in the form of runoff. Under different rainfall intensity, the runoff rate is over 90%. Therefore, a low intensity but longer duration of rainfall is more likely to cause large scale runoff, which can lead to urban waterlogging.
3. Under extreme wet condition, the surface soil is nearly saturated, besides the runoff rate is over 50%. Under different rainfall intensity, rainfall mainly infiltrates into the slope, and its runoff rate is similar to normal climatic condition.

4 Conclusion

In this paper, three typical previous climatic conditions are considered, and the mechanism of rainfall induced landslide under different climatic conditions is studied:

(1) Under normal climatic condition, the effect of different intensity rainfall on slope stability is not significant under the same rainfall amount. More than half of the rainfall infiltrates into the slope.
(2) Under extreme drought condition, rainfall is difficult to infiltrate into the slope. During the rainfall process, the slope has remained stable during the rainfall process, but it is prone to cause debris flow. The long duration rainfall will result in the shallow sliding.
(3) Under extreme wet condition, the ground water table rises, and the surface soil is nearly saturated. Therefore, a low intensity but longer duration of rainfall is more likely to cause deep sliding.

Based on two-dimensional rainfall infiltration, this paper analyses the stability of rainfall-triggered landslides considering previous climatic conditions. Further study will consider the stability of unsaturated slope under the three-dimensional rainfall infiltration and dynamic change of rainfall intensity with time.

References

1. Leung, Y.K., Yeung, K.H., Ginn, E.W.L., Leung, W.M.: Climate change in Hong Kong, Hong Kong Observatory Technical Note No. 107 (2004)
2. Wang, S.: Extreme Geo-disasters and risks. J. Eng. Geol. **19**(3), 289–296 (2011)
3. Zhang, L.M.: Analysis of geo-hazards caused by climate changes. In: Tenth International Symposium on Landslides and Engineered Slopes (2008)
4. Griffiths, D.V., Lu, N.: Unsaturated slope stability analysis with steady infiltration or evaporation using elasto-plastic finite elements. Int. J. Numer. Anal. Meth. Geomech. **29**(3), 249–267 (2005)
5. Fredlund, D.G., Xing, A.: Equations for the soil-water characteristic curve. Can. Geotech. J. **31**(4), 521–532 (1994)
6. Fredlund, D.G., Xing, A., Huang, S.: Predicting the permeability function for unsaturated soils using the soil-water characteristic curve. Can. Geotech. J. **31**(4), 159A (1994)
7. Fredlund, D.G., Morgenstern, N.R.: Constitutive relations for volume change in unsaturated soils. Can. Geotech. J. **13**(3), 261–276 (2011)
8. National Standard of People's Republic of China GB/T 20481-2006: The grade of meteorological drought. China Standard Press, Beijing (2006)

Study on Features and Genetic Mechanism of Debris-Bedrock Interface Landslide

Sheng-Rui Su[✉], Chi Ma, and Min Guo

Department of Geological Engineering, Chang'an University, Xi'an, China
shengruisu@163.com

Abstract. The debris-bedrock interface landslides are widely distributed in the southwest and northwest region of China, but the features and genetic mechanism are not revealed thoroughly in the existing researches. On the basis of analyzing the futures of debris-bedrock interface landslides from literatures and the potential geological disasters investigation in Langao County, Shaanxi Province, this paper generalized the features of debris-bedrock interface landslides. According to laboratory tests results, the physical and mechanical properties of soil-rock mixture materials and rock-soil interface are discussed. It is found that the soil-bedrock interface is the weak structural plane which leads to the debris-bedrock interface landslide by comparison the strength between soil-rock mixture materials and soil-rock interface. Taking the Zushimiao landslide in Langao County which is a typical debris-bedrock landslide, the genetic mechanism is revealed based on site investigation, laboratory test and numerical simulations. It is thought that: the soil-bedrock interface as weak structural plane is controlling factor, the surface water infiltrating and groundwater seeping are triggering factors, and the debris-bedrock interface landslides occurred by combination of the two factors.

1 Introduction

Due to complex geological condition and active geological structure, China is one of the countries that most affected by geological disasters. Landslides are the main geological disasters, which restrict the development of national economy and threaten property safety. The debris-bedrock interface landslides are widely distributed in the southwest and northwest region of China, according to the survey, it is found that 96.63% of the Three Gorges reservoir area and 92.3% of the southern Shaanxi mountainous landslide belonged to this type (Zhu 2010; He 2013).

The structural surface which is widely developed in the slope has a great impact on the formation of landslides and then tends to be a potential slip surface (Cheng 2003). As a kind of discontinuous and low intensity interface, the soil-rock interface always is a key part and weak link of slope damage in engineering (Yang 1990). At home and abroad the study on the mechanical properties of the structural surface has a long history and rich research results: a series of strength formulas were summed up on the basis of a large number of experimental and theoretical analysis (Jeage 1971; Sun 1998). The main factors that affect the shear strength characteristics of structural surface were also

© Springer Nature Singapore Pte Ltd. 2018
A. Farid and H. Chen (Eds.): GSIC 2018, *Proceedings of GeoShanghai 2018 International Conference: Geoenvironment and Geohazard*, pp. 240–250, 2018.
https://doi.org/10.1007/978-981-13-0128-5_28

summed up (Xia and Sun 2002). The contact surface between clay and the structure is in accordance with the Mohr-Coulomb intensity criterion (Potyondy 1961).

In the aspect of formation mechanism and stability evaluation, the authors have done a lot of researches on the debris-bedrock interface landslides. The shape of bedrock surface has a great influence on the stability and resurrection of the debris-bedrock interface landslides by comparing the quantitative calculation with monitoring values (Yang 1995). Rainfall and the change of groundwater level are main causes of landslide instability (Chen et al. 2015).

The features and genetic mechanism of debris-bedrock interface landslides are not revealed thoroughly in the existing researches. As the Zushimiao landslide in Langao County is a typical debris-bedrock landslide, the futures and genetic mechanism are revealed based on site investigation, laboratory test and numerical simulations.

2   Basic Characteristics of Debris-Bedrock Landslides

Based on the potential geological disasters investigation, 318 debris-bedrock interface landslides are found in Langao County (Zhang 2012), this paper generalized the features of debris-bedrock interface landslides (Ma 2016).

(1) This kind of landslides which mostly shallow surface are generally developed at a height of 600 to 1200 m, with 20° to 40° gradient and 180° to 270° slope direction.

(2) The slip body of such landslide consists mainly gravel soil or gravel silty clay, and this type of landslide developed along the underlying bedrock as a fractured structure.

(3) The development of such landslides usually pass through four stages: the creep stage, the slip stage, the slippery stage and the stable stage, however, after the landslide stabilizes, the slope will slip again and repeat the destruction of the previous experience when the various predisposing factors can form a greater impact again.

(4) This kind of landslides are mainly controlled by the soil-rock interface, which failure modes are simple, mainly including creep-cracking and slippage-push failure modes.

In the following, the genetic mechanism of debris-bedrock landslides is revealed by taking the Zushimiao landslide.

3   The General Situation of Zushimiao Landslide

The Zushimiao landslide is located on the west side of the county town of Langao County, Ankang City. The trailing edge of the landslide is a steep, the leading edge is located above the retaining wall, and the main slip direction is about 61°. The landslide's leading edge and trailing edge elevation is 427.0 m and 455.0 m. The upper part of the landslide is wide and the lower part is narrow, the rear was curved, and the overall shape likes a tongue. The landslide is 62 m long and 26 m wide, its thickness is about 4 m, and the total volume of landslide is about 5.9×10^3 m³, it belongs to a small landslide. The sliding belt is near the debris-bedrock interface, its thickness is about 1.2 m, it has obvious signs of deformation. The sliding bed mainly is strong weathered tuffaceous conglomerates with fractured structure.

After two rains from August 8 to 15 and August 29 to August 1, 2003, a partial collapse of the slope occurred in the middle part of the landslide, cracks and deformation appeared at the trailing edge and houses on the slope. Since the front edge of the slope is steep, the landslide cut off from the upper part of the retaining wall, resulting in the trees skewed in landslide and the Ankang-Langao highway roadbed sank, the sink range is about 8–12 cm, and the length of the sink is about 9.3 m.

4 Experimental Study on Physical and Mechanical Properties of Soil-Rock Interface

Nature of the soil and rock mass is the key factor of the debris-bedrock interface landslides, therefore, the study on the physical and mechanical properties of gravel and gravel-bedrock interface have a great significance for the investigation of the futures and genetic mechanism of debris-bedrock interface landslides.

4.1 Physical Properties Test Results

The contents of the experiment mainly include density tests and water content determination tests. The basic physical parameters of gravel soil are shown in Table 1.

Table 1. Physical index of gravel soil

Natural density (g/cm³)	Dry density (g/cm³)	Saturation density (g/cm³)	Natural water content (%)	Saturated water content (%)
1.98	1.65	2.11	19.6	28.97

4.2 Mechanical Properties Test Results

The tests include gravel soil and soil-rock interface shear tests, the influence of water content on the mechanical indexes of gravel soil and debris-bedrock interface are studied emphatically.

4.2.1 Shear Test Results of Gravel Soil

In this test, the samples were tested with different moisture content, which were dry, slightly wet, moist and saturated respectively. First, the original samples were tested and get the C = 23.25 kPa, φ = 19.86°. According to the four statistics (Table 2), the internal friction angle gradually decreases with the increase of water content.

Table 2. Mechanical properties of gravel soil

Condition	Water content (%)	C (kPa)	φ (°)	tanφ
Dry	0.54	10.27	36.13	0.73
Slightly wet	10.86	83.77	26.84	0.51
Moist	20.30	76.43	20.01	0.36
Saturated	28.97	62.69	15.11	0.27

It is found that the water content ω is approximately linear with the internal friction angle φ through experiments (Fig. 1), and the experimental data are linearly fitted to obtain the following relationship:

$$\varphi = -74.07\omega + 35.757$$

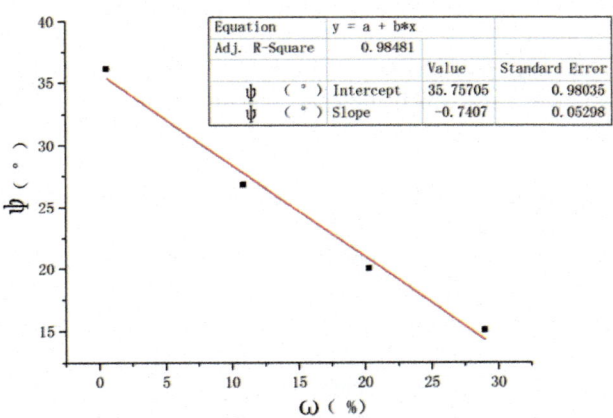

Fig. 1. Relationship between internal friction angle of gravel soil and moisture content

4.2.2   Shear Test Results of Soil-Rock Interface

This test employed the same conditions with gravel soil shear test. The friction angle decreased with the increase of the water content of the samples (Table 3).

It is also found that the water content at the soil-rock interface is linear with the internal friction angle (Fig. 2):

$$\varphi = -112.39\omega + 38.210$$

Table 3. Mechanical properties of soil-rock interface

Condition	Water content (%)	C (kPa)	φ (°)	tanφ
Dry	0.43	44.44	38.78	0.89
Slightly wet	11.70	71.78	24.27	0.43
Moist	18.70	30.72	15.54	0.25
Saturated	28.60	26.83	7.45	0.19

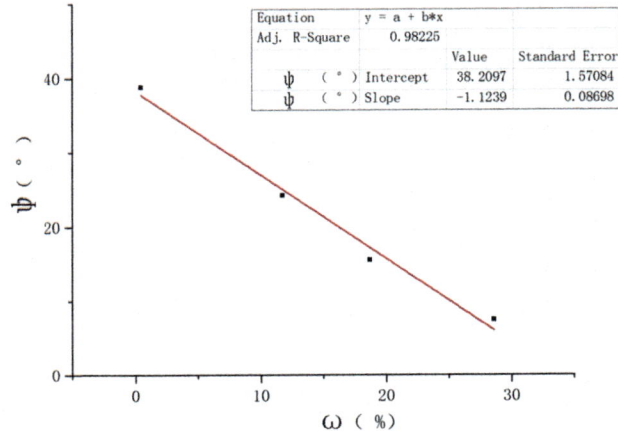

Fig. 2. Relationship between internal friction angle of soil-rock interface and moisture content

Except dry condition, the internal friction angles of the debris-bedrock interfaces are smaller than gravel soil, which indicates that the debris-bedrock interface is the weak structural plane of landslide deformation and failure.

4.2.3   Shear Test Results of Soil-Rock Interfaces of Different Fine Particle Content

In order to study the effect of fine particle matter (≤ 2 mm) on the stability of landslide near the soil-rock interface, extremely fine, fine and slightly fine particles were added on the basis of natural gradation. The soil shear strength decreases with the increase of fine particles (Table 4). For debris-bedrock interface landslides, due to the soil-rock interface is the dominant seepage surface and the impermeable interface, under the action of seepage, the fine particulate matter near the interface is easy to gather, so the interface is easy to become the weak structural plane.

Table 4. Mechanical properties of soil-rock interface in different grain group

Condition	Water content (%)	C (kPa)	φ (°)
Slightly fine (2–1 mm)	18.7	16.80	14.06
Fine (0.5–0.25 mm)	18.7	30.50	12.45
Extremely fine (≤ 0.075 mm)	18.7	51.02	10.07

5   Numerical Simulation Analysis of the Genetic Mechanism of the Debris-Bedrock Landslides

Based on the field investigation and indoor tests, the formation process of Zushimiao landslide is simulated by using the MIDAS/GTS finite element analysis software. The genetic mechanism of the debris-bedrock interface landslides were revealed by simulating the deformation and the strain distribution rule under different working conditions.

5.1   The Establishment of Numerical Simulation Model

According to the actual situation, the model was established as follows: The overall direction of the model (X direction) is 110 m, and the vertical direction (Y) is 54 m. The geological model was defined as: debris (gravel soil), soil-rock interface (weak zone) and bedrock.

The mesh division of the model is shown in Fig. 3.

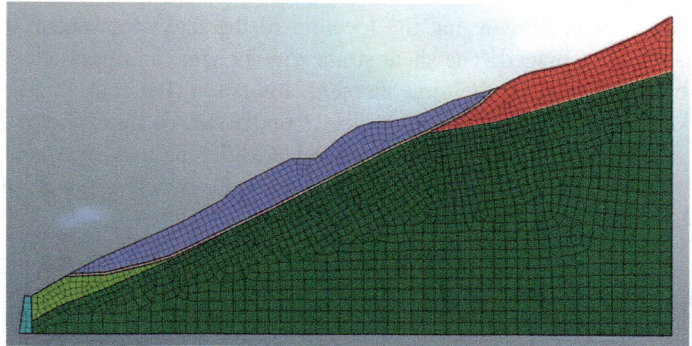

Fig. 3. Model meshing figure

According to Tables 1, 2, 3 and 4 and the stratigraphic lithology in the area where the Zushimaio landslide is located, the physical and mechanical parameters used in this simulation are shown in Table 5.

Table 5. Physical and mechanics index of model materials

Material	Elastic modulus (MPa)	Poisson's ratio	Natural unit weight (KN/m^3)
Retaining wall	4000	0.28	26
Bedrock	5000	0.25	30
Slip body	30	0.30	19.6
Sliding surface	18	0.35	20
Material	Saturated unit weight (KN/m^3)	C (MPa)	φ (°)
Retaining wall	27	0.060	40
Bedrock	31	0.100	45
Slip body	21	According to Tables 4.1–4.3	
Sliding surface	22		

5.2   Definition of Boundary Conditions

The specific boundary conditions are as follows: (1) The slope of the landslide is set as the free surface; (2) The direction of the gravitational field is Y direction without considering the tectonic stress field and the temperature field; (3) According to the actual situation, deformation and displacement can occur only on the slope side of the rock and soil, therefore, the bottom and the sides of the model are fixed.

5.3   Numerical Simulation Results

5.3.1   Results of Natural State

The stability of the slope was calculated by Midas/NX Slope Stability (SRM) module without considering seepage state. The stability coefficient of the landslide is 1.0575 under natural state.

Many scholars have shown that the location of the maximum shear strain is the place that easy to be damaged. The shear strain zone become an important criterion for the destruction of the slope in the finite element simulation (Pei et al. 2010). The shear strain zone occurs at the debris-bedrock interface under natural state, and the maximum shear strain is concentrated in the middle of the landslide, the maximum shear strain is 1.465 (Fig. 4). The debris-bedrock interface is the weak structural plane of the landslide, and it's also the place that easy to be damaged.

Fig. 4. Maximum shear strain clouds in natural state

5.3.2   Results of Seepage State

In order to analyze the deformation characteristics of landslide under the conditions of rainfall infiltration and surface water seepage, the boundary condition of the model is changed, the stress-seepage complete coupling analysis module is used to simulate the slope. Details are as follows: the maximum precipitation in the study area is 134.9 mm/d (2010.07.18), which is chosen as the simulated rain intensity. The rainfall infiltration boundary is set on the slope, and the node flow rate $Q = 0.00562$ m^3/h according to the maximum rain intensity. Set drainage boundary in the front of landslide, the water pressure is 0; the analysis time is set to 24 h and the analysis step is 24.

The shear strain first appear at the slip zone and then spread to the trailing edge of the landslide gradually along with the rainfall time, it forms a shear strain zone through the landslide finally (Figs. 5, 6 and 7). $T = 3.800\mathrm{e}{-}001$ h, the shear strain is mainly located in the lower part of the sliding surface, the shear strain is small, the maximum shear strain is 0.911; $T = 7.526\mathrm{e}{+}000$ h, the shear strain zone of the slope continues to expand, but does not penetrate to the trailing edge of the landslide, the maximum shear

Fig. 5. Maximum shear strain clouds at T = 3.8000e–001 h

Fig. 6. Maximum shear strain clouds at T = 7.526e+000 h

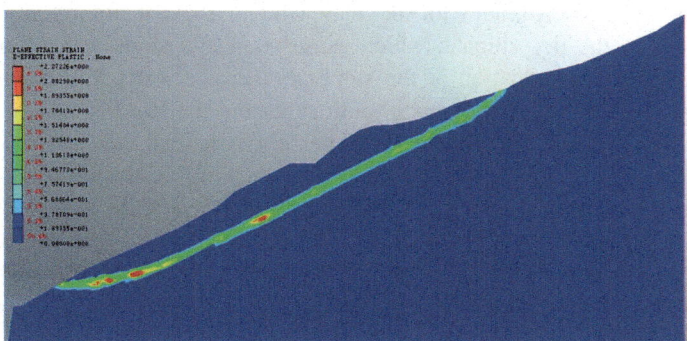

Fig. 7. Maximum shear strain clouds at T = 2.4e+001 h

strain is 1.522; T = 2.4e+001 h, the shear strain zone continues to increase, the maximum is 2.272, and the shear strain zone is formed at the soil-rock interface, the landslide occurs along the debris-bedrock interface.

6 The Genetic Mechanism of Debris-Bedrock Interface Landslide

Zushimiao landslide is a typical debris-bedrock interface landslide. Combined with the results of site investigation, laboratory tests and numerical simulations, the formation of the landslide has a close relationship to the physical-mechanical properties of the debris-bedrock interface, the composition of the landslide, rainfall and groundwater.

6.1 Sliding Surface Properties

The sliding surface of Zushimiao landslide is the soil-rock interface, the underlying bedrock is tuffaceous conglomerate, and the upper part is a thicker debris layer. Through laboratory tests, it can be concluded that the shear strength index decreases with the increase of the water content both in the gravel soil and the soil-rock interface. As the debris-bedrock interface is a weak rock layer and the leaching of groundwater, the fine particulate matter near the interface is easy to gather,which reducing the internal friction angle of the debris-bedrock interface,with far bigger falls for saturated state. Through the numerical simulation results, it is found that the strain zone is most likely occured near the debris-bedrock interface. It is also proved that the soil-rock interface is the weak structural surface that is most likely to form the slip surface.

6.2 Material Composition of Landslide Body

First of all, the sliding surface of the Zushimiao landslide is concave in morphology, this topography is conducive to the stacking of the slope sediments, which provides the material condition for the formation of the landslide. The main components of the landslide are gravel soil, the gravel size is concentrated in 1.25–10 mm, which contain some stones, and graded well.

Secondly, the gravel is filled with silty clay, which has expansibility. So it is easy to produce cracks in the dry climate, which causes rain and slope water flow into the cracks. In the initial deformation of the slope, the surface water infiltration is aggravated by the cracks of the trailing edge, both sides and the middle.

Finally, there is a large difference in permeability between the sliding surfaces,the upper loose soil permeability is good and the lower tuff permeability is poor. The increase of water content in the slope leads to the gain of saturation and weight of the slope, and the declining force of the landslide increases,in addition, stagnant water is formed near the concave sliding surface, which further causes the landslide to slide.

6.3 The Role of Water

According to the investigation, the blockage of the ditches along the central part of Zushimiao landslide caused the infiltration of rainwater into the slope during rainfall. In addition,there is a large area of farmland on the slope, the water diversion irrigation resulting in continued infiltration of groundwater, making the slope a higher level of the groundwater, resulting in the increase of soil moisture content, pore water pressure and unit weight increase, as a result, the sliding force increases and the anti-skid force decreases, at the same time, due to the difference of water permeability, greater dynamic water pressure and buoyancy force are produced at the soil-rock interface. The results of numerical simulation show that with the continuous infiltration of rainfall, the shear strain zone of landslide increases continuously until it through the landslide, which has a great influence on the stability of the slope.

6.4   The Genetic Mechanism of Debris-Bedrock Interface Landslide

To sum up, because of the special geological structure and material composition of the Zushimiao landslide, together with more rainfall in Langao County, the hydrogeo-logical conditions of slope can be changed easily, meanwhile, the soil-rock interface is the weak structural plane, the dominant seepage surface and the impermeable interface, so the fine particulate matter near the interface is easy to gather. Coupled with the drainage ditch of the highway in the middle of the slope is not smooth, there are many cracks in the slope, and the structure of the gravel soil is loose, which can cause a large amount of rainwater to infiltrate the slope. The above reasons lead to the increase of slope weight, pore water pressure and the landslide slip force, at the same time, the saturated zone is formed near the debris-bedrock interface and the shear strength is greatly reduced. These factors contribute to the upper soil slide to the lower part of the slope along the soil-rock interface.

7   Conclusions

(1) The slip body of debris-bedrock interface landslide is generally gravel soil or gravel silty clay, sliding surface is soil-rock interface. This kind of landslides are generally developed at a height of 600 to 1200 m, a gradient of 20° to 40°, and a slope direction of 180° to 270°, and mostly are shallow surface landslides.

(2) The internal friction angle of the debris-bedrock interface is smaller than gravel soil, and it decreases with the increase of water content. It indicates that the soil-rock interface is the weak structural plane which leads to the debris-bedrock interface landslide.

(3) By changing the grading of gravel soil, it is considered that the seepage of groundwater can enrich the fine particulate matter at soil-rock interface, which can decrease the shear strength of soil-rock interface. It is the important reason that the soil-bedrock interface becomes weak structural plane.

(4) The soil-bedrock interface as weak structural plane is controlling factor, the surface water infiltrating and groundwater seeping are triggering factors, and the debris-bedrock interface landslides occurred by combination of the two factors.

References

Chen, S.X., Xu, X.C., Xu, H.B.: Features and stability analysis of rainfall induced colluvial landslides. Rock Soil Mech. **26**, 6–10 (2015)

Cheng, Y.G.: Current situation and developments of landslide stud in china in recent twenty years. J. Geol. Hazards Environ. Preserv. **14**(4), 1–5 (2003)

He, H.: Research on micropile reinforcing mechanism and application of expansive soil landslides in shallow accumulative layer. Xi'an University of Science and Technology (2013)

Jeager, J.C.: Friction of rocks and stability of rock slopes. Geotechnique **21**, 91–134 (1971)

Ma, C.: Study on features and cause mechanism of debris-bedrock interface landslide. Chang'an University (2016)

Pei, L.J., Qu, B.N., Qian, S.G.: Uniformity of slope instability criteria of strength reduction with FEM. Rock Soil Mech. **31**(10), 3337–3341 (2010)

Potyondy, J.G.: Skin friction between various soils and construction materials. Technique **11**(4), 339–353 (1961)

Sun, G.Z.: Structure mechanics of rock mass. Science Press, Beijing (1998)

Xia, C.C., Sun, Z.Q.: Engineering Mechanics of Rock Joints. Tongji University Press, Shanghai (2002)

Yang, J.B.: Influence of bedrock surface feature on reviviscence of accumulation landslide. J. Catastrophol. **10**(2), 38–42 (1995)

Yang, Y.Y.: Damage Fracture Mechanics Model of Jointed Fractured Rock Mass and Its Application in Rock Mass Engineering. Tsinghua University (1990)

Zhang, N.B.: The Evaluating of the Susceptibility about Geological Disaster in Langao County Based on GIS. Chang'an University (2012)

Zhu, D.P.: Revival Mechanism and Deformation Prediction of Typical Accumulative Landslide in the Three Gorges Reservoir. China University of Geoscience (2010)

Environmental Geotechnics

Study on Degradation in Axial Bearing Capacity of a Cast-in-Situ Pile Caused by Sulfate Attack in Saline Area

Gaowen Zhao and Jingpei Li[(✉)]

Key Laboratory of Geotechnical and Underground Engineering of Ministry
of Education, Tongji University, Shanghai 200092, China
89099@tongji.edu.cn

Abstract. Sulfate ions attack cast-in-situ piles of bridges and roads and hence deteriorate their bearing capacities. A theoretical model is proposed for evaluating the bearing capacity of a cast-in-situ pile under sulfate corrosive condition. Evolution rules of the side resistance, the end resistance, the bearing capacity, the strength of pile body, as well as the compressive rigidity and the flexural rigidity are studied systematically. The effects of pile diameter and length on the bearing capacities are analyzed and compared in detail. Studies show that the side resistance of pile is influenced by both radial compressive stress redistribution and concrete deterioration caused by accumulating of sulfate corrosion products. And it is also influenced by length and diameter of the pile. The end resistance and the effective area decrease with the increase of sulfate corrosion depth. The strength of pile body decreases rapidly during sulfate corrosion process. Both the compressive and flexural rigidities show remarkable drops with the increase of corrosion depth. The results show that increasing the pile diameter can enhance the pile's resistance against corrosion and slow the decline in its bearing capacity.

Keywords: Sulfate saline soil · Cast-in-situ pile · Corrosion mechanism
Bearing properties · Evolution rules

1 Introduction

Degradation of concrete exposed to sulfate environment has been noticed since the early years of the 19th century and widely studied in recent years [1, 2]. Sulfate ions react with the hydration products of cement and the reaction products are mainly ettringite and gypsum [3]. It is also proved that the sulfate attack is caused by the reactions of sulfate and the cement hydration products [4]. The corrosion involves several chemical reactions that may produce secondary products such as ettringite, gypsum, and thaumasite [5–7].

Over the last decade, considerable efforts have been devoted to corrosion mechanism and models [8, 9]. Study of Santhanam [10] showed that expansion increased dramatically and the same rate of expansion was maintained until failure. Skaropoulou [11] studied the influencing factors on the thaumasite form in concrete sulfate corrosion. MA [12] argued that destruction of concrete in saline soil was mainly caused by

© Springer Nature Singapore Pte Ltd. 2018
A. Farid and H. Chen (Eds.): GSIC 2018, *Proceedings of GeoShanghai 2018*
International Conference: Geoenvironment and Geohazard, pp. 253–264, 2018.
https://doi.org/10.1007/978-981-13-0128-5_29

salt crystallization and corrosion products. Santhanam [13] claimed that sulfate attack caused the loss of strength and cohesion in concrete. Generation of corrosion products were widely studied and discussed in literature deals with corrosion mechanisms [14]. The bearing behavior of a cast-in-situ pile in saline area is influenced by sulfate corrosion, and such effect should be considered in the design of the cast-in-situ pile in sulfate saline area. However, studies on the bearing properties of the cast-in-situ pile subjected to sulfate attack in saline areas are rarely reported.

This paper analyzes the chemical reactions and corrosion mechanism of concrete that servicing in the sulfate saline soil. The side resistance, the end resistance, the bearing capacity, the strength of pile body, as well as the compressive rigidity and the flexural rigidity calculate models of cast-in-situ pile are proposed to evaluate the change of bearing behavior. A case study is combined to analyze the bearing behavior evolution rules of the cast-in-situ pile. Unfortunately, there is still no testing data from practical projects, and thus, more field tests are needed for further studies and verification in the future.

2 Sulfate Attack Reactions

Chemical attack is the main cause of the cast-in-situ pile failure [15–17]. The reactions are as follows [2]:

$$3CaO \cdot Al_2O_3 + 3(CaSO_4 \cdot 2H_2O) + 26H_2O \rightarrow 3CaO \cdot Al_2O_3 \cdot 3CaSO_4 \cdot 32H_2O \quad (1)$$

$$3CaO \cdot Al_2O_3 \cdot CaSO_4 \cdot 12H_2O + 2(CaSO_4 \cdot 2H_2O)$$
$$\rightarrow 3CaO \cdot Al_2O_3 \cdot 3CaSO_4 \cdot 32H_2O \quad (2)$$

$$4CaO \cdot Al_2O_3 \cdot 13H_2O + 3(CaSO_4 \cdot 2H_2O) + 14H_2O$$
$$\rightarrow 3CaO \cdot Al_2O_3 \cdot 3CaSO_4 \cdot 32H_2O + CaO \cdot H_2O \quad (3)$$

In the reactions from (1) to (3), H_2O can be supplied timely and the product $CaO \cdot H_2O$ can dissolve in environmental solution in concrete. So the effect of H_2O and $CaO \cdot H_2O$ on volume change can be neglected. Therefore, sulfate attack reactions are simplified into a unified reaction as

$$\omega CA + \xi CS \xrightarrow{\ Corrosion\ } \psi CACS \quad (4)$$

where CA refers to aluminate. CS refers to gypsum and CACS stands for ettringite. ω, ζ and ψ are the stoichiometric coefficients of reactants and products, respectively.

As shown in Table 1, the products molar volume involved in the previous chemical reactions far surpass the molar volume of reactants [7].

The volume swelling ratio can be calculated based on the corrosion reactions and Table 1. The corresponding equations can be described as

$$\left(\frac{\Delta V}{V}\right)_i = \frac{\psi m_e}{\omega m_{CA} + \xi m_g} - 1 \quad (5)$$

Table 1. Volume of reactants and resultant involved in corrosion reactions

Reactant	Name of compound	Molar volume (cm^3/mol)
$3CaO \cdot Al_2O_3$	Tricalcium aluminate	88.8
$3CaO \cdot Al_2O_3 \cdot CaSO_4 \cdot 12H_2O$	Monosulfoaluminate	313.0
$4CaO \cdot Al_2O_3 \cdot 13H_2O$	Calcium aluminate hydrate	276.2
$3CaO \cdot Al_2O_3 \cdot 3CaSO_4 \cdot 32H_2O$	Ettringite	725.1
$CaSO_4 \cdot 2H_2O$	Gypsum	74.2
H_2O	Water	18.0

$$\left(\frac{\Delta V}{V}\right)_\infty = \sum_{i=1}^{n} \left(\frac{\Delta V}{V}\right)_i \tag{6}$$

where ΔV is the volume change of concrete reacted with sulfate ions. V is the initial volume of reactant, i denotes the ordinal number of corrosive reactions. m_{CA}, m_g, m_e are the molar volume of aluminate compound, gypsum and ettringite in corrosion reaction i, respectively. n is the total number of corrosion reactions.

Study of Tixier [2] showed that void in concrete cannot be fully filled by corrosive reaction products. The void volume that can be filled with expansive products corresponds to a fraction of concrete porosity (φ), defined as the fraction of capillary porosity (f). Then, the final deformation ratio can be calculated by:

$$\alpha = \left(\frac{\Delta V}{V}\right)_\infty - f\phi \tag{7}$$

where [3] the parameter, ϕ, takes the following form

$$\phi = \max\left(V_c \frac{\frac{w}{c} - 0.39\beta}{\frac{w}{c} + 0.32}, 0\right) \tag{8}$$

where f is a parameter ranging from 0.05 to 0.45, which to represent the filling degree of the voids in concrete by the reaction products [2]. w/c is the water-cement ratio of concrete. V_c is the volumetric fraction of cement in the concrete. β is the degree of hydration of the cement. It is assumed that the swelling of reaction products only take place in the radial direction and damaged concrete of pile losses the ability of vertical bearing capacity. R is the radius of the cast-in-situ pile and l is the depth of damaged concrete in cast-in-situ pile, then one can obtain

$$R_v = R - l \tag{9}$$

$$R_c = R + \Delta R \tag{10}$$

$$\Delta R = \alpha l \tag{11}$$

where R_v is the radius of the cast-in-situ pile bearing vertical load after sulfate attack. R_c is the final radius of the pile after sulfate attack. ΔR is the radius increment of the pile.

3 Degradation Models of a Cast-in-Situ Pile

3.1 Pile Strength in Saline Environment

It is assumed that the concrete is still continuous and isotropous medium after sulfate corrosion. The corrosion damage ratio (D) can be defined as [18]

$$D = \frac{A_f}{A} \tag{12}$$

where A_f and A are the damaged and undamaged cross-sectional area of cast-in-situ pile. Du [18] proposed the calculated models for strength and elasticity modulus of the sulfate attack-damaged concrete, which can be written as

$$f_c' = (-0.8129D + 1.0291)f_c \tag{13}$$

$$E' = (-1.2824D + 1.0255)E_0 \tag{14}$$

where f_c' is the compressive strength of the concrete after sulfate attack. f_c is the compressive strength of the non-corroded concrete. E' is the elasticity modulus after sulfate attack and E_0 is the initial elasticity modulus of concrete. Then, the ultimate bearing capacity of axial compressive reinforced concrete pile, N_k, can be given by [19]

$$N_k = \varphi_c f_c' A_{ps} + 0.9f_y' A_s' \tag{15}$$

Combining Eqs. (10) and (15), the compressive strength of the pile body after corrosion, N, can be expressed as

$$N = \varphi_c(-0.8129D + 1.0291)f_c A_{ps} + 0.9f_y' A_s' \tag{16}$$

where φ_c is the technological coefficient of a cast-in-situ pile, the value of which ranges from 0.60 to 0.90. f_c is the design compressive strength of concrete. A_{ps} is the cross-section area of the pile. f_y' is the design compressive strength of longitudinal steel bar. A_s' is the cross-sectional area of longitudinal steel bars. Hence, the compressive rigidity (EA) and the flexural rigidity (EI) of a circular cast-in-situ pile can be described as

$$EA = E'A = (-1.2824D + 1.0255)E_0 \pi R_v^2 \tag{17}$$

$$EI = E'I = \frac{(-1.2824D + 1.0255)E_0 \pi R_v^4}{4} \tag{18}$$

where A is the cross-section area and I is the moment of inertia of cast-in-situ pile.

3.2   Side Resistance of Corrosive Pile

Assuming that soil around pile is homogeneous and the distribution of soil pressure is linearly along lengthways of pile. The horizontal stress and the side friction can be given by

$$\sigma_h = k_h \gamma z \tag{19}$$

$$\tau_f = \sigma_h \mu \tag{20}$$

where σ_h is the horizontal stress. τ_f is the side friction. μ is the friction coefficient. k_h is the lateral pressure coefficient. γ is the specific weight of soil and z is the distance from the calculated point to the ground surface.

The radial stress increment and the total radial stress caused by swelling products can be written as

$$\Delta \sigma_h = E_h \Delta R \tag{21}$$

$$\sigma_u = k_h \gamma z + E_h l \alpha \tag{22}$$

where E_h is the lateral compressive modulus of soil. $\Delta \sigma_h$ is the radial stress increment and σ_u is the final radial stress around the cast-in-situ pile.

3.3   Vertical Bearing Property Models

It is assumed that the damaged concrete loses the vertical bearing ability after being damaged by corrosion reactions. The vertical bearing capacities of the piles after sulfate attack can be expressed as

$$Q_{uk} = \varphi_{si} Q_{sk} + \varphi_b Q_{pk} = \varphi_{si} Q_{sk} + \varphi_b q_{pk} A_p \tag{23}$$

$$Q_{sk} = \varphi_{si} \int_{S_p} \mu \sigma_u dz = \varphi_{si} \int_{S_p} \mu (k \gamma z + E_h l \alpha) dz \tag{24}$$

$$Q_{pk} = \varphi_b q_{pk} A_p = \varphi_b q_{pk} \pi R_v^2 = \varphi_b q_{pk} \pi (R - l)^2 \tag{25}$$

where Q_{uk} is the vertical bearing capacity. Q_{sk} is the side resistance. Q_{pk} is the end resistance. A_p is the cross-sectional area of pile. S_p is the lateral surface area of pile, which can be calculated by $S_p = 2\pi R_v dz$. q_{pk} is the ultimate end resistance coefficient. φ_{si} and φ_b are the size effect coefficient of side resistance and end resistance [19]. $\varphi_{si} = \varphi_b = 1$ when pile radius less than 0.4 m. The size effect coefficient can be described by $\varphi_{si} = (0.8/d)^{1/5}$ and $\varphi_b = (0.8/d)^{1/4}$ when the radius of pile exceeds 0.4 m.

Based on the chemical corrosion reactions (1)–(3) along with the Eqs. (17)–(19), the bearing capacity of a pile in sulfate saline environment can be calculated as

$$Q_{uk} = \varphi_{si} \int_{S_p} \mu(k_h \gamma z + E_h l \alpha) dz + \varphi_b q_{pk} \pi (R - l)^2 \tag{26}$$

in which, E_h, k_h, μ and γ can be obtained in field or lab test. q_{pk} can be obtained in investigation report. α can be obtained according to Eq. (5).

3.4 Selection of Sulfate Diffusion Model

Most of previous studies on sulfate diffusion models primarily consider the diffusion coefficient as constant. Diffusion of sulfate ions in concrete can be described by the Second Fick's law. Thus, the diffusion model given by Eq. (27) is selected to determine the relationship between the time and the diffusion depth of sulfate attack in concrete [20].

$$C(l, t) = (C_S - C_0)\left[1 - erf\left(\frac{l}{2\sqrt{D_e t}}\right)\right] + C_0 \tag{27}$$

where l is the diffusion depth of sulfate ions in concrete. t is corrosion time. C_s is the original concentration of sulfate ions around the pile. C_0 is the initial sulfate concentration in concrete. D_e is the diffusion coefficient of sulfate ions in concrete. C_s, C_0, and D_e in the diffusion model can be obtained from lab test results.

4 Analysis of the Proposed Models

4.1 Selection of Parameters and Calculating Method

In this paper, diffusion model expressed by Eq. (27) is used to determine the relation between the time and the corresponding diffusive depth. In the model, calculating steps of time and diffusion depth are assumed to be 0.05 a (from 0 to 20 years) and 0.01 mm, respectively. Four working conditions are considered in the case study: $L = 30$ m with $D = 0.8$ m, $L = 30$ m with $D = 1.2$ m, $L = 50$ m with $D = 0.6$ m, $L = 50$ m with $D = 1.0$ m. It is assumed that the initial compressive strength of pile concrete is 24.29 MPa, the initial compressive modulus of pile concrete is 30 GN/m^2 and $\mu = 0.38$. The reactions between the sulfate ions and steel bars are not involved so $A_{ps} = A$. It is also proposed that $\varphi_c = 0.7$, $E_h = 30$ MPa, $\gamma = 19$ kN/m^3, $k_h = 1.0$, $q_{pk} = 1500$ kPa, $\alpha = 0.018$ [34], $\varphi_{si} = (0.8/d)^{1/5}$, $\varphi_b = (0.8/d)^{1/4}$, $D_e = 8.8 \times 10^{-8}$ mm^2/s, $C_s = 800$ mol/m^3, $C_0 = 0$ mol/m^3.

4.2 Results and Analyses

4.2.1 Side and End Resistance

Figure 1 shows the variation of the sulfate diffusion depth with the corrosion time from 0 to 20 years. Diffusion of sulfate ions is relatively fast from the first year to the 7th year and then keeps constant at a lower rate in the following 13 years.

Figure 2 indicates the variation of the end resistance with the corrosion time, in which the ordinate is the ratio of the real-time end resistance and the initial end resistance.

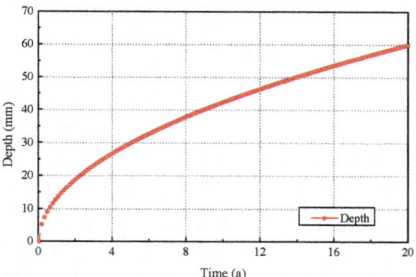

Fig. 1. Variation of diffusion depth with corrosion time

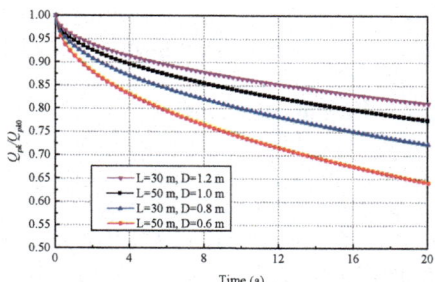

Fig. 2. Variation of end resistance with corrosion time

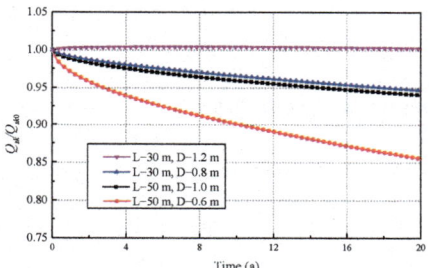

Fig. 3. Variation of side resistance with corrosion time

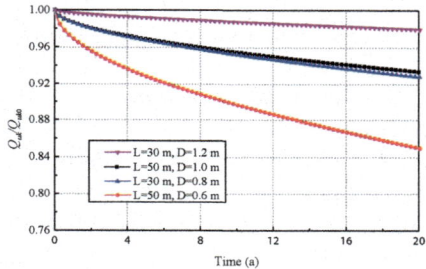

Fig. 4. Variation of bearing resistance with corrosion time

End resistance decreases apparently during the first to the 20th year. The end resistance under each working condition show drops of 94.4 kN ($L = 50$ m, $D = 0.6$ m), 152.5 kN ($L = 50$ m, $D = 1.0$ m), 127.8 kN ($L = 30$ m, $D = 0.8$ m) and 175.9 kN ($L = 30$ m, $D = 1.2$ m) respectively in the 7th year. After 20 years of sulfate corrosion, drops of 152.6 kN ($L = 50$ m, $D = 0.6$ m), 251.2 kN ($L = 50$ m, $D = 1.0$ m), 209.1 kN ($L = 30$ m, $D = 0.8$ m) and 291.1 kN ($L = 30$ m, $D = 1.2$ m) are observed. New cracks are formed because of the formation of ettringite caused by sulfate attack. The expanded cracks and new cracks make it easier for sulfate ions to diffuse further into internal concrete. Therefore, the effective cross-sectional area that bearing the vertical load becomes smaller. It can be seen in the Fig. 2 that the falling range of end resistance accounts for 19.0% of previous end resistance when $L = 30$ m and $D = 1.2$ m. While a loss of 36.0% is noticed when $L = 50$ m and $D = 0.6$ m.

The ordinate in Fig. 3 is the ratio of the real-time side resistance to the initial side resistance. The changing tendency of side resistances are quite different in each working condition. A sharp drop of 2477.6 kN, accounting for 14.5% of the initial value, is indicated when $L = 50$ m and $D = 0.6$ m. While, a slight increase of 0.2% is observed when $L = 30$ m and $D = 1.2$ m. Generation of new cracks and the swelling of surface concrete cause the growth of radial compressive stress of pile-soil, leading to an increase of frictional resistance. However, the effective friction area of pile decreases because the damaged concrete is incompact and the compressive strength can be neglected.

4.2.2 Bearing Capacity

The bearing capacity of a cast-in-situ pile is calculated using Eq. (26) and the results are shown in Fig. 4. The ordinate in Fig. 4 is the ratio of the real-time bearing capacity to the original bearing capacity. The bearing capacity of the pile decreases from 17469.9 kN all the way to 14839.6 kN after 20 years when $L = 50$ m and $D = 0.6$ m. The falling range occupies 15.1% of the initial bearing capacity. However, a bearing capacity drop of 2.1% appears when $L = 30$ m and $D = 1.2$ m. Sulfate corrosion reactions cause the redistribution of the radial compressive stress and decrease the effective friction area of pile-soil interface. The effective cross-sectional area, bearing the vertical load, decreases because the outer concrete layer is damaged. The end resistance of the pile drops sharply and hence the bearing capacity of pile shows a downtrend by a large decrement from the first year to the 20th year.

4.2.3 Effect of Pile Length on Bearing Capacity Degradation

The bearing behavior of the cast-in-situ pile effected by pile length is studied and the results are shown in Table 2. The loss ratio in Table 2 is the increment ratio, which is the specific value of the bearing capacity increment and the initial bearing capacity. Table 2 shows that the bearing capacity of the cast-in-situ pile is mainly controlled by the decrease of the side resistance. The side resistance is influenced by both the decease of effective friction area and the increase of radial compressive stress. The variation of

Table 2. Effects of pile length on bearing properties before and after corrosion

D/m	L/m	SR/kN	ER/kN	BC/kN	EA/MN	EI/MNm^2	Loss ratio/%
0.8	15	+88.9	−209.1	−120.2	−9653.2	−450.6	−4.3
	35	−748.3		−957.4			−8.0
	55	−2673.6		−2882.8			−10.2
	75	−5687.2		−5896.3			−11.4
	95	−9789.0		−9998.1			−12.1
1.2	15	+297.3	−291.1	+6.2	−14870.9	−1678.9	+0.1
	35	−189.5		−480.6			−2.8
	55	−1679.7		−1970.8			−5.0
	75	−4173.3		−4464.4			−6.2
	95	−7670.4		−7961.5			−6.9
1.6	15	+483.9	−366.0	+118.0	−20088.7	−4178.6	+2.0
	35	+293.6		−72.3			−0.3
	55	−844.0		−1210.0			−2.4
	75	−2929.0		−3294.9			−3.6
	95	−5961.3		−6327.3			−4.4

+: increase of the index value; −: decrease of the index value. In the table, D is the pile diameter, L is the pile length, SR is the side resistance, ER is the end resistance, BC is the bearing capacity.

bearing capacity is mainly influenced by the loss of end resistance when the pile is relatively short. The effect on side resistance by the increase of radial compressive stress is getting weak with the increase of pile length. The decrease of effective friction area caused by sulfate attack leads to a remarkable decrease in the side resistance. The bearing capacity shows a sharp reduce when the length of pile is getting longer. Table 2 also indicates that the loss ratio of bearing capacity shows an increase of around 7% when the pile length varies from 15 m to 95 m.

4.2.4   Effect of Pile Diameter on Bearing Capacity Degradation

The effect of pile diameter on the bearing behavior of a cast-in-situ pile is studied and the results are shown in Table 3. Table 3 indicates that the diameter of pile has a pivotal influence on the bearing behavior of a pile. Taking the pile with $L = 55$ m as example, the loss ratio of bearing capacity is 10.2% when $D = 0.8$ m while it becomes 2.4% when $D = 1.6$ m. The similar rules are observed when pile lengths are 15 m, 35 m, 75 m and 95 m. The bearing capacity loss ratio of the cast-in-situ pile shows a remarkable decrease when the diameter varies form 0.8 m to 1.6 m. Hence, the increase of pile radius in the given pile length shows a significant improvement in developing the anti-corrosion ability of a cast-in-situ pile.

Table 3. Effects of pile diameter on bearing properties before and after corrosion

L/m	D/m	SR/kN	ER/kN	BC/kN	EA/MN	EI/MNm^2	Loss ratio/%
15	0.8	+88.9	−209.1	−120.2	−9653.2	−450.6	−4.3
	1.2	+297.3	−291.1	+6.2	−14870.9	−1678.9	+0.1
	1.6	+483.9	−366.0	+118.0	−20088.7	−4178.6	+2.0
35	0.8	−748.3	−209.1	−957.4	−9653.2	−450.6	−8.0
	1.2	−189.5	−291.1	−480.6	−14870.9	−1678.9	−2.8
	1.6	+293.6	−366.0	−72.3	−20088.7	−4178.6	−0.3
55	0.8	−2673.6	−209.1	−2882.8	−9653.2	−450.6	−10.2
	1.2	−1679.7	−291.1	−1970.8	−14870.9	−1678.9	−5.0
	1.6	−844.0	−366.0	−1210.0	−20088.7	−4178.6	−2.4
75	0.8	−5687.2	−209.1	−5896.3	−9653.2	−450.6	−11.4
	1.2	−4173.3	−291.1	−4464.4	−14870.9	−1678.9	−6.2
	1.6	−2929.0	−366.0	−3294.9	−20088.7	−4178.6	−3.6
95	0.8	−9789.0	−209.1	−9998.1	−9653.2	−450.6	−12.1
	1.2	−7670.4	−291.1	−7961.5	−14870.9	−1678.9	−6.9
	1.6	−5961.3	−366.0	−6327.3	−20088.7	−4178.6	−4.4

+: increase of the index value; −: decrease of the index value. In the table, D is the pile diameter, L is the pile length, SR is the side resistance, ER is the end resistance, BC is the bearing capacity.

4.2.5   Bearing Strength of Pile Body

The bearing behavior of a cast-in-situ pile is influenced by both the pile-soil interface properties and the bearing strength of pile body. The ordinate in the Fig. 5 represents the ratio of the real-time strength to the initial strength of pile body. In Fig. 5, the axial

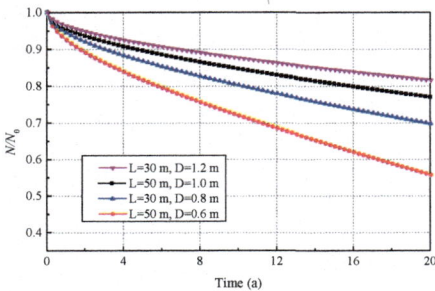
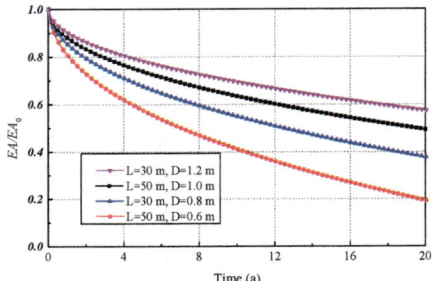

Fig. 5. Variation of compressive strength of pile body with corrosion time

Fig. 6. Variation of compressive rigidities with corrosion time

bearing ability of the pile decreases gradually as the sulfate ions diffuse deeper. A drop of 2196.7 kN is found as the corrosion depth reaches 60.0 mm for the working condition that $L = 50$ m and $D = 0.6$ m. The loss of bearing strength accounts for 44.4% of initial bearing strength. The loss ratio of bearing strength becomes 18.5% when $L = 30$ m and $D = 1.2$ m. Thus, the damage caused by sulfate ions corrosion is significant, especially for the bearing strength of the cast-in-situ pile.

4.2.6 Rigidity of a Cast-in-Situ Pile

(1) Compressive rigidity

The compressive rigidities of piles are calculated by Eq. (17) and the results are shown in Fig. 6. The Y-axis in Fig. 6 represents the ratio of the real-time compressive rigidity to the original compressive rigidity. The initial compressive rigidities of the piles are 8698.6 MN, 24162.8 MN, 15464.2 MN and 34794.4 MN for the working condition $L = 50$ m with $D = 0.6$ m, $L = 50$ m with $D = 1.0$ m, $L = 30$ m with $D = 0.8$ m, $L = 30$ m with $D = 1.2$ m, respectively. However, the values become 1654.3 MN, 11900.7 MN, 5811.0 MN and 19923.5 MN after 20 years. Remarkable drops of 7044.3 MN, 12262.1 MN, 9653.2 MN and 14870.9 MN are presented for each working condition in Fig. 6. The loss of compressive rigidity of each working condition occupy 81.0% ($L = 50$ m, $D = 0.6$ m), 50.7% ($L = 50$ m, $D = 1.0$ m), 62.4% ($L = 30$ m, $D = 0.8$ m) and 42.7% ($L = 30$ m, $D = 1.2$ m) of the original rigidity. The value of compressive rigidity relates to the resistance ability of compressive and tensile deformation, especially the compressive deformation for piles.

(2) Flexural rigidity

The flexural rigidity of a pile is influenced by the concrete properties and the section characteristics of the pile. The flexural rigidity is calculated based on Eq. (18) and the results are shown in Fig. 7, in which the ordinate is the ratio of the real-time flexural rigidity to the initial flexural rigidity. One can observe from Fig. 7 that the flexural rigidity of the pile drops form 618.6 MN/m^2 to 168.0 MN/m^2 after 20 years when $L = 30$ m and $D = 0.8$ m. The results of other working conditions show the

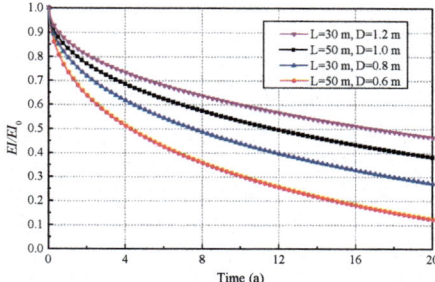

Fig. 7. Variation of flexural rigidities with corrosion time

similar trend with the drops of 172.0 MN/m^2 ($L = 50$ m, $D = 0.6$ m), 934.1 MN/m^2 ($L = 50$ m, $D = 1.0$ m) and 1678.9 MN/m^2 ($L = 30$ m, $D = 1.2$ m). The significant decreases take place because the pile concrete is deteriorated and the section area is getting smaller. Flexural rigidity stands for the resistance ability against bending deflection when a horizontal load is applied to the pile. The bearing capacity of piles suffer serious damage after 20 years of sulfate attack. The horizontal stability of the entire structure against static and dynamic load reduces in a large degree.

5   Conclusions

In this study, vertical bearing behavior calculating models are proposed to clarify the degradation behavior of a cast-in-situ pile in sulfate saline environment. Based on the analysis results, the following conclusions can be drawn:

(1) The side resistance of a cast-in-situ pile is influenced by both the decrease of effective friction area and the radial stress redistribution caused by sulfate attack. End resistance decreases because of the decline of the effective cross-sectional area.

(2) Pile diameter plays an important role in decreasing the loss ratio of bearing capacity in sulfate saline environment. Increase of pile length has no significant effects on the bearing capacity loss ratio of the pile.

(3) The compressive and the flexural rigidity of the cast-in-situ pile show a remarkable decrease with the increase of corrosion depth. The ability of resistance to deformation is deteriorated seriously after sulfate attack.

Acknowledgments. This research was supported financially by the National Natural Science Foundation of China (No. 41772290) and the Project of Shaanxi Province (No. 20170522).

References

1. Al-Amoudi, O.S.B.: Attack on plain and blended cements exposed to aggressive sulfate environments. Cement Concr. Compos. **24**(3), 305–316 (2002)
2. Tixier, R., Mobasher, B.: Modeling of damage in cement-based materials subjected to external sulfate attack. I: formulation. J. Mater. Civ. Eng. **15**(4), 305–313 (2003)
3. Al-Amoudi, O.S.B.: Sulfate attack and reinforcement corrosion in plain and blended cements exposed to sulfate environments. Build. Environ. **33**(1), 53–61 (1998)
4. Tixier, R., Mobasher, B.: Modeling of damage in cement-based materials subjected to external sulfate attack. II: comparison with experiments. J. Mater. Civ. Eng. **15**(4), 314–322 (2003)
5. Oliveira, I., Cavalaro, S.H.P., Aguado, A.: New unreacted-core model to predict pyrrhotite oxidation in concrete dams. J. Mater. Civ. Eng. **25**(3), 372–381 (2012)
6. Chinchón-Payá, S., Aguado, A., Chinchón, S.: A comparative investigation of the degradation of pyrite and pyrrhotite under simulated laboratory conditions. Eng. Geol. **127**, 75–80 (2012)
7. Sun, C., Chen, J., Zhu, J., et al.: A new diffusion model of sulfate ions in concrete. Constr. Build. Mater. **39**, 39–45 (2013)
8. Idiart, A.E., López, C.M., Carol, I.: Chemo-mechanical analysis of concrete cracking and degradation due to external sulfate attack: a meso-scale model. Cement Concr. Compos. **33**(3), 411–423 (2011)
9. Bary, B.: Simplified coupled chemo-mechanical modeling of cement pastes behavior subjected to combined leaching and external sulfate attack. Int. J. Numer. Anal. Methods Geomech. **32**(14), 1791–1816 (2008)
10. Santhanam, M., Cohen, M.D., Olek, J.: Mechanism of sulfate attack: a fresh look: part 1: summary of experimental results. Cem. Concr. Res. **32**(6), 915–921 (2002)
11. Skaropoulou, A., Kakali, G., Tsivilis, S.: Thaumasite form of sulfate attack in limestone cement concrete: the effect of cement composition, sand type and exposure temperature. Constr. Build. Mater. **36**(36), 527–533 (2012)
12. Ma, H., Li, Z.: Multi-aggregate approach for modeling interfacial transition zone in concrete. ACI Mater. J. **111**(2), 189–199 (2014)
13. Santhanam, M., Cohen, M.D., Olek, J.: Sulfate attack research-whither now? Cem. Concr. Res. **31**(6), 845–851 (2001)
14. Taylor, H.F.W., Famy, C., Scrivener, K.L.: Delayed ettringite formation. Cem. Concr. Res. **31**(5), 683–693 (2001)
15. Gao, R., Li, X., Xu, Q., et al.: Concrete deterioration under alternate action of chemical attack environments. In: International Conference on Multimedia Technology (ICMT), pp. 1008–1012. IEEE (2011)
16. Whittaker, M., Black, L.: Current knowledge of external sulfate attack. Adv. Cem. Res. **27**(9), 532–545 (2015)
17. Campos, A., López, C.M., Aguado, A.: Diffusion-reaction model for the internal sulfate attack in concrete. Constr. Build. Mater. **102**, 531–540 (2016)
18. Du, J., Liang, Y., Zhang, F.: Mechanism and Performance Degradation of Underground Structure Attacked by Sulfate. China Railway Press, Beijing (2011). (in Chinese)
19. JGJ 94-2008: Technical Code Building Pile Foundations (2008). (in Chinese)
20. Tumidajski, P.J., Chan, G.W., Philipose, K.E.: An effective diffusivity for sulfate transport into concrete. Cem. Concr. Res. **25**(6), 1159–1163 (1995)

Study on Pollution in Shanghai Contaminated Site Based on Resistivity CPTU

Cong Yan[1], Guojun Cai[1(✉)], Xuepeng Li[1], Min Chen[1], Songyu Liu[1],
Xinrong Mao[2], Jun Lin[1], and Hanliang Bian[1]

[1] Institute of Geotechnical Engineering, Transportation College of Southeast
University, Nanjing 210096, China
220153347@seu.edu.cn, focuscai@163.com
[2] Test Center of Geotechnical Engineering of Shanghai City,
Shanghai 200436, China

Abstract. The resistivity of soil is the basic parameter that characterizes the conductivity of soil in contaminated sites. It is very important to evaluate the resistivity of soil in contaminated sites. As normal CPTU's development, resistivity CPTU (RCPTU) takes both the advantages of CPTU and traditional resistivity method in pollution evaluation. In this paper, RCPTU was used to test the 5 boreholes in the contaminated site of Shanghai Chemical plants and obtain the resistivity values of different depth. Combined with the indoor test data, the relationship between the resistivity of contaminated soil and heavy metal pollution was analyzed. As resistivity is also influenced by porosity ratio, the relationship between relative density and site pollution is also analyzed based on data from RCPTU results.

Keywords: RCPTU · Resistivity · Heavy metal pollution · Relative density

1 Introduction

CPTU is a new international in situ testing technology risen in 1980s which has been in the developed countries to promote the use of a large number of projects. The technique is to add pore water pressure sensor to the taper cone according to the special nature of soil engineering based on the traditional static cone penetration technology. It has the advantages of high test precision, large amount of data collected, continuous testing, no sampling, fast and convenient, low interference and low cost which make it particularly suitable for highways, railways, subways and other large-scale project survey and design. Resistivity sensor is a new type of sensor used in CPTU application. Through the detection of resistivity CPTU, the resistivity value of each position in the soil can be obtained. Also the trend of changes of the resistivity of different direction in the soil can be obtained.

© Springer Nature Singapore Pte Ltd. 2018
A. Farid and H. Chen (Eds.): GSIC 2018, *Proceedings of GeoShanghai 2018*
International Conference: Geoenvironment and Geohazard, pp. 265–272, 2018.
https://doi.org/10.1007/978-981-13-0128-5_30

2   Relative Theories

In 1942, Archie [1] developed a resistivity model for saturated and cohesive soils by experimenting with the relationship between resistivity and its structure. The famous Archie equation was proposed:

$$\rho = a\rho_w n^{-m} \tag{1}$$

Where ρ is the soil resistivity; ρ_w is the pore water resistivity; n is the soil porosity; a is the soil parameters; m is the cementing factor.

Kalinski and Kelly [2] extend Archie's model for unsaturated soils and propose the following equation:

$$\rho = a\rho_w n^{-m} S_r^{-p} \tag{2}$$

Where s_r is saturation; p is saturation index.

In view of the fact that the surface conductivity of soil particles cannot be neglected, Waxman and Smits [3], through the two parallel resistance test study, proposed for the surface conductivity of a good cohesive soil resistivity model:

$$\rho = \frac{a\rho_w n^{-m} S_r^{-p}}{s_r + \rho_w BQ} \tag{3}$$

Where B is the electric conductivity of the electric charge in the double electric layer opposite to the surface electrical energy of the soil particles; Q is the cation exchange capacity in the pores of the unit soil; BQ is the electrical conductivity of the surface layer of the soil particle.

3   Experimental Study

3.1   Site Description

The test and sampling site is located in a chemical plant site in Taopu area, Shanghai which has a flat landform. There used to be a river of 5 to 10 m wide in the northwest-southeast direction of the site intersected by another river in the northeast. There are ponds and puddles in the site. The water in the south is basically no color or lighter in color. The middle of the water was light red to red. The color of the northern water body is darker and darker with a noticeable smell.

According to the multi-functional resistivity CPTU test in the area, the soil layers from the top to bottom can be divided into: miscellaneous fill, silty clay and silt silty clay. Groundwater is mainly phreatic water in shallow soil. According to the geotechnical investigation data, the phreatic water depth of the Taopu area is about 0.5–2.3 m. The annual change rate of phreatic water level is about 1.0 m.

3.2   Test Hole Situation

A total of 6 holes were drilled at the experimental site, and RCPTU 1, RCPTU 2, RCPTU 3, RCPTU 4, RCPTU 5 and RCPTU 6, respectively. Multipurpose RCPTU drilling information is shown in Table 1.

Table 1. Multifunctional RCPTU test hole basic information

Hole number	Drilling depth (m)	Hole depth (m)
RCPTU 1	11.0	1.5
RCPTU 2	10.0	1.5
RCPTU 4	12.55	1.5
RCPTU 5	12.0	2.3
RCPTU 6	15.0	2.0

3.3   CPTU Test Equipment

Compared to conventional CPTU, the core of the RCPTU device is a resistivity sensor mounted on the rear of a standard CPTU probe. Early resistors are used to estimate the porosity and density of the in situ soil. Recently, the resistivity sensor has been used to delineate the scope of contaminated soil or water [4]. At the beginning of 2005, Institute of Geotechnical Engineering, Transportation College of Southeast University introduced the United States Hogentogler original multi-function digital RCPTU system, equipped with the latest resistivity test probe [5]. The structure of the Hogentogler resistivity CPTU probe is shown in Fig. 1.

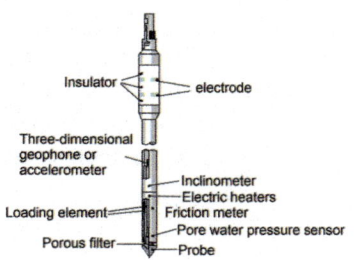

Fig. 1. RCPTU probe structure

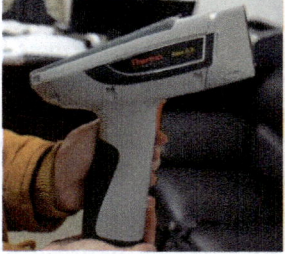

Fig. 2. Handheld XRF instrument

3.4   Testing Equipment

The heavy metal ions in the soil can be determined by the XRF test. XRF technical principle is the use of radioactive radiation X-ray excitation to detect the inner layer of electrons. At this point in order to maintain energy balance, the external electronic components get into the inner layer of electronic vacancies. Due to the external electronic energy is high, the external electrons get into the inner orbit releasing a specific fluorescent energy and the amount of the measured object can be inferred by the size of the energy. According to the processing of the collected data, the test equipment is divided into wavelength dispersion X-ray fluorescence and energy dispersive X-ray fluorescence instrument. The XRF analyzer used in this test was a Niton XL-31 handheld XRF instrument (shown in Fig. 2).

4 Experimental Data Analysis

Polluted soil's conductive capacity is affected by many factors. The influence factors of the resistivity of compacted silty clay were studied [6]. The influence of different factors on the conductivity of compacted silty clay was obtained, which are shown in Table 2.

Table 2. Influence of different variables on conductivity of saturated soil [6]

Variable	Variable change	Effect on conductivity
Porosity ratio	increase↑	↑(***)
Saturation	increase↑	↑(***)
Salinity	increase↑	↑(***)
In the pore water		
Granularity	increase↑	↓(**)
Grading	increase↑	↑(**)
Temperature	increase↑	↑(**)
Activity	increase↑	↑(**)

Note: (***) indicates very important, (**) indicates important

It can be seen that the three factors that have the greatest influence on soil resistivity are its porosity ratio, saturation and salinity. Because the groundwater level is below 1.5 m and shallow soil resistivity data discretization degree is large, the data below the depth of 1.8 m is taken for analysis. The effects of relative density and ion concentration on resistivity are mainly considered.

4.1 Relationship Between Relative Density and Resistivity

Because the continuous in-situ porosity ratio of soil is difficult to obtain and the cost is high, relative density is often used to express the compact state of soil in CPTU test. Studies have shown that if the porosity gets smaller, the conductivity of the soil will be stronger and the resistivity will be smaller [1]. Correspondingly, considering the relative density as a reference index, if the relative density gets greater, the conductivity of the soil will be smaller and the resistivity will increase. The results of Arulmoli [7] show that the sand soil resistivity in the main gradation is not totally linearly related, but its resistivity and relative density are monotonic.

The definition of relative density is:

$$D_r = \frac{e_{max} - e}{e_{max} - e_{min}} \tag{4}$$

Since the in-situ porosity ratio data is difficult to obtain, the relative density is estimated using the formula based on the CPTU in situ test data. According to the calibration tank test [8], considering the compressive effect, it is recommended to use the following formula to estimate the relative density of the normal consolidated sand:

$$D_r = -98 + 66 \log \frac{q_c}{\sqrt{\sigma'_{v0}}} \tag{5}$$

Where q_c is measured cone tip resistance, σ'_{v0} is effective overburden stress, Units are t/m2.

The results of the relative densities of each hole are shown in Fig. 3:

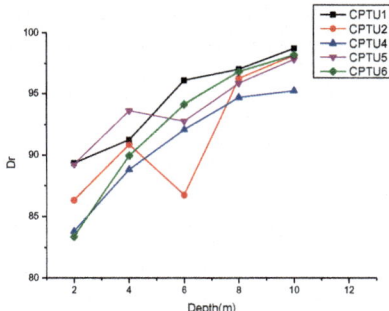

Fig. 3. Relative density of test holes

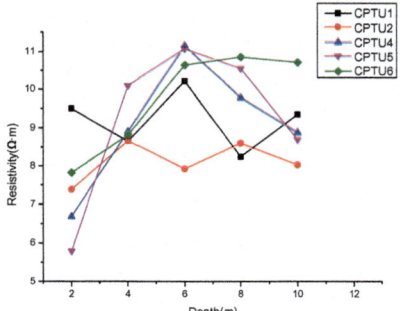

Fig. 4. Resistivity of test holes

As the site of the soil have different degrees of pollution, it is difficult to obtain the original undisturbed soil resistivity of the most benchmark resistivity. In view of the relative density and the change of soil resistivity, although the influence of pollutants on soil resistivity is present, it is assumed that the influence of the main metal pollutants on the soil resistivity will reduce the earth resistivity compared with the original level. With the increase of the depth and the increase of the relative density, the general trend of the earth resistivity is still increasing. It can be concluded that there is a positive correlation between soil resistivity and relative density of soil (Fig. 4).

4.2   Relationship Between Metal Ion Pollution and Resistivity

Many types of products make the site a lot of inorganic and organic pollutants. The main detection of heavy metal pollutants are as follows (with the corresponding ion valence) (Table 3):

Table 3. Metal pollutants detected

Type	Zr	U	Th	Ti	As	Fe	Sr	Zn	Cu	Ni	Co	Ca	Mn	Pb	Rb	K
Valence	4	4	4	4	3	3	2	2	2	2	2	2	2	2	1	1

For these heavy metal contaminants, they exist in the form of ions in the pore water of the soil. Arulanandan and Smith [9] evaluated the effect of the type and amount of electrolyte on the current diffusion characteristics of the soil. The inorganic pollutants are ionized in the pore water, which will greatly improve the conductivity of the pore water and reduce the resistivity of the contaminated soil. At the same time, due to the presence of double layers on the surface of the soil particles, the cationic and anions in the electric double layer have the electrical conductivity under the action of the electric field, which will reduce the resistivity of the contaminated soil. The conductivity of ions containing contaminated soils is related to the ionic content and the valence state of ions (Table 4).

Table 4. Sum of the ion charge and concentration product at different depths (e·ppm)

Depth (m)	CPTU1	CPTU2	CPTU6	CPTU4	CPTU5
2	85328		85581		91358
3	66217	90304	74523	63335	71938
4	92262		75428	78445	
5	58282	65857		70910	69294
6	66257		70941	58093	
7		62804			62732
8					
9		71356			

The increase in ion concentration will obviously increase the conductivity of contaminated soil and reduce the resistivity. The metal ions with different valence carry different charges, so that they have different effects on the resistivity of contaminated soil under the same pollution concentration. In general, the higher the ion valence is, the more obvious the resistivity decreases. The main metal ions detected in the site are iron (Fe) and titanium (Ti) and their valence is different. Although the other ions are at least one order of magnitude lower than the major contaminants, they are detected and the total amount is comparable to the major contaminants. Therefore, considering the concentration of pollutants and the amount of charge, the effect of pollutants on resistivity is expressed as the product of ion valence and concentration.

According to the data of the resistivity and corresponding product of the ion charge and concentration, the resistivity is negatively correlated with the ion charge concentration (Fig. 5). The larger the charge concentration of pollutants is, the smaller the resistivity will be, which is consistent with the existing soil resistivity studies. Since the resistivity of undisturbed soil is not measured in this field, the exact relationship is still to be further studied.

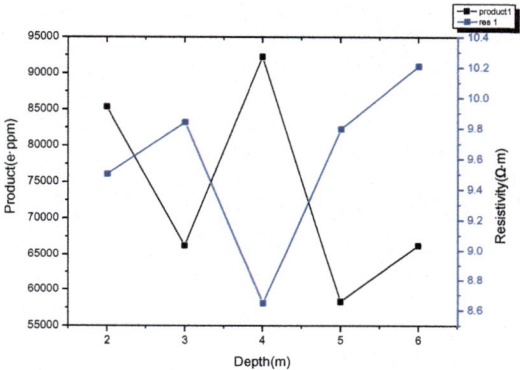

Fig. 5. Typical relationship between charge-concentration product and resistivity (cptu1)

5 Conclusion

a. The soil bulk density has a significant effect on the soil resistivity. There is a significant positive correlation between the relative density and the resistivity in this test site. The resistivity of the soil will increase if the relative density of the soil get larger, which is consistent with the existing research results. For traditional test methods, it have to use drilling sampling to obtain the soil density data, which will cost much time and money and the result is scattered. CPTU test method used in this site can directly get the continuous data without the trouble of drilling. The accuracy of the test result shows that CPTU test system provides a reliable basis for measuring soil densities.

b. The heavy metal ion pollution in this field is mainly iron ion pollution, followed by titanium ion pollution. These two pollutants are also the most common raw materials and pollutants in the dyeing industry, consistent with the production of the original polluting enterprises. The relationship between soil resistivity and contaminant ion charge concentration product shows that: In the metal ion contamination site, the contamination of the site can be assessed by testing the resistivity sensor combined with the common CPTU test after a small number of indoor tests have identified the major pollutant species. In the absence of additional test holes and equipment, the indoor test workload is greatly reduced and the continuous data obtained is better than the single point data of the indoor test.

References

1. Archie, G.E.: The electrical resistivity log as an aid in determining some reservoir characteristics. Trans. AIME **146**(01), 54–62 (1942)
2. Kalinski, R., Kelly, W.: Electrical-resistivity measurements for evaluating compacted-soil liners. J. Geotec. Eng. **120**(2), 451–457 (1994)
3. Waxman, M., Smits, L.: Electrical conductivities in oil-bearing shaly sands. Soc. Petrol. Eng. J. **8**(02), 107–122 (1968)
4. Cai, G., Liu, S.: Geoenvironmental investigation based on resistivity piezocone tests (RCPTU) for contaminated site. Environ. Monit. Manag. Technol. **22**(5), 48–52 (2010)
5. Cai, G., et al.: Analysis of formation characteristics of marine clay based on resistivity cone penetration test (RCPT). Chin. J. Geotech. Eng. **30**(4), 529–535 (2008)
6. Rinaldi, V.A., Cuestas, G.A.: Ohmic conductivity of a compacted silty clay. J. Geotech. Geoenvironmental Eng. **128**(10), 824–835 (2002)
7. Arulmoli, K., Arulanandan, K., Seed, H.B.: New method for evaluating liquefaction potential. J. Geotech. Eng. **111**(1), 95–114 (1985)
8. Jamiolkowski, M., Lo Presti, D., Manassero, M.: Evaluation of relative density and shear strength of sands from CPT and DMT. In: Soil Behavior and Soft Ground Construction, pp. 201–238 (2003)
9. Arulanandan, K., Smith, S.S.: Electrical dispersion in relation to soil structure. J. Soil Mech. Found. Div. **99**(sm2), 1113–1133 (1973)

Experimental Study on the Strength Characteristics of Lead Contaminated Soil with NZVI Treatment

Wan-Huan Zhou[1(✉)], Yong-Zhan Chen[1], Fuming Liu[1,2], and Shuping Yi[2]

[1] University of Macau, Macau, China
hannahzhou@umac.mo
[2] Southern University of Science and Technology, Shenzhen, China

Abstract. This investigation focuses on the shear strength of lead-contaminated soil before and after its stabilization by nanoscale zero valent iron (NZVI)—one of the magnetic nanoparticles (MNPs) having significant stabilization and remediation properties for heavy metal and organics. Being highly toxic carcinogens, lead ions are considered an extensive heavy metal contaminant in groundwater and soil, hazardous to the nervous system and organs of human beings. NZVI has shown an aptitude for effective stabilization when used to treat soil contaminated with heavy metals. In this study, a series of laboratory tests were conducted on lead-contaminated soil subjected to NZVI treatment in different percentages. Microstructures and elements were analyzed to evaluate degradation and morphological variations before and after the treatment, and vane shear tests were conducted to measure the shear strength of the soil with additives of lead nitrate and NZVI. From a microstructural perspective, the morphology of the treated soil indicates that the particle skeleton became more conjoint with use of NZVI. The experimental data show that addition of lead nitrate has a significant negative influence on shearing. It was found that the shear strength of polluted soil was reduced significantly even as that of a treated specimen increased. The investigation indicated that application of NZVI is a promising approach to reusing contaminated ground from the perspectives of both environmental and geotechnical engineering.

Keywords: NZVI treatment · Contaminated soil · Shear strength

1 Introduction

Heavy metals are the most toxic inorganic pollutants in the environment. Some of these ions can be very toxic even at very low concentrations. Recent decades, the production of heavy metals such as lead, copper, and zinc has increased exponentially. Chief among these, lead is an important environmental health concern, being a non-biodegradable toxic metal whose use is widespread in the world for many industrial applications.

Magnetic nanoparticles (MNPs) exhibit noteworthy properties, among them high BET specific surface area, high dispersibility, and high reactivity, among other

© Springer Nature Singapore Pte Ltd. 2018
A. Farid and H. Chen (Eds.): GSIC 2018, *Proceedings of GeoShanghai 2018*
International Conference: Geoenvironment and Geohazard, pp. 273–279, 2018.
https://doi.org/10.1007/978-981-13-0128-5_31

unexpected beneficial properties. Recently, nanoscale zero valent iron (NZVI) has been demonstrated to be a high-efficiency reductant. Furthermore, because of its nanoscale particle size, it also shares the advantages of MNPs. The corrosion of the nanoparticles in aqueous systems, caused by the presence of oxygen and Fe3+, accelerates the formation of iron oxide and oxyhydroxide compounds on the surface of NZVI. These particles exhibit an outstanding ability to remove organic and inorganic contaminants [1]. Such metallic and organic contaminants include Pb^{2+}, Ni^{2+}, Ba^{2+}, Co^{2+}, ClO^{4-}, CrO_4^{2-}, and AsO_4^{3-} from water and wastewater [2–4].

When considering soil enhancement through the use of NZVI reagents, the geotechnical properties of treated soil are worthy of attention. Previous research has focused mostly on solving foundation problems and on the removal of contaminants [5–8]. Hence it is necessary to clarify the effects of NZVI treatment on contaminated soil, which has been the subject of little research [9, 10].

This study investigates NZVI-induced enhancement of lead-contaminated soil. Scanning electron microscopy was also employed to visualize particle space and size variation. Furthermore, vane shear tests and triaxial tests were used to provide an understanding of the geotechnical properties of contaminated and NZVI-treated soil.

2 Experimental Program

2.1 Sample Preparation

Clay soil was selected from a foundation pit of construction site close to the Hong Kong–Zhuhai–Macau bridge in Praia de Areia Preta, Macau SAR. In this investigation, the Macau soil was used as a material in context of which to study the morphological and geotechnical properties of lead-contaminated soil and NZVI-treated soil. The selected natural soil was sieved through 1 mm standard screen hole and mixed with distilled water until reaching twice its liquid limit (LL). The resulting slurries were loaded to effective vertical pressure of about 230 kPa with a double draining condition. When deformation of the pre-consolidation was stable, the samples were trimmed for vane shear tests using a cylindrical tube 38 mm in diameter.

2.2 Contaminant and Treated Agent

Lead nitrate, $Pb(NO_3)_2$, an analytical reagent (AR, 99.8%), was obtained from the Xi-Long Chemical Co. Ltd., China. Nano zero-valent irons (NZVI) were commercially synthesized and supplied by Xiang-Tian Corp. (Shanghai, China) as the treated agent. To prevent oxidation of NZVI, the particles were sealed in nitrogen-filled packaging for immediate use. They were characterized as follows: average particle size 50 nm, purity 99.9%, BET specific surface area (SSA) 23 m^2/g, bulk density 0.45 g/cm^3, density 7.9 g/cm^3.

The contaminated agent of soil sample was contaminated by lead nitrate and then divided into four types of sample: natural soil, Pb-polluted soil, Pb + 1% NZVI-treated soil, and Pb + 10% NZVI-treated soil. To achieve this, the soil was artificially contaminated with lead ions at 500 mg/kg dry weight, left standing for 48 h.

2.3 Basic Properties of Soil Samples

Atterberg limits inform the fundamental classification and assessment of soil particles, which are associated with moisture adsorption, state, and geotechnical properties. The Atterberg limits comprise liquid limit (LL), plastic limit (PL), and plastic index (PI) as defined by the ASTM standard test method, ASTM D4318.

Particle size distribution (PSD) classification of all specimens followed ASTM D2487. To aid measurement of PSD, different type soil samples were blended with a dispersant—distilled water—by mechanical and ultrasonic stirring for at least 300 s. After the solution was totally homogenized, the grain size distribution was evaluated by Malvern Mastersizer (Malvern, UK).

2.4 Scanning Electron Microscope

The microstructures and morphologies of the clay and the additive mixtures were analyzed by means of a scanning electron microscope (SEM, Zeiss, Germany). Microstructural changes to soil caused by the addition of lead and NZVI significantly affected the geotechnical properties of the soil. Prior to use, the soil particles were dehydrated at 35° centigrade for 12 h to completely remove free water from the surface of particles. The samples that had a conductive coating were transferred to the chamber for sputtering with gold before being put into the scanning electron microscope for examination, thereby preventing the charging of artifacts under electron beam exposure and allowing high-resolution pictures to be obtained.

2.5 Vane Shear Tests

The laboratory vane shear test at atmospheric pressure is a standardized investigation method for determining the shear strength of fine-grained soils and is described in ASTM D4648. The trend of the undrained strength gain of contaminated and treated specimens was measured using a laboratory vane shear test (manufactured by Wyke-ham Farrance, Ltd., Slough, UK) on the 38 mm diameter specimens. The rotation rate was controlled by an electrical motor at a constant rotational speed of 6° per min.

3 Results

3.1 Atterberg Limits Results

Table 1 shows the particle size distribution of different specimens.

Table 1. Particle size distribution for soil samples

Particle Size Distribution (PSD)	Natural soil	Soil + Pb	Pb + 1% NZVI	Pb + 10% NZVI
Sand (0.075–2 mm)	9.76	3.83	0.00	15.60
Silt (0.002–0.075 mm)	70.09	73.56	73.16	63.61
Clay (<0.002 mm)	20.15	22.61	26.84	20.78

The particle size distribution of contaminated soil tends more toward fine particles than in natural soil. Furthermore, with the addition of 1% NZVI, the PSD result indicates the appearance of smaller particles. For 10% NZVI-treated soil, the result indicates that larger particles appeared (Fig. 1).

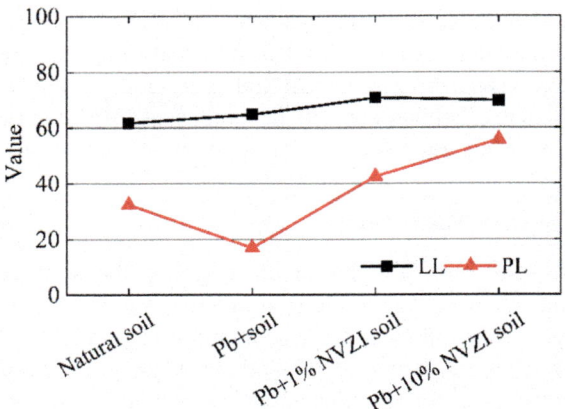

Fig. 1. The unified soil classification system of samples

The experimental data show that the LL of lead-polluted soil increased with the addition of lead ions and that the PL presented a sharply reduced tendency. However, LL and PL increased with the addition of NZVI. The PL of NZVI-treated soil has a significant increment, and the observed increase in Plasticity Index (PI) for the contaminated soil was evident in increased water content and clay content. NZVI-treated soil showed an opposite trend. It can be found that the nature of each samples changes visibly from results of Atterberg tests.

3.2 Morphology Results

Figure 2 shows SEM images of stabilized contaminated soils with mixed Pb contamination, Pb + 1% NZVI-treated soil and Pb + 10% NZVI-treated soil.

D-values can be thought of as the diameters of particles that divide a sample's mass into a specified percentage when the particles are arranged by mass, ascending. On examination of the image sequence in order of increasing size, smaller sizes of contaminated soil may be caused by lead ions.

After the NZVI additive was mixed in, the SEM was then analyzed to demonstrate the effects of the reaction concentration of NZVI. Larger and conjoint structures were observed. In particular, incidence of clastic soil particles gradually lessened. Meanwhile, it was evident that polymerized skeleton was in the majority, as shown in the image associated with 10% NZVI.

Structural and surface reconstructions in samples of pristine NZVI were analyzed by SEM. SEM images of the treated soil sample showed that schistose and laminar texture like structures formed depending on the NZVI reaction, with their size larger

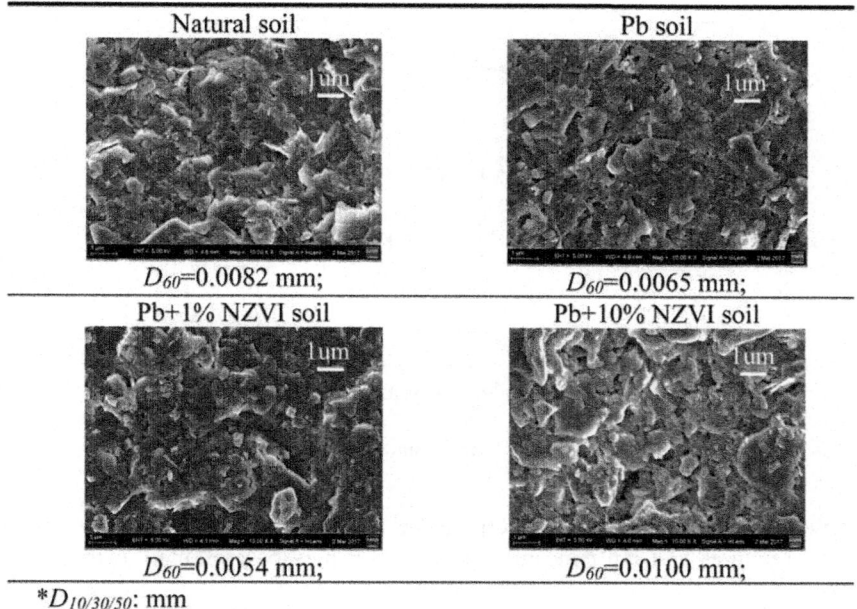

Natural soil	Pb soil
D_{60}=0.0082 mm;	D_{60}=0.0065 mm;
Pb+1% NZVI soil	Pb+10% NZVI soil
D_{60}=0.0054 mm;	D_{60}=0.0100 mm;

*$D_{10/30/50}$: mm

Fig. 2. SEM images of reaction specimens of Pb(II) remediation by NZVI. (scale: 1 um)

than that seen in soil particles. Similar results were reported in the literature. [11] Variations in morphology should be ascribed to the formation of ferric oxides/hydroxides in response to addition of NZVI. Such SEM results are consistent with the particle size distribution curves already mentioned.

3.3 Vane Shear Tests Results

Figure 3 compares vane shear strength (from five repeated tests for each case) of specimens having a natural status and having a Pb-contaminated status. To evaluate the effects of the NZVI used, and to verify the results obtained in a laboratory, a second test series was carried out.

It was revealed that the vane shear strength of natural soil decreased sharply with the introduction of lead nitrate, from 17.24 kPa to 7.25 kPa. Upon NZVI treatment, a significant strength increase from 7.25 to 32.4 kPa (average vane shear strength) in the presence of 1% (dry weight ratio) addition was observed. With 10% NZVI addition, vane shear strength increased to 69.3 kPa.

The following reasons could explain this behavior: (1) Lead ions were immobilized. NZVI was added to the contaminated soil, chiefly for reduction and adsorption of Pb^{2+} and to eliminate the negative effects of lead nitrate. Because the standard reduction potential of Pb^{2+} (−0.13 V) is greater than that of Fe^{2+} (−0.44 V), insoluble lead may be observed after reaction [11]. (2) Pb^{2+} ions may through co-precipitation be incorporated into the lattice points of ferrites ($Pb_xFe_xO_4^{3-}$) during their formation [12].

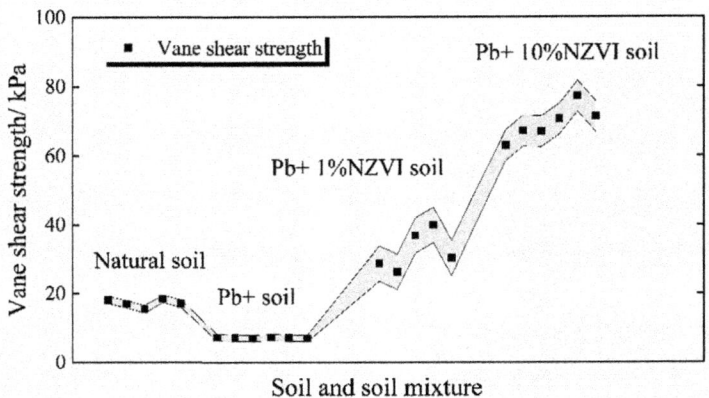

Fig. 3. Vane shear strength of each sample

The new materials (e.g. ferrites, PbO) can be attached to the soil particles. (3) The conjoint structures and ion concentration variation may be attributed to the strength enhancement.

4 Conclusion

This study presents certain results obtained when studying the effects of nanoscale zero valent iron (NZVI) treatment on the microstructure and geotechnical properties of lead-contaminated soil. The following items are sketched.

The microstructure of each kind of sample was significantly affected by the additive. With the addition of a strong oxidant, 1% NZVI, the lead ions absorbed by the soil particles reacted, threatening to disturb interparticle forces. Hence the particle size of 1% NZVI sample is smaller than for contaminated soil and natural soil. Because of the aggregation effect under excess NZVI conditions, the results of 10% NZVI treatment were marked by larger particle size and conjoint structures.

Lead contamination has a negative effect on the strength of soil, but NZVI can enhance vane shear strength dramatically, with increased presence of NZVI consistent with increased vane shear strength. Conjoint aggregation and variation of soil nature may be shown to be the primary reason for this enhancement.

Based on these results, addition of NZVI to lead-contaminated soil represents a potential approach to enhancing and treating lead-contaminated soil.

Acknowledgments. The authors wish to thank the financial support from the Macau Science and Technology Development Fund (FDCT) (125/2014/A3), the University of Macau Research Fund (MYRG2017-00198-FST, MYRG2015-00112-FST).

References

1. Stefaniuk, M., Oleszczuk, P., Ok, Y.S.: Review on nano zerovalent iron (nZVI): from synthesis to environmental applications. Chem. Eng. J. **287**, 618–632 (2016)
2. Moazeni, M., Ebrahimi, A., Rafiei, N., Pourzamani, H.R.: Removal of lead ions from aqueous solution by Nano Zero-Valent Iron (nZVI). Heal. Scope **6**(2), e40240 (2017) (In press)
3. Mueller, N.C., et al.: Application of nanoscale zero valent iron (NZVI) for groundwater remediation in Europe. Environ. Sci. Pollut. Res. **19**, 550–558 (2012)
4. Li, S., Wang, W., Liang, F., Zhang, W.: Heavy metal removal using nanoscale zero-valent iron (nZVI): theory and application. J. Hazard. Mater. **322**, 1–9 (2016)
5. Zhou, W.-H., Zhao, L.-S., Garg, A. & Yuen, K.-V. Generalized analytical solution for the consolidation of unsaturated soil under partially permeable boundary conditions. Int. J. Geomech. **17**(9), 04017048 (2017) (Accepted)
6. Zhou, W.H., Lok, T.M.H., Zhao, L.S., Mei, G.Xiong, Li, X.B.: Analytical solutions to the axisymmetric consolidation of a multi-layer soil system under surcharge combined with vacuum preloading. Geotext. Geomembranes **45**, 487–498 (2017)
7. Zhou, W.-H., Zhao, L.-S.: One-dimensional consolidation of unsaturated soil subjected to time-dependent loading with various initial and boundary conditions. Int. J. Geomech. **14**, 291–301 (2014)
8. Chen, R.P., Zhou, W.H., Wang, H.Z., Chen, Y.M.: One-dimensional nonlinear consolidation of multi-layered soil by differential quadrature method. Comput. Geotech. **32**, 358–369 (2005)
9. Nasehi, S.A., Uromeihy, A., Morsali, A., Nikudel, M.R.: Use of nanoscale zero-valent iron to improve the shear strength parameters of gas oil contaminated clay. Geopersia **5**, 161–175 (2015)
10. Nasehi, S.A., Uromeihy, A., Nikudel, M.R., Morsali, A.: Use of nanoscale zero-valent iron and nanoscale hydrated lime to improve geotechnical properties of gas oil contaminated clay: a comparative study. Environ. Earth Sci. **75**, 1–20 (2016)
11. Xi, Y., Mallavarapu, M., Naidu, R.: Reduction and adsorption of Pb^{2+} in aqueous solution by nano-zero-valent iron - A SEM, TEM and XPS study. Mater. Res. Bull. **45**, 1361–1367 (2010)
12. Tu, Y., Chang, C., You, C., Wang, S.: Treatment of complex heavy metal wastewater using a multi-staged ferrite process. J. Hazard. Mater. **209–210**, 379–384 (2012)

A Three-Dimensional Dual-Permeability Numerical Flow Model in Bioreactor Landfills

Zhen-Bai Bai, Shi-Jin Feng$^{(\boxtimes)}$, and Shi-Feng Lu

Department of Geotechnical Engineering,
Tongji University, Shanghai 200092, China
fsjgly@tongji.edu.cn

Abstract. To optimize the landfill operation and to ensure optimal distribution of water content in recirculation, a three-dimensional dual-permeability numerical flow model is proposed to simulate vertical wells recirculation in bioreactor landfills. Compared with the traditional single-porosity model, the dual-permeability model considering the highly heterogeneity of waste materials divides the porous media into two interacting pore domains, and can be used to reflect the preferential movement in the heterogeneous media. The governing equations are solved by a self-define solver using finite volume method in an open source platform OpenFOAM. The accuracy of the proposed model is verified by the numerical software Hydurs. A typical three-dimensional landfill cell with a vertical leachate recirculation well is adopted to show the performance of proposed model. The result indicates that the preferential movement of leachate makes a significant impact on the water content distribution in recirculation, and the single-porosity model will underestimate the volume and radius of impact zones in vertical wells for leachate recirculation.

Keywords: Dual-permeability model · Bioreactor landfill · Recirculation

1 Introduction

The solid waste industry attempts to control, monitor, and optimize the waste stabilization processes rather than simply contain the waste. Increasing water content by leachate recirculation or addition of other liquids has been the most widely used [1]. The major purpose of recirculation is to accelerate landfill settlement, reduce the contaminant life span of the landfill site, and minimize long-term environmental risk and liability [2]. In order to reasonably design the leachate recirculation and collection systems, it is important to understand the waste hydraulic properties for leachate flows. Extensive researches have revealed that the traditional single-porosity model cannot reasonably reflect the hydraulic properties of MSW, as preferential flow is observed in numerous in-situ tests. The dual-permeability model divides the structure of waste into two subdomains that can reflect the heterogeneity of municipal solid waste (MSW) and preferential flow of leachate.

Only Hydrus 2D (a finite element software) can consider the highly nonlinear dual-permeability model in numerical simulation at the present stage. However, for the vertical well recirculation, a typical three-dimensional problem, it simplified the actual

© Springer Nature Singapore Pte Ltd. 2018
A. Farid and H. Chen (Eds.): GSIC 2018, *Proceedings of GeoShanghai 2018*
International Conference: Geoenvironment and Geohazard, pp. 280–288, 2018.
https://doi.org/10.1007/978-981-13-0128-5_32

conditions into a plane model and will cause noticeable errors. Therefore, the aim of this study is to develop a method for three-dimensional dual-permeability in bioreactor landfills to simulate the vertical well recirculation. The governing equations are solved by a self-define solver using finite volume method in an open source platform Open-FOAM [3]. A typical bioreactor landfill cell with a vertical leachate recirculation well is adopted to show the performance of proposed model.

2 Methodology

2.1 Governing Equation

To describe the leachate flow in MSW, the single-porosity model, which simplifies the waste as a homogeneous material, was firstly adopted in recent decades. However, this assumption has been challenged by the observation of preferential flow in numerous landfill laboratory and field experiments. MSW should be considered as a notably heterogeneous and complex material in any modeling effort [4], so the dual-permeability model was developed to describe the heterogeneous waste material.

The main assumption of the dual-permeability model is that the medium consists of two subdomains: the fracture (macropores) and matrix (micropores) with different hydraulic characteristics. The fracture represents the connected large pores in the waste, while the matrix represents the relatively homogeneous materials. The governing equations of the dual-permeability model were derived by Gerke and van Genuchten from the nonlinear Richards equation as follows [5]:

$$\text{Fracture: } C(h_f)\frac{\partial h_f}{\partial t} = \nabla \cdot [K_f(h_f) \cdot \nabla(h_f + z)] - \frac{\Gamma_w}{w} \tag{1}$$

$$\text{Matrix: } C(h_m)\frac{\partial h_m}{\partial t} = \nabla \cdot [K_m(h_m) \cdot \nabla(h_m + z)] + \frac{\Gamma_w}{1 - w} \tag{2}$$

For convenience, the subscript i represents fracture (subscript f) or matrix (subscript m) domains; where h_i (m) is the pressure head, K_i (m/s) is the hydraulic permeability, and $C(h_i)$ is the specific moisture capacity; w is the ratio of the volume of fracture domain to the volume of total domain. The hydraulic properties of two domains follow the Mualem-van Genuchten model, which has been widely applied to MSW. The specific moisture capacity is defined as

$$C(h) = \frac{\partial \theta}{\partial h} \tag{3}$$

where θ is the water content, and the relationship between θ and h is expressed as:

$$\theta(h) = \begin{cases} \theta_r + \frac{\theta_s - \theta_r}{[1 + (\alpha|h|)^n]^m} & \text{if } h < 0 \\ \theta_s & \text{otherwise} \end{cases} \tag{4}$$

where θ_r is the residual water content, θ_s is the saturated water content, α the inverse of air-entry pressure, n the pore-size distribution index, and $m = 1 - 1/n$. K_i is defined as:

$$K = K_s k_r(\theta) = K_s (\frac{\theta - \theta_r}{\theta_s - \theta_r})^{0.5} \left\{ 1 - \left[1 - (\frac{\theta - \theta_r}{\theta_s - \theta_r})^{1/m} \right]^m \right\}^2 \tag{5}$$

where K_s (m/s) is the saturated permeability; and k_r is the relative permeability.

Γ_w is the transfer term, which describes the transfer of liquid between the two domains:

$$\Gamma_w = \frac{\beta}{a^2} \gamma K_{sa} k_{ra} (h_f - h_m) \tag{6}$$

where β is a dimensionless geometry-dependent coefficient; a is the characteristic length of the matrix; γ is a dimensionless scaling coefficient; K_{sa} is the saturated hydraulic permeability of the fracture-matrix interface; k_{ra} is the relative permeability function of the fracture-matrix interface evaluated in terms of both h_f and h_m as follows:

$$k_{ra} = 0.5 \left[k_{rf}(h_f) + k_{rf}(h_m) \right] \tag{7}$$

As the dual-permeability structure consists of two subdomains, a certain variable contains one total value and two local values at any point in time and space. The total value is the weighted average of the two local values:

$$X = wX_f + (1 - w)X_m \qquad X = h, \theta, K \ldots \tag{8}$$

2.2   Finite Volume Discretization in OpenFOAM

The model is built and solved based on an open source platform OpenFOAM (open source field operations and manipulations). The computational space domain is divided into numerous control volumes, which is bounded by a set of flat faces and can be any shape. Time is split into a set of time steps Δt. For equation discretization, the generalized Gauss divergence theorem is adopted to convert the volume integral into surface integral. Consequently, each term in the partial differential equations can be linearized as explicit or implicit form:

$$\underbrace{\oint_{\partial Vp} [C(h_f) \frac{\partial h_f}{\partial t}]}_{\text{implicit}} = \underbrace{\oint_{\partial Vp} [K_f(h_f) \cdot \nabla h_f]}_{\text{implicit}} + \underbrace{\oint_{\partial Vp} [K_f(h_f) \cdot \nabla z]}_{\text{explicit}} - \underbrace{\oint_{\partial Vp} \frac{\Gamma_w}{w}}_{\text{explicit}}$$

$$\underbrace{\oint_{\partial Vp} [C(h_m) \frac{\partial h_m}{\partial t}]}_{\text{implicit}} = \underbrace{\oint_{\partial Vp} [K_m(h_m) \cdot \nabla h_m]}_{\text{implicit}} + \underbrace{\oint_{\partial Vp} [K_m(h_m) \cdot \nabla z]}_{\text{explicit}} + \underbrace{\oint_{\partial Vp} \frac{\Gamma_w}{1 - w}}_{\text{explicit}} \tag{9}$$

The implicit parts contain the unknown variables of the current time level, while the explicit parts can be calculated from the variable values of the old time level. Then the

Modified Picard iteration for nonlinear equations is applied to get the solution, which will be introduced in the next section.

2.3   Solution Procedures

The solution procedures with modified Picard iteration are shown in Fig. 1. In this model, the primary variables are h_f and h_m, the others are related variables, which can be explicitly calculated based on h_f and h_m. For a specific iteration at a new time step from t to $t + 1$, the related variables are initialized first. Equation (9) can be rewritten as follows:

$$
\begin{aligned}
C(h_f^t)\frac{h_f^{t+1,n+1} - h_f^t}{\Delta t} &= \nabla \cdot [K_f^t(h_f^t) \cdot \nabla(h_f^{t+1,n+1})] + \nabla \cdot [K_f^t(h_f^t) \cdot z] - \frac{\Gamma_w^t}{w} \\
C(h_m^t)\frac{h_m^{t+1,n+1} - h_m^t}{\Delta t} &= \nabla \cdot [K_m^t(h_m^t) \cdot \nabla(h_m^{t+1,n+1})] + \nabla \cdot [K_m^t(h_m^t) \cdot z] + \frac{\Gamma_w^t}{1 - w}
\end{aligned}
\tag{10}
$$

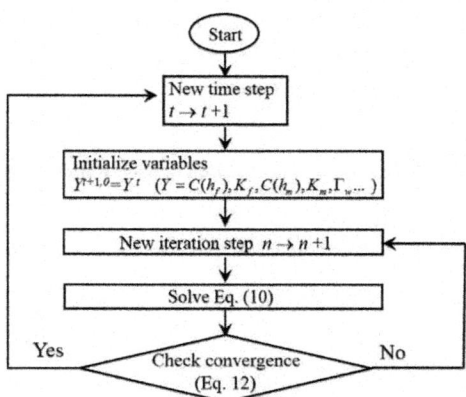

Fig. 1. Solution procedures of the solver developed in this study.

The superscripts t and n denote time level and iteration level, respectively. The equations for all the control volumes are assembled for the whole domain and a matrix equation about the primary unknown variables can be established as

$$
\begin{cases}
\mathbf{A}_f \mathbf{h}_f^{t+1,n+1} = \mathbf{r}_f \\
\mathbf{A}_m \mathbf{h}_m^{t+1,n+1} = \mathbf{r}_m
\end{cases}
\tag{11}
$$

where $\mathbf{A}_f$ and $\mathbf{A}_m$ are coefficient matrixes about the primary unknown variables; $\mathbf{r}_f$ and $\mathbf{r}_m$ are column vectors calculated explicitly by substituting the related variables. The convergence condition of the modified Picard iteration is that the maximum difference in two pressure heads between two successive iteration levels become less than a tolerance value:

$$
\begin{cases}
\max \left| h_f^{t+1,n+1} - h_f^{t+1,n} \right| \leq \delta_f \\
\max \left| h_m^{t+1,n+1} - h_m^{t+1,n} \right| \leq \delta_m
\end{cases}
\tag{12}
$$

where δ_f and δ_m are set as 10^{-9} m in this study. If convergence requirements are achieved at the present iteration, the calculation turns into the next time level; otherwise, continuing the next iteration until meeting the convergence criteria.

2.4 Model Verification

To ensure the accuracy of the finite volume method, a recirculation case in landfill was conducted using the proposed model and the finite element software Hydrus 1-D, which has a built-in the dual-permeability model. The depth of the bioreactor landfill is 30 m, and the pressure heads at the upper and lower boundaries are −0.4 m and 0 m, respectively. The related parameters are given in Table 1. Figure 2 shows the comparison of the pressure head and water content between the proposed model and the Hydrus 1-D at different time. The agreement between the two numerical methods indicates the precision of the proposed model.

Table 1. Parameters for numerical simulation

Single-porosity model					
θ_r	θ_s	α (m^{-1})	n	K_s (m/s)	SF[a]
0.15	0.69	2	1.5	5×10^{-6}	10
Dual-permeability model					
Matrix					
θ_{rm}	θ_{sm}	α_m (m^{-1})	n_m	K_{sm} (m/s)	SF[a]
$\varepsilon_r/(1-w)$	$(\varepsilon - w)/(1-w)$	2	1.5	5×10^{-6}	10
Fracture					
θ_{rf}	θ_{sf}	α_f (m^{-1})	n_f	K_{sf} (m/s)	w
0	1	10	1.5	10^{-3}	0.1
Transfer term					
a (m)	β	γ	K_{sa}(m/s)	Residual porosity	
0.1	3	0.4	10^{-6}	$\varepsilon_r = 0.15$ Total porosity $\varepsilon = 0.69$	

Note [a]: SF is the scaling factor between horizontal and vertical Ks

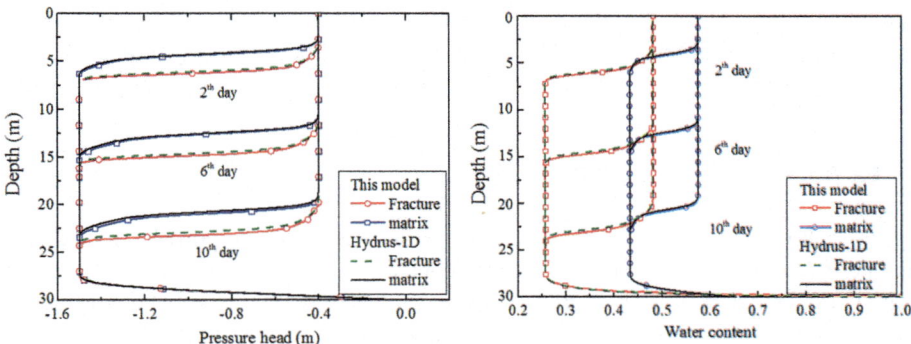

Fig. 2. Comparisons with Hydrus- 1D: (a) pressure head; (b) water content.

3   Three-Dimensional Model Simulation

3.1   Model Description

A typical three-dimensional landfill cell with a vertical leachate recirculation well is adopted to simulate the leachate flow. The surface of bioreactor landfill is usually covered by topsoil with a green landscape. The leachate is added through the vertical well and the recycle layer is the leachate collection and removal system placed between the low-permeability liner system and the waste. It is assumed that the size of the simulated landfill cell is $20 \times 20 \times 20$ m (Fig. 3). The leachate recirculation well is simplified as quarter squares with an area of 0.2×0.2 m in the simulated landfill cell. The depth of the vertical well is 3 m from the top boundary. The left, right, front and back boundaries are set to be symmetrical boundaries.

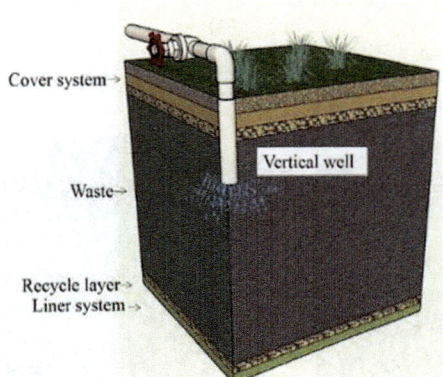

Fig. 3. Conceptual model of a typical bioreactor landfill.

3.2   Parameters Selection

The parameters of the proposed model are summarized in Table 1, and the hydraulic parameters determined by Audebert et al. were obtained from a full-scale landfill recirculation test, which are particularly suitable for the simulation in this study [6]. It also gave the parameters of the single-porosity model corresponding to the dual-permeability model.

4   Results and Discussion

Figure 4 shows the water content contours with time in single-porosity model and dual-permeability model. The initial water content of waste is set to 0.3, and as the recirculation, the impact zones increase. Given that the MSW is a notably heterogeneous and complex material, the dual-permeability model is more appropriate to simulate the properties of waste [6–8]. Under the same recirculation pressure, the impact zones in dual-permeability model expands sharply with the greater impact zones. So, the single-porosity model will underestimate the impact zones in recirculation.

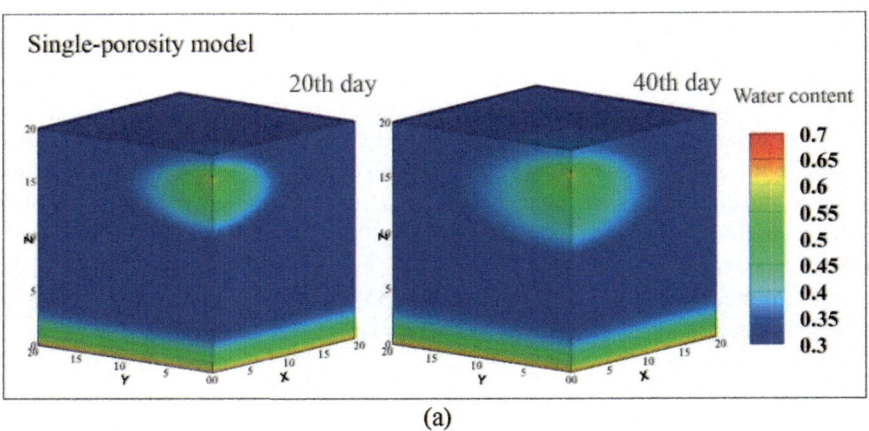

(a)

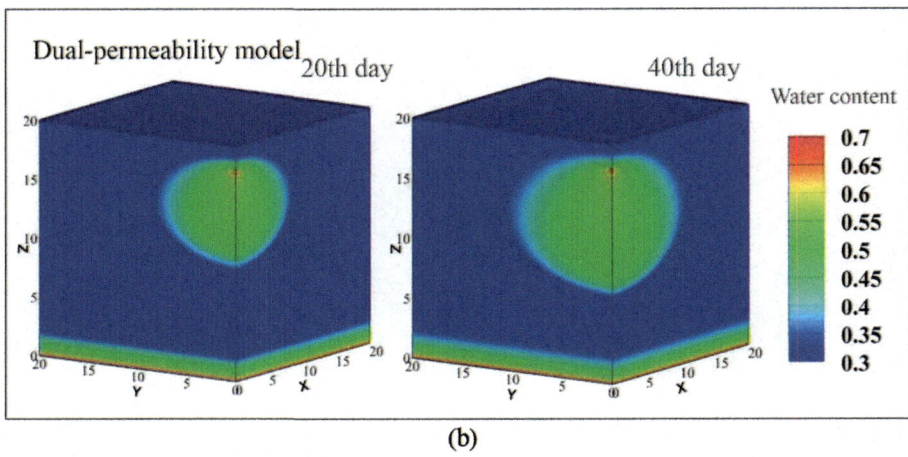

(b)

Fig. 4. Water content contours with time: (a) single-porosity model; (b) dual-permeability model.

The volume of impact zone and the radius of impact zone are two important parameters to design of recirculation system. To further demonstrate the performance of the three-dimensional dual-permeability model, Fig. 5(a) shows the differences of volume of impact zones with time in two models. When the water content of waste exceeds 0.5, it can be recognized as impact zones. As can be seen, volume of impact zones of dual-permeability model is three times lager than that of single-porosity model. The difference of the two models is mainly caused by the preferential pathways in waste, which means that if the big openings of the waste particles (fracture) are neglected, it will lead to the noticeable error in the recirculation simulation. Figure 5(b) shows the radius of impact zone using dual-permeability model on 40th day, and the maximum radius can reach 10 m in the bioreactor landfill, which means that 20 m is a suitable distance between two vertical wells.

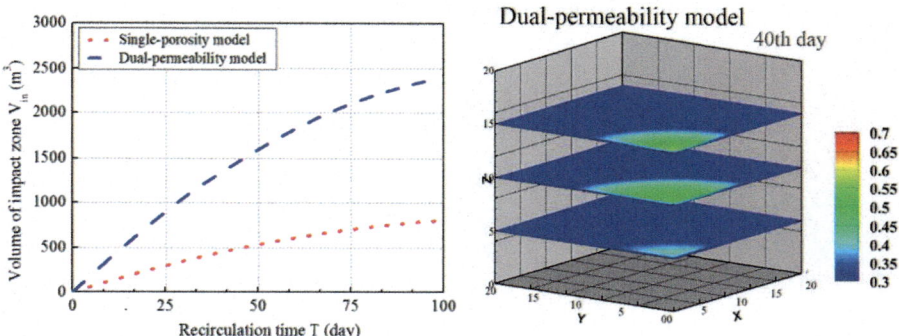

Fig. 5. (a) The volume of impact zones with time; (b) The radius of impact zones 40th day.

5   Conclusion

In this study a three-dimensional dual-permeability numerical flow model is developed to describe the leachate flow in bioreactor landfills. The model is solved in OpenFOAM with modified Picard iteration. The accuracy and capacity of the proposed model are verified by the Hydrus-1D. A typical three-dimensional landfill cell with a vertical leachate recirculation well is adopted to simulate the leachate flow. The results indicated that the traditional single-porosity model can lead to relatively noticeable errors, while the dual-permeability model is reasonably close to the actual properties of waste materials.

References

1. Benson, C.H., Barlaz, M.A., Lane, D.T., Rawe, J.M.: Practice review of five bioreactor/recirculation landfills. Waste Manag. **27**(1), 13–29 (2007)
2. Warith, M.: Bioreactor landfills: experimental and field results. Waste Manag. **22**(1), 7–17 (2002)

3. Jasak, H.: Error analysis and estimation for the finite volume method with applications to fluid flows. Ph.D. Thesis, University of London Imperial College, London, the UK (1996)

4. Capelo, J., de Castro, M.A.: Measuring transient water flow in unsaturated municipal solid waste - a new experimental approach. Waste Manag. **27**(6), 811–819 (2007)

5. Gerke, H.H., Genuchten, M.T.V.: A dual-porosity model for simulating the preferential movement of water and solutes in structured porous media. Water Resour. Res. **29**(2), 305–319 (1993)

6. Audebert, M., Clément, R., Moreau, S., Duquennoi, C., Loisel, S., Touze-Foltz, N.: Understanding leachate flow in municipal solid waste landfills by combining time-lapse ERT and subsurface flow modelling – Part I: analysis of infiltration shape on two different waste deposit cells. Waste Manag. **55**, 165 (2016)

7. Han, B., Scicchitano, V., Imhoff, P.T.: Measuring fluid flow properties of waste and assessing alternative conceptual models of pore structure. Waste Manag. **31**(3), 445–456 (2011)

8. Tinet, A.J., Oxarango, L., Bayard, R., Benbelkacem, H., Stoltz, G., Staub, M.J., et al.: Experimental and theoretical assessment of the multi-domain flow behaviour in a waste body during leachate infiltration. Waste Manag. **31**(8), 1797–1806 (2011)

Microstructure and Triaxial Shear Testing of Compacted Clay Presented in Landfill Leachate

Shicheng Xu, Haijun Lu$^{(\boxtimes)}$, Jixiang Li, Xiaofan Liu,
and Weiwei Wang

School of Civil Engineering and Architecture,
Wuhan Polytechnic University, Wuhan 430023, China
lhj_whpu@163.com

Abstract. In order to reasonably predict failure law of compacted clay presented in landfill leachate as well as help provide theoretical basis for curbing and restoration of contaminated soil, triaxial shear, SEM, XRD and laser particle size tests were used to determine strength and microstructure of compacted clay presented in landfill leachate in this paper. The results show that when the leachate concentration was low, the stress-strain curve of compacted clay belonged to typical strain-softening behavior. With the increase of leachate concentration, the stress-strain curve of soil began to show typical strain-hardening behavior. The concentration of leachate was negatively related to the cohesion and internal friction angle of soil sample. In the aspect of soil microstructure, with the increase of leachate concentration, a series of microstructural changes, such as a decrease of secondary minerals, an increase of pore, the overhead pores developed into cracks and so on, had result in a decrease of the strength of compacted clay. The particle size curve of compacted clay presented in landfill leachate was about single peak curve and the value of $D_v(50)$ tend to stable.

Keywords: Compacted clay · Landfill leachate · Triaxial shear test
Microstructure

1 Introduction

In 2015, Chinese City has 640 landfill disposal facilities and the total of landfill disposal reached 115 million tons, which more than 8 million tons in 2014. The organic matter in garbage disposal and storage is degraded to produce a lot of landfill leachate and methane [1]. The materials for most landfill liners are made up of remolded compacted clay. After longterm influence of leachate corrosion, the strength of landfill liner has changed, which lead to the destruction of landfill liners. The properties of compacted clay presented in landfill leachate are different from that of clay. The strength criterion may not be suitable for compacted clay polluted by landfill leachate. The complexity and randomness of soil microstructure lead to the complexity and randomness of soil macro-mechanical strength.

© Springer Nature Singapore Pte Ltd. 2018
A. Farid and H. Chen (Eds.): GSIC 2018, *Proceedings of GeoShanghai 2018*
International Conference: Geoenvironment and Geohazard, pp. 289–297, 2018.
https://doi.org/10.1007/978-981-13-0128-5_33

Scholars have done a lot of research on the microstructure of contaminated soil. Zhao, Nayak et al. [2, 3] found that aggregate structure was destroyed by leachate, causing an increase of smaller pores and a decrease of larger pores, the permeability of clay increased. Cuisinier et al. [4] found that the infiltration and mechanical characteristics of soil with a long period of alkaline solute diffusion changed significantly. Khamehchiyan et al. [5] discovered that the petroleum pollution reduced the strength of sand and clay and the seepage resistance ability. Yong et al. [6] found that the double layer swelling force and the buffering capacity had an important influence on the microstructure of the montmorillonite clay mixture. Compared with the existing research of contaminated soil, the research on microstructure of soil under the coupling of organic pollutants and stress have less been studied. So it is necessary to do further research.

Consolidated-undrained triaxial shear, X-ray diffraction, scanning electron microscopy and laser particle size tests were used to determine mechanical strength characteristics of compacted clay and the corresponding microstructure characteristics under different concentration of leachate and different stress.

2 Test Materials and Methods

2.1 Test Materials

The clay were drawn from a construction site near Huangpi District in Wuhan, Hubei province, China (about 2.5 m in depth). The basic physical and chemical characteristics of compacted clay are summarized in Tables 1 and 2. The clay were broken and layered compaction in tamping cylinder to obtain cylindrical sample (height = 140 mm, diameter = 70 mm). The leachate used in the test was derived from the solid waste landfill site in Wuhan city. Its basic characteristics are shown in Table 3.

Table 1. The basic physical characteristics of compacted clay

ρ_{dmax} (g/cm^3)	Specific surface area (m^2/g)	W_{opt} (%)	W_L (%)	W_p (%)	I_p (%)	Particle size distribution (%)			
						>0.05 (mm)	0.05~0.005 (mm)	0.005~0.002 (mm)	<0.002 (mm)
1.67	275.4	21.2	48.8	26.1	22.7	42	34	16	8

Table 2. The basic chemical characteristics and composition of compacted clay mineral

pH	Soluble salt (%)	Organic (%)	Mineral composition (%)					
			Quartz	Illite	Montmorillonite	Muscovite	Albite	Kaolinite
5.5	0.4	1.5	23.41	23.10	20.94	21.09	6.69	4.76

Table 3. Physical and chemical characteristics of landfill leachate

Contaminants	Parameter values / (mg·L^{-1})	Determination method	Contaminants	Parameter values / (mg·L^{-1})	Determination method
P-PO$_4$	22.4	HJ 669-2013	TOC	671.3	HJ 501-2009
DO	3.90	GB/T 7489-87	BOD$_5$	1472.6	HJ 505-2009
N-NO$_3$	17.64	HJ/T 346-2007	Cl$^-$	2055.8	GB/T 11896-89
TP	26.7	HJ 671-2013	Na$^+$	1858.3	GB/T 11904-89
N-NH^{4+}	2813.20	HJ 666-2013	COD	4131	HJ/T 399-2007
N-NO$_2$	0.50	HJ/T 197-2005	SO$_4^{2-}$	5.8	GB/T 13196-91

2.2   Test Methods

2.2.1   Triaxial Shear Test

The prepared cylinder specimens were placed in vacuum saturated cylinder. A leachate to water ratio of 1:0, 1:1, 1:2, 1:5 (diluted 0, 1, 2, 5 times) were respectively pumped into the vacuum saturated cylinder after vacuumed 24 h. Test was performed after 30d.

Based on the "Specification of soil test" (SL237-1999) [7], triaxial shear test used CKC pneumatic cyclic triaxial apparatus under digital closed-loop control (United States Geotechnical Engineering Equipment Co.). In order to obtain the strength parameters of compacted clay through Mohr-Coulomb theory, tests were conducted with three confining pressures of 50, 100, and 150 kPa [8]. The maximum volumetric and deviatoric strains were 5% and 15%, respectively and the shear rate was 0.1%/min. When the specimen was bulging greatly without shear failure or the deviatoric strain is more than 15%, stop triaxial shear test.

2.2.2   XRD Test

About 400 g soil samples from 150 kPa confining pressure inside the soil as the test object after the triaxial shear test finished. The soil samples were air-dried, sieved through a 200 mm sieve. The sieved samples were placed on the 15 mm × 15 mm plate. Based on the "General Rule of Modern Instrument Analytical Method and Metrology Examine Regulation" [9], EMPYREAN X-ray diffraction (Holland PANalytical Co.) was used to determine mineral composition of clay at 45 kV and 40.0 mA with CuKα radiation and the emission slit was fixed at 0.19 mm. The scanning speed was 6 °/min (2θ). The scanning range from 5° to 85° (2θ) at 0.04 °/min (2θ).

2.2.3   SEM Test

The clay samples used in SEM test was the same as that of the XRD test (size = 1 mm × 1 mm × 1 mm). Soil samples were pretreated with a vacuum conductive film with a thickness of about 10 nm on the surface of soil samples. Based on the "General Rule of Modern Instrument Analytical Method and Metrology Examine Regulation", the equipment used in this test was S-3000N scanning electron microscope (Hitachi, Japan). SEM images with magnifications of 5000× were obtained.

2.2.4   Laser Particle Size Test

The clay samples used in laser particle size test were the same as that of the XRD, too. According to the operation manual of instrument and "Specification of soil test" (SL237-1999), the sieved 300 g soil samples were tested by Malvern Mastersizer 3000 laser particle size analyzer (Malvern Instruments Ltd., UK) with range from 0.01– 10000 μm. The test used dry injection and Mie diffraction model.

3   Test Results and Discussion

3.1   Triaxial Shear Test

Figure 1 shows the stress-strain curves of compacted clay polluted by landfill leachate. When the diluted times > 5, the stress-strain curve of soil belong to typical strain-softening behavior. With the increase of leachate concentration, the stress-strain curve of soil began to show typical strain-hardening behavior [10]. The stress-strain curve of confining pressure 50 kPa still showed strain-softening behavior with leachate to water ratio of 1:1. The change with cohesion and internal friction angle of compacted clay polluted by landfill leachate was showed in Fig. 2. The leachate concentration was negatively related to cohesion and internal friction angle of compacted clay.

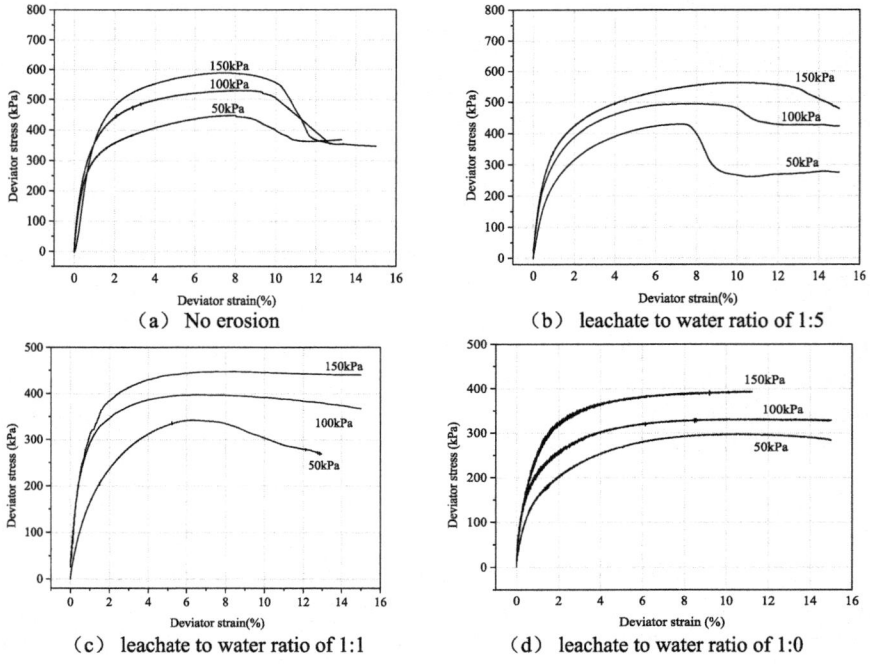

Fig. 1. The stress-strain curves of compacted clay polluted by landfill leachate

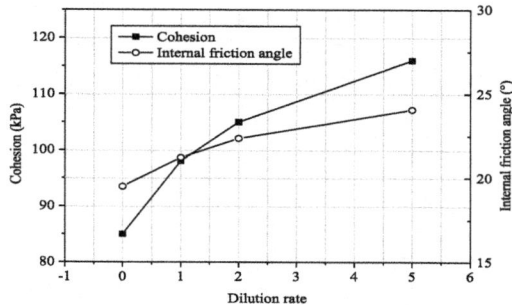

Fig. 2. Change with cohesion and internal friction angle of compacted clay polluted by landfill leachate

The phenomenon can be explained as follows: The clay was layered and compacted in a tamping tube. Therefore, soil particles were closely connected with soil particles. When leachate concentration was low, soil particles were less dispersed and there was still strong force between soil particles to resist the deformation and failure. So the stress-strain curve exhibited strain-softening behavior. When leachate concentration was higher, bound water membrane in the soil surface became thicker and the electrical characteristics of soil were modified by metal ions in leachate, which makes soil particles become more dispersed. Therefore, soil exhibited strain-hardening behavior. When leachate to water ratio was 1:1, soil still has a certain ability to resist damage under 50 kPa due to compacted structure. The greater the confining pressure, the more obvious the damage of the soil structure. So soil still showed strain-softening behavior under confining pressure 50 kPa, while soil showed strain-hardening behavior under confining pressure 150 kPa.

The soil strength is determined by cohesion and internal friction angle [11]. As leachate concentration increased, organic matter and metal ions in leachate destroy cementation force of soil, which result in a decrease of cohesion. Soil particles was broken, which were rearranged and dislocated between particles. These changes mean that the contact area of soil particles decreased as well as internal friction angle.

3.2   XRD Test

The XRD pattern of compacted clay polluted by landfill leachate was shown in Fig. 3. The main minerals of soil polluted by landfill leachate include Quartz, Albite, Montmorillonite, Kaolinite, Muscovite, Illite. The diffraction peaks of Montmorillonite and Illite gradually disappear with the increase of concentration of leachate in 15° and 45°.

The percentage of mineral content of compacted clay obtained by quantitative analysis was shown in Table 4. The content of primary minerals in soil was equal to that of secondary minerals, secondary mineral content accounts for 48.8%. With the increase of leachate concentration, the contents of primary minerals such as Quartz and Albite were little changed, while the contents of secondary minerals such as Illite and Montmorillonite were decreasing continuously. The content of Illite and Montmorillonite in soil decreased by 12% and 11.29% with leachate to water ratio of 1:0, respectively, compared with no erosion.

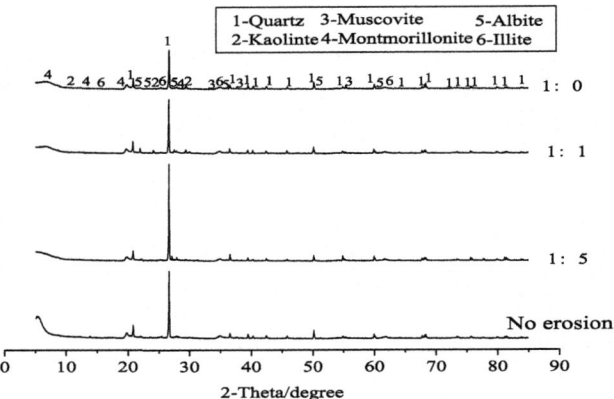

Fig. 3. The XRD pattern of compacted clay polluted by landfill leachate

Table 4. Content of mineral of compacted clay polluted by landfill leachate (%)

Mineral types	Ratio			
	No erosion	Leachate to water ratio of 1:5	Leachate to water ratio of 1:1	Leachate to water ratio of 1:0
Quartz	23.41	23.30	23.78	23.62
Illite	23.10	18.20	17.22	16.25
Montmorillonite	20.94	17.80	17.25	15.79
Muscovite	21.09	18.14	18.94	20.99
Albite	6.69	18.33	17.69	17.55
Kaolinite	4.76	4.24	5.13	5.79

After the leachate erosion, chemical composition and stable crystal lattice structure of primary minerals have not changed, which make it has strong resistance for leachate. However, secondary minerals have active crystal lattice with high dispersion and strong adsorption. The organic matter and metal ions in leachate reacted with the secondary minerals to form inorganic salts, which reduce the content of secondary minerals and lead to obvious change of soil engineering properties, for example plasticity and dilatability significantly increased. Therefore, the content of primary mineral remains relatively stable, the clay mineral content of soil decreased with the increase of leachate concentration. These changes were similar to cohesion decreased with the increase of leachate concentration. Therefore, the decrease of strength was related to the decrease of the content of secondary minerals [12].

3.3 SEM Test

Figure 4(a) shows that soil structure not eroded by leachate was mainly plate-like and honeycombed. The structure was compact, but a few pores are still visible. The contact between soil particles was mainly surface-surface contact and there are some

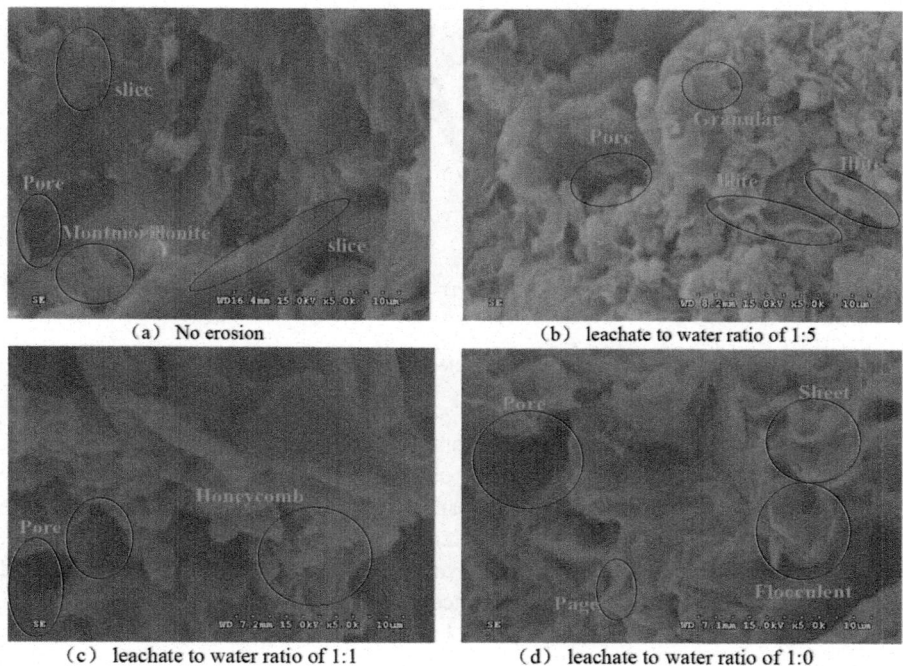

(a) No erosion

(b) leachate to water ratio of 1:5

(c) leachate to water ratio of 1:1

(d) leachate to water ratio of 1:0

Fig. 4. SEM images (magnified 5000 times) of compacted clay polluted by landfill leachate

side-surface contact and point-surface contact. Figure 4(b) and (c) show that soil surface became rough and there were many particles on the soil surface. Some of the particles began to agglomerate and form larger particles, which may be the product of the reaction between the organic matter in the leachate and soil mineral. Most of soil structures were unstable flocculent structures. There were many overhead pores in the soil and the number of pores increases obviously. The contact between soil particles was point-side and side-side contact. Figure 4(d) shows that there were many pores on the soil surface and some of the overhead pores have developed into cracks.

The change of soil microstructure affect the strength and deformation of soil. With the increase of the leachate concentration, a series of microstructure changes, including an increase of pore, unstable flocculent structures and so on, result in a decrease of soil strength as well as cohesion and internal friction angle.

3.4   Laser Particle Size Test

Figure 5 shows particle size curve of compacted clay polluted by landfill leachate. The particle size curve of soil was a single peak curve. From the composition of soil particles, the silt (5 μm–75 μm) were the dominant particle size and there were small amount of sand (75 μm–3000 μm). With the increase of leachate concentration, the volume fraction of sand decreased gradually. $D_v(50)$ indicates the size distribution of the soil particles. The value of $D_v(50)$ was stable, which range from 20 to 40 μm.

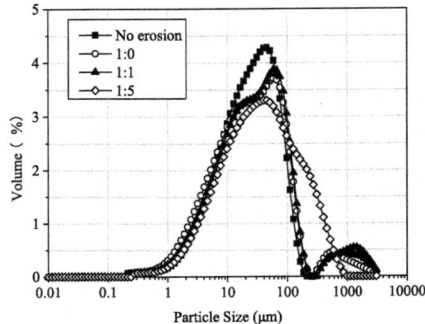

Fig. 5. Particle size curve of compacted clay polluted by landfill leachate

As leachate concentration increased, compacted clay was eroded more vigorous. Quantitative results of laser particle size test combined with SEM show that an increase of leachate concentration result in smaller particles, which reflect that a decrease of the force between particles and particles, a decrease of the contact area between particles. Therefore, cohesion, internal friction angle and strength of compacted clay decreased obviously with the increase of leachate concentration.

4 Conclusion

In order to explain relationship between the macro mechanical strength and the microstructure of compacted clay presented in landfill leachate, the mechanical strength and microstructures were tested by triaxial shear, XRD, SEM and laser particle size test. The following conclusions are drawn:

(1) When leachate concentration was low, the stress-strain curve belong to typical strain-softening behavior. With the increase of leachate concentration, the stress-strain curve began to show strain-hardening behavior. The leachate concentration was negatively related to cohesion and internal friction angle of compacted clay.

(2) The content of Illite and Montmorillonite decreased by 12% and 11.29% with leachate to water ratio of 1:0, respectively, compared with no erosion. The decrease of strength was related to the decrease of the content of secondary minerals.

(3) With the increase of the concentration of leachate, such as the increase in the number of pores, small granular agglomerate big granular, the cementation material was eroded and the overhead pores developed into cracks, a series of microstructure changes result in a decrease of soil strength as well as cohesion and internal friction angle.

(4) The particle size curve of soil was a single peak curve. With the increase of leachate concentration, the value of $D_v(50)$ decreased and tend to stable gradually.

Acknowledgments. This research was financially supported by the "National Natural Science Foundation of China (11672216 and 11602183)," the "Natural Science Fund of Hubei Province (2016CFB636)."

References

1. He, X.S., Yu, H., Xi, B.D., Cui, D.Y., Pan, H.W., Li, D.: Difference of contaminant composition between landfill leachates and groundwater and its reasons. Environ. Sci. **35**(4), 1399–1406 (2014)
2. Zhao, Y., Xue, Q., Huang, F.X., Hu, X.T., Li, J.S.: Experimental study on the microstructure and mechanical behaviors of leachate-polluted compacted clay. Environ. Earth Sci. **75**(12), 1–9 (2016)
3. Nayak, S., Sunil, B.M., Shrihari, S.: Hydraulic and compaction characteristics of leachate-contaminated lateritic soil. Eng. Geol. **94**(3–4), 137–144 (2007)
4. Cuisinier, O., Masrouri, F., Pelletier, M., Villieras, F., Mosser, R.: Microstructure of a compacted soil submitted to an alkaline PLUME. Appl. Clay Sci. **40**(1–4), 159–170 (2008)
5. Khamehchiyan, M., Charkhabi, A.H., Tajik, M.: Effects of crude oil contamination on geotechnical properties of clayey and sandy soils. Eng. Geol. **89**(3–4), 220–229 (2007)
6. Yong, R.N., Ouhadi, V.R., Goodarzi, A.R.: Effect of Cu^{2+} ions and buffering capacity on smectite microstructure and performance. J. Geotech. Geoenviron. Eng. **135**(12), 1981–1985 (2009)
7. SL 237-1999: Specification of Soil Test. Nanjing Hydraulic Research Institute. China Water Power Press, Beijing (1999)
8. Afolagboye, L.O., Talabi, A.O., Ajisafe, Y.C., Alabi, S.: Geotechnical assessment of crushed shales from selected locations in Nigeria as materials for landfill liners. Geotech. Geol. Eng. **35**, 1847–1858 (2017)
9. Liu, Z.M.: General Rule of Modern Instrument Analytical Method and Metrology Examine Regulation. Science Technology Reference Press, Beijing (1997)
10. Yang, J., Lu, H.J., Zhang, X., Li, J.X., Wang, W.W.: An experimental study on solidifying municipal sewage sludge through skeleton building using cement and coal gangue. Adv. Mater. Sci. Eng. **2017**, 1–13 (2017)
11. Zhang, Q., Lu, H.J., Liu, J.Z., Wang, W.W., Zhang, X.: Hydraulic and mechanical behavior of landfill clay liner containing SSA in contact with leachate. Environ. Technol. **39**, 1307–1315 (2017)
12. Li, J.S., Xue, Q., Wang, P., Liu, L.: Influence of leachate pollution on mechanical properties of compacted clay: a case study on behaviors and mechanisms. Eng. Geol. **167**(18), 128–133 (2013)

Investigation of Local Processes and Spatial Scale Effects on Suffusion Susceptibility

Chuheng Zhong[1], Van Thao Le[1,2], Fateh Bendahmane[1],
Didier Marot[1], and Zhenyu Yin[3(✉)]

[1] Université de Nantes, Institut de Recherche
en Génie Civil et Mécanique, ST-Nazaire, France
[2] The University of Danang - University of Science and Technology,
Da Nang, Vietnam
[3] Ecole Centrale de Nantes, Institut de Recherche
en Génie Civil et Mécanique, Nantes, France
zhenyu.yin@ec-nantes.fr

Abstract. Many failures of earth structures are caused by the internal erosion occurring in these structures and their foundations. Suffusion, one of four internal erosion types, is a selective erosion of fine particles which move through the matrix formed by the coarser particles. In literature, most investigations on suffusion took it as a single erosion process. However, the suffusion is a complex process due to the combination of three processes: detachment, transport and possible filtration of finer fraction. The influence of the local processes on suffusion susceptibility, especially the filtration process, is not well established. The objectives of this study are investigating the filtration process by verifying results of filtration tests with the basic filtration equation and analyzing the influence of spatial scale effects on the filtration process by performing tests with two different-sized devices. The filtration tests results show the consistency with the basic filtration equation on suspended particle concentration. And suffusion tests indicate the significant effect of specimen size on filtration process. The interpretative method based on the energy expended by the seepage flow and the cumulative loss dry mass is more appropriate with filtration process than those based on the geometric shape of the particles.

Keywords: Filtration process · Suffusion · Spatial scale effect

1 Introduction

Internal erosion resulted from a seepage within an earth structure is an intricate phenomenon which is one of the most common origins of failure of levees and soil dams. Among 11192 surveyed dams, 136 reveal dysfunctions, around 46% show internal erosion, 48% show overtopping and 5.5% show sliding [1]. Nowadays, there are four forms of internal erosion that are identified: concentrated leak erosion, backward erosion, suffusion and contact erosion [2]. This paper focus on suffusion and there are three criteria which have to be satisfied for suffusion existing: geometric criterion, stress criterion and hydraulic criterion: (1) the size of fine particles must be smaller than the size of the voids between the coarser particles, (2) the fine particles are not enough

© Springer Nature Singapore Pte Ltd. 2018
A. Farid and H. Chen (Eds.): GSIC 2018, *Proceedings of GeoShanghai 2018*
International Conference: Geoenvironment and Geohazard, pp. 298–305, 2018.
https://doi.org/10.1007/978-981-13-0128-5_34

to cram the space between the coarser particles and otherwise the effective stresses will load the fine particles, (3) the rate of flow through the soil must be great enough to move the fine particles through the constrictions between the larger particles [3]. According to the two first criteria, proposals of various geometric assessment methods exist in the literature, mostly based on the particle size distribution [3–5]. With the purpose to take also into account the influence of the relative density, some criteria based on constriction size distribution are proposed [6]. The third criterion means that suffusion leads to detachment and transport of the fine particles, which will result in the filtration of some detached particles.

More and more researchers are interested in the role of filtration process in suffusion because the decrease of the hydraulic conductivity which is resulted from the clogging caused by the filtration of some detached particles is visible [7–9]. The basic filtration equation is expressed as follow [10]:

$$C(L) = C_0 e^{-\lambda L} \tag{1}$$

Where C_0 is the initial particle mass concentration, C is the particle concentration after flow through a filter with length L and λ is the filter coefficient. To eliminate the influence of filtration on soil fabric, which limits the geometric evaluation methods, it may be better to consider variations of both differences of hydraulic head and flow rate [11]. Both rigid wall cylinder and specific triaxial cell are selected to design the devices to study suffusion, and this leads to various tested specimen sizes [3–5, 7, 12]. Considering few research about filtration in the context of suffusion, this paper aims to compare results from filtration tests and Eq. (1). By comparing results of suffusion tests performed with two different-sized devices, the potential spatial scale influence on filtration and discussed.

2 Laboratory Experiments

2.1 Main Characteristics of Testing Devices

Since two different apparatuses are necessary to study the spatial scale effects, a large device, which is composed of a rigid wall cylinder cell with an inner diameter of

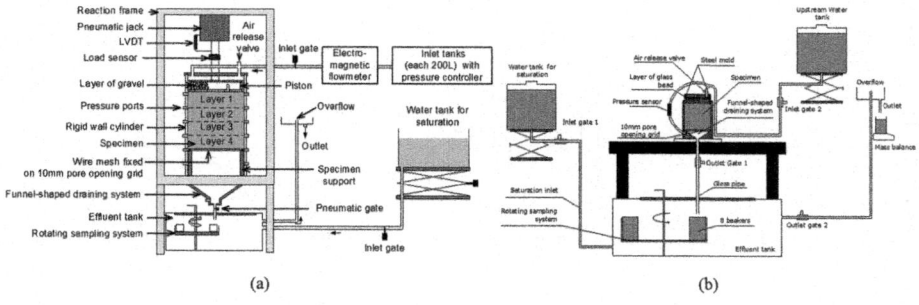

(a)   (b)

Fig. 1. Sketch of the experimental benches (a) oedopermeameter; (b) triaxial erodimeter

280 mm (see Fig. 1(a)) and a small device, which is modified from a triaxial cell and compatible with the specimens whose sizes are 50 mm in diameter and height up to 100 mm (see Fig. 1(b)) were designed to perform suffusion tests with a flow in downward direction. The larger one, named as oedopermeameter, is also used to investigate the local filtration process, due to the local pressures ports. Each draining system is connected to a collecting system which is consist of effluent tank containing a rotatable support with eight containers to catch the eroded particles. The confining pressures are replaced by a steel mold for the triaxial erodimeter in order to test specimens in the same oedometric condition with both devices. A differential pressure transducer between the top cap and base pedestal of triaxial erodimeter is used to measure differential pore water pressure across the specimen. Twelve pressures ports are installed along the rigid wall of oedopermeameter cell, a pressure port is placed on piston base plate and a fourteenth port is located below the specimen on the funnel-shaped draining system. In the literature, there are more detailed descriptions for both devices [7, 12].

2.2 Testing Materials

Grain size distribution of tested materials is shown as Fig. 2. In filtration tests, gravel was selected as the filter (number 1) and Fontainebleau Sand (number 2) played the role of suspended particles. On the other hand, two types of gradations were selected to realize suffusion tests, gap graded and widely graded. One gap graded soil is composed of sand and gravel (number 5). Two soils are widely graded, one cohesionless composed of sand and gravel (number 4) and one clayey soil (number 3) composed of 25% of Kaolinite Proclay and 75% of Fontainebleau sand. All these selected soils for suffusion tests are indeed internally unstable.

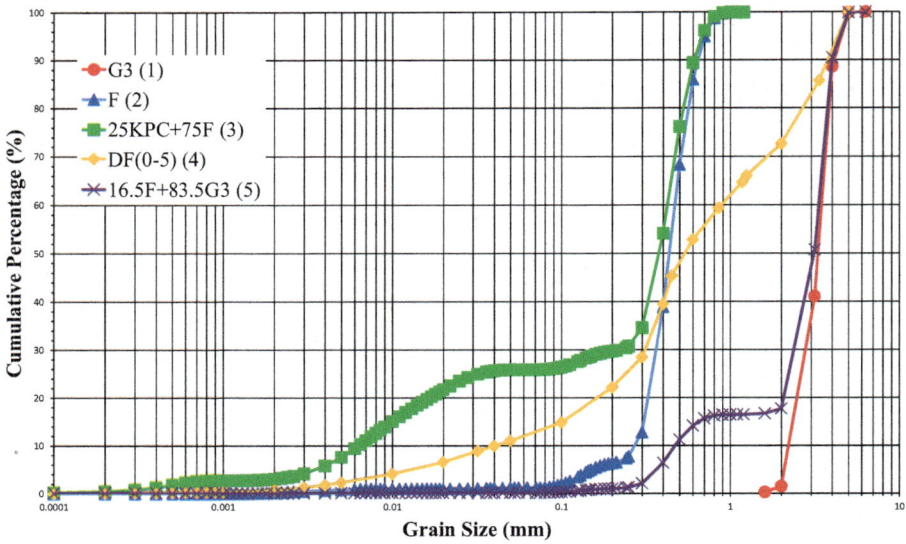

Fig. 2. Grain size distribution of tested soils

2.3   Specimen Preparation and Testing Program

To perform a filtration test, a binary mixture of sand and gravel (1:1) prepared with a water content of 3% was placed over a 200 mm high bed of gravel. Two different testing programs were used: (1) the binary mixture filled an impervious tube with a diameter of 98 mm (depicted in Fig. 3(a)), (2) the binary mixture filled the gap between the oedopermeameter cylinder and an impervious internal plastic bucket with an average diameter of 210 mm (depicted in Fig. 3(b)). All samples were saturated first with CO_2 and then with downward water seepage. Specimens were saturated for both devices under the same moistening velocity and a beaker was used to catch the loss of particles during the saturation phase. The history of hydraulic loading has a significant influence on the development of suffusion [13], so the specimens were systematically tested under a multi-staged hydraulic gradient and each stage lasted thirty minutes. A beaker was used to catch the eroded particles during each hydraulic gradient stage and the corresponding dry masses were measured.

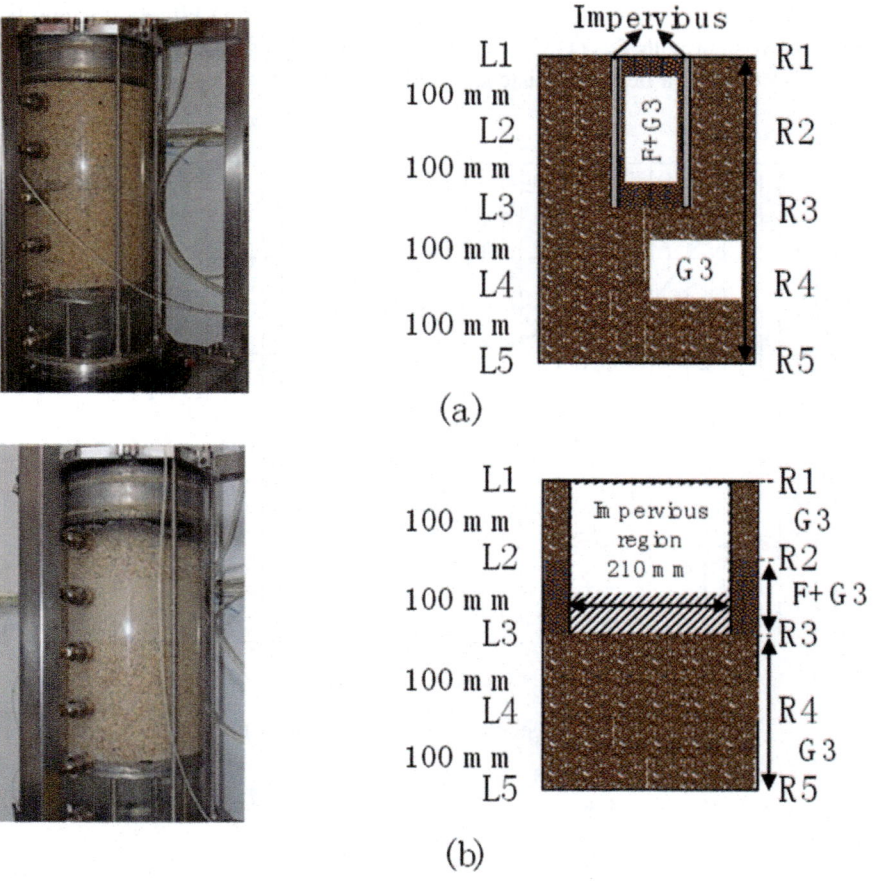

Fig. 3. Schematics and pictures of the filtration tests performed with oedopermeameter

3   Results and Discussion

With the objective to improve the readability, for filtration tests, C and G means the tests with the binary mixture in the center (Fig. 3(a)) and in the gap (Fig. 3(b)) respectively and the number is the specimen number. On the other hand, the first number of each suffusion test name is related to the gradation (Fig. 2). The letter indicates the used apparatus: O for oedopermeameter test and T for triaxial erometer test, and the last number details the specimen number.

3.1   Post-test Particle Size Distributions of Specimens

For filtration tests, we focus on the area between pressure ports L3-R3 and L5-R5 because the filtration process exists in the downstream. For all specimens, it can be noted that the sand content decreases along the seepage length in Fig. 4. If we use the sand content to present the particle mass concentration, this result is in agreement with Eq. (1) but not a complete exponential decrease. Ideally, the filtration process ends when the result reaches the state described as Eq. (1). For whole specimen, Filtration-G1 and Filtration-G-2 have a higher sand content than Filtration-C and they are more consistent with Eq. (1). In consequence, Eq. (1) results and the experimental results seem to have a better consistency with more fine particles within specimens.

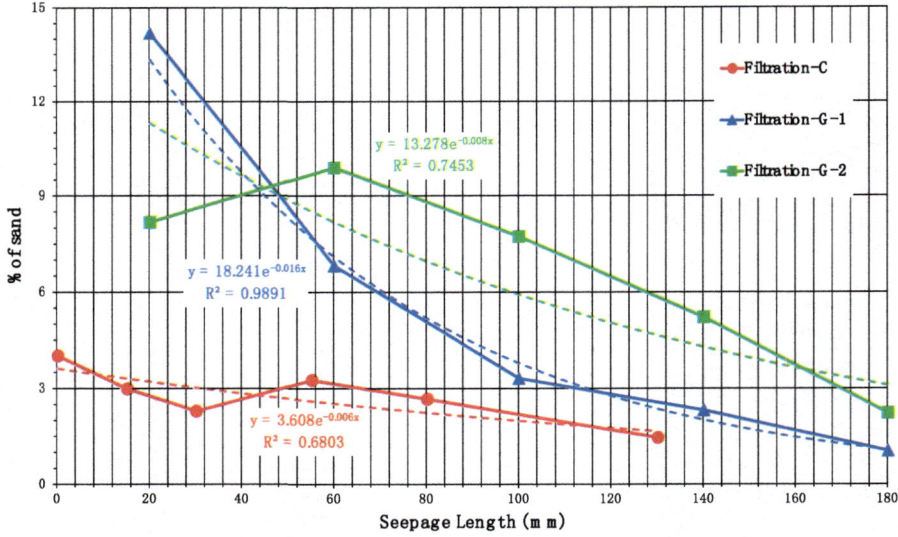

Fig. 4. Sand percentage along seepage length after filtration tests

3.2   Hydraulic Conductivity

As mentioned above, if the filtration process achieves the final state, it stops, which means the process of the soil fabric change caused by the filtration ends. It leads to the constant hydraulic conductivity. And Fig. 5 indicates that the hydraulic conductivity of Filtration-G-2 is more stable than Filtration-C, in other words, the filtration process in Filtration-G-2 is closer to complete. Comparing the hydraulic conductivity of the downstream (L4R4 to L5R5) and the upstream at the end, it seems that fine content maybe decrease the hydraulic conductivity. The curves of hydraulic conductivity of L2R2-L3R3 and L3R3-L4R4 are very closed, but the sand contents in these two layers are very different (initial sand content is 50% in L2R2-L3R3). Therefore, a small amount of fine content has a great impact on hydraulic conductivity but the influence cannot increase infinitely.

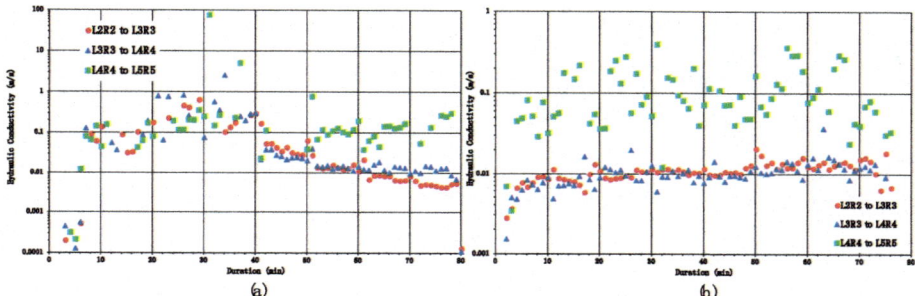

Fig. 5. Time evolution of hydraulic conductivity for (1) Filtration-C; (2) Filtration-G-2

3.3   Spatial Scale Effects

Based on the variation of hydraulic conductivity shown as Fig. 6, which decreases first then increases, the filtration process seems to be the major process before 30 min for Suffusion-5-T and the detachment process becomes significant after 30 min. In contrast to it, the filtration is always the main process for Suffusion-O-T because the hydraulic conductivity of specimen on oedopermeameter continue to decrease. Filtration process seems to reach the final state later in the big device than in the small device. Table 1 displays all values of the critical hydraulic gradient [14] and erosion resistance index [11] for suffusion test. Considering the critical hydraulic gradient, it is not available for oedopermeameter in some cases and there is a big difference between the results from two devices. Filtration, as a local process, it has a greater influence on local parameter as the critical hydraulic gradient than on global parameter, e.g. one soil on both devices with the same erosion resistance index. Suffusion is a combination of three processes: detachment, transport and filtration, and the spatial scale mainly affected the filtration process. Moreover, it needs more time to achieve the final state of filtration on a larger scale device.

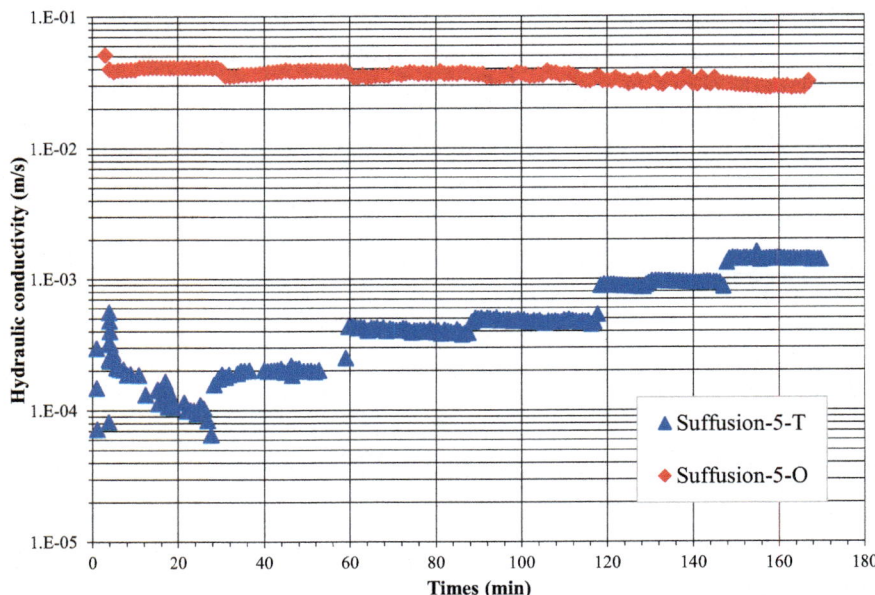

Fig. 6. Time evolution of hydraulic conductivity for Suffusion-5-T and Suffusion-5-O

Table 1. Critical hydraulic gradient and erosion resistance index

Specimen reference in paper	Critical hydraulic gradient	Erosion resistance index
Suffusion-3-O-1	Determination not possible	Moderately resistant
Suffusion-3-O-2	Determination not possible	Moderately resistant
Suffusion-3-T	4.000	Moderately resistant
Suffusion-4-O	Determination not possible	Moderately resistant
Suffusion-4-T	1.200	Moderately resistant
Suffusion-5-O	0.075	Moderately erodible
Suffusion-5-T	0.130	Moderately erodible

4   Conclusion

According to the filtration tests performed with two different programs, the content of filtered fine particles decreases along the seepage length. The basic filtration equation indicates the final state of filtration process. In suffusion process, the filtration process ends when the fine content is closed to the value computed by the basic filtration equation.

The results of suffusion tests on two devices with different sizes show the significant effect of specimen size on filtration process. It seems to need more time to reach the final state of filtration process with a greater specimen, as a local process of suffusion, which leads to the sensitivity of suffusion to spatial scale. In response to the impact of the filtration process, the interpretative method based on the energy expended by the seepage flow and the cumulative loss dry mass is a good way, which is more appropriate with filtration process than those based on the geometric shape of the particles. Therefore, the same soil has the same classification on two devices with this method.

Mechanical coupling with the evolution of fines content [15–17] will be carried out in the near future.

References

1. Foster, M., Fell, R., Spannagle, M.: The statistics of embankment dam failures and accidents. Can. Geotech. J. **37**(5), 1000–1024 (2000)
2. Bonelli, S.: Erosion in Geomechanics Applied to Dams and Levees, 1st edn. Wiley-ISTE, London (2013)
3. Wan, C.F., Fell, R.: Assessing the potential of internal instability and suffusion in embankment dams and their foundations. J. Geotech. Geoenviron. Eng. **134**(3), 401–407 (2008)
4. Chang, D.S., Zhang, L.M.: Extended internal stability criteria for soils under seepage. Soils Found. **53**(4), 569–583 (2013)
5. Kenney, T.C., Lau, D.: Internal stability of granular filters. Can. Geotech. J. **22**(2), 215–225 (1985)
6. Indraratna, B., Israr, J., Rujikiatkamjorn, C.: Geometrical method for evaluating the internal instability of granular filters based on constriction size distribution. J. Geotech. Geoenviron. Eng. **141**(10), 04015045 (2015)
7. Bendahmane, F., Marot, D., Alexis, A.: Experimental parametric study of suffusion and backward erosion. J. Geotech. Geoenviron. Eng. **134**(1), 57–67 (2008)
8. Marot, D., Bendahmane, F., Rosquoet, F., Alexis, A.: Internal flow effects on isotropic confined sand-clay mixtures. Soil Sediment Contam. **18**(3), 294–306 (2009)
9. Nguyen, H.H., Marot, D., Bendahmane, F.: Erodibility characterisation for suffusion process in cohesive soil by two types of hydraulic loading. La Houille Blanche **6**, 54–60 (2012)
10. Iwasaki, T., Slade, J.J., Stanley, W.E.: Some notes on sand filtration [with Discussion]. J. (American Water Works Association) **29**(10), 1591–1602 (1937)
11. Marot, D., Rochim, A., Nguyen, H.H., Bendahmane, F., Sibille, L.: Assessing the susceptibility of gap-graded soils to internal erosion: proposition of a new experimental methodology. Nat. Hazards **83**(1), 365–388 (2016)
12. Sail, Y., Marot, D., Sibille, L., Alexis, A.: Suffusion tests on cohesionless granular matter. Eur. J. Environ. Civil Eng. **15**(5), 799–817 (2011)
13. Rochim, A., Marot, D., Sibille, L., Le, V.T.: Effects of hydraulic loading history on suffusion susceptibility of cohesionless soils. J. Geotech. Geoenviron. Eng. **143**(7), 04017025 (2017)
14. Skempton, A.W., Brogan, J.M.: Experiments on piping in sandy gravels. Géotechnique **44**(3), 449–460 (1994)
15. Chang, C.S., Yin, Z.-Y.: Micromechanical modeling for behavior of silty sand with influence of fine content. Int. J. Solids Struct. **48**(19), 2655–2667 (2011)
16. Yin, Z.-Y., Zhao, J., Hicher, P.Y.: A micromechanics-based model for sand-silt mixtures. Int. J. Solids Struct. **51**(6), 1350–1363 (2014)
17. Yin, Z.-Y., Huang, H.W., Hicher, P.Y.: Elastoplastic modeling of sand-silt mixtures. Soils Found. **56**(3), 520–532 (2016)

Construction Efficiency Analysis of Shield Tunnelling in Ji'nan R1 Yu-Wang Section

Yongliang Huang[✉], Yongjun Wang, Qi Ding, and Lianyong Sun

Jinan Rail Transit Group Co. Ltd., Jinan 250010, Shandong, China
hyl@stu.ouc.edu.cn

Abstract. Based on the shield tunnel practice of Yu-Wang section in Ji'nan Rail Transit Line R1, the energy and resource consumption, and environmental impaction on the shield tunnel were researched to estimate the engineering quantity accurately and resolve estimation difficulties when bidding for rail tunneling projects. The consumption of electric energy, water resources, soil conditioning agent and grouting fluid during the construction process were quantified according to an investigation of shield construction parameters from the first 100 rings of the launching section. It is shown that the power consumption is relatively stable during segments installation, while it is mainly related to formation factors (shear wave velocity), the cutter head working time and construction parameters for the tunnelling stage. There is a positive correlation between the ring power consumption with formation shear wave velocity and the cutter head working time of each ring. And a clear linear correlation between the total power and the advancing speed, the cutter torque and the power of the cutter head is severally presented. The cutter head working time is around 20 min to 40 min and the power consumption is between 270 kWh to 580 kWh for each ring. The cumulative industrial water consumption is mainly between 1100 L to 4050 L per ring, and it is positively correlated with the cutter head working time. The total amount of solution and mixed liquid of foam in each ring are arranged 800 L to 2700 L and 60 L to −220 L, respectively. There is a significant positive correlation between them and the cutter head working time in each ring. The cumulative grouting amount of each ring is between 3100 L to 7000 L. Nevertheless, there is no obvious correlation between the cutter head working time.

Keywords: Rail transit · Shield construction
Power and materials consumptions · Efficiency analysis

1 Introduction

Constructions of rail transit by shield tunnelling would cause a series problem of resource consumption, environmental pollution. Many case histories have been reported during the past decades which helped with the understanding of the efficiency analysis associated with shield tunnelling for rail transit. Based on the construction technology and experience, the green construction technology is applied to shield tunnelling [1–3], and researched green mud slurry, energy consumption simulation and assembly trajectory of assembling machine hydraulic system, green construction

© Springer Nature Singapore Pte Ltd. 2018
A. Farid and H. Chen (Eds.): GSIC 2018, *Proceedings of GeoShanghai 2018*
International Conference: Geoenvironment and Geohazard, pp. 306–314, 2018.
https://doi.org/10.1007/978-981-13-0128-5_35

evaluation [4–9]. Despite the large number of green shield tunnelling reported in the literature, it is still lack of green implementation effect of shield construction technology in resource consumption.

To address the limitation described, in this study a shield tunnel in Ji'nan Rail Transit Line R1 was heavily instrumented and closely monitored. The observed resource consumptions (i.e., Electric energy, water and material consumption) are analyzed in detail. The case reported in this paper provide feedback for the green construction of track traffic shield, and provide reference for shield construction in the late stage.

2 Site

2.1 A Brief Description of the Tunnel

Figure 1 shows a plan view of the site. The length of shield section researched in this paper is 1303 m, and the buried depth of shield structure from interval working well to Wangfuzhuang Station is about 13.7 m to 24.0 m, respectively. The shield tunnelling is started from Yu-Wang interval work well and received at Wangfuzhuang Station. In order to facilitate data analysis, the data of the first 100 rings of the shield are analyzed in this paper. For reducing the tunnel block ratio and the operation energy consumption, the inner and outer diameter of shield tunnel in Ji'nan Rail Transit Line R1 are increased to 5.80 m and 6.40 m, respectively. But the length of each segment is still 1.2 m.

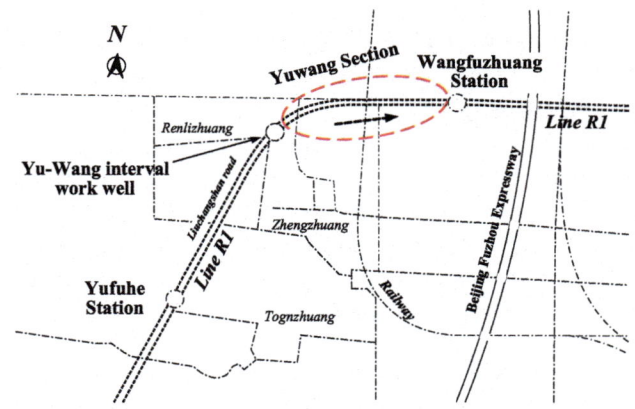

Fig. 1. Plan view of site (not to scale).

2.2 Geology

Ji'nan is located in the middle-west part of Shandong Province, south of Taishan and north across the Yellow River. The layer in researched site can be classified into three categories, the artificial accumulation layer, e.g., the Quaternary Holocene flooding layer, and the fourth upper Pleistocene flooding layer. And further subdivided into 12 layers according to the stratigraphic lithology and their physical and mechanical properties. Figure 2 shows the vertical profile of the first 100 ring shield through the formation, of which the BH1-BH6 is prospecting boreholes. The physical mechanics of the BH3 is shown in Fig. 3.

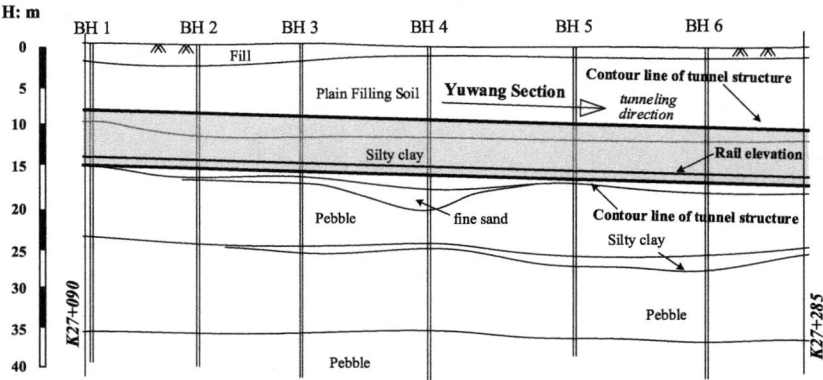

Fig. 2. Soil profile of the site.

3   Interpretation of Measured Results

3.1   Energy Consumption Analysis of Shield Construction

Energy Consumption System
The shield machine diameter, length weight and minimum turning radius of Yu-Wang tunnel is 6.68 m, 85 m, 500 t and 250 m, respectively. The main bearing diameter of the shield used in this section is 3217 mm, and the maximum thrust load is 4086 t. The effective service life of more than 10000 H (i.e., hour). There are 9 groups of hydraulic drive satisfied the requirement of sand layer in the formation with higher torque requirement, which rated torque and turnaround torque is 7070 kNm and 8610 kNm respectively.

There are two main energy consumption operation modes of tunnelling and segment installation under working condition show in Table 1. However, the other conditions also consume energy, but the overall is relatively small.

Analysis of Energy Consumption
Geological condition and cutter operating time are two of the main factors that affect the energy consumption of the shield tunnelling. Equivalent soil shear wave velocity is selected as stratum parameter to analyze in this paper, which is easy to acquire and most closely related to shield cutter cutting soil and driving.

Figure 4 shows the relationship between the shear wave velocity of the equivalent soil layer and the cumulative electricity consumption of the ring. With the increase of

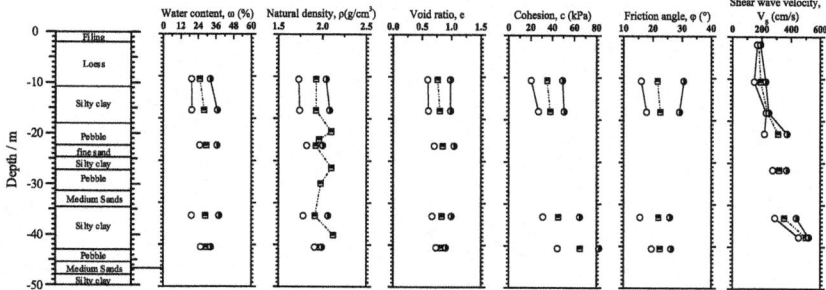

Fig. 3. Soil properties obtained from BH3.

Table 1. Technical parameters of main energy consumption system of shield.

Project	Power	Unit
Cutter head drive	945	kW
Screw conveyor	250	kW
Advance	75	kW
Segment erector	55	kW
Grouting	55	kW
Auxiliary pump	22	kW
Main drive slippage pump	55	kW
Articulated pump	18.5	kW
Control oil pump	11	kW
Air compressor 1	55	kW
Air compressor 2	55	kW
Bentonite pump	18.5	kW
Bentonite mixing	12	kW
Booster water pump	11	kW

shear wave velocity of the equivalent soil layer, the formation hardness increases, and the power consumption of each ring of the shield also increases. This means that there is a certain linear relationship between the ring power consumption and the formation. The greater the relative hardness of the stratum, the greater the total energy consumption of the ring.

Figure 5 shows the relationship between the total power consumption of the ring and the working time of the cutter head. The working time of each ring the cutter head of the first 100 rings is concentrated around from 20 to 40 min. As shown in Fig. 5, the ring power consumption of the shield increases with the increase of the cost per ring, which shows a positive correlation.

Moreover, the construction parameters also affect the overall ring power consumption of the shield machine to a certain extent. The overall power consumption of the shield machine is the sum of the working power of each brake motor. As the actual power is closer to the rated power, the efficiency of the system is higher. The power of

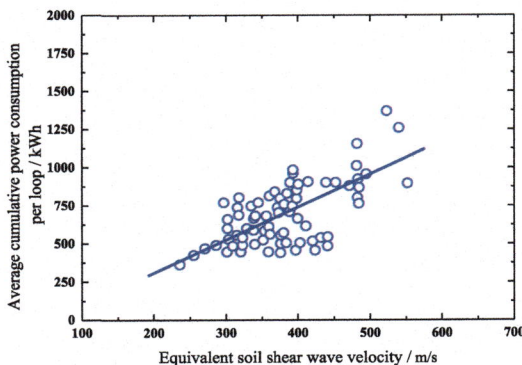

Fig. 4. Relationship between cumulative energy consumption and equivalent shear wave velocity.

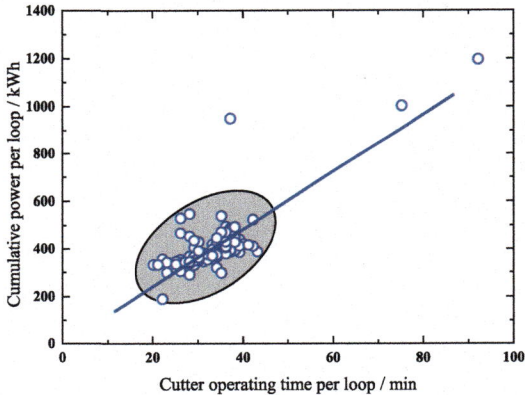

Fig. 5. The relationship between the total power consumption of the ring and the working time of the cutter head.

the cutter head is mainly affected by the speed of the cutter head and the torque of the cutter head.

It can be seen from Fig. 6 that the power of cutter head is mainly about 440 to 530 kW, while the torque of cutter head is mainly between 1350 to 1750 kNm, and the cutter torque is significantly related to the power of the cutter head. According to the theory of reducing useless power, in order to improve the utilization of cutter energy, the power of cutter head should been increased by increasing the torque of cutter head within the limits of the rated power.

The thrust speed of shield machine is mainly between 0 to 5.5 cm/min, and the total power is about 330 to 590 kW (in Fig. 7). The total power of shield propulsion is obviously related to the speed of propulsion. So it is obvious that the total operating power increases with the propulsion speed increasing, which could increase the use efficiency of useful energy sources.

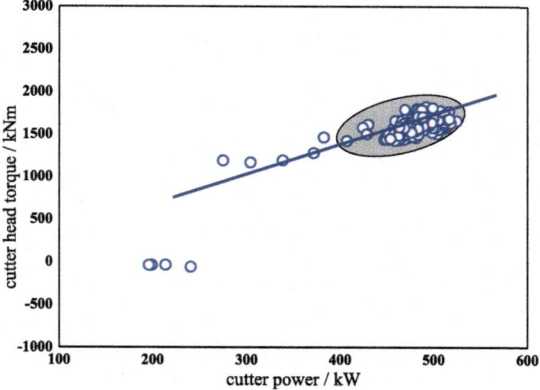

Fig. 6. The relation between cutter head torque and cutter power.

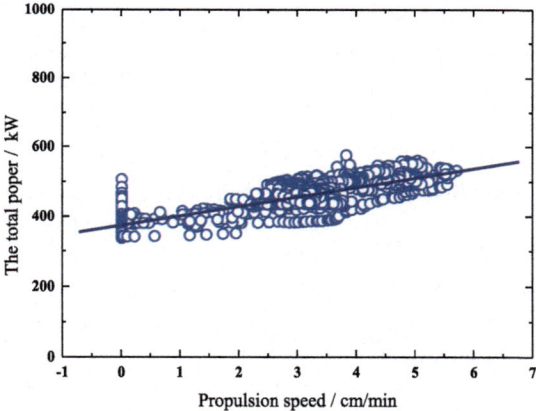

Fig. 7. Relationship between total power and propulsion speed.

3.2 Analysis of Water Consumption

The industrial water consumption of earth pressure balanced shield (EPBs) is mainly concentrated in shield cooling system. The relationship between the cumulative industrial water consumption and the working time of the ring cutter is described in Fig. 8. The cumulative industrial water consumption in each ring is mainly around 1100 to 4050 L, and there is a significant linear correlation between them and the working time of the ring cutter head. The overall industrial water consumption is relatively less.

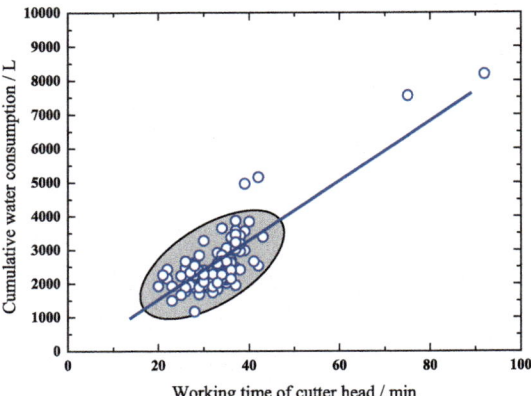

Fig. 8. Relationship between cumulative water consumption and working time of cutter head.

3.3   Analysis of Material Consumption

There are many kinds of materials consumed during the construction of shield tunnelling, but soil amendment (i.e., foam) and grouting are emphatically discussed in this paper.

Soil Amendment

Figure 9 shows the relationship between the cumulative amount of foam solution and the mixture of foam liquid with the working time of ring cutter. The total amount of foam solution and the mixture of foam liquid per ring is mainly between 800 to 2700 L and 60 to 220 L, respectively. There is a linear correlation between the amount of foam and the amount of foam used in each ring with the working time of the ring cutter head. But the amount and effect of the use of a large extent based on the builder's judgment while tunnelling, there is no index of operation. There is a certain space for optimization and improvement.

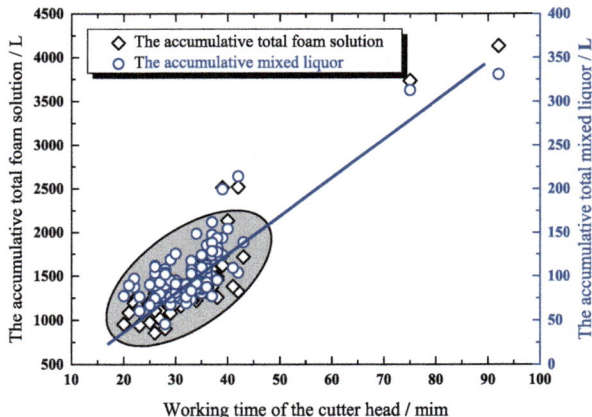

Fig. 9. The relationship between the accumulative total foam solution/mixed liquor and the working time of the cutter head.

Grouting Fluid

Grouting fluid is used to control formation deformation which greatly depends on the formation factor. The relationship between the cumulative grouting amount of each ring and the working time of the ring cutter as show in Fig. 10. The cumulative grouting amount of each ring is concentrated between 3100 to 7000 L. There is no obvious correlation between the accumulated liquid content of the shield ring and the working time of the ring cutter. That's because the complex formation factors affect the grouting amount of the grouting fluid.

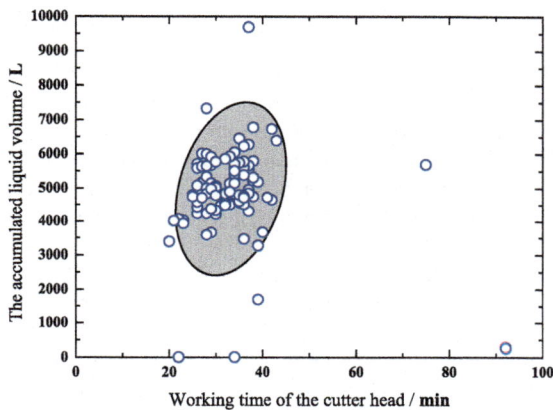

Fig. 10. The relationship between the accumulated liquid volume and the working time of the cutter head.

4   Conclusions

In this study, the energy, water and material consumption of shield tunnelling in Ji'nan R1 Yu-Wang Section were analyzed. The following conclusions may be drawn:

(a) There are two main energy consumption operation modes of tunnelling and segment installation under working condition. The working time of each ring is between 20 to 40 min, which is mainly related to formation conditions, cutter working time and construction parameters.

(b) There is a clear correlation between the power consumption of each ring with the shear wave velocity and the working time of the ring cutter head. The larger the relative hardness of the stratum is, the greater the total energy consumption of the ring is. Similarly, the cutter torque is linearly related to the power of the cutter head, and the total power of shield propulsion is obviously related to the speed of propulsion.

(c) The water consumption of shield construction mainly lies in the consumption of industrial water in shield cooling system. There is a significant linear correlation between the cumulative industrial water consumption and the working time of the ring cutter head.

(d) There is a linear correlation between the amount of foam and the amount of foam used in each ring with the working time of the ring cutter head. Nevertheless, there is no obvious correlation between the accumulated liquid content of the shield ring and the working time of the ring cutter.

Acknowledgements. The research work described herein was funded by the Nature Science Foundation of Shandong Province (Grant No. ZR2016EEQ 13).

References

1. Chai, Q.S.: Research on the Application of Green Construction Technology Subway Shield Tunneling Construction. Master thesis, Nanchang University, China (2016)
2. Wen, F.Q., Zhang, L.Q., Pi, Y.H., et al.: Application of green construction technology in Metro Shield Construction. In: 3rd National Symposium on Intelligent Urban and Rail Transportation, p. 4. Urban Science Research Association of China, Suzhou (2016)
3. Zhang, P.: Environmental protection technology for metro tunnel constructed by shield. Tunnel Constr. **32**(S2), 98–100 (2012)
4. Li, D.S.: Green treatment technology of super large diameter shield slurry under geological condition of full section clay. J. Shijiazhuang Tiedao University (Nature Science Edition) **27**(S1), 72–75 (2014)
5. Zhang, Z.X., Hu, X.Y., Huang, X.: Experimental study on environment friendly slurry for slurry shield tunnelling. J. Tongji University (Nature Science Edition) **38**(11), 1574–1578 (2010)
6. Wang, T.: A study on preparation of environment friendly slurry and its adaptation during slurry shield tunnelling in high permeability stratum. Master thesis, Beijing Jiaotong University, China (2015)
7. Cui, J.Z.: Design and simulation of segment erector hydraulic control system and energy saving analysis of system assembly path. Master thesis, Shanghai JiaoTong University, China (2014)
8. Pi, Y.H.: Study on calculating carbon emissions of shield tunnel construction. Master thesis, Nanchang University, China (2016)
9. Li, Y., Chen, C.: The fuzzy comprehensive evaluation of the green construction of slurry shield cross-river tunnel section. Constr. Technol. **44**(S1), 280–284 (2015)

Volumetric Shrinkage of Compacted Lateritic Soil Treated with Bacillus *pumilus*

K. J. Osinubi[1], Adrian O. Eberemu[1(✉)], T. Stephen Ijimdiya[1],
John E. Sani[2], and S. E. Yakubu[3]

[1] Department of Civil Engineering, Ahmadu Bello University, Zaria, Nigeria
aeberemu@yahoo.com
[2] Department of Civil Engineering, Nigeria Defence Academy, Kaduna, Nigeria
[3] Department of Micro Biology, Ahmadu Bello University, Zaria, Nigeria

Abstract. Lateritic soil was treated with Bacillus *pumilus* in stepped concentration of 0/ml, 1.5×10^8/ml, 6.0×10^8/ml, 12×10^8/ml, 18×10^8/ml and 24×10^8/ml, respectively. Specimens were prepared at −2, 0, 2 and 4% relative to optimum moisture content (OMC) and compacted with British Standard light (BSL) (or standard Proctor) energy. The specimens extruded from the compaction moulds were air-dried in the laboratory to determine the volumetric shrinkage of the material when used to construct liners and covers for the containment of municipal solid waste (MSW). High volumetric shrinkage strain (VSS) values were recorded in the initial 5 days of air-drying but became relatively constant after 15 days. VSS values of specimens increased with higher Bacillus *pumilus* concentration and moulding water content in the range from the dry side (–2%) to wet side (+4%) relative to OMC. A compaction plane that satisfied the regulatory VSS value $\leq 4\%$ was obtained at 24×10^8/ml Bacillus *pumilus* concentration recommended for the optimal treatment of lateritic soil to be used as liner or cover in engineered MSW landfill.

Keywords: Acceptable zone · Bacillus pumilus · Compaction
Liner · Lateritic soil · Volumetric shrinkage strain

1 Introduction

The current trend in soil improvement techniques involves the use of sustainable and environmentally friendly modifiers/stabilizers [1, 2]. Researchers over time have shown that the use of the conventional soil modifiers/stabilizers (i.e., cement, lime, etc.) often times modify the soil pH which in turn contaminate soil and groundwater [1, 3].

Microbial-Induced Calcite Precipitation (MICP) is a novel soil improvement technique. It is a green and sustainable method which uses biological and chemical processes to improve soil engineering properties [4].

Stabilizing agents such as cement, fly ash, asphalt, bagasse ash, lime, aqueous polymers, blast furnace slag, etc. have been successfully used to reduce shrinkage cracking in clayey barrier soils, but the used Bacillus *pumilus* is not documented. This study focused on the use of Bacillus *pumilus* for the treatment of lateritic soil and evaluated its effect on the volumetric shrinkage strain of compacted soil.

© Springer Nature Singapore Pte Ltd. 2018
A. Farid and H. Chen (Eds.): GSIC 2018, *Proceedings of GeoShanghai 2018*
International Conference: Geoenvironment and Geohazard, pp. 315–324, 2018.
https://doi.org/10.1007/978-981-13-0128-5_36

2   Materials and Methods

2.1   Soil

The lateritic soil used in the study was collected by method of disturbed sampling from a depth of 1.5 m at an erosion site in Abagana (68°24′31″N and 27°52′11″E), Njikoka Local Government Area of Anambra state, South East Nigeria.

2.2   Cementation Reagents

The cementation reagents used in the study comprise 20 g of urea (CO $(NH_2)_2$) and 2.8 g of calcium chloride ($CaCl_2$). It also contained 10 g ammonium chloride (NH_4Cl), 3 g nutrient broth and 2.12 g sodium bicarbonate ($NaHCO_3$) per litre of de-ionized water [3, 6–8].

2.3   MICP Treatment

The air-dried lateritic soil that passed through BS #4 sieve was prepared at moulding water content −2, 0, 2 and 4% relative to OMC and 1/3 pore volume of Bacillus *pumilus* solution in stepped concentration as recommended by [9]. The mixture was allowed to stand for 12 h for saturation of the Bacillus *pumilus* within the soil mass before compaction in the moulds using BSL energy [10]. Cementation reagent was introduced on the surface of the soil and allowed to percolate until the specimen was saturated. The procedure was repeated at six hours interval for two days.

2.4   Volumetric Shrinkage

Volumetric shrinkage strain was determined by extruding compacted specimens from the compaction moulds and then air-drying them in the laboratory with an ambient room temperature of 29 ± 2 °C for 30 days. Vernier caliper was used to measure the diameters and heights of specimens treated with 0/ml, 1.5×10^8/ml, 6.0×10^8/ml, 12×10^8/ml, 18×10^8/ml and 24×10^8/ml Bacillus *pumilus* concentration, respectively, at 5 days interval for 30 days. The averages of the values recorded were used in the computation of volumetric shrinkage strain of specimens for each Bacillus *pumilus* concentration.

3   Discussion of Results

3.1   Index Properties of the Natural Soil

The physical properties of the lateritic soil are presented in Table 1. The soil belongs to the SC group in the Unified Soil Classification System [11] or A-4(3) soil group of the AASHTO soil classification system [12]. The oxide composition of the lateritic soil used is shown in Table 2.

Table 1. Physical properties of the lateritic soil used

Property	Quantity
Natural moisture content (%)	11.8
Percentage Passing #200 Sieve (Wet Sieving)	35.3
Liquid Limit (%)	34.4
Plastic Limit (%)	8.3
Plasticity Index (%)	26.1
Linear Shrinkage (%)	8.7
Specific Gravity	2.65
AASHTO classification	A-4(3)
USCS	SC
Colour	Reddish brown
Dominant Clay Mineral	Kaolinite

Table 2. Oxide composition of lateritic soil used

Oxide	Concentration %
SiO_2	56.5
Al_2O_3	19.00
CaO	0.33
TiO_2	2.89
V_2O_5	0.061
Cr_2O_3	0.051
Fe_2O_3	15.41
MnO	0.075
CuO	0.056
ZrO_2	0.290
L.O.I	4.54

3.2   Effect of Air-Drying

The variation of mass with drying time for specimens prepared at −2, 0, +2 and +4 relative to OMC for different Bacillus *pumilus* concentration is shown in Fig. 1. Large reduction in mass was recorded within the first five days of air-drying, but the mass of specimens became constant by the 15th day. The reduction in mass was proportional to the moulding water content; it reduced faster for specimens prepared at lower moulding water content (i.e., on the dry side of OMC) than those prepared at higher moulding water content (i.e., on the wet side of OMC). Similar trends were reported by [13–15].

The relationship between VSS and air-drying period for specimens prepared at −2, 0, +2 and +4 of OMC for different Bacillus *pumilus* concentrations is depicted in Fig. 2. The rates of change in mass were affected by the Bacillus *pumilus* concentrations used. Lower VSS values were recorded at higher Bacillus *pumilus* concentration because of the higher quantity of calcite deposited that clogged the voids of the specimens. Similar trends were also reported by [13–15].

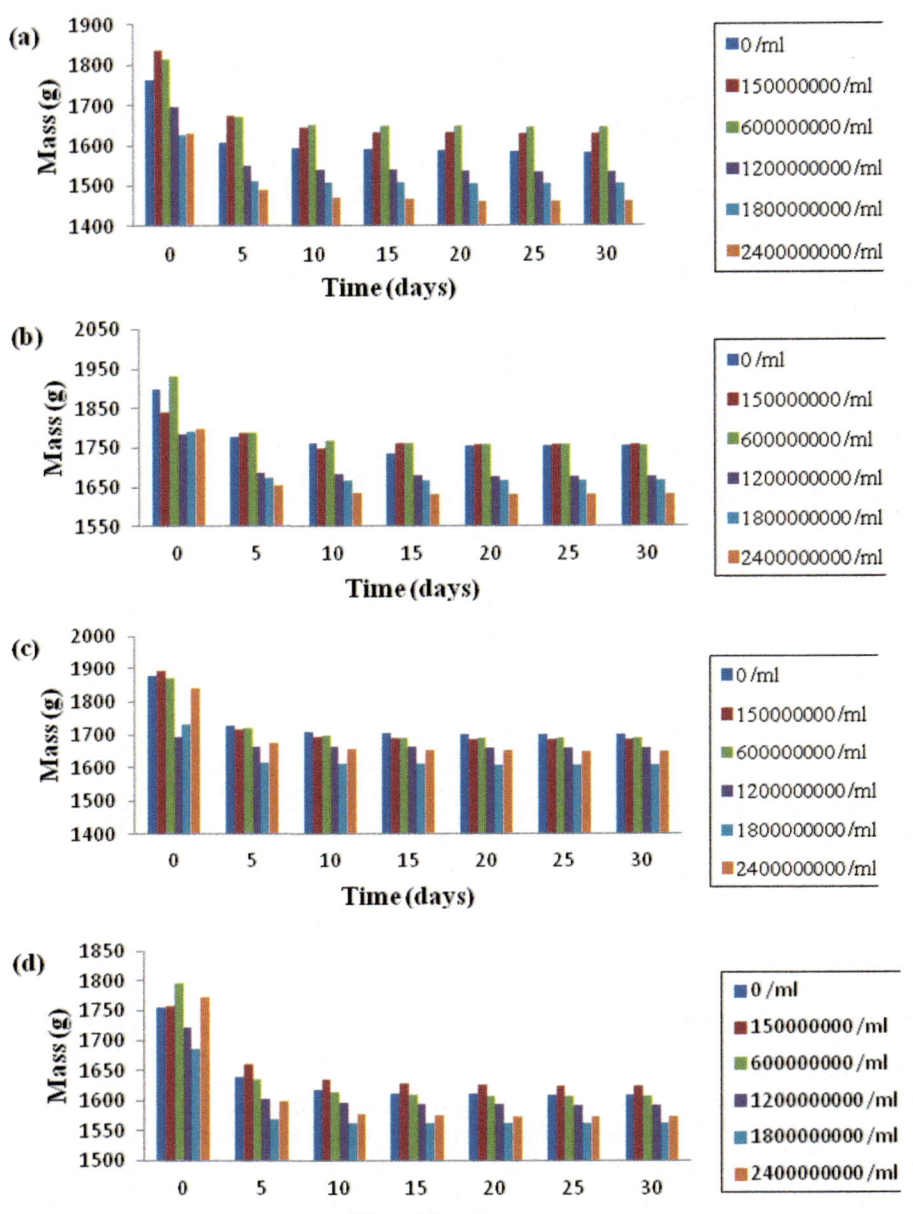

Fig. 1. Variation in mass of lateritic soil with air-drying time for varying Bacillus *pumilus* concentration (a) −2% OMC (b) OMC (c) +2% OMC (d) +4% OMC

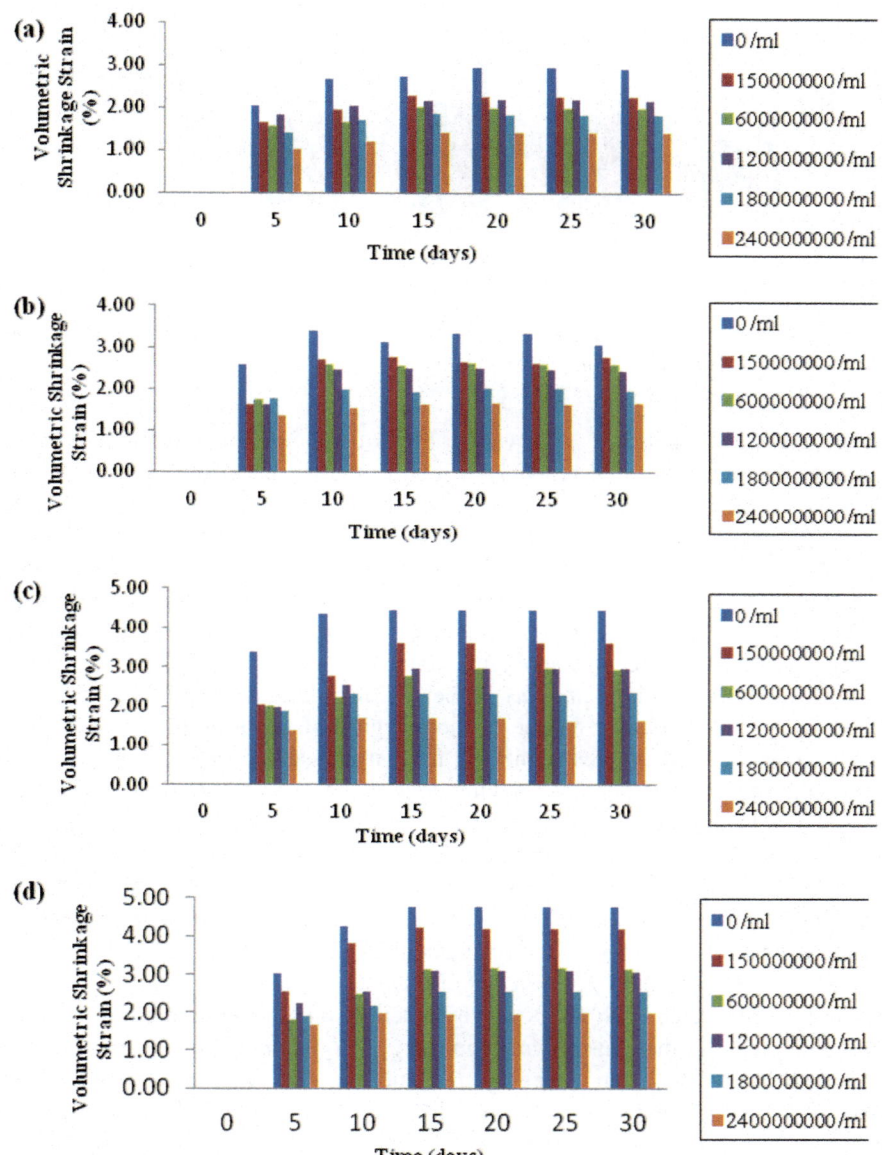

Fig. 2. Relationship between volumetric shrinkage strain of lateritic soil - Bacillus *pumilus* mixtures prepared at: (a) −2% OMC (b) OMC (c) +2% OMC (d) +4% OMC

3.3 Influence of Moulding Water Content

The histogram depicting the relationship between VSS and moulding water content at varying Bacillus *pumilus* concentration is presented in Fig. 3.

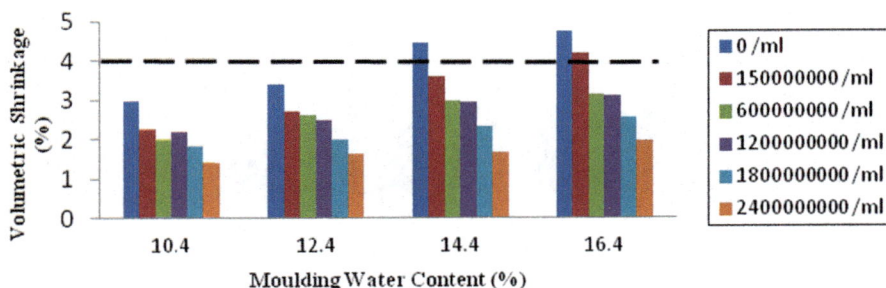

Fig. 3. Relationship between volumetric shrinkage strain of lateritic soil and moulding water content at varying Bacillus *pumilus* concentration

The observed increase in VSS values of air-dried specimens with increase in moulding water content could probably be due to the presence of more water in the voids of the compacted soil that were prepared at higher moulding water content. Haienes [16] reported that, VSS is proportional to the amount of water getting out of the voids of the soil when drying occurs in saturated soils. For the moulding water content between −2 and +4% of the OMC considered in the study, VSS values for the natural soil below the maximum regulatory 4% were recorded only for specimens prepared at −2% and OMC with moulding water content in the range 10.4–13.4%.

At 1.5×10^8/ml Bacillus *pumilus* concentration it was observed that VSS value reduced. This could probably be due to bio-clogging of voids by the calcite formed when B. *pumilus* released urease enzyme from its metabolic process. This enzyme triggered the MICP process by hydrolyzing urea [9] The ammonium ($NH4^+$) produced increased the pH of the environment which favoured the bicarbonate (HCO_3^-) to precipitate calcium ion (Ca^{2+}) from the calcium chloride added, producing calcium calcite ($CaCO_3$) [17]. VSS values below the maximum regulatory 4% were recorded for specimens prepared at moulding water content in the range 10.4–16.0% and the percentage reductions in VSS values at this moulding water content are 30.9%, 25.5%, 24.2% and 13.1% for −2, 0, 2, 4% relative to OMC, respectively.

Lateritic soil treated with 6.0×10^8/ml Bacillus *pumilus* concentration recorded VSS values below the maximum regulatory 4% at moulding water contents between 10.4 and 16.4%. The percentage reduction in VSS values at moulding water content of −2, 0, +2 and +4 relative to OMC are 48%, 30.3%, 50.7% and 51.3%, respectively. On the other hand, at 12.0×10^8/ml Bacillus *pumilus* concentration, the regulatory criterion was achieved at moulding water content between 10.4 and 16.4%. The percentage reduction in VSS values at moulding water contents of −2, 0, +2 and +4% relative to the OMC are 35.8%, 36.5%, 51.2% and 53.7%, respectively.

At 18.0×10^8/ml Bacillus *pumilus* concentration satisfactory VSS values were obtained at moulding water content between 10.4 and 16.4%. The percentage reduction in VSS at moulding water content of −2, 0, +2 and +4% relative to the OMC are 60.9%, 70.0%, 93.9% and 87.8%, respectively. Finally, at 24.0×10^8/ml Bacillus *pumilus* concentration, satisfactory VSS values were recorded at moulding water content between 10.4 and 16.4%. The percentage reduction in VSS at moulding water content of −2, 0, +2 and +4% relative to the OMC are 111.4%, 108.6%, 167.1% and 144.8%, respectively.

3.4   Influence of Moulding Water Content Relative to Optimum

The relationship between VSS of lateritic soil - Bacillus *pumilus* mixtures and moulding water content relative to optimum is shown in Fig. 4.

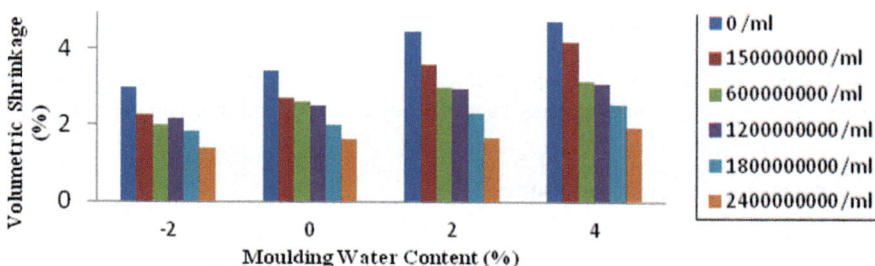

Fig. 4. Relationship between volumetric shrinkage strain of lateritic soil Bacillus *pumilus* – mixtures and moulding water content relative to optimum

The VSS values increased with increasing moulding water content relative to optimum. The natural lateritic soil specimens prepared at moulding water content between −2 and +4% relative to the OMC recorded VSS values below the maximum regulatory 4% at moulding water 1.3% relative to OMC (see Fig. 4). At 1.5×10^8/ml Bacillus *pumilus* concentration satisfactory VSS values were achieved at 3.7%, while for 6.0×10^8/ml, 12.0×10^8/ml, 18.0×10^8/ml and 24.0×10^8/ml Bacillus *pumilus* concentration, satisfactory VSS values were achieved for specimens prepared with 4% moulding water content relative to OMC.

3.5   Influence of Bacillus Pumilus Concentration

The relationship between the VSS of lateritic soil and Bacillus *pumilus* concentration for specimens prepared at −2, 0, +2 and +4% relative to OMC is presented in Fig. 5. The VSS values decreased with increase in Bacillus *pumilus* concentration regardless of the moulding water content at which specimens were prepared. This could be attributed to the higher concentration of Bacillus *pumilus* that produced more urease enzyme from the hydrolysis urea that triggered the MICP process [9]. The ammonium (NH_4^+) produced increased the pH of the environment and enabled the bicarbonate (HCO_{3^-}) to precipitate calcium ion (Ca^{2+}) from the calcium chloride added. This resulted in the formation of calcite ($CaCO_3$) in the voids of the soil [17].

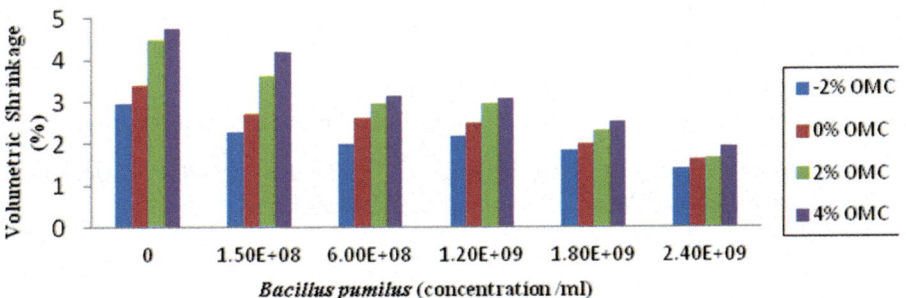

Fig. 5. Relationship between volumetric shrinkage strain of lateritic soil and Bacillus *pumilus* concentration at varying moulding water content relative to optimum moisture content

3.6 Acceptable Compaction Planes

Compaction planes as described by [5] on which VSS values below the maximum regulatory 4% were recorded for each Bacillus *pumilus* concentration was developed by relating the moulding water content to the average dry density of specimens recorded during the compaction process. The compaction planes on which the regulatory VSS criterion is satisfied are depicted in Fig. 6.

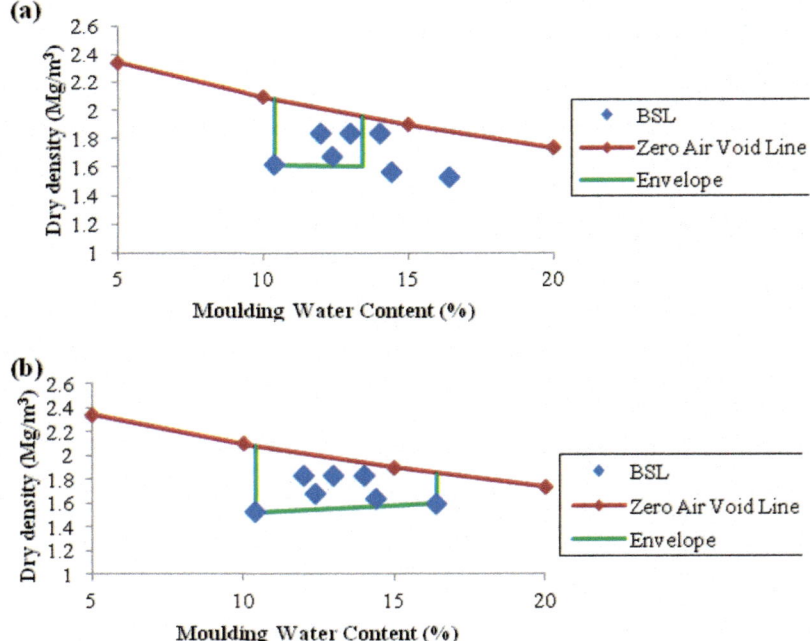

Fig. 6. Satisfactory compaction planes for volumetric shrinkage strain of lateritic soil - Bacillus *pumilus* concentration mixture (a) 0 (i.e., Natural soil) (b) 24.0 × 10[8]/ml treatment

The satisfactory compaction plane for the natural lateritic soil was achieved at moulding water content between 10.4 and 13.4% (see Fig. 6a). It was also achieved between 10.4 and 16.2% for 1.5×10^8/ml Bacillus *pumilus* concentration. Satisfactory compaction planes were obtained for treatment of lateritic soil with 6.0×10^8/ml, 12.0×10^8/ml, 18.0×10^8/ml and 24.0×10^8/ml Bacillus *pumilus* concentration when specimens were at moulding water content between 10.4 and 16.4%. The compaction plane for the optimally treated soil is shown in Fig. 6b. The plots show that beyond 1.5×10^8/ml Bacillus *pumilus* concentration, larger satisfactory compaction planes were achieved with the optimally treated soil having wider satisfactory compaction plane when compared with that of the natural lateritic soil.

4 Conclusion

A lateritic soil belonging to SC group in the Unified Soil Classification System or A-4 (3) in the AASHTO classification system was injected with up to 24.0×10^8/ml Bacillus *pumilus* concentration to evaluate the influence of air-drying on the compacted material when used as a liner and cover in engineered municipal solid waste landfill. Soil samples were prepared at moulding water contents of −2, 0, 2 and 4% of their respective OMC using British Standard light (or standard Proctor) energy. Volumetric shrinkage strain (VSS) values of specimens peaked within the first five (5) days of air-drying and became were constant by the 15th day.

Generally, VSS increased with greater moulding water content as well as the moulding water content relative to optimum. On the other hand, VSS generally decreased with higher Bacillus *pumilus* concentration in the lateritic soil irrespective of the moulding water content of the specimens. From the results of this study, an optimal 24×10^8/ml Bacillus *pumilus* concentration is recommended for the treatment of lateritic soil to be used as liner and cover in engineered landfills.

References

1. Karol, R.H.: Chemical Grouting and Soil Stabilization, 3rd edn. M. Dekker, New York (2003)
2. Basha, E.A., Hashim, R., Mahmud, H.B., Muntohar, A.S.: Stabilization of residual soil with rice husk ash and cement. Constr. Build. Mater. **19**, 448–453 (2005)
3. DeJong, J.T., Fritzges, M.B., Nüsslein, K.: Microbially induced cementation to control sand response to undrined shear. J. Geotech. Geoenviron. Eng. **132**, 1381–1392 (2006)
4. Achal, V., Pan, X., Özyurt, N.: Improved strength and durability of fly ash-amended concrete by microbial calcite precipitation. Ecol. Eng. **37**(4), 554–559 (2011)
5. Daniel, D.E., Benson, C.H.: Water content density criteria for compacted soil liners. J. Geotech. Eng. ASCE **116**(12), 1811–1830 (1990)
6. Stocks-Fischer, S., Galinat, J.K., Bang, S.S.: Microbiological precipitation of $CaCO_3$. Soil Biol. Biochem. **31**(11), 1563–1571 (1999)

7. Stoner, D.L., Watson, S.M., Stedtfeld, R.D., Meakin, P., Griffel, L.K., Tyler, T.L., Pegram, L.M., Barnes, J.M., Deason, V.A.: Application of stereo lithographic custom models for studying the impact of biofilms and mineral precipitation on fluid flow. Appl. Environ. Microbiol. **71**(12), 8721–8728 (2005)

8. Qabany, A.A., Mortensen, B., Martinez, B., Soga, K., Dejong, J.: Microbial carbonate precipitation correlation of s-wave velocity with calcite precipitation. In: Geo-Frontiers 2011, pp. 3993–4001 (2011)

9. Soon, N., Lee, L., Khun, T., Ling, H.: Factors affecting improvement in engineering properties of residual soil through microbial-induced calcite precipitation. ASCE J. Geotech. Geoenvironmental Eng. **140**(5), 1–11 (2014)

10. BS 1377: Method of Testing Soils for Civil Engineering Purpose. BSI, British Standard Institute, London (1990)

11. ASTM: Annual Book of Standards, vol. 04.08, American Society for Testing and Materials, Philadelphia (1992)

12. AASHTO: Standard Specifications for Transport Materials and Methods of Sampling and Testing. 14th edn. American Association of State Highway and Transport Officials (AASHTO), Washington, D.C (1986)

13. Albrecht, B.A., Benson, C.H.: Effect of desiccation on compacted natural clay. J. Geotech. Geoenvir. Engrg. ASCE **127**(1), 67–75 (2001)

14. Osinubi, K.J., Eberemu, A.O.: Desiccation induced shrinkage of compacted lateritic soil treated with blast furnace slag. Geotech. Geol. Eng. **28**, 537–547 (2010)

15. Osinubi, K.J., Eberemu, A.O.: Volumetric shrinkage of compacted lateritic soil treated with bagasse ash. J. Solid Waste Technol. Manag. **37**(3), 210–220 (2011)

16. Haienes, W.: The volume changes associated with variations of water content in soils. J. Agric. Sci. **13**, 296–310 (1923)

17. Harkes, M.P., Van Paassen, L.A., Booster, J.L., Whiffin, V.S., Van Loosdrecht, M.C.M.: Fixation and distribution of bacterial activity in sand to induce carbonate precipitation for ground reinforcement. Ecol. Eng. **36**(2), 112–117 (2010)

Effect of pH and Grain Size on the Leaching Mechanism of Elements from Recycled Concrete Aggregate

Yi-Bo Zhang[1], Jian-Nan Chen[1(✉)], Matthew Ginder-Vogol[2], and Tuncer B. Edil[3]

[1] Faculty of Geosciences and Environmental Engineering, Southwest Jiaotong University, Chengdu, China
jchen@swjtu.edu.cn
[2] Geological Engineering, University of Wisconsin-Madison, Madison, WI, USA
[3] Civil and Environmental Engineering, University of Wisconsin-Madison, Madison, WI, USA

Abstract. Using recycled concrete aggregate (RCA) as base course in pavement construction also requires the investigation on the potential environmental risk. In this study, leaching characteristics of major elements including aluminum (Al), calcium (Ca), iron (Fe) and magnesium (Mg), trace and minor elements including arsenic (As) and barium (Ba) from RCA under different pH conditions (pH $\sim$ 2–14) were investigated by laboratory batch leaching tests in accordance with the United States Environmental Protection Agency (USEPA) leaching method 1313. Geochemical speciation modelling by Visual Minteq was applied to determine leaching mechanisms of these elements. Results showed that leaching of Fe followed a cationic leaching pattern, where the elemental concentrations leached from RCA decreased with an increase in pH. Leaching of Al present the amphoteric pattern, and leaching behavior of Ca, Mg, As and Ba showed different leaching patterns, which is less pH dependent. Maximum leaching concentrations of the majority elements were measured at extreme acidic (pH $\sim$ 2). The fine particles tended to leach more trace elements than sand and gravel-sized particles at pH > 10. Leaching of Al from all RCA samples are controlled by dissolution/precipitation of (hydr)oxides mineral solid phases. Leaching of Ba is controlled by witherite from pH of 8 to 10, and it seems to be controlled by barite at extremely alkaline pH condition (pH = 13).

Keywords: Recycled concrete aggregate · Leaching · Grain size
pH · Geochemical modeling

1 Introduction

Roadways including highways and railways are frequently rehabilitated due to the end of their service lives, and the existing roadway surfaces need to be removed and preplaced with new construction materials. Increasing roadway reconstruction along with virgin aggregate in the United States (U.S.) causes increasing demand for more

© Springer Nature Singapore Pte Ltd. 2018
A. Farid and H. Chen (Eds.): GSIC 2018, *Proceedings of GeoShanghai 2018*
International Conference: Geoenvironment and Geohazard, pp. 325–334, 2018.
https://doi.org/10.1007/978-981-13-0128-5_37

construction materials. The Federal Highway Administration (FHWA 2004) reported that the demand of aggregates are two billion tons in the United States and the quantity of quarried aggregates will increase to 2.5 billion tons by 2020. In 2016, two major construction material (sand and gravel) valued at $8.9 billion was produced, and it is estimated that about 44% of them was used as concrete aggregates; which 25% for road base and coverings and road stabilization (USGS 2017). Since the construction sector consumes huge amount of natural resources and produces significant quantity of waste, recycling and reusing of concrete aggregates present preeminent affection on reduction of energy consumption (Alagusic et al. 2016). Moreover, environmental friendly benefits such as reduce of greenhouse gas emissions also can be achieved.

However, beneficial use of recycled materials requires to assess environmental impacts first, which usually be identified by leaching characteristics (Garrabrants et al. 2014; Komonweeraket et al. 2014; Zhang et al. 2016a, b; Chen et al. 2016). Recycled concrete aggregates (RCA) is a cement-based material, so highly alkaline effluent and the leaching of heavy metals have been reported (Sadecki et al. 1996). The impact of pH on leaching behavior of elements from RCA can be determine by pH-dependent leaching test. Engelsen et al. (2009; 2010) assessed leaching characteristics of the major constituents, minor and trace elements from RCA, and geochemical speciation modelling was applied by framework ORCHESTRA. Chen et al. (2012) have investigated the leaching behavior of trace elements, including copper (Cu) and zinc (Zn) and minor heavy metals, including oxyanion chromium (Cr), and the pH and grain size effect also been considered. However, to gain a comprehensive investigation on leaching behavior and mechanism affect by the pH and grain size, more elemental leaching and geochemical modeling need to be performed.

In this study, pH-dependent leaching test along with consideration of grain size (gravel, sand and fine particle) were conducted on RCA sample collected from state of Colorado, U.S. The impact of pH and grain size on leaching behavior of elements from RCA were investigated, and the controlling leaching mechanisms were determined via geochemical modeling. Visual MINTEQA2 (United States Environmental Protection Agency, U.S.) geochemical modeling is used to determine the dominant oxidation states of the leached elements and to predict the controlling leaching mechanisms.

2 Materials

The RCA sample were collected from multiple concrete demolition sources at state of Colorado (CO), and provided by the State Departments of Transportation (DOT) as part of a Pooled Fund project. RCA sample had been stockpiled for more than one year, and mainly contained feldspar, quartz, and calcite (Chen et al. 2013). To explore the effect of grin size on leaching characteristics, three grain-size fractions of RCA sample were prepared by sieving process. Gravel fraction refers to gravel-sized particles (<75 mm, >4.75 mm), Sand fraction refers to sand-sized particles (<4.75 mm, >0.075 mm), and Fine fraction refers to fine particles (<0.075 mm). Physical properties and grain size distribution of RCA sample were summarized in Table 1.

Table 1. Physical properties and grain size distribution of RCA

Properties	Test methods	RCA sample from Colorado
Physical properties		
In-situ water content (%)	ASTM D2216	3.9
Optimum water content (%)	ASTM D1557	11.9
Maximum dry unit weight (kN/m^3)	ASTM D1557	18.9
Specific gravity (G$_s$)	AASHTO T85	2.6
Absorption (%)	AASHTO T85	5.8
Void ratio (e)	AASHTO T85	0.36
Unified soil classification system (USCS)		SM
Hydraulic conductivity (k, m/s)	ASTM D5856	1.6×10^{-5}
Grain size distribution		
75–4.75 mm (wt%)	ASTM D2487	40.9
4.75–0.075 mm (wt%)		46.3
<0.075 mm (wt%)		12.8

Total elemental analysis was determined by acid digestion according per EPA Method 3050B. Tests were conducted in triplicate. A 1:1 nitric acid digestion of 1-g of solid sample was performed at 90–95 °C for 2 h, and 30% H_2O_2 was added to start a peroxide reaction. The total carbon (TC), total inorganic carbon (TIC), and total organic carbon (TOC) were determined with a SC144 DR sulfur and carbon analyzer (LECO Inc., St. Joseph, MO, USA). Table 2 summarizes the chemical composition of the RCA sample. More detailed information regarding the index properties and chemical properties of RCA samples can be found in Chen et al. (2013).

Table 2. Chemical compositions of RCA

Properties	Test methods	RCA sample from Colorado
Chemical composition		
Total carbon (%)	LECO carbon analyzer	1.9
Total organic carbon (%)		0.3
Total inorganic carbon (%)		1.5
Major elements		
Calcium (Ca, %)	EPA method 3050 B	8.5
Iron, (Fe, %)		1.1
Aluminum, (Al, %)		1.5
Magnesium (Mg, %)		0.3
Trace elements		
Arsenic (As, mg/kg)		6.5
Barium (Ba, mg/kg)		88.8
Material pH		
Bulk specimens	Accumet	12.1
Gravel-sized (75–4.75 mm)	AR 50	12.1
Sand-sized (4.75–0.075 mm)	pH meter	11.9
Fines (<0.075 mm)		11.8

3   Methods

3.1   pH-Dependent Leaching Tests

pH-dependent batch tests were performed according to US EPA method 1313 (U.S. EPA 2012). Tests were conducted at a liquid to solid (L/S) ratio of 10:1. A pH range of 2 to 13 was used for the pH-dependent leaching tests, with target pH values of 13, 12, 10.5, 9, 8, 7, 5.5, 4 and 2. A pre-test titration was conducted to determine the equilibrium time and the acid/base addition required for each batch. Batches were agitated in an end-over-end tumbler at a speed of 30 rpm. After equilibration, the solution was allowed to settle for 5 min. pH, electrical conductivity (EC), and oxidation-reduction potential (ORP) were recorded. The suspended solution was filtered through a 0.45-μm membrane disk filter using a 20-mL plastic syringe, and then stored in sealed 15-mL high-density polyethylene tube bottles.

All filtered samples were preserved using nitric acid to pH < 2 and stored at 4 °C prior to chemical analysis. Concentrations of major and trace elements, including Al, Ca, Fe, Mg, Ba and As, were quantified to investigate leaching characteristics. Concentrations of Al, Ca, Fe, Mg and Ba were determined by inductively coupled plasma optical emission spectrometry (ICP-OES, Vista-MPX CCD Simultaneous ICP-OES, Varian Inc., CA, U.S.). Concentrations of As were determined by inductively coupled plasma mass spectrometry (ICP-MS, VG PlasmaQuad PQ2 Turbo Plus ICP-MS, Thermo Fisher Scientific Inc., MA, U.S.).

3.2   Geochemical Modeling

MINTEQA2, a geochemical modeling software developed by USEPA, has been used for the calculation of the equilibrium composition of natural aqueous systems and aqueous solutions in laboratory setting (Allison et al. 1991). In this study, MINTEQA2 was used to determine the predominant oxidation states and leachate controlling mechanisms of the leached elements from RCA samples with fractionated particle. The controlling mechanism for elements release from the RCA samples (e.g., whether solubility control or not) were investigated via geochemical modeling. Major and trace element concentrations from pH-dependent leaching tests, leachate pH, electrical conductivity (EC) and leachate Eh collected in leaching tests were used as an input in the MINTEQA2 geochemical modeling program.

4   Results and Discussion

4.1   Effects of pH and Grain Size on Leaching of Major Alkaline Elements

Elemental leaching behavior of recycled construction material under the influence pH has been reported that follow three distinct patterns, including cationic, oxyanionic, and amphoteric patterns (Komonweeraket et al. 2015a, b; Zhang et al. 2016a, b). Leaching concentrations of elements decrease consistently as pH increases in the cationic pattern. In the oxyanionic pattern, the leachate concentrations of elements increase with pH and reach a relatively high concentration at alkaline condition (pH > 10). In the amphoteric pattern, minimum leaching concentrations of elements can be observed neutral condition (pH = 7), and elemental leaching concentration increase at acidic and basic pH conditions. Figure 1 shows leaching behavior of major alkaline elements as a function of pH from RCA with different fractions. Elements of Al, Ca, Fe and Mg leaching from RCA samples were considered as major alkaline elements, due to the relative higher concentrations (>0.1%) in total elemental analysis (Table 2).

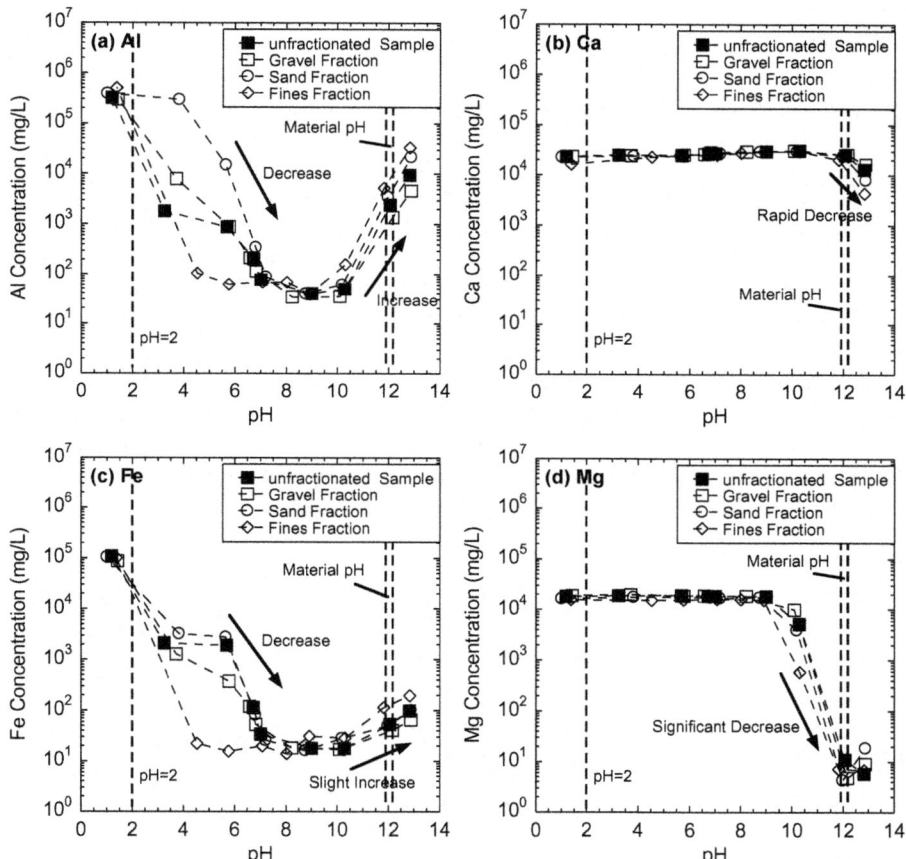

Fig. 1. Concentrations of major alkaline elements as a function of pH from RCA with different fractions: (a) Al, (b) Ca, (c) Mg and (d) Fe

Figure 1(a) shows concentrations of Al of different fraction RCA samples as a function of pH, thus indicating the amphoteric pattern. All RCA samples exhibit similar leaching behavior for Al, while the sand fraction sample present relative higher concentration (more than 10 times) than other fraction samples at pH of 2 to 7. The minimum leaching concentration of Al can be observed at around pH = 9 with approximately 5 mg/L, and the maximum concentration is about 3–5 × 10⁵ mg/L, which occurs at pH of 2 and 13, respectively. Leaching concentrations of Ca from RCA samples are similar for different fraction samples (Fig. 1b). The leaching concentrations are 2–3×10^4 mg/L with nearly no change for pH ranging from 2 to 12, while the concentration of Ca rapidly decreases when pH increases beyond 12. Figure 1(c) and (d) shows that leaching behavior of Fe and Mg from RCA sample are similar to leaching behavior of Al and Ca. Leaching of Fe from sand fraction sample is higher than other samples during pH of 2 to 7, and minimal concentration of Fe is about 50 mg/L at pH = 9. No change for concentration of Mg can be detected at pH of 2–9. However, the concentration of Mg rapidly decreases from 2×10^4 mg/L to 10 mg/L at pH of 12.

4.2   Effects of pH and Grain Size on Leaching of Trace Element and Heavy Metal

Trace metal element and heavy metal (As and Ba), which showed that leaching behavior are sensitive to the changes in pH of the aqueous solutions, are chosen to be analyzed. Figure 2 shows concentrations of trace element and heavy metal in leachate from the fraction RCA samples under different pH conditions. Leaching behavior of As from all samples follows a similar trend (Fig. 2a). Leaching behavior of As for all three fraction of the RCA samples present relatively independent of pH until pH = 10 (Fig. 2a), following with a significant drop from 500 µg/L to 10 µg/L. The minimum release of As occurred at natural pH condition (pH = 12), and the concentration is about 10 µg/L, which is same with the US EPA maximum contaminant level (MCL) for drinking water. It means that the concentration of As from all RCA samples exceed the EPA MCL, which could put potential environmental risk.

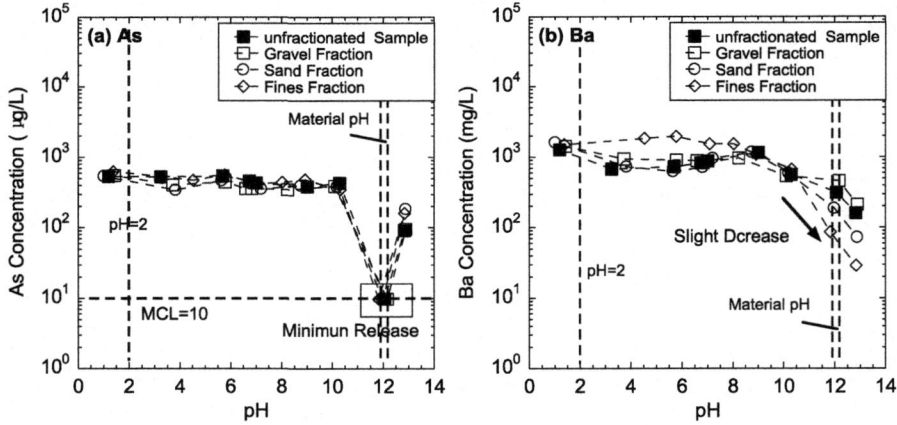

Fig. 2. Concentrations of (a) As and (b) Ba as a function of pH from RCA with different fractions:

Leaching concentration of Ba for all fraction of the RCA samples present relatively independent of pH until pH = 9 (Fig. 2b), following with a significant drop at alkaline conditions (pH > 10). Similar leaching behavior has also been reported for recycled construction material, such as MSWI fly ash (Zhang et al. 2016a, b) and coal fly ash mixtures (Komonweeraket et al. 2014). It can be seen that the fine fraction RCA samples release more Ba than other particle size fraction samples at pH < 9, while the opposite case occurred when pH > 10.

4.3 Geochemical Modeling

Apul et al. (2005) reported that saturation index (SI) and aqueous phase equilibrium composition of all leachates with respect to solids or minerals can be computed by allowing aqueous complexation reactions at fixed pH. It was assumed that the equilibrium achieved between the leachate and solubility-controlling elements, and the reaction system was assumed under influence of atmospheric CO_2 with 25 °C as well. Two equilibrium mechanism including solubility (dissolution-precipitation) and sorption has been reported that control the leaching of metals from coal combustion byproducts (Fruchter et al. 1990; Mudd et al. 2004; Komonweeraket et al. 2015a). The dominant oxidation states of selected elements including Al and Ba are Al^{3+} and Ba^{2+}, respectively. Log activities of the concerned elements were calculated by MINTEQA2, and log activities corresponding with leachate pH were graphed as Log activity diagrams (as shown in Fig. 3). These graphs can be used to determine whether elements of interest were controlled by mineral solubility. When the calculated activities of elements are close to the stability/solubility line of the mineral or solid, release of these element could be controlled by mineral solubility, otherwise, the elements are not solubility controlled.

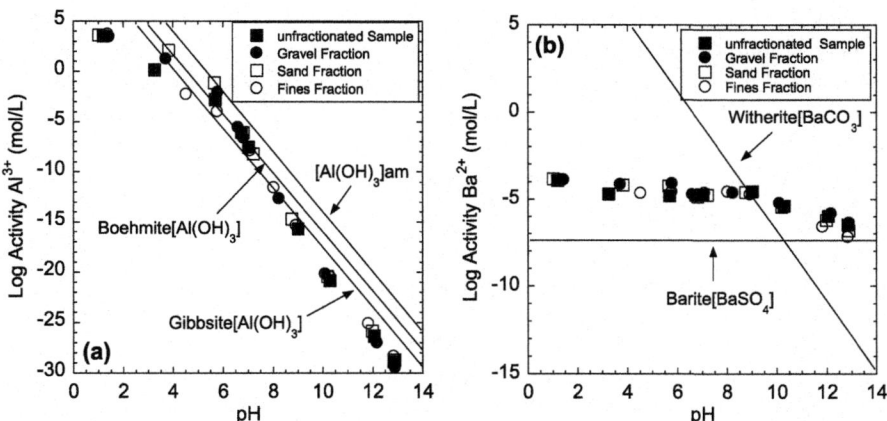

Fig. 3. Log activity of elements vs pH: (a) Al^{3+} and (b) Ba^{2+}.

Figure 3(a) shows that leaching of Al from all RCA samples are controlled by dissolution/precipitation of (hydr)oxides mineral solid phases. It can be observed that leaching of Al is controlled by gibbsite at pH < 6 and pH > 8, while the dissolution/precipitation of Boehmite is controlling the leaching of Al at neutral pH condition. Previous studies found that leaching of Al was controlled by amorphous Al

(OH)$_3$ and gibbsite on coal and MSWI fly ash samples (Komonweeraket et al. 2015b; Zhang et al. 2016a, b). Two compounds including Barite [Ba(SO$_4$)(s)] and witherite [BaCO$_3$(s)] has been indicated that control the leaching of Ba in aqueous solutions from coal combustion byproducts (Fruchter et al. 1990; Mudd et al. 2004). Figure 3(b) shows that leaching of Ba is controlled by witherite from pH of 8 to 10, and it seems to be controlled by barite at extremely alkaline pH condition (pH = 13).

5   Conclusions

pH-dependent leaching tests was performed on RCA samples with different fractions, leaching behavior of major and trace elements under different pH conditions were investigated. Leaching mechanisms of Al and Ba were explored via MINTEQA2. Conclusions of the current study are summarized below:

(1.) The leaching concentrations of major and trace elements including heavy metal, were significantly influenced by pH. Leaching of Al present the amphoteric pattern with minimal concentration at neutral pH (=7) and an increase in concentration at acidic and basic pH conditions. Leaching behavior of Ca, Mg, As and Ba showed different leaching patterns, which is less pH dependent, and maximum releases occurred at acidic conditions (pH = 2). Leaching of As followed a cationic leaching pattern where the concentrations of elements decreased with an increase in pH of the effluent solutions. The fine particles tended to leach more trace elements than sand and gravel-sized particles at pH > 10.

(2.) Geochemical modeling analyses indicated that leaching of Al and Ba from RCA samples were solubility controlled. Leaching of these elements depended on the dissolution/precipitation of their (hydr)oxides and carbonate. Results of geochemical modeling analyses also showed that the leaching controlling mechanism of elements leached from different fraction RCA samples was not different.

(3.) Recycled concrete aggregate (RCA) is one of alternative material for roadway construction, while more laboratory and numerical analysis still need to be performed to confirm that it can be reused environmental friendly.

Acknowledgments. The National Science Foundation of China (NSF-China, 41701347), Sichuan Science and Technology Project (2016HH0084) and the Fundamental Research Funds for the Central Universities (A0920502051708-89) provided support for Dr. Jian-Nan Chen and Dr. Yi-Bo Zhang. This study was financially supported by the TPF-5 (129) Recycled Unbound Materials Pool Fund administered by The Minnesota Department of Transportation and Federal Highway Administration (FHWA) Recycled Materials Resource Center (RMRC). The findings and opinions in this paper are solely those of the authors. Endorsement by sponsors is not implied and should not be assumed.

References

Alagusic, M., Milovanovic, B., Pecur, I.B.: Recycled aggregate concrete – sustainable use of construction and demolition waste and reduction of energy consumption. In: Advances in Cement and Concrete Technology in Africa (2016)

Allison, J.D., Brown, D.S., Novo-Gradac, K.J.: Minteqa2/Prodefa2, a Geochemical Assessment Model for Environmental Systems: Version 3.0 User's Manual, p. 115. U.S. Environmental Protection Agency EPA/600/3-91/021, March 1991

Apul, D.S., Gardner, K.H., Eighmy, T.T., Fällman, A., Comans, R.N.: Simultaneous application of dissolution/precipitation and surface complexation/surface precipitation modeling to contaminant leaching. Environ. Sci. Technol. 39(15), 5736–5741 (2005)

Chen, J., Bradshaw, S.L., Benson, C.H., Tinjum, J.M., Edil, T.B.: pH-dependent leaching of trace elements from recycled concrete aggregate. In: GeoCongress 2012, pp. 3729–3738. American Society of Civil Engineers (2012)

Chen, J., Soleimanbeigi, A., Zhang, Y., Edil, T.B., Li, L.: Leaching characteristics of recycled asphalt shingles mixed with industrial byproducts as structural fills. In: Geo-Chicago 2016, pp. 400–410. American Society of Civil Engineers (2016)

Chen, J., Tinjum, J.M., Edil, T.B.: Leaching of alkaline substances and heavy metals from recycled concrete aggregate used as unbound base course. Transp. Res. Rec.: J. Transp. Res. Board 2349(1), 81–90 (2013)

Engelsen, C.J., van der Sloot, H.A., Wibetoe, G., Justnes, H., Lund, W., Stoltenberg-Hansson, E.: Leaching characterisation and geochemical modelling of minor and trace elements released from recycled concrete aggregates. Cem. Concr. Res. 40(12), 1639–1649 (2010)

Engelsen, C.J., van der Sloot, H.A., Wibetoe, G., Petkovic, G., Stoltenberg-Hansson, E., Lund, W.: Release of major elements from recycled concrete aggregates and geochemical modelling. Cem. Concr. Res. 39(5), 446–459 (2009)

Federal Highway Administration (FHWA): Transportation Applications of Recycled Concrete Aggregate. FHWA State of the Practice National Review (2004). https://www.fhwa.dot.gov/pavement/recycling/applications.pdf

Fruchter, J.S., Rai, D., Zachara, J.M.: Identification of solubility-controlling solid phases in a large fly ash field lysimeter. Environ. Sci. Technol. 24(8), 1173–1179 (1990)

Garrabrants, A.C., Kosson, D.S., DeLapp, R., van der Sloot, H.A.: Effect of coal combustion fly ash use in concrete on the mass transport release of constituents of potential concern. Chemosphere 103(May), 131–139 (2014)

Komonweeraket, K., Cetin, B., Aydilek, A., Benson, C.H., Edil, T.B.: Geochemical analysis of leached elements from fly ash stabilized soils. J. Geotech. Geoenviron. Eng. 141(5), 4015012 (2015a)

Komonweeraket, K., Cetin, B., Aydilek, A.H., Benson, C.H., Edil, T.B.: Effects of pH on the leaching mechanisms of elements from fly ash mixed soils. Fuel 140, 788–802 (2015b)

Komonweeraket, K., Cetin, B., Benson, C.H., Aydilek, A.H., Edil, T.B.: Leaching characteristics of toxic constituents from coal fly ash mixed soils under the influence of pH. Waste Manag. 38, 174–184 (2014)

Mudd, G.M., Weaver, T.R., Kodikara, J.: Environmental geochemistry of leachate from leached brown coal ash. J. Environ. Eng. 130(12), 1514–1526 (2004)

Sadecki, R.W., Busacker, G.P., Moxness, K.L., Faruq, K.C., Allen, L.G.: An investigation of water quality in runoff from stockpiles of salvaged concrete and bituminous paving. MN/PR-96/31. Minnesota Department of Transportation (1996)

U.S. Environmental Protection Agency (U.S. EPA): Method 1313: liquid-solid partitioning as a function of extract pH using a parallel batch extraction procedure. U.S. (2012). http://www.epa.gov/epawaste/hazard/testmethods/sw846/pdfs/1313.pdf

United States Geological Survey (USGS): Mineral Commodity Summaries. U.S. Geological Survey (2017)

Zhang, Y., Cetin, B., Likos, W. J., Edil, T.B.: Impacts of pH on leaching potential of elements from MSW incineration fly ash. Fuel **184**, 815–825 (2016a)

Zhang, Y., Chen, J., Likos, W.J., Edil, T.B.: Leaching characteristics of trace elements from municipal solid waste incineration fly ash. In: GeoChicago, Chicago, USA (2016b)

Stress-Strain Relation and Strength Prediction Method of a KMP Stabilized Zn, Pb and Cd Contaminated Site Soil

Wei-Yi Xia[1], Ya-Song Feng[1], Martin D. Liu[2], and Yan-Jun Du[1(✉)]

[1] Jiangsu Key Laboratory of Urban Underground Engineering
and Environmental Safety, Institute of Geotechnical Engineering,
Southeast University, Nanjing 210096, China
duyanjun@seu.edu.cn
[2] University of Wollongong, Northfields Ave 2., Wollongong, Australia

Abstract. This paper presents a systematic laboratory investigation of the effects of KMP content and curing time on the mechanical properties of a Zn, Pb and Cd contaminated site soil. Some basic parameters quantifying the stress-strain relation and strength properties of KMP stabilized site soils are evaluated. They are the values of unconfined compressive strength (q_u), failure strain (ε_f), and Young's modulus (E_{50}). Furthermore, some empirical equations are proposed, which are useful for engineering applications.

Keywords: Solidification/Stabilization · Heavy metals contaminated soil
Phosphate · Strength prediction · Stress-strain relationship

1 Introduction

Heavy metal contamination can have considerable influence on the mechanical properties of the soils, and the influence can be unfavorable for the redevelopment of the contaminated sites [1]. The process of stabilization/solidification (S/S) has been standardized and forms an important part of the contaminated soil remediation technology. The S/S is a process that involves the mixing of waste with a binder to reduce the contaminant leachability from both chemical and physical means, and to convert a hazardous waste into an environmentally acceptable material for land disposal or construction use. A new phosphate-based binder-KMP for heavy metal contaminated soils is developed recently [2, 3]. The KMP consists of oxalic acid-activated phosphate rock (APR), monopotassium phosphate (KH_2PO_4), and reactive magnesia (MgO). Previous studies find that the KMP stabilized soils have low leachability, ecotoxicity, alkalinity, but high strength [2]. However the results are mainly obtained from artificial soils spiked with Pb and/or Zn contaminants, the applicability of these features observed for heavy metal contaminated site soils need to be investigated conclusively. This study aims to investigate the properties of unconfined compressive strength (q_u), failure strain (ε_f), and Young's modulus (E_{50}) of a KMP stabilized Pb, Zn and Cd contaminated site soil from a Pb–Zn smelter. A series of unconfined compression tests (UST) are carried out. The effects of curing time, KMP content on q_u, E_{50}, ε_f as well as

© Springer Nature Singapore Pte Ltd. 2018
A. Farid and H. Chen (Eds.): GSIC 2018, *Proceedings of GeoShanghai 2018
International Conference: Geoenvironment and Geohazard*, pp. 335–345, 2018.
https://doi.org/10.1007/978-981-13-0128-5_38

the relationships of q_u versus E_{50} and q_u versus ε_f of the KMP stabilized soils are studied. Empirical formula for predicting q_u is proposed, and predictions made by using the formula are validated against experimental data.

2   Materials and Methods

2.1   Raw Materials

Soil samples are collected from the surroundings of the "Northwest" Pb–Zn smelter, located in Baiyin City, Gansu, China. This smelter is constructed in 1996 as a base for non-ferrous metals smelting in China. The basic physicochemical properties including heavy metal concentrations of the soil are listed in Table 1. The KMP consists of APR, KH_2PO_4, and MgO with a mass ratio of 1:1:2. Commercial phosphate rock cores $(Ca_{10}(PO_4)_6F_2)$ are crushed and ground to pass through a sieve with opening size of 0.075 mm. The KH_2PO_4 (chemical analytical reagent) and MgO (industrial reagent) are obtained from Nanjing Chemical Reagent Co., Ltd., and Sinopharm Chemical Reagent Co., Ltd., respectively. Prior to the KMP preparation, the sieved phosphate rock powder is mixed with 1.0 mol/L oxalic acid at a liquid solid ratio of 2:1 for 5 min using an electronic mixer. The mixture is equilibrated in a top-sealed vessel at 28 ± 1 ° C for 4 days. After having been dried in the oven at 60 °C, the mixture is pulverized and passed through a 0.075 mm sieve to achieve desired APR [2]. The chemical compositions of the phosphate rock and MgO are measured by X-ray fluorescence method using ARL9800XP+ XRF spectrometry and are listed in Table 2.

Table 1. Physicochemical properties of the contaminated soil

Index	Value	Test method
Natural water content, w_n (%)	46.2	GB/T 50123-1999
pH	6.04	ASTM D 4972
Plastic limit, w_P (%)	17.2	ASTM D 4318
Liquid limit, w_L (%)	33.3	ASTM D 4318
Soil type	CL	ASTM D 2487
Specific gravity, G_s	2.73	GB/T 50123-1999
Heavy metal concentration (mg/kg)		
Zinc, Zn	17300	GB/T 17138-1997
Lead, Pb	9710	GB/T 17141-199
Cadmium, Cd	425	GB/T 17141-199
Grain size distribution (%)		
Clay (<0.005 mm)	6.54	GB/T 50123-1999
Silt (0.005–0.075 mm)	43.95	GB/T 50123-1999
Sand (0.075–1 mm)	49.51	GB/T 50123-1999

Table 2. Oxide chemistry of phosphate rock and MgO tested

Oxide chemistry	Phosphate rock (%)	MgO (%)
Calcium oxide (CaO)	45.93	0.23
Aluminium oxide (Al_2O_3)	1.23	0.28
Magnesium oxide (MgO)	–	87.95
Phosphorus oxide (P_2O_5)	25.10	0.03
Potassium oxide (K_2O)	–	0.011
Silicon oxide (SiO_2)	6.14	0.28
Sulphate oxide (SO_3)	–	0.45
Fluorine (F)	2.35	–
Chlorine (Cl)	–	0.28
Loss on ignition (950 °C)	13.12	9.87

2.2   Sample Preparation and Testing Methods

In this study, there are four curing times, i.e., 3, 7, 14, and 28 days, and four KMP contents, i.e., 0% (i.e., untreated soil), 6%, 8% and 10% (dry weight basis). To prepare stabilized soils, a predetermined volume of deionized water is added into the air-dried soil, and the water content reaches 22%. The soil and the deionized water are mixed thoroughly by using an electronic mixer for about 5 min to obtain homogenous slurry. Later, the KMP powder is poured into the slurry and the mixture is mixed thoroughly to achieve homogeneity. The final mixture is then compacted into a $\Phi50 \times H100$ mm cylindrical iron mold using a hydraulic jack to achieve a dry density of 1.51 g/cm³, which is the same with the maximum dry density of the untreated soil. Then the soil specimen is extruded from the mold, wrapped by a polythene bag, and cured under controlled ambient condition (22 °C and relative humidity of 95%).

Unconfined compression tests (UCT) are performed as per ASTM D4219 [4], by fixing the strain rate at 1%/min. From these tests, q_u, E_{50}, ε_f of soils are obtained. ε_f is the strain corresponding to q_u and is an important parameter to delineate the ductile and brittle behavior of cemented soils and other materials. E_{50} is the secant modulus given as:

$$E_{50} = \frac{\sigma_f}{2\varepsilon_{1/2}} \tag{1}$$

where σ_f = the peak strength q_u, $\varepsilon_{1/2}$ = the strain corresponding to the state with $\sigma = 0.5\,\sigma_f$. Triplicate samples are tested, and the mean values are given in the result-figures.

3   Test Results and Analyses

3.1   Unconfined Compressive Strength

It is evident that the q_u increases with KMP contents and curing times (Fig. 1). For examples, after 7 days, q_u increases approximately by 1.7 and 1.8 times when KMP content increases from 6% to 8% and 6% to 10%, respectively. The values of q_u corresponding to 28 days are found to be about 10% to 30% higher than those for 7 days. When the KMP content is higher than 8%, the effect of KMP content on q_u becomes very moderate. This can be attributed to the limited water content in soil specimens, indicating that for a relatively high KMP content ($\geq 8\%$), a further increase in initial water content would result in an improvement of the strength of stabilized soils.

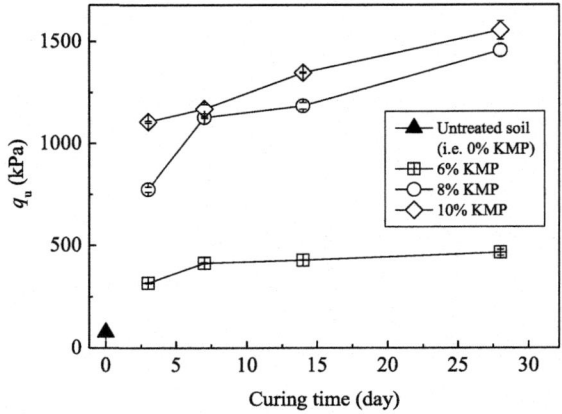

Fig. 1. Change of q_u with curing time and KMP content

3.2 Failure Strain

Figure 2 reveals the results of ε_f. It is seen that the increment in KMP content leads to substantial increase in the brittleness of the stabilized soils, characterized by the decrease in ε_f. It should also be noted that ε_f reduces with curing time. A power-function relationship between curing time and ε_f is found, which is given as:

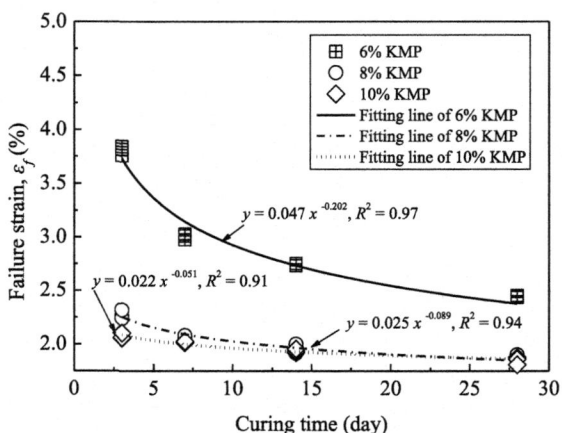

Fig. 2. ε_f values of the soils stabilized with the KMP

$$\varepsilon_f = a \bullet t^b \qquad (2)$$

where t is the curing time. For the soils tested in this study, a summary of the values of parameter a and b is given in Table 3.

Table 3. Values of parameters a and b for Eq. (2)

KMP content	a	b	R^2
6%	0.047	−0.202	0.97
8%	0.025	−0.089	0.94
10%	0.022	−0.051	0.91

3.3   Young's Modulus E_{50}

Figure 3 summarizes the variation of E_{50}. For different KMP contents and curing times, the E_{50} values measured are in a range of 18.55 to 110.76 MPa for this series of tests. It is seen that KMP stabilized soils have a strong deformation resistant ability. The results also show that the values of E_{50} increases with curing time and KMP content. Furthermore, the rate of E_{50} increment generally decreases over time.

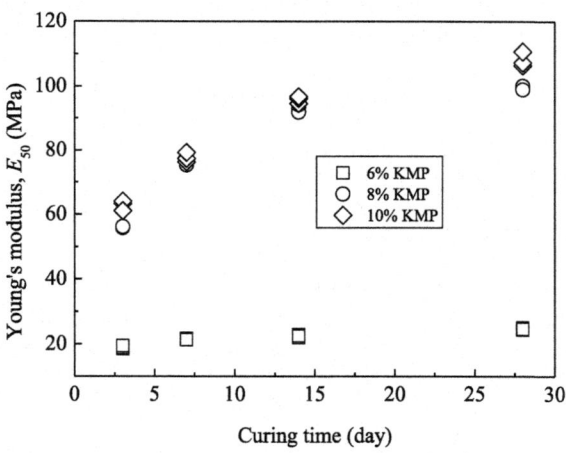

Fig. 3. E_{50} values of the soils stabilized with the KMP

3.4   Relationship Between ε_f and Q_u

Figure 4 illustrates the relationship between q_u and ε_f. Zhu [5] proposes that the reduction of ε_f with q_u can be described by an exponential function for cement solidified sediments. However, a power function is suggested in this paper as follows:

$$\varepsilon_f = 0.34 \bullet q_u^{-0.40}, R^2 = 0.87 \tag{3}$$

The fair correlation coefficient, R ($R^2 = 0.87$) suggests that the derived power function may be used for engineering estimation for the KMP stabilized contaminated site soils. Ductile behavior is associated with low strength and higher failure strain, and brittle behavior vice versa. It is seen that increments in KMP content and curing time are the main parameters to cause the stabilized soils brittle. The existence of the ε_f–q_u relation can be attributed to similar impacts as the KMP content and curing time on q_u and ε_f (see Figs. 1 and 2). This observation shows that the relationships between ε_f and q_u are not sensitive to the KMP content or curing time.

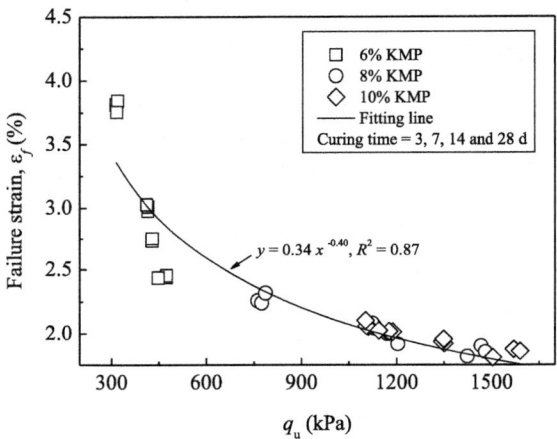

Fig. 4. Relationships between ε_f and q_u of the stabilized soils

3.5   Relationship Between E_{50} and Q_u

Figure 5 illustrates the relationship between E_{50} and q_u. It is seen clearly that the E_{50} (i.e. stiffness) of the soils increases with soil strength q_u. The relationship is basically linear. Therefore the relationship between E_{50} and q_u is expressed by

$$E_{50} = \eta \bullet q_u \tag{4}$$

in which η is a dimensionless parameter. For the soils tested in this study, 12 pairs of E_{50} and q_u corresponding to the curing times of 3, 7, 14, and 28 days at different KMP contents, are found to follow well Eq. (4). The value of η is obtained using the Least-Squares-Fitting method, and is equal to 68.2 ($R^2 = 0.99$).

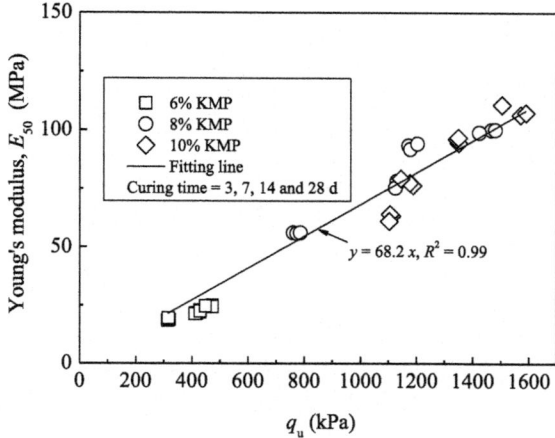

Fig. 5. Relationships between E_{50} and q_u of the stabilized soils

3.6   Proposed Method for q_u Estimation

Semi-empirical equations are proposed to provide a useful means to estimate the unconfined compressive strength q_u of the KMP stabilized soils. For the purpose of engineering application, the method is kept simple. The strengths q_u are dependent on KMP content (a_w) and curing time (t), and the corresponding experimental data are shown in Figs. 6 and 7, respectively. It is seen that the value of q_u increases with the KMP content monotonically at a specific curing time (see Fig. 6), indicating a linear relationship between q_u and a_w. In addition, the q_u–t relationship at a given KMP content is nonlinear and can be expressed by a power equation (see Fig. 7).

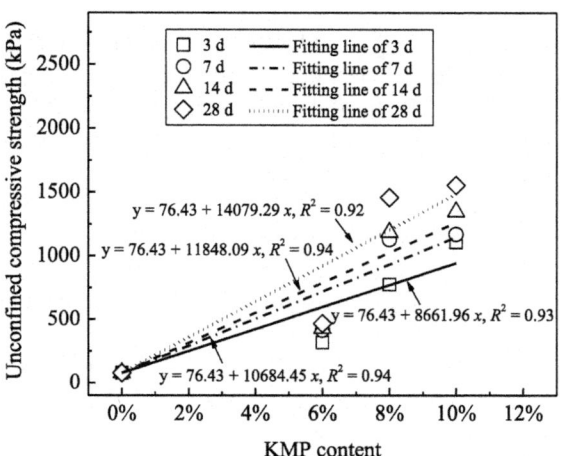

Fig. 6. Change of q_u with KMP content

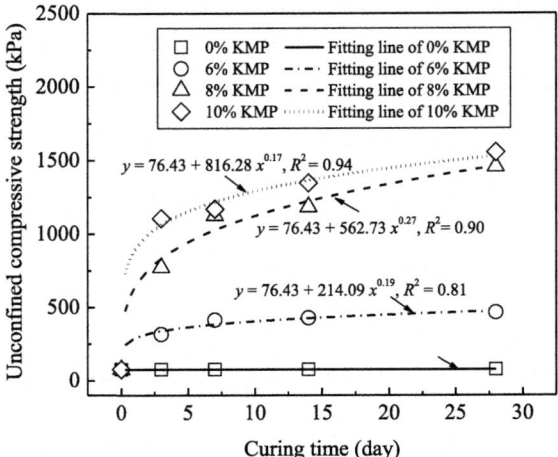

Fig. 7. Change of q_u with curing time

The strength of stabilized soils can be divided into two parts: the strength of the untreated soil and the strength increment from the cementation effect of the binder stabilization [6]. Therefore, the following linear equation for q_u of KMP stabilized soils is suggested.

$$q_u(a_w, t) = q_{u,0(t)} + k_{(t)} \bullet a_w \quad (5)$$

where $q_u(a_w, t)$ is the strength of the soil at a given KMP content a_w after t days of curing; $q_{u,0(t)}$ is q_u of the untreated soil after t days of curing, it is assumed that $q_{u,0(t)} = q_{u,0}$, a constant over time, equal to 76.43 kPa for the soil reported in this paper; and $k_{(t)}$ describes the rate of strength increment with KMP content, and it is time dependent.

The relationships and the fitting lines between q_u and the KMP content at different curing times are shown in Fig. 6. The values of parameter $k_{(t)}$ of different curing times are listed in Table 4 ($k_{(0)} = 0$ kPa for 0 day of curing). It is found that the value of $k_{(t)}$ varies with curing time t as a power function (see Fig. 8), which is consistent with the analysis based on Fig. 7. It can be expressed as

$$k_{(t)} = 6924.51 \bullet t^{0.21}, R^2 = 0.99 \quad (6)$$

Substituting Eq. (6) into Eq. (5) gives the following empirical equation:

$$q_u(a_w, t) = q_{u,0} + k_{(t)} \bullet a_w = 76.43 + 6924.51 \bullet t^{0.21} \bullet a_w \quad (7)$$

Equation (7) can be used to estimate q_u for the stabilized Pb, Zn and Cd contaminated soils with a given KMP content of a_w and a specified curing time of t. To examine the validity, the predicted values of q_u by using Eq. (7) and the measured values of q_u are plotted in Fig. 9.

Table 4. Fitting parameters $k_{(t)}$ for Eq. (5)

Curing time, t (days)	$k_{(t)}$ (kPa)	R^2
0	0	l
3	8661.96	0.93
7	10684.45	0.94
14	11848.09	0.94
28	14079.29	0.92

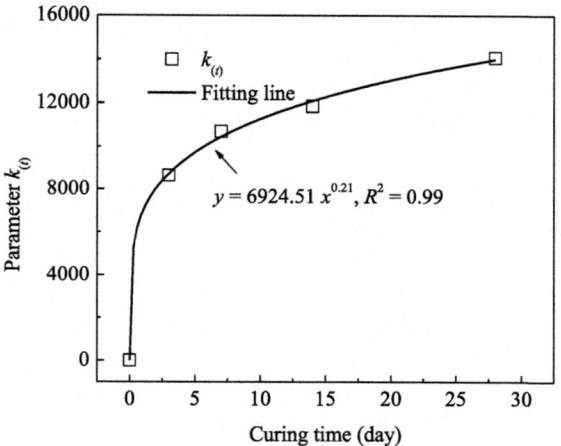

Fig. 8. Relationship between parameter $k_{(t)}$ and curing time

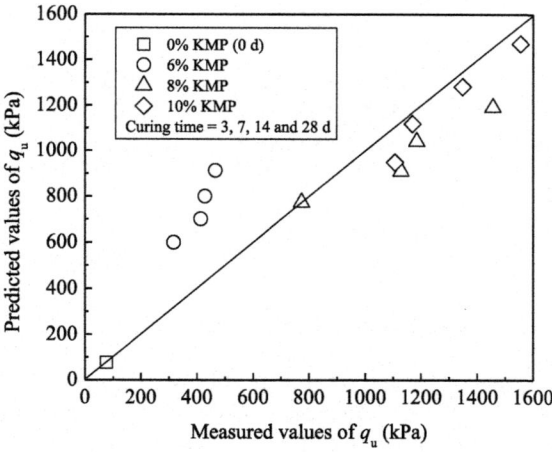

Fig. 9. Strength q_u measured and predicted for the stabilized soils

It can be seen that the predicted values for q_u are slightly higher than the measured when the measured values are less than 500 kPa (i.e., KMP content $\leq$ 6%) and are slightly lower than the measured when the measured values exceed 1100 kPa (i.e., KMP content > 6%). Based on the comparisons, it is suggested that Eq. (7) may be employed to estimate the strength of KMP stabilized soils contaminated by heavy metals with the effect of KMP content and curing time.

4 Conclusions

An experimental investigation is made of the stress-strain behavior and strength of a KMP stabilized contaminated site soil. The effects of binder content and curing time on the peak strength q_u, deformation modulus E_{50} and failure strain ε_f are studied. A method for estimating strength q_u of the stabilized soils based on KMP content and curing time has been proposed. Based on this study, the following conclusions can be drawn:

(1) Strength q_u increases dramatically with KMP content and curing time. The stabilized soils exhibit brittle characteristic when the soil has a KMP content greater than 8% with the curing time more than 3 days.
(2) Parameters ε_f and E_{50} of stabilized soils are dependent on KMP content and curing time. ε_f decreases with KMP content and curing time, however, E_{50} increases with KMP content and curing time.
(3) The relationship between curing time and ε_f can be described by a power function. Failure strain ε_f decreases with q_u, and their relationship is of a power function. Young's modulus E_{50} increases with q_u for studied soils, and a linear relationship exists between the two parameters.
(4) An empirical equation is proposed for the estimation of q_u of the stabilized soils with different KMP contents and curing times. The proposed equation is found to give acceptable estimation of the strength of KMP stabilized heavy metal contaminated soils.

Acknowledgements. This research is financially supported by the Environmental Protection Scientific Research Project of Jiangsu Province (Grant No. 2016031), National Natural Science Foundation of China (Grant No. 41330641 and 41472258),and National High Technology Research and Development Program of China (Grant No. 2013AA06A206).

References

1. Zha, F., Xu, L., Cui, K.: Strength characteristics of heavy metal contaminated soils stabilized/solidified by cement. Rock Soil Mech. **33**(3), 652–656 (2012)
2. Du, Y., Wei, M., Reddy, K.R., Jin, F., Wu, H., Liu, Z.: New phosphate-based binder for stabilization of soils contaminated with heavy metals: leaching, strength and microstructure characterization. J. Environ. Manag. **146**, 179–188 (2014)

3. Du, Y., Wei, M., Reddy, K.R., Wu, H.: Effect of carbonation on leachability, strength and microstructural characteristics of KMP binder stabilized Zn and Pb contaminated soils. Chemosphere **144**, 1033–1042 (2016)
4. ASTM.: Standard test method for unconfined compressive strength index of chemical-grouted soils. ASTM standard D4219. American Society for Testing and Materials, West Conshohocken, Pa (2002)
5. Zhu, W., Zhang, C., Gao, Y., Fan, Z.: Fundamental mechanical properties of solidified dredged marine sediment. J. Zhejiang Univ. Eng. Sci. **39**(10), 1561–1565 (2005)
6. Liu, M.D., Indraratna, B., Horpibulsuk, S., Suebsuk, J.: Variations in strength of lime-treated soft clays. Proc. Inst. Civ. Eng. **165**(4), 217–223 (2012)

The Effect of Particle Size on the Electrochemical Corrosion Behavior of X70 Steel in NaCl Contaminated Sandy Environment

Bin He[1] and Xiaohong Bai[1,2(✉)]

[1] College of Architecture and Civil Engineering, Taiyuan University
of Technology, Taiyuan 030024, China
bxhong@tyut.edu.cn
[2] Shanxi Province Key Laboratory of Geotechnical and Underground
Engineering, Taiyuan 030024, China

Abstract. Corroded soil is a complex material consisting of soil particles, pore fluid, and pore gas. The behavior of buried pipelines in a corrosive soil is influenced by not only fluid, gas, but also particle sizes. However, the effect of soil particle size on the electrochemical corrosion behavior of steel pipelines in a corrosive soil system remains unknown. An indoor experiment is conducted. Three sorts of uniform sands and two sorts of grade sands contaminated by sodium chloride are used as a contaminated environment in this experiment. Steel disc buried in the contaminated sand is a type of pipeline steel, grade API 5L X70 (for short X70). The results from uniform sands tests show that the corrosion behavior of X70 steel is greatly affected by the particle size and distribution. The corrosion rate decreases first then increases with the soil particle size increasing for the uniform sandy environment. This implies that there is a critical particle size. The corrosion rate of X70 steel in the poor gradation sandy soil environment is significantly larger than that in the well grade sandy soil environment. The experimental results provide that the electrochemical impedance spectroscopy (EIS) is an effective way to evaluate the corrosion of soil environment.

Keywords: Sodium chloride contaminated sandy · Particle-size
EIS · X70 pipeline steel

1 Introduction

Buried pipelines have been commonly used for energy delivery purposes, for example, the West–East Gas Pipeline Project operated in 2004 was developed to transmit natural gas from the western part to the east in China [1]. The majority of the pipelines in this project are buried in sandy soils in Northwest China which mainly are salinized sandy soils [2]. With the increase of the service life, the corrosion of pipeline steels in soils, especially in contaminated soils, becomes an important issue, as it can affect the normal operation of the pipelines and result in safety and economic impacts. Many researchers

© Springer Nature Singapore Pte Ltd. 2018
A. Farid and H. Chen (Eds.): GSIC 2018, *Proceedings of GeoShanghai 2018*
International Conference: Geoenvironment and Geohazard, pp. 346–353, 2018.
https://doi.org/10.1007/978-981-13-0128-5_39

have investigated extensively in previous studies [3–9]. There are various factors affecting the corrosion of buried metals, including soil moisture and types, chemical constituents, environmental pH value, soil porosity, and others. However, the effect of particle size of sandy soil on the electrochemical corrosion behavior of buried metal (such as steel pipeline) is not fully understood since the research in this area is rare.

In order to uncover the effect of sand particle size on the electrochemical corrosion behavior and process of buried pipeline steel, a series of tests in door are conducted by using electrochemical test system and electrochemical impedance spectroscopy (EIS).

2   Materials and Experimental Methods

2.1   Materials

Soils Used in This Study. The soil used was commercially available quartz sand (China ISO). For the purpose studying particle size effect on the electrochemical corrosion behavior of buried pipeline steel, three sorts of uniform sands with different particle sizes were used named as 1#sand, 2#sand and 3#sand, respectively, shown in Fig. 1. Meanwhile, two sorts of grade sands, the continuous gradation (CG) and the gap gradation (GG), were used too, shown in Fig. 1.

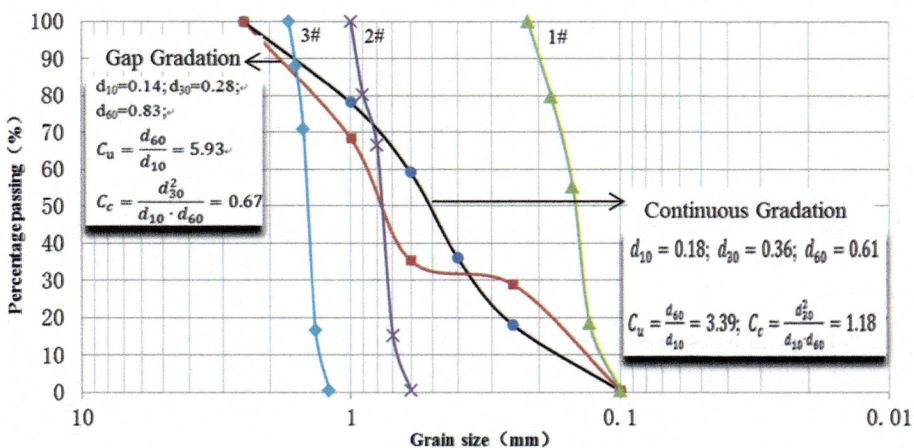

Fig. 1. Particle-size distribution curves of two typical grad sands, where the black line represents the curve for the continuous gradation (CG) and the red line represent the curve for the gap gradation (GG).

Metal Used in This Study. Based on the No. 1 West-East Gas Pipeline Project in China, the metal studied in this paper was pipeline steel X70. Its main chemical composition and parameters are listed in Tables 1 and 2 respectively.

Table 1. Main chemical compositions of X70 pipeline steel (wt%)

C	Si	Mn	P	S	Cr
0.0645	0.201	1.906	0.0119	<0.0005	0.021
Nr	Mo	Cu	Co	V	Fe
0.021	0.234	0.012	0.013	0.011	Balance

Table 2. Parameters of X70 pipeline steel

Parameter	Value
Yield strength σ_y (MPa)	559.73
Ultimate tensile strength σ_u (MPa)	681.45
Young's module E (MPa)	2.06E5
Poisson's ratio υ	0.3

2.2 Experimental Methods

Preparation of Contaminated Soil Environment. The sand to be used was first ultrasonically cleaned in de-ionized water and dried naturally for more than 24 h. Then, the contaminated soil was made by mixing the clean sand and NaCl solution with a concentration of 1.5wt%. The moisture of the contaminated sand was 18%. The contaminated sand was manually compacted inside an experimental box using a proctor compactor, controlling the dry density of 1.65 g/cm^3 which means the porosity of the sand equals to 0.38.

Preparation of Steel Specimen. The steel specimen is a plate of 15 mm × 15 mm × 2 mm. The plate was enlaced with copper conductor around the borders. One surface of the plate was polished first using a series of silicon carbide emery papers, cleaned with ethanol, followed by rinsing with de-ionized water, degreased with alcohol, and dried naturally. The working area about 10 m × 10 m was chosen and must be kept clean in this surface. The other part except working area of the plate was covered with epoxy resin.

Electrochemical Measurement. A CS350 electrochemical test system was employed to measure electrochemical data during whole experimental process. The testing conditions were room temperature of 20 ± 1° and moisture of 45 ± 2%. The conventional three electrodes system was adopted to measure potentio-dynamic polarization curves. Three electrodes system includes a platinum counter electrode, a saturated calomel electrode (SCE) as reference electrode, and the working electrode of the X70 disc which was buried in the NaCl contaminated sand as described above.

It must be noticed that the X70 disc was placed in the position and the working area faced to the center of the box, seen Fig. 2. All potentials were reported with respect to SCE. Furthermore, top of the box was sealed with the waterproof breathable layer in order to ensure constant water content throughout the experimental process.

The potentio-dynamic polarization curves were recorded by a constant scanning rate of 0.167 mV/s, and the open-circuit potential was stable for 30 min before

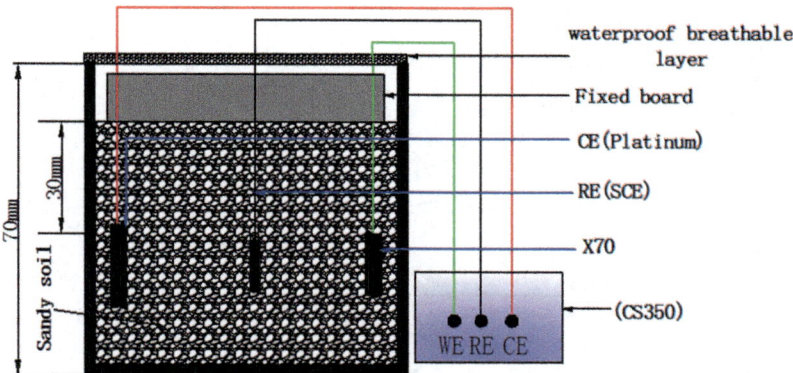

Fig. 2. The schematic diagram of the electro-chemical system designed for the study of the behavior of X70 disc in NaCl contaminated sandy environments

recording the polarization curves. The experiments were performed for 7, 28, 60 and 90 days, respectively, and the results analyzed using the fit program CView2 software combining with Tafel extrapolation method

Electrochemical impedance spectroscopy (EIS) measurements were conducted at the open circuit potential sine signal of 10 mV and in 10^5 Hz and 10^{-2} Hz frequency range. In order to reach stable conditions, 30 min was set before test. The EIS results were fitted with the fit program ZSimpWin software.

3 Results and Discussions

3.1 Corrosion Behaviors of X70 Disc in Uniform Sandy Environment

Linear Polarization Resistance (Rp). Figure 3 gives the change curves of linear polarization resistance (Rp) of X70 disc with buried times in the NaCl contaminated uniform sandy environments. In general, the Rp of X70 decreases linearly with buried time increasing for given sand particle size. But, Rp increases first, and then decreases with the particle size increasing for given buried time. Based on the results of Choi et al. [10], the polarization resistance is inverse proportion with the corrosion current density, while the corrosion current density is proportional to the corrosion rate of metal. This fact indicates that, as the buried time increasing, the corrosion rate of X70 increased for given particle size; and as the particle size increasing from 0.15 mm to 1.5 mm it decreased first, and then increased for given buried time.

Polarization Curves. Figure 4 shows the polarization curves of X70 discs buried in the NaCl contaminated uniform sandy environments for 7 days and 60 days with different particle sizes. Table 3 lists the parameters Ecorr and Icorr which were obtained by using the fit program CView2 software combine with Tafel extrapolation method.

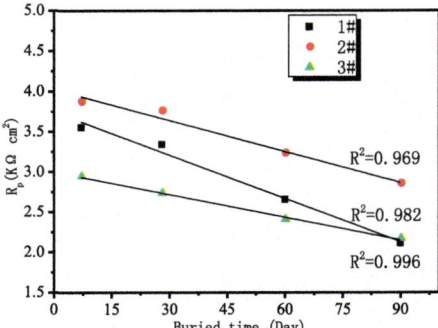

Fig. 3. Linear polarization resistances of X70 disc vs. buried time in uniform sandy environments contaminated by NaCl.

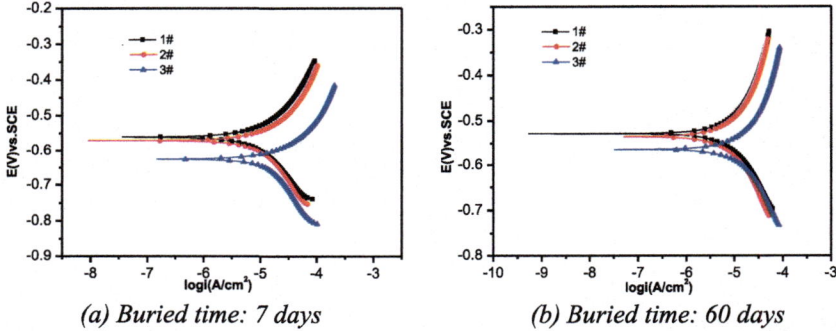

| *(a) Buried time: 7 days* | *(b) Buried time: 60 days* |

Fig. 4. Polarization curves of X70 disc after buried 7 days and 60 days in NaCl contaminated uniform sandy environments, respective.

In Fig. 4, it can be seen that the anodic polarization curves of X70 discs for 1# and 2# sandy environments are very similar, but different from 3# sandy environment. This may imply that the anodic corrosive procedures of X70 discs in both environments of 1# and 2# sands are same, but different from that in 3# sandy environment. This also can be proved from the parameters Ecorr and Icorr listed in Table 3. For 1# and 2# sandy environments, the Ecorr and Icorr are very closer, however, Icorr for 3# sandy environment is greater than those for 1# or 2# sandy environment at same buried time.

Based on above discussions, it may conclude that there is critical particle size, when the particle size in sandy environment is smaller than it, the corrosion rate of X70 disc decreases with particle size increasing, but increases with particle size increasing when the particle size in sandy environment is greater than it. For the different particle size zones, the corrosion mechanisms may differ.

Table 3. Fitting parameters of X70 disc from polarization curves

Buried time	Sand number	Ecorr, (V)	Icorr, (μA/cm^2)
7 days	1#	−561	6.789
	2#	−572	6.091
	3#	−627	10.320
	CG	−0.572	8.042
	GG	−0.660	30.894
28 days	1#	−554	8.132
	2#	−575	8.042
	3#	−592	10.198
	CG	−0.575	8.689
	GG	−0.671	42.682
60 days	1#	−528	8.797
	2#	−536	8.689
	3#	−565	11.386
	CG	−0.536	6.091
	GG	−0.628	30.116
90 days	1#	−537	10.112
	2#	−572	10.050
	3#	−552	15.541
	CG	−0.572	10.050
	GG	−0.628	48.180

Notes: E_{corr} is corrosion potentials; I_{corr} is corrosion currents.

3.2   Corrosion Behaviors of X70 Disc in Grade Sandy Environment

Polarization Curves. Figure 5 shows the typical polarization curves of X70 discs buried in NaCl contaminated graded sandy environments for 7 days and 28 days. Table 3 also lists the parameters Ecorr and Icorr of X70 disc buried in graded sandy environments which were obtained by using the fit program CView2 software combine with Tafel extrapolation method.

From Fig. 5, it can be seen that the catholic branches were promoted not obvious in these two types of sandy environments while the anodic branches were promoted distinctly. This fact indicated that the anodic metal dissolution process was promoted in the GG sandy environment, and the catholic oxidation-reduction reaction process at the metal surface was also promoted, which contributed to the negative shift of the corrosion potential.

In addition, the corrosion current density (Icorr) for GG sandy environment is greater than that for CG sandy environment, in Table 3. Consequently, the results imply that the corrosion process of X70 pipeline steel is considerably dependent on the soil particle size distribution in NaCl contaminated sandy environment.

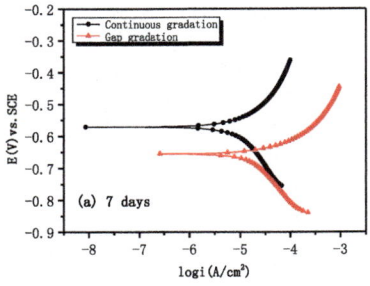 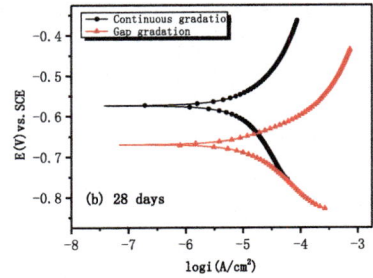

Fig. 5. Polarization curves of X70 disc buried in NaCl contaminated sandy environments with CG or GG for 7 days (a) and 28 days (b).

Electrochemical Impedance Spectroscopy (EIS). Based on the results of EIS, the equivalent circuit can be obtained by using the ZSimpWin software. Then electrochemical parameters can be calculated and are listed in Table 4. Rf represents the resistance of the corrosion product film on the metal surface, and Rct represents the charge transfer resistance. From the data listed in Table 4, both Rf and Rct of X70 disc buried in NaCl contaminated GG sandy environment are much greater than those of X70 disc buried in NaCl contaminated CG sandy environment. As known that Rf and Rct are the resistances of corrosion process of carbon steel [11], the polarization resistance, the sum of Rf and Rct, is used to evaluate the corrosion rate of X70 disc buried in contaminated soil environment. The (Rf + Rct) values in CG sandy environment are obviously higher than those in GG sandy environment, as shown in Table 4. It indicates that the corrosion rate of X70 pipeline steel buried in NaCl contaminated GG sandy environment increases obviously than that in CG sandy environment.

Table 4. Rf and Rct of X70 pipeline steel fitted from EIS results

Buried time	Sandy environment	R_f ($\Omega \cdot cm^2$)	R_{ct} ($\Omega \cdot cm^2$)	$R_f + R_{ct}$ ($\Omega \cdot cm^2$)
7 days	CG	361.1	1800.0	2161.1
	GG	135.4	381.1	516.5
28 days	CG	752.4	718.3	1470.7
	GG	171.2	335.3	506.5
60 days	CG	869.5	2643.0	3512.5
	GG	320.5	493.9	814.4
90 days	CG	1002.6	1089.0	2091.6
	GG	198.3	463.9	662.2

4 Conclusions

The results from uniform sands tests show that the corrosion behavior of X70 steel is greatly affected by the particle size. The corrosion rate decreases with the soil particle size increasing from 0.15 mm to 0.8 mm, then increases with the further increasing particle size from 0.8 mm to 1.5 mm. This implies that there is a critical particle size. For the different particle size zones, the corrosion mechanisms may differ. The corrosion rate of X70 steel in sodium chloride contaminated uniform sands increases with increments of buried time.

The results from two grade sands tests show that the corrosion behavior of X70 steel is affected by the soil particle-size distribution, too. The corrosion rate of X70 steel in the poor gradation sandy soil environment (GG) is significantly larger than that in the well grade sandy soil environment (CG).

The experimental results prove that the electrochemical corrosion behavior of metal materials in contaminated soil is an effective way to evaluate the corrosion of soil environment.

Acknowledgements. The authors thank the supports of National Natural Science Foundation of China (51178287, 51208333) and Youth Foundation of Taiyuan University of Technology (1205-04020203).

References

1. Piao, S.L., Fang, J.Y., Liu, H.Y.: NDVI-indicated decline in desertification in China in the past two decades. Geophys. Res. Lett. **32**(6), 1–4 (2005)
2. Zhou, H., Li, Q., Zhu, R.: Saline soils' characteristic of engineering geological in Ku'erle, Xinjiang Province. West-China Explor. Eng. **2000**(3), 37–38 (2000)
3. Jin, M., Meng, X., Huang, H., et al.: Corrosion mechanics of carbon steel in four types of soils. J. Huazhong Univ. Sci. Technol. (Nature Sci.) **30**(7), 104–107 (2002)
4. Benmoussa, A., Hadjel, M., Traisne, M.: Corrosion behaviour of API 5L X-60 pipeline steel exposed to near-neutral pH soil simulating solution. Mater. Corros. **57**(10), 771–777 (2006)
5. Yan, M., Wang, J., Hana, E., Ke, W.: Local environment under simulated disbanded coating on steel pipelines in soil solution. Corros. Sci. **50**(5), 1331–1339 (2008)
6. Alamilla, J.L., Espinosa-Medina, M.A., Sosa, E.: Modeling steel corrosion damage in soil environment. Corros. Sci. **51**(11), 2628–2638 (2009)
7. Jeannin, M., Calonnec, D., Sabot, R., Refait, P.: Role of a clay sediment deposit on the corrosion of carbon steel in 0.5 mol L^{-1} NaCl solutions. Corros. Sci. **52**(6), 2026–2034 (2010)
8. Noor, E.A.: Comparative analysis for the corrosion susceptibility of Cu, Al, Al–Cu and C-steel in soil solution. Mater. Corros. **62**(8), 786–795 (2011)
9. Cole, I.S., Marney, D.: The science of pipe corrosion: a review of the literature on the corrosion of ferrous metals in soils. Corros. Sci. **56**(2012), 5–16 (2012)
10. Choi, Y.S., Kim, J.G., Lee, K.M.: Corrosion behavior of steel bar embedded in fly ash concrete. Corros. Sci. **48**(7), 1733–1745 (2006)
11. Zhang, G.A., Zeng, Y., Guo, X.P., et al.: Electrochemical corrosion behavior of carbon steel under dynamic high pressure H2S/CO_2 environment. Corros. Sci. **65**, 37–47 (2012)

A Hydro-Chemo-Mechanical Analysis
of the Slip Surface of Landslides in the Three
Gorges Area of China

Ashok Gaire[1], Yu Zhao[2], and Liang-Bo Hu[1(✉)]

[1] University of Toledo, Toledo, OH 43606, USA
Liangbo.Hu@utoledo.edu
[2] Institute of Mountain Hazards and Environment,
Chinese Academy of Sciences, Chengdu 610041, Sichuan, China

Abstract. Landslides in the Three Gorges area of China are often triggered by rainfalls of significant acidity which potentially lead to wetting effects by rainfall precipitation and mineralogical changes induced by adverse chemical reactions. This paper presents a hydro-chemo-mechanical analysis of slope stability in this context. It first surveys typical landslide events that occurred in the Three Gorges area where mineralogical changes were identified and further simulated in the laboratory; both field and laboratory evidences strongly suggest that these changes contributed to the weathering and strength decline of sliding masses. Subsequently numerical simulations are conducted to investigate possible scenarios of strength evolution due to rainfall induced wetting as well as mineralogical changes. The factor of safety of potential slip surfaces is computed under different combinations of hydrological and chemical scenarios. The results of a parametric study indicate the potential consequences of hydro-mechanical and chemo-mechanical processes that may play an important role in the landslides in the Three Gorges area.

Keywords: Landslides · Mineral transformation · Rainfall · Three Gorges

1   Introduction

Chemical weathering of geological sediments has been recognized to potentially play a significant role in landslides. The mineralogical history and evolution in geomaterials may induce significant changes in soil strength (e.g., Moon 1993; Shuzui 2001), and as a consequence they become more susceptible to landslides when subject to a triggering event such as rainfall. Landslides in the Three Gorges areas of China occur rather frequently (Liu et al. 2004). Of particular interest is the potential weathering effect of rainfall with significant acidity, not unusual in the Three Gorges area, typically leading to intricate mineral evolution and transformation. Zhao et al. (2011) studied a typical landslide in this area potentially trigged by acid rainfall. Their field study suggested that at the location of landslides illites or mixed layer illite/smectite of indigenous slopes were altered into smectite and subsequently kaolinite neo-formed from that transitory smectite. The experimental simulation of the remolded soils also revealed a complex pattern of strength change with time.

© Springer Nature Singapore Pte Ltd. 2018
A. Farid and H. Chen (Eds.): GSIC 2018, *Proceedings of GeoShanghai 2018*
International Conference: Geoenvironment and Geohazard, pp. 354–362, 2018.
https://doi.org/10.1007/978-981-13-0128-5_40

Whist the fundamental understanding of mineralogical changes due to weathering remains an intriguing subject under extensive research, it is also of great interest, especially from the practice perspective, to analyze their potential consequences on the stability of slope surfaces. This paper presents a hydro-chemo-mechanical analysis of slope stability in this context. It first surveys a landslide event that occurred in the Three Gorges area where mineralogical changes were identified and further simulated in the laboratory. Subsequently numerical simulations are conducted to investigate possible scenarios of strength evolution due to rainfall induced wetting as well as mineralogical changes, focusing on the evolution of the factor of safety of potential slip surfaces under different combinations of hydrological and chemical scenarios.

2 Geo-Hydro-Chemical Effects and Strength Evolution During Acid Rainfall Scenarios

2.1 Mineralogical Alterations and Associated Effects

There are potentially various processes involved in the water-mineral interaction that may have a strong impact on the behavior of geological materials and possibly promote the vulnerability to landslides. Weathering effects possibly involved in landslides include the dissolved carbonates and loss of cementitious bonds (Anson and Hawkins 2002), lithological structure changes (Udvardi et al. 2016), and weakening or destruction of intergranular bonds (Gajo et al. 2015), and so on. Of particular interest is the smectitization of illite through leaching of potassium, which has been observed in several investigations (Anson and Hawkins 2002; Zhao et al. 2011; Chai et al. 2013) and believed to have potentially contributed to the alteration of shear strength.

In the 2003 Qianjiangping landslide (Fig. 1), and later another similar event, Diaojiaozui landslide, both of which occurred after an intense rainfall period which was characterized by strong acidity, a field investigation focusing on geochemical changes on the slip surface and on the no-slip surface of vicinity (Zhao et al. 2011) indicated that significant amount of illite seemed to be altered to smectite on the weakened slip surface but not in the vicinity where slides did not occur. A key reaction involved may be expressed as:

$$\text{illite + I/S (mixed layer) + kaolinite + chlorite + } H_2O \text{ + acid } \Rightarrow$$
$$\text{smectite (w/adsorbed } H_2O\text{) + kaolinite + potassium + silicic acid}_{(soluble)} \tag{1}$$

The scenario also predicts that at a particular condition alteration of I/S to smectite becomes replaced or dominated by a reaction of neoformation of kaolinite on the basis of the previously produced smectite. This reaction takes place, if there is a supply of H^+, whereas the base cations and silica are leached out or precipitated (e.g. McBride 1994). It is probable that such effects have to be accumulated over extend periods of time.

Subsequent laboratory study on remolded soils subject to an acidic solution as an attempt to simulate the chemical effects of acid rainfall showed evident mineralogical alterations and strength changes. SEM images show the evolution of the material structure during its exposure to acidic solution; for example, after 30 days (Fig. 2) each

Fig. 1. (a) A landslide that occurred in Qianjiangping, Hubei, China, July 2003. (b) A slip surface of the landslide. (c) Samples collected from the slip surface for subsequent investigations (Zhao et al. 2011).

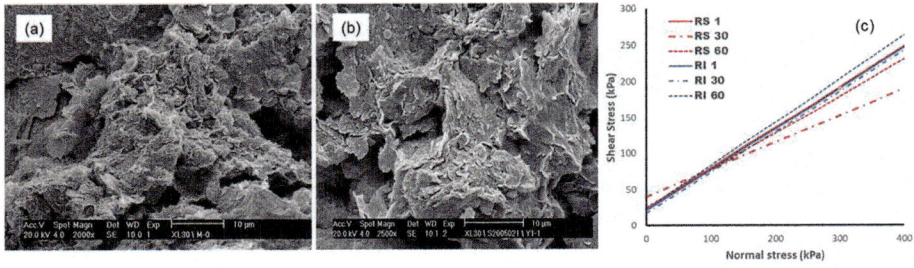

Fig. 2. SEM image of mineral structure in (a) a remolded smectite soil in its original state; (b) a remolded illite soil after 30 days of soaking in an acidic solution. (c) Strength evolution of the remolded smectite (RS) and illite (RI) soils after 1-day, 30-day and 60-day exposure to the acidic solution.

sample exhibited relatively small clusters of smectite filling most pores between the silt and sand particles.

The strength evolution (Fig. 2c) clearly suggests the changes in the cohesion and angle of fiction of each remolded soil, but does not exhibit a consistent pattern or directly offer the support for the proposed smectitization of illite *in situ* from the strength aspect. However, the inner mechanisms of such mineralogical alterations have not been well understood even from a purely geochemical point view; nor does the evolution of the mechanical strength manifest in a consistent manner. Most likely such mechanisms have to be studied from a multi-scale perspective to elucidate the inter-action of complex chemo-mechanical processes.

In the present paper a concept of mineralogically affected strength evolution is explored and subsequently used in the analysis of slope stability. A generic formulation of the yield condition is,

$$f = \tau - c'\left(\varepsilon_q^{pl}, \zeta_s, \zeta_k, \zeta_i\right) - \sigma' \tan \phi'\left(\varepsilon_q^{pl}, \zeta_s, \zeta_k, \zeta_i\right) \tag{2}$$

where τ and σ' are the shear and effective normal stress components, c' and ϕ' are effective cohesion and friction angle, and ε_q^{pl} is plastic shear strain. Both the cohesion and angle of friction can be strain and mineralogically dependent. The parameter for each mineral can be expressed as $\zeta_k = \Delta m_k/m^0$, i.e., the mass or molar fraction of the change of the mineral, Δm_k, to the overall total mass or mol of the material, m^0. The subscript k, is herein confined to one of smectite, kaolinite and illite.

Based on the observations made in Zhao et al. (2011), only the alteration in the angle of friction is considered in the present study; in addition, the strain hardening/softening effect is ignored for the sake of simplicity for the conventional slope stability analysis. The following form of function is proposed to represent the

$$\phi' = \phi'_0(1 + \lambda_s\zeta_s + \lambda_k\zeta_k + \lambda_i\zeta_i) \tag{3}$$

Each coefficient λ represents the dependence on each major mineral considered.

2.2 Evolution of Shear Strength During Saturation and Desaturation Phases

Conventional formulation for the saturation dependent shear strength is employed in the present study. A typical failure envelop can be described in a modified form of Mohr-Coulomb failure criterion for both partially and fully saturated soils (e.g., Vanapalli et al. 1996),

$$\tau = c' + (\sigma - u_a + \chi(u_a - u_w))\tan\phi' \tag{4}$$

where σ is the total normal stress, u_a is the pore air pressure, u_w is the pore water pressure. The term, $(\sigma_n - u_a)$ is the net normal stress, and $(u_a - u_w)$ is the matric suction; thus the term, $\sigma - u_a + \chi(u_a - u_w)$ forms an expression of the effective stress, the coefficient χ can be replaced by a function of saturation, herein it adopts a formulation of effective saturation:

$$\chi = \frac{\theta_w - \theta_r}{\theta_s - \theta_r} \tag{5}$$

where θ_w is the volumetric water content, θ_s is the saturated water content and θ_r is the residual water content. Under the saturated case ($\chi = 1$), Eq. (4) returns to the conventional form for saturated soils.

Equations (3) and (4) represent the shear strength of soils simulated in the present study, taking onto account both the effects of mineralogical changes and saturation. In the following section a parametric study is undertaken focusing on a variety of scenarios to assess the potential stability of slopes subject to the effects acid rainfall as hypothesized in the preceding section.

Parameters adopted in the present study regarding the strength change are: $\lambda_s = -6.70$, $\lambda_i = 20.05$ and $\lambda_k = -8.51$. The calibrations are based on the results reported in Zhao et al. (2011) and some typical parameter values are offered here. It was observed that the content of chlorite remains almost constant (10% of total clay) and thus did not contribute to the strength change. Table 1 provides the alteration in the angle of friction after significant smectitization of illite. Table 2 offers the resultant changes due to the subsequent formation of kaolinite from newly produced smectite. All the values are based on Eq. (3) and later used in the analysis of slope stability.

Table 1. Evolution of friction angle with transformation of illite to smectite.

Smectite (%)	Kaolinite (%)	Illite (%)	Chlorite (%)	Friction angle (kPa)
54	11	25	10	31.0
59	11	20	10	26.3
64	11	15	10	21.6
69	11	10	10	16.9
74	11	5	10	12.2
79	11	0	10	7.6

Table 2. Evolution of friction angle with subsequent neo-formation of kaolinite from smectite.

Smectite (%)	Kaolinite (%)	Illite (%)	Chlorite (%)	Friction angle (kPa)
54	11	25	10	31.0
35	35	20	10	24.7
37.5	37.5	15	10	19.9
40	40	10	10	15.0
42.6	42.4	5	10	10.17
45	45	0	10	5.3

3   Numerical Simulations

An idealized homogeneous, isotropic slope with an angle of inclination 40° to the horizontal plane, height of 30 m and length of 70 m is analyzed in the present study (Fig. 3). Typical scenarios are explored based on the hypothesis discussed in the preceding section: considerable mineralogical changes are induced and accumulated after prolonged periods of acid rainfall, resulting in alteration of the shear strength; when later confronting with an intense rainfall, its stability further reduces when water content and saturation rises and may become susceptible to landsliding. Primary parameters used in this study include the saturated hydraulic conductivity, relevant fitting parameters for soil-water characteristic curve and unsaturated permeability, in addition to the strength parameters discussed above.

The first, simple, scenario explored is focused on the mineralogical effects only; the results of potential slope stability due to accumulated smectitization of illite and subsequent formation of kaolinite are shown in Fig. 4. Two cases discussed in the preceding section are considered; first, transformation (from illite) into the smectite, which reduces the FOS (factor of safety) substantially; secondly, the neo-formation of kaolinite from the produced smectite, which shoes even further reduction in the FOS due to the lowered shear strength.

The FOS, as conventionally defined in typical geotechnical analysis, refers to the ratio of the summation of resisting shear stresses to the summation of mobilized shear stresses. In the present simulations the well-known limit equilibrium method is used. It should be noted that the most vulnerable sliding surfaces are identified by first conducting a conventional slope stability analysis and found to be located a few meters below the inclined surface (BC). Since moisture variation due to rainfall is more significant in the upper part of the slope, shallow circular slip surfaces are assumed in computing the factor of safety in the present simulations.

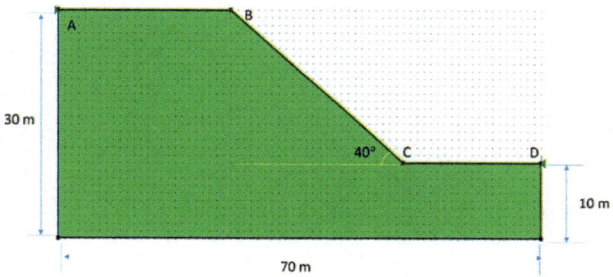

Fig. 3. Geometry of the simulated slope model.

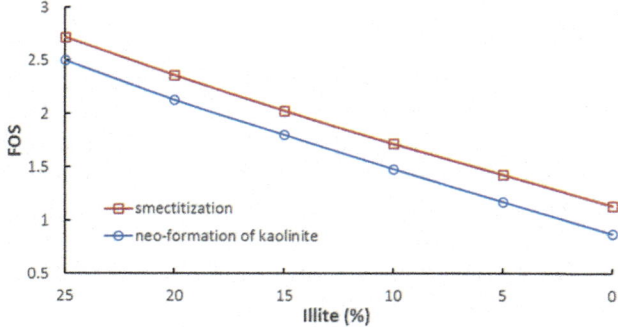

Fig. 4. Evolution of the factor of safety of potential slope stability with the transformation of illite.

Subsequently a rainfall scenario is simulated as a rainfall flux of 1×10^{-6} m/s is imposed on the exposed surface (from A to D) for 5 days. The analysis of the seepage and the slope stability is conducted via the numerical software SLOPE/W. It should be noted that water ponding is not considered, hence further progression to the rise of water pressure after full saturation in the entire domain is beyond the scope of the present investigation, although it can be addressed under properly formulated boundary values problems as well.

The evolution of the slop stability is shown in Fig. 5. Four soils at different illite clay contents as a result of mineral transformation are examined. Their original strengths prior to the rainfall are listed in Table 2, resulting in the substantial differences in the original factor of safety. The progression of rainfall induces further changes in the soil strength and as a consequence causes the reduction in the factor of safety as shown in Fig. 5. It is noted that although the changes appear smaller for the soils with lower illite content, the percentages are actually quite comparable.

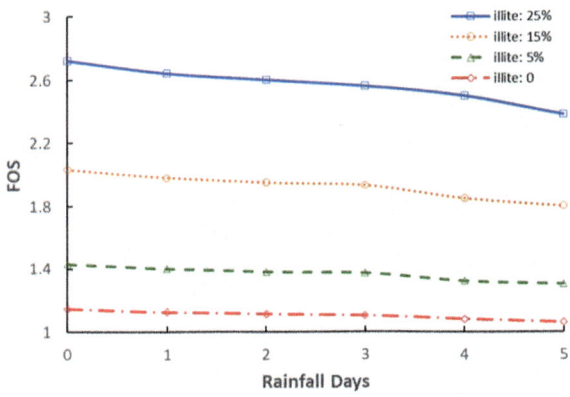

Fig. 5. Evolution of the factor of safety of potential slip surface at different initial illite contents subject to rainfall.

The third scenario explored is the simultaneous, combined effects of mineral transformation and rainfall. In the absence of kinetics of mineral transformation, a constant rate is assumed. Figure 6 shows the evolution of factor of safety for two cases: complete transformation of illite to smectite in 5 days, and formation of kaolinite immediately from transformed smectite in the same period. The slope is slightly more vulnerable in the second case since the strength is reduced further. As compared with Fig. 4, it suggests that although the effect of rainfall is still significant, the mineral transformation exerts more influence in the present study.

It should be noted that the presented results are largely based on the strength evolution as described in the preceding which renders considerable alteration in shear strength based on the laboratory studies of remolded soils; it is of great uncertainty that the range of strength evolution *in situ* under complex field conditions and environments can be replicated in simulated laboratory tests. Nonetheless, the framework employed in the present study offers a numerical tool to assess the stability of the potential sliding masses under different scenarios.

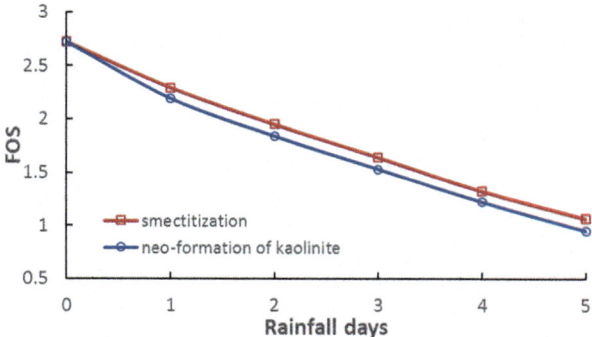

Fig. 6. Evolution of the factor of safety of potential slip surface subject to the combined effects of rainfall and mineral transformation.

4 Conclusion

The present study is focused on the effects of potential mineral transformation during acid rainfalls. Two typical processes are considered, illite or mixed layer I/S of indigenous slopes is altered into smectite and subsequently kaolinite is neo-formed from that transitory smectite. The variability in the internal friction angle with the progression of the mineral transformation, is employed to describe the shear strength evolution and subsequently simulate the slope stability in a numerical example. The reduction in the factor of safety of the potential slip surfaces is numerically simulated, taking into account both the effect due to mineral transformation and the saturation induced strength reduction during the rainfall.

Potential coupled effects of chemical, hydraulic and mechanical interactions deserve more attention; in the present study, their interactions are treated as a sequential process rather than a simultaneous one. This would demand more extensive efforts to better understand the mechanisms behind the mineral transformation during acid rain scenarios, which most likely involve multi-scale considerations to adequately address multi-physics phenomena occurring across the scales.

References

Anson, R.W.W., Hawkins, A.B.: Movement of the Soper's Wood landslide on the Jurassic Fuller's earth, bath, England. Bull. Eng. Geol. Environ. **61**(4), 325–345 (2002)

Chai, B., Yin, K., Du, J., Xiao, L.: Correlation between incompetent beds and slope deformation at Badong town in the Three Gorges reservoir. China. Environ. Earth Sci. **69**(1), 209–223 (2013)

Gajo, A., Cecinato, F., Hueckel, T.: A micro-scale inspired chemo-mechanical model of bonded geomaterials. Int. J. Rock Mech. Min. Sci. **80**, 425–438 (2015)

Liu, J.G., Mason, P.J., Clerici, N., Chen, S., Davis, A., Deng, H., Liang, L.: Landslide hazard assessment in the Three Gorges area of the Yangtze river using ASTER imagery: Zigui-Badong. Geomorphology **61**(2), 171–187 (2004)

McBride, M.B.: Environmental Chemistry of Soils. Oxford University Press, NewYork (1994)

Moon, V.G.: Microstructure control on the geomechanical behaviour of ignimbrite. Eng. Geol. **35**(1), 19–31 (1993)

Shuzui, H.: Process of slip-surface development and formation of slip-surface clay in landslide in tertiary volcanic rocks, Japan. Eng. Geol. **61**(4), 199–219 (2001)

Vanapalli, S.K., Fredlund, D.G., Pufahl, D.E., Clifton, A.W.: Model for the prediction of shear strength with respect to soil suction. Can. Geotech. J. **33**(3), 379–392 (1996)

Udvardi, B., Kovacs, I.J., Szabo, C., Falus, G., Ujvari, G., Besnyi, A., Bertalan, E., Budai, F., Horvath, Z.: Origin and weathering of landslide material in a loess area: a geochemical study of the Kulcs landslide, Hungary. Environ. Earth Sci. **75**, 1299 (2016)

Zhao, Y., Cui, P., Hu, L.B., Hueckel, T.: Multi-scale chemo-mechanical analysis of the slip surface of landslides in the Three Gorges, China. Sci. China Ser. E: Technol. Sci. **54**(7), 1757–1765 (2011)

Study on Comprehensive Testing of Lossless Precision Testing for Leakage of Waterproof Curtain in Foundation Pit

Liu Tao[1,2(✉)], Sun Wenjing[1,2], and Chen Jian[3]

[1] Shandong Provincial Key Laboratory of Marine Environment and Geological Engineering, Ocean University of China, Qingdao 266100, China
liutao09_ouc@126.com
[2] Laboratory for Marine Geology, Qingdao National Laboratory for Marine Science and Technology, Qingdao 266061, China
[3] China Railway 14th Construction Bureau Co., LTD, Jinan 250014, China

Abstract. As the rapid development of ultra-deep foundation pit, the effective detection technology for the foundation pit envelope leaks is still an urgent need. Resistivity method is an important way to carry on anti-seepage detection. However, there is no relevant theoretical formula to guide the specific construction for practical application. The main objective of this note is to build a new theoretical model of detection range of the micro-logging method. The numerical simulations of transverse and longitudinal cracks detection have been carried out by the AM and AMN methods, respectively. The calculation formula of micro-logging electric detection's sensitivity has been proposed and the influence of electrode tube distance, the electrode pole distance and crack size have been analyzed. Finally, the results of the practical application in the project illustrate reliability and accuracy of the micro-logging theory, indicating that the equivalent simplification of the testing process is realized.

Keywords: Micro-logging method · Anti-seepage test · Diaphragm wall
Theoretical analysis

1 Introduction

The leakage during foundation pit is one of the most common, but the most challenging problems faced by foundation pit designers and constructors. It is easy to cause the deformation of the foundation pit, resulting in the destabilization of the foundation pit. The leakage will impair project schedule and stop all forward progress. For properly controlling leakage in time, the most important task is to find the specific location leakage in the foundation pit.

To ensure the smooth construction of foundation pit, it is of great significance to pre-inspect water barrier effect of the curtain before excavation. Some detecting methods, such as light dynamic penetration, pore water pressure and pumping test, can only indirectly reflect quality of the local curtain construction. However, these methods cannot accurately evaluate overall effect of the curtain. In recent years, the DC resistivity method to detect seepage of diaphragm wall of dams has been used in hydraulic

© Springer Nature Singapore Pte Ltd. 2018
A. Farid and H. Chen (Eds.): GSIC 2018, *Proceedings of GeoShanghai 2018*
International Conference: Geoenvironment and Geohazard, pp. 363–372, 2018.
https://doi.org/10.1007/978-981-13-0128-5_41

engineering, which has been widely applied because of its advantages of non-destructive, high efficiency and accuracy (Zhu and Ren 2004). For non-destructive testing methods, lots of laboratory and field investigation have been conducted, and some theoretical models have been proposed (Wang et al. 2001; Liu et al. 2015). Moon and Jeong (2011) investigated the variation of water inflow rate into the tunnel as a highly pervious zone near the tunnel using image well method. Flow-field fitting method can detect inlet of water leakage (Huang et al. 2013; Zou 2009). However, internal cracks involving washout of the fine particles in dam foundation lead to formation of the leakage pathways overtime. The strength of current through intact concrete and the internal crack are similar, because neither of them has a connected passage for ions propagation. Thus, flow-field fitting method is not able to distinguish between internal crack and intact concrete by calculating ground resistance. The temperature tracer requires a lot of deep holes in outside of the pit, and is subject to the need for strong seepage to use this limitation (Duan et al. 2016). It is difficult and cumbersome to do qualitative and quantitative analysis using ultrasonic testing method (Feng et al. 2016). Jianan (2011) tried different damage detection and non-destructive detecting methods in a reservoir project and researched the applicability of each test method to the specific anti-seepage curtain and the correlations of these methods. Gao (2013) used finite difference method and model testing method to analyze the characteristics of high density electrical detection profiles of three typical foundation pit waterproofing leakages, and gave the typical anomaly characteristics.

Through the summary and comparison to the above methods, these methods all detect the soil on both sides of the waterproof curtain or diaphragm wall and get the seepage trends of soil water to ensure the position of precipitation funnel and find the leakage point. The method belong to indirect testing that can't detect the wall directly. When we use these methods in the field testing process, the construction site is limited, such that the methods have weak operability and inaccuracy because of the unfitness with the construction process. To take partial advantage of the flow-field fitting method into account, this paper proposed a micro-logging method to detect the leakage. The theoretical formula of the detection range with the micro-logging method is deduced. The influences of different factors on the sensitivity of crack detection in different size are discussed in detail. The feasibility of this method is further proved by the anti-seepage test of Xujiaping Station Project in Wuhan Metro Line 8.

2 Proposed Method

2.1 Principle of Micro-Logging Method

Two power supply points A and B are used to establish the electric field with the measurement of potential difference of the other two points M and N (see Fig. 1).

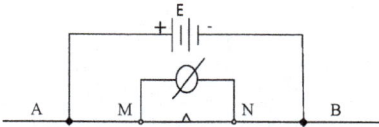

Fig. 1. The principle of micro - logging method.

According to Fig. 1, the potentials between M and N can be calculated as:

$$U_M = I\rho/[2\pi(1/AM - 1/BM)] \tag{1}$$

$$U_N = I\rho/[2\pi(1/AN - 1/BN)] \tag{2}$$

where ρ is the earth resistivity; I is the supply current; U is the measurement potential of M and N point; AM, BM, AN and BN are electrode distance. Potential difference between M and N can be calculated as:

$$U_M = I\rho/[2\pi(1/AM - 1/BM - 1/AN + 1/BN)] \tag{3}$$

where $\rho_{MN} = K \cdot \Delta U_{MN}/I$, $K = 2\pi/(1/AM - 1/AN - 1/BM + 1/BN)$ is defined as the electrode device coefficient, which is a physical quantity associated with the electrode position. In this paper, the AMN and AM detection methods are used to detect the ground wall defect (see Fig. 2).

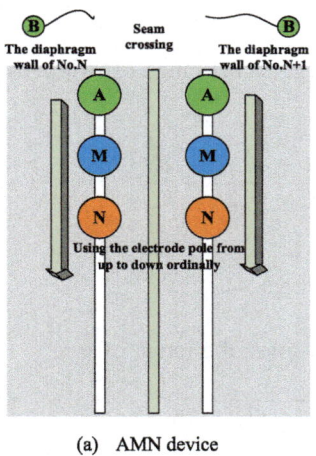

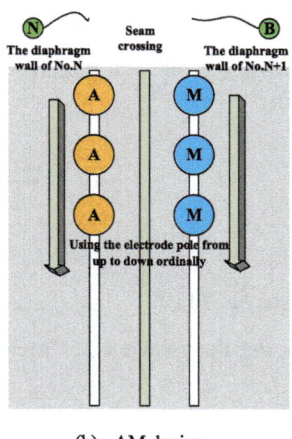

(a) AMN device   (b) AM device

Fig. 2. Two different detection devices.

2.2 Theoretical Calculation of Detection Range Using Micro-Logging Method

Two electrodes are equally placed on both sides of connecting wall. For simplicity, assuming the cross section of the crack is rectangular in which length is dy, width is dx, electrode tube distance is $X1$, electrode pole distance is $H1$, concrete resistivity is $R1$,

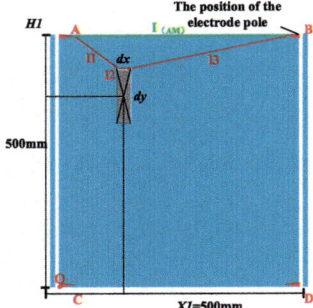

Fig. 3. Theoretical model of electrode tube detection.

water resistivity in crack is $R1$, point O is origin of the coordinates, and the coordinate of the crack is (x, y).

For AM device, we assume that the resistance value in path 1 is less than that in path 2 (see Fig. 3). It can be recognized that the crack can be detected because current always propagates along the path with the least resistance, which can be expressed by:

$$R_1 \cdot (l1 + l2) + R_2 l2 = R_1 \cdot X1 \tag{4}$$

Due to the symmetry, the left $(x \in X1/2)$ of the model can be analyzed instead of analyzing the entire model. The whole model can be divided into two parts:

(1) $x \in [0, dx/2]$

Replacing the relevant parameters into Eq. (4) gives the range of y (ignoring R_2 as $R_2 \ll R_1$):

$$y < \frac{-x^2 - xdx + 2xX1 + dxX1 - dx^2/4}{2X1} + \frac{dy}{2} \tag{5}$$

(2) $x \in [dx/2, X1/2]$

Replacing the relevant parameters into Eq. (4) gives the range of y:

$$y < \sqrt{\frac{x^2 dx^2}{X1^2} + 2xdx - \frac{2x^2 dx}{X1} - \frac{xdx^2}{X1}} + \frac{dy}{2} \tag{6}$$

Then, the total area of the crack can be calculated as:

$$S = \frac{3dx^2}{2} - \frac{7dx^3}{12X1} + dyX1 - \frac{2t_2\sqrt{b^2 + 4at_2^2} - 2t_1\sqrt{b^2 + 4at_1^2}}{a} + \frac{b^2(c_2 + 0.5\sin 2c_2 - c_1 - 0.5\sin 2c_1)}{a\sqrt{-4a}} \tag{7}$$

where $a = dx^2/X1^2 - 2dx/X1$; $b = 2dx - dx^2/X1$; $t_1 = \sqrt{adx^2/4 + bdx/2}$;
$t_2 = \sqrt{aX1^2/4 + bX1/2}$; $c_1 = \arcsin(t_1/\sqrt{-b^2/4a})$; $c_2 = \arcsin(t_2/\sqrt{-b^2/4a})$. In
practical engineering, pole distance of the electrical method should be controlled to
cover full seams area, expressed as

$$H1\max = \begin{cases} -x^2 - xdx + 2xX1 + dxX1 - (dx^2/4)/X1 + dy, & x \in [0, dx/2] \\ 2\sqrt{x^2dx^2/X1^2 + 2xdx - 2x^2dx/X1 - xdx^2/X1} + dy, & x \in [dx/2, X1/2] \end{cases}$$

$$(8)$$

For AMN device, the theoretical range of the detection can be obtained by the same
way.

2.3 Numerical Analysis Using AM Method

Results of Transverse Crack Detection. In this program, MATLAB software is used
to verify the proposed theoretical model. For the electrode with distance of 500 mm
and pole distance of 500 mm, transverse crack analysis with 20 mm length and
100 mm width can explain the problem well (see Figs. 4 and 5).

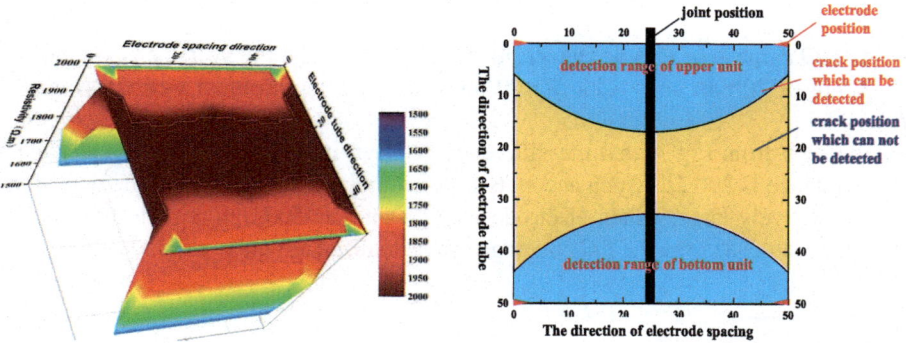

Fig. 4. Generalization analysis of resistivity of
transverse crack in AM method.

Fig. 5. AM method for transverse cracks can
detect the scope of the map.

It can be seen from Fig. 4 that the difference between maximum resistance and minimum resistance is 500 Ω, which accounts for one-third of the minimum, and the
detection range is relatively large. When electrode tube distance is 500 mm and electrode pole distance is set to 320 mm, full coverage can be realized on the ground seam.

Results of Longitudinal Crack Detection. For the electrode with the distance of
500 mm and the pole distance of 500 mm, it can be seen from Fig. 6 that the difference
between maximum resistance and minimum resistance is 200 Ω, which accounts for
one-tenth of the minimum. The larger the difference, the easier it is to be measured,
indicating that the AM method has a strong perceptual ability to transverse cracks and

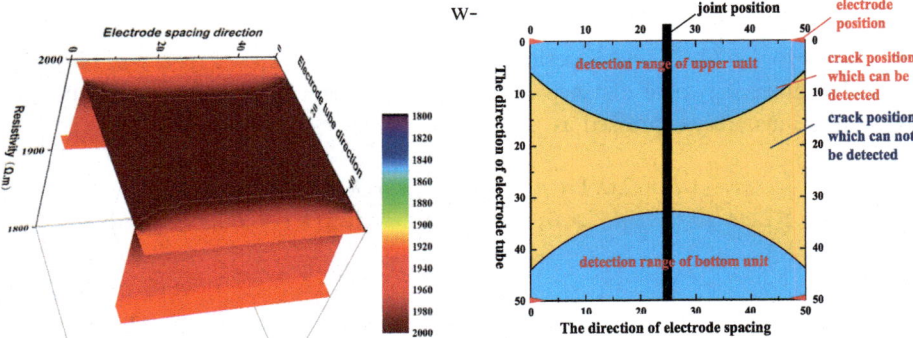

Fig. 6. Generalization analysis of longitudinal fracture resistivity based on AM method.

Fig. 7. AM method for longitudinal crack detection range.

eak perceptual ability to longitudinal cracks in the 500 mm × 500 mm unit, and the detection range is relatively small. When the electrode tube distance is 500 mm and the electrode pole distance is set to 240 mm, the full coverage can be realized on the ground seam (Fig. 7).

2.4 Numerical Analysis Using AMN Method

Results of Transverse Crack Detection. The results of the transverse crack analysis with length of 20 mm and width of 100 mm can solve the problem well (see Figs. 8 and 9).

It can be seen from Fig. 8 that the difference between maximum resistance and minimum resistance is 200 Ω, which accounts for one-tenth of minimum, and the detection range is relatively large. When electrode tube distance is 500 mm and electrode pole distance is set to 240 mm, the full coverage can be realized around middle secant in unit.

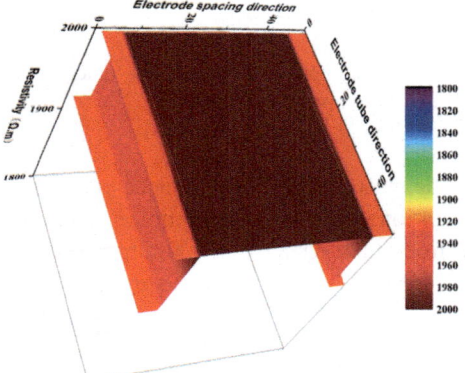

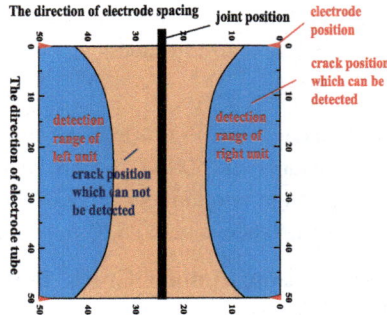

Fig. 8. Generalization analysis of transverse crack resistivity based on AMN method.

Fig. 9. The AMN method is used to detect the range of transverse cracks.

Results of Longitudinal Crack Detection. For electrode with distance of 500 mm and pole distance of 500 mm, it can be seen from Fig. 10 that the difference between maximum resistance and minimum resistance is 500 Ω, which accounts for one-third of the minimum, indicating that the AMN method has a strong perceptual ability to longitudinal cracks and weak perceptual ability to transverse cracks in 500 mm × 500 mm unit, and the detection range is relatively large (Fig. 11). When the electrode tube distance is 500 mm and the electrode pole distance is set to 320 mm, respectively, the full coverage can be realized around the middle secant in the unit.

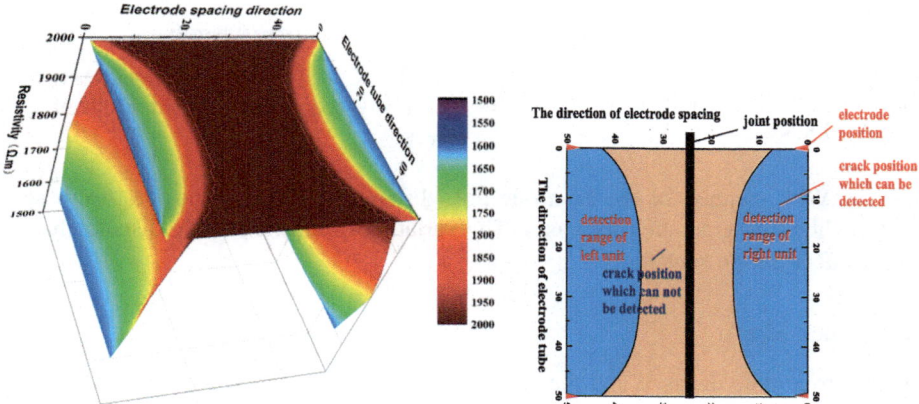

Fig. 10. Generalization analysis of longitudinal crack resistivity based on AMN method.

Fig. 11. AMN method for longitudinal crack detection range.

3 Detection Sensitivity Calculation

To quantitatively study the influence of different factors on the detection range, the concept of detection sensitivity is proposed as follows:

$$\eta = S/S_{total} \tag{9}$$

where S is the area of the detectable range; and S_{total} is the total cross-section area. The effects of different influencing factors on the sensitivity of detection were analyzed for the AM method.

3.1 Electrode Tube Distance and Electrode Pole Distance

When the electrode tube distance is set to 500 mm, corresponding electrode pole distance changes from 300 mm to 700 mm. Sensitivity of each case is calculated in 20 groups, and corresponding curves of sensitivity with electrode tube distance and electrode pole distance are plotted in Fig. 12. With increase of the electrode tube distance, the sensitivity of the AM method is close to the linear trend with a small

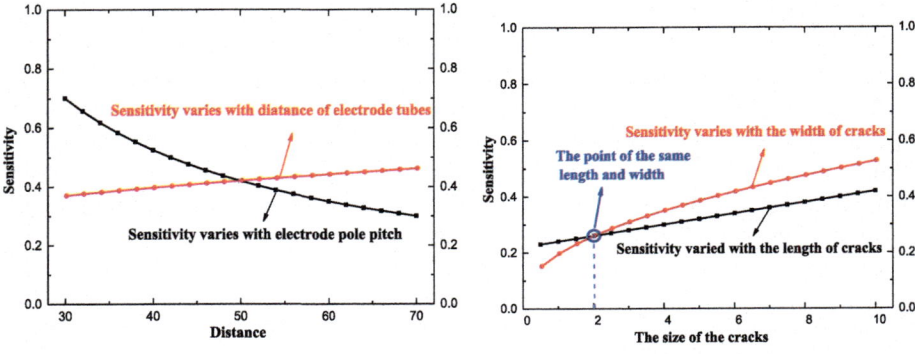

Fig. 12. Curve of the relationship between sensitivity and the distance of electrode tube and pole distance.

Fig. 13. Curve of the relationship between sensitivity and the length and width of the crack.

increase. With increase of the electrode pole distance, sensitivity shows a decreased trend, andthe rate gradually decreases. The variation of electrode pole distance has a greater effect on the sensitivity.

3.2   Crack Size

The electrode tube distance and the electrode pole distance are set to 500 mm in this program. When crack length is set to 20 mm, the corresponding crack width changes from 5 mm to 100 mm. When crack width is set to 20 mm, the corresponding crack length changes from 5 mm to 100 mm. Sensitivity of each case is calculated in 20 groups, and corresponding curves of sensitivity with crack length and crack width are plotted in Fig. 13. With the increase of crack length, sensitivity curve of the AM method shows an approximate linear increasing trend, however, the increment is slow. With increase of crack width, sensitivity increases significantly, the speed of which is higher than the increase speed of the sensitivity caused by the change of the crack length. It is obvious that the larger the crack width, the slower the increase speed. This proves that the change of the crack width has a greater effect on the effect for the AM method.

4   Field Test

4.1   Project Overview

The site layer is a typical soft soil composite stratum. The upper part is the soft and permeable silty sand layer; the lower part is a strong weathered conglomerate, weak cementation conglomerate and medium cementation conglomerate. The groundwater is widespread and uneven. In precipitation process, it is easy to be filled with water through the soil. However, the construction is very difficult which is called "cancer stratum" of foundation pit construction.

4.2   Verification Process

As shown in Figs. 14 and 15, the WG24-25, WG25-26 and WG26-27 diaphragm wall have been selected in axis-48, and 6 electrode tubes were laid.

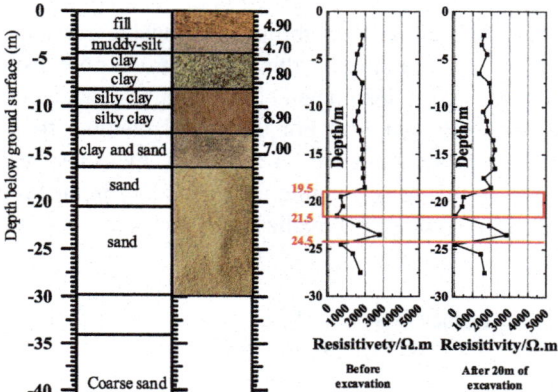

Fig. 14. The detective result of the AMN resistivity method in the seams of WG26-27.

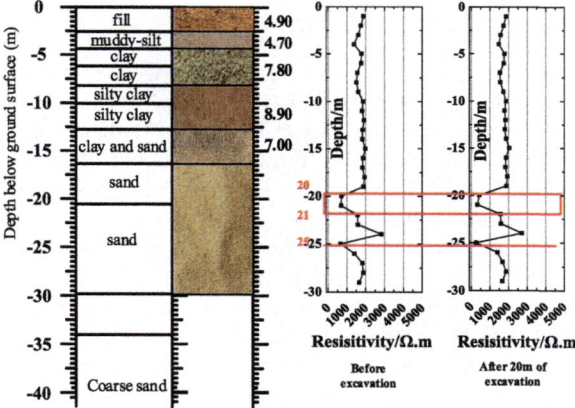

Fig. 15. The detective result of the AM resistivity method in the seams of WG26-27.

According to actual conditions, the electrode tube distance and electrode pole distance are set to 500 mm and 1000 mm, respectively, and 30 needle electrodes were laid in each electrode tube. The results indicate that significant low resistance and suspected leakage area are detected in depth of 19.5–21.5 m and 24–25 m position by the AM method and the AMN method. As a result, the construction unit took the plugging measures in time to prevent this. Field test results further verify the feasibility and effectiveness of the AM method and the AMN method proposed in this paper.

5 Conclusion

Theoretical formula of detection range using the micro-logging method was deduced in this program. Then detailed comparisons of the numerical results with the theoretical data were presented. The influences of different parameters on the sensitivity of crack detection were also analyzed. In seepage detection of the diaphragm wall, the AM and AMN methods can obtain a good effect through reasonable parameters selecting. For the AM method, the influence of crack width and electrode pole distance on detection range is greater than that of crack length and electrode tube distance. The electrode pole distance should be chosen in practice. For the AMN method, the crack length and electro distance have a greater influence on detection. Field test results indicate that reliability and accuracy of the micro-logging method theory in the seepage detection of the diaphragm wall is effective and realizes the equivalent simplification of the detection process.

Acknowledgements. The financial supports of the National Natural Science Foundation of P.R. China (No. 41427803 and 41672272) are greatly acknowledged.

References

Duan, Q., Shi, J., Wu, D.: Design and realization of foundation pit leakage detection system based on flow field fitting principle. J. Cent. S. Univ. (Sci. Technol.) **47**(12), 4108–4114 (2016)

Feng, J.: Studies on Selecting Soft Soil Foundation Anti-Seepage Measures and Related Testing Methods. Huazhong University of Science and Technology, Wuhan (2011)

Gao, Z.: Hidden Leakage Diseases Prospecting of Waterproof Curtain of Foundation Pit Using Geophysical Methods in Soft Soil Region. Ocean University of China, Qingdao (2013)

Huang, T., Li, P., Duan, Q., Zhang, L., Shi, J.: Development of Foundation Pit Leakage Detector based on Flow-field Fitting Method (2013)

Liu, T., Nie, Y., Hu, L., et al.: Model tests on moisture migration based on high-density electrical resistivity tomography method. Chin. J. Rock Mech. Eng. **33**(1), 234–238 (2015)

Moon, J., Jeong, S.: Effect of highly pervious geological features on ground-water flow into a tunnel. Eng. Geol. **117**(3), 207–216 (2011)

Wang, C., Dong, H., Liu, Z.: An investigation into the monitoring and forewarning technique for dyke ridden trouble under the condition of high water level. Geophys. Geochem. Explor. **4** (25), 294–299 (2001)

Zhu, D.B., Ren, Q.W.: Numerical and physical simulation of resistivity of transverse profiles of vertical cut-off walls. J. Hobai Univ. **32**(4), 410–414 (2004)

Zou, S.J.: The Research on Theory and Application of the Flow-Fitting Method Foe Detection of Piping and Leakage in Dykes and Dams. Central South University, Changsha (2009)

Study on Electrical Parameters and Microstructure of the Life Source-Contaminated Soil

Junchao Zang[1,2(✉)], Xinyu Xie[1,2], and Lingwei Zheng[1,2]

[1] Research Center of Coastal and Urban Geotechnical Engineering,
Zhejiang University, Hangzhou 310058, China
zjc001zju@163.com
[2] Ningbo Institute of Technology, Zhejiang University, Ningbo 315100, China

Abstract. In this study, the life source polluted soils at various seepage depths (with different pollution degrees) were taken as the research object to perform the experiment tests of 150 samples for the determination of basic physical, electrochemical indicators and microstructure characters. The interaction between life source pollutants and clay promotes the increase of acidity, salinity, TDS, electro-conductibility, as well as the contents of such elements as Al, Fe, Mg, Ca, K and S with a critical depth exists at 23 cm. Such parameters as void ratio, clay content, and saturated permeability coefficient of the polluted soils are positively correlated to the seepage depth above the critical depth, while dry density are negatively correlated to seepage depth. Besides, a series of laboratory tests was conducted in a Miller Soil Box to investigate the electro-osmosis reinforcement properties of life source-contaminated soil. This study determined that the rank of the primary factors affecting the electro-osmosis treatment effect was voltage, energizing time, initial moisture content, initial pore water salinity and electrode material. The energy consumption coefficient increased by 100% after 20 h.

Keywords: Drained consolidation · Electro-osmosis · Orthogonal design
Contaminated soil · Ion transport

1 Background

Contaminated soil contains organic contamination and inorganic pollutants, and the former is caused by heavy metal ions. Most of the existing related studies focus on the gathering and detoxification of heavy metal ions around electrodes [1]. Currently, studies on life source-contaminated soil focus on engineering properties, settlement, seepage, stabilization and diffusion [2, 3]. However, there are a few studies on the indoor drained consolidation of life source-contaminated soil. Due to the acceleration of urban construction, increasing amounts of life source-contaminated soil have developed in construction sites and the properties of the contaminated clay change significantly, including the relative concentration, liquid and plastic limit, void ratio, etc. Kaniraj, S.R. performed a series of electro-osmotic consolidation experiments on organic soil and found that it was considerably different from normal clay [4].

Furthermore, significant research has been devoted to experimental studies and the engineering applications of electro-osmosis. Esrig created the theory of electro-osmosis

© Springer Nature Singapore Pte Ltd. 2018
A. Farid and H. Chen (Eds.): GSIC 2018, *Proceedings of GeoShanghai 2018*
International Conference: Geoenvironment and Geohazard, pp. 373–385, 2018.
https://doi.org/10.1007/978-981-13-0128-5_42

drainage [5]. Gong studied the effects of voltage, electrode materials, salinity and other factors on electro-osmotic drainage [6]. Li conducted a thorough inquiry of the effects of voltage on electro-osmotic drainage [7, 8]. However, we cannot determine the electro-osmosis reinforcement properties of life source-contaminated soil based on the study of normal clay. This article presents the indoor drained consolidation of life source-contaminated soil and optimal parameter combinations as well as sorting electro-osmosis reinforcing factors. The objective of this report is to provide guidance regarding the treatment of life source-contaminated soil in practice.

2   Experimental Design and Process

2.1   Clay Sample Preparation

In practice, it is hard to get the life source contaminated soil of landfill. So that, this article prepared life source-contaminated soil through device in Fig. 1 and simulated pollutant seepage process of uncontaminated clay.

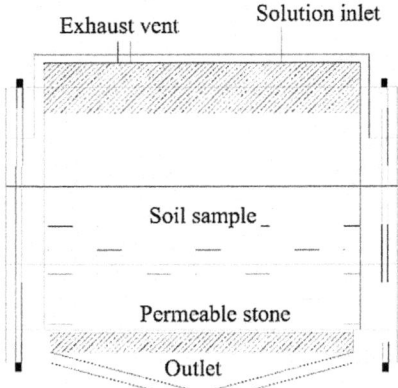

Fig. 1. Preparation device for the life source-contaminated soil.

Basic parameters of uncontaminated soil are listed in Table 1. The composition and content of garbage are listed in Table 2.

Table 1. Basic parameters of uncontaminated soil

Parameter	Plastic limit $\omega_P/\%$	Liquid limit $\omega_L/\%$	Plasticity index $I_P/\%$	Natural water content/%	Natural density $\rho/(\mathrm{g/cm^3})$	Optimal water content/%	Maximum dry density/ $(\mathrm{g/cm^3})$
Value	17.20	34.25	17.05	23.00	1.64	22.00	1.69

Element type and content/%							
Na	Mg	Al	Si	K		Ca	Fe
0.501	0.894	8.301	29.717	1.716		0.896	3.887

Table 2. Main compositions and content of garbage

Components	Organic fractions (%)			Inorganic fractions (%)			Mixed fractions (%)
	Food	Vegetable	Leaves	Paper	Slag	Ash	Soil
Value	15.00	61.00	4.50	3.40	4.40	8.30	3.40

2.2 Basic Properties of Life Source Contaminated Soil at Various Seepage Depths

Changes of Soil Water Content with the Depth of Seepage

In order to reflect the true natural water content of the compacted clay layer in the model, only the average of the soil water content test data in the initial experimental stage is analyzed.

The variation of water content in different seepage depths is not monotonic, there is a critical depth. The water content reduced gradually when the critical depth is exceeded. Changes of soil water content with the depth of seepage is shown in Fig. 2.

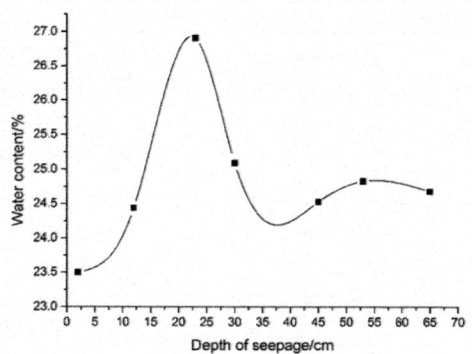

Fig. 2. Changes of soil water content with the depth of seepage.

Changes of Plasticity Index with the Depth of Seepage

The plastic limit and liquid limit of the contaminated soil are determined by the photoelectric liquid and plastic limit joint tester, taking the depth of 2 mm corresponding to the water content as the plastic limit, taking 10 mm corresponding to the water content as the liquid limit, the difference between the two parameter is plasticity index.

Plastic limit, liquid limit and plasticity index show a decreasing trend with the increase of seepage depth. But the growth rate decreases with the depth of seepage, which can reflect the influence degree of contaminants to clay in the model device decreases with the increase of seepage depth. Changes of plasticity index with the depth of seepage is shown in Fig. 3.

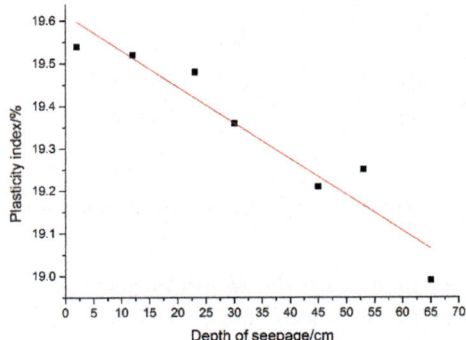

Fig. 3. Changes of plasticity index with the depth of seepage.

Changes of PH with the Depth of Seepage

The pH of the life source contaminated soil was tested using a soil-water mixed solution with a ratio of 1: 5. The soil of different seepage depth was taken through 2 mm sieve, weighing 10 g. Then 50 ml of deionized water was added and allowed to stand for 24 h. The supernatant was measured using a pH meter.

As can be seen from Fig. 4, pH is positively correlated with seepage depth, shallow soil is weak acid, deep soil is neutral, and the pH of the leaching solution in the shallow and deep contaminated soils increased slowly.

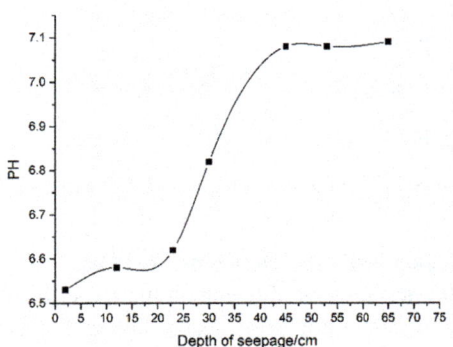

Fig. 4. Changes of PH with the depth of seepage.

Changes of Resistivity with the Depth of Seepage

The resistivity of soil is a combination of factors such as soil composition, structure and external conditions [9–12]. This paper mainly considers the relationship between the pollutant content and the resistivity of different soil samples with different seepage depth. Resistivity test using DUK-2A quadrupole sounding high-density electrical measurement system, the electrode arrangement for the four-pole device. Take the different depth of life of the source of pollution of soil samples, which is 200 cm³ ring knife size.

On the whole, with the increase of the depth of seepage, the resistivity of life source contaminated soil increased gradually. In general, the resistivity of soil is negatively correlated with water content and is positively correlated with dry density. Considering the characteristics of water content and dry density of soils with different seepage depth, water content increased gradually and dry density decreases with the depth of seepage depth. In theory, the resistivity of soil should be accelerated under the influence of double factors. The regularity of shallow soil layer is opposite to the theoretical trend while the rate of increase in the resistivity of the deep soil layer is gentle. Based on this, it is speculated that the resistivity of soil will no longer be dominated by water content and dry density, and the pollution degree of soil determines the change trend of soil resistivity. The smaller the depth of seepage, the higher the degree of pollution. Changes of resistivity with the depth of seepage is shown in Fig. 5.

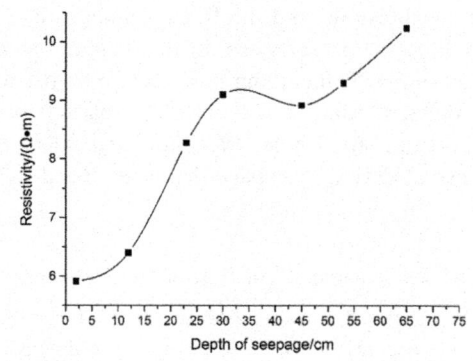

Fig. 5. Changes of resistivity with the depth of seepage.

Changes of Permeability Coefficient with the Depth of Seepage

Infiltration characteristics research south-55 variable water head permeameter, getting saturated infiltration coefficient of life source contaminated soil. Soil sample should be putted in saturated box in advance to vacuum pumping about 2 h. In order to reduce the error of evaporation, adds a controlled trial to test the role of reading period room temperature water head loss values. Permeability coefficient of different flow depth are shown in Fig. 6, in which k is filtration coefficient, d is seepage depth.

It can be observed from Fig. 6 that life source contaminated soil has a low saturated permeability which varies with depth. A large value in the vicinity of 23 cm seepage depth and corresponds to critical depth, permeability and reduced gradually reach a stable trend when the depth exceeds.

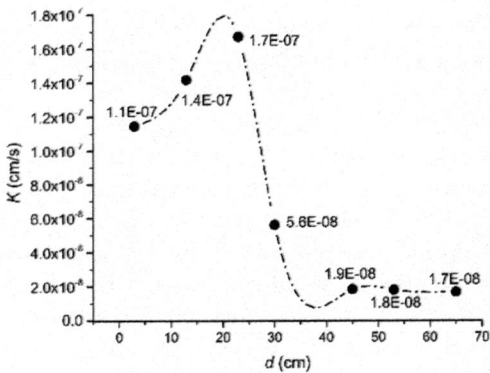

Fig. 6. Permeability coefficient of different flow depth.

Quantitative Analysis of Microstructure

In this paper, SEM images are enhanced, denoised, segmented and binarized by MATLAB. The microstructure parameters of life source contaminated soil was calculated use the appropriate threshold and the fractal dimension of the microstructure image was extracted by Fractal Fox software. In this paper, we mainly analyzed the microstructure parameters of 500X, including pore area, pore perimeter, apparent pore ratio and fractal dimension. According to the size of the particle, large hole (>20 μm), medium size holes (5–20 μm), small holes (2–5 μm) and micro-holes (<2 μm). The microstructural parameters of different seepage depth are listed in Table 3.

Table 3. Quantitative Analysis of Microstructure

Depth of seepage/cm	Pore area/um^2	Circumference of pore/um	Apparent porosity	Fractal dimension	Correlation coefficient
3	21062	16278	0.0239	1.8189	0.9368
23	25649	19748	0.0292	1.6551	0.9416
53	10494	8033	0.0117	1.8717	0.9353
65	7566	6441	0.0084	1.916	0.9339

The total amount of voids increases first and then decreases with the increase of seepage depth, the maximum value appears at a seepage depth of 23 cm. The total amount of voids at a seepage depth of 45 cm or 65 cm is half of seepage depth of 23 cm or 3 cm in which is small medium and small pore.

2.3   Design of the Experimental Apparatus

There are some differences in the nature of soil in different seepage depth, the soil at the depth of 23 cm is selected for electroosmotic consolidation test to investigate the electro-osmosis reinforcement properties of life source-contaminated soil. This article sampled and measured their basic properties, as shown in Table 4.

Table 4. Basic parameters of the life source contaminated soil

Parameter	Plastic limit $W_P/\%$	Liquid limit $W_L/\%$	Friction angle $\psi/°$	Cohesion c/kPa	Unit weight $\gamma/$ (kN/m^3)	Dry density $\rho_d/(g/cm^3)$
Value	23.4	42.6	20.7	19.1	19.1	1.50

The test box model consists of a Miller Soil Box and a sedimentation measuring device [13] The inner edge of the polymethyl methacrylate box is 186 by 100 by 97 mm. The settlement combination measuring device includes a mounting bracket, adjustable hanging rod and dial indicator. The dial indicator is fixed to the adjustable hanging rod, and the lower end is pressed against the Plexiglas sheet, which is 168 by 94 mm. The experiment uses a plate electrode, which is 97 by 97 by 2 mm. The outside of the cathode plate is wrapped with geotextiles, and several holes are drilled evenly on the negative plate using a drilling machine. The displacement is obtained using electronic scales (Fig. 7).

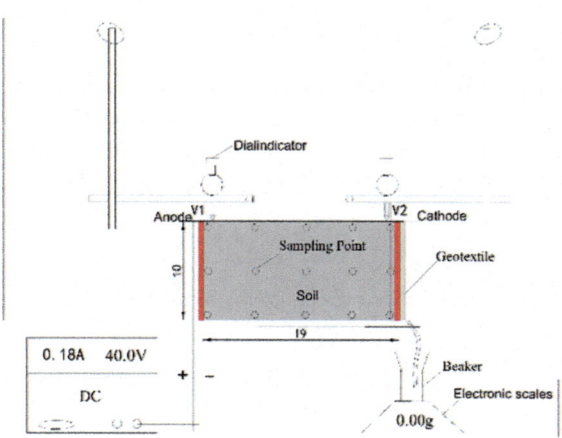

Fig. 7. Schematic diagram of electro-osmosis test device.

2.4 Design of the Test Plan

To determine the optimal level of electro-osmosis for the life source-contaminated soil, a reasonable range of the different factors using 9 groups of preliminary experiments was determined. Consequently, the conditions for formal electro-osmosis are listed in Table 5.

Table 5. Factors and level of the formal electro-osmosis

Factors	A: Power voltage (V)	B: CaCl$_2$ supplementation (g/L)	C: Electrode material	D: Power-on time (h)	E: Initial moisture content (%)
Level	A1:30.0	B1:0.0	C1:Iron	D1:15.0	E1:42.0
	A2:40.0	B2:0.4	C2:Copper	D2:20.0	E2:48.0
	A3:50.0	B3:0.8	C3: Aluminium	D3:25.0	E1:42.0
	A4:58.0	B4:1.2	C4:Graphite	D4:30.0	E2:48.0

The primary experimental process can be given as follows: (1) the geotextile is placed in a suitable position; (2) the soil specimen is compacted by layer; (3) the electric potential arrow is placed approximately 5 mm away from the electrode plates; (4) the power supply is connected, and the voltage is adjusted; (5) the voltage, electric current and displacement are measured every hour.

3 Analysis of the Orthogonal Test

3.1 Range of Parameters for Electro-Osmosis

To thoroughly evaluate the drainage effects and energy consumption [14–16], the percentage of the water content reduction (ratio of changes before and after electro-osmosis), shear strength and energy consumption of the unit displacement was selected, and the range of different factors [17] was analyzed. The test results are listed in Table 6.

Table 6. Summarized results of the formal electro-osmosis

Experiment number	Experiment results		
	ω_r (%)	Ω_u (10^{-3}kW $\times$ h/L)	τ (kPa)
Optimal groups	F9, F6	F1	F9, F6

The effective potential decreases with time during the electro-osmotic consolidation process, which varies with different experimental conditions. We can obtain a more accurate consumption of energy using a unit energy consumption.

The factors and level of displacement using the unit energy consumption are illustrated in Fig. 8.

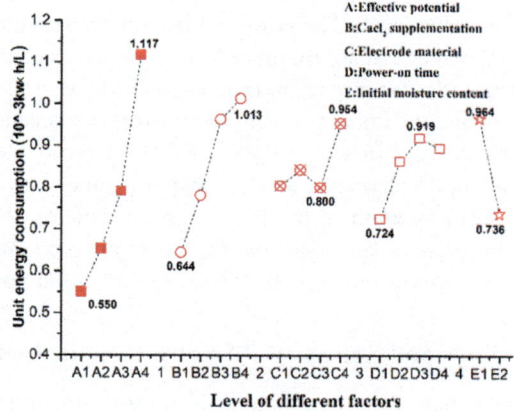

Fig. 8. Factors and level of displacement by unit energy consumption.

Based on the range analysis, the rank of the effect of each factor on the energy consumption per unit displacement of electro-osmosis can be given as follows: power voltage > CaCl$_2$ supplementation > initial moisture content > power-on time > electrode material.

The unit energy consumption of 40 V increased by 2% compared to 30 V whereas the unit energy consumption of 58 V increased by 50% compared to 50 V. The power consumption increased with the power voltage, and considering the drainage effect, it was determined that we could obtain enough displacement through less energy consumption when the range of voltage was controlled between 40 V–50 V.

The power-on time and electrode material have similar reinforcement effects; the power consumption of the aluminum electrode is significantly less than that of the other three electrode materials.

Using the comprehensive analysis method for the orthogonal experiment, this article created a comprehensive evaluation index using different percentages attached to different indicators. The calculation results are provided in Table 7.

Table 7. Summarized results of the electro-osmotic reinforcement using the comprehensive range and the optimal level

Factors	A: Power voltage	B: CaCl$_2$ supplementation	C: Electrode material	D: Power-on time	E: Initial moisture content
Comprehensive range	21.3 V	8.6 g/L	7.641	9.2 h	8.953%
Sequence	1	4	5	2	3
Optimal level	50 V	0.2 g/L	Iron	20 h	48%

Based on a comprehensive analysis of the above three indicators, the primary conclusions can be stated as follows: (1) The power voltage plays a dominant role, reaching an optimal power supply voltage of approximately 50 V considering the drainage effects and energy consumption; (2) The power-on time is approximately 20 h; (3) The initial moisture content has the largest impact on the electro-osmosis because water is easy to discharge under a high initial moisture content; (4) The F6 and F9 test groups have similar conditions with optimal levels, and the experimental data indicates that the drainage volume and shear strength of the F6 test group remains at the optimal level. Additionally, the electro-osmotic effects of the F9 test group are similar to the optimal level due to the high initial moisture content caused by the initial poor drying effects.

3.2   Study on the Effects and Efficiency of Electro-Osmotic Consolidation

Firstly, this article evaluates the electro-osmosis reinforcement effects considering parameter drainage, drainage rate and soil settlement. Secondly, this article selects a coefficient of energy consumption to evaluate the electro-osmotic efficiency. Thirdly, this article analyzes the variation in the corresponding evaluation parameters in the electro-osmotic consolidation process comparatively.

Settlement
Figure 9 illustrates the soil specimens' anode and cathode settlement graph for all of the test groups; the settlement of the anode is greater than that of the cathode with a gap of approximately 1–2 mm. The settlement of the initial moisture content of 42% exceeded that of the case of 48%.

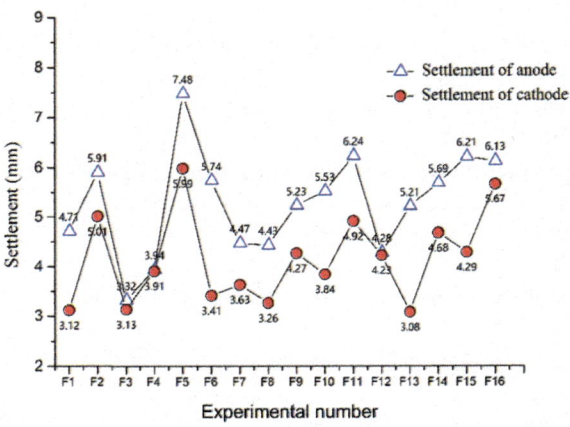

Fig. 9. Anode and cathode settlement graph for sixteen groups of soil samples.

Apart from considering the impact of the electrode material and supply voltage, this article analyzes the consolidation process using different combinations of tests.

We can conclude that the soil settlement growth decreases with time, and its curve shape is similar to the preloading method's consolidation settlement curve, which has a similarly curved shape with a logarithmic function. In the first few hours, the soil settlement value and the increasing rate of the cathode exceed that of the anode. Lastly, this analysis obtained a gap of approximately 0.65 mm—1.08 mm. This result occurs because the water around the cathode drains out first. Consequently, the soil near the cathode generates the settlement first. When water from the anode is transported to the cathode, the sedimentation rate of the anode exceeds that of the cathode. The time that the sedimentation rate anode exceeds that of the cathode decreases as the voltage increases.

Energy Consumption Coefficient

The electro-osmotic drainage reinforcement efficiency is the required power energy of the unit volume, and the energy utilization ratio of the electro-osmotic drainage consolidation process comprises the part used in soil consolidation, which are different from each other [18, 19]. The introduction of the coefficient of energy consumption C can reflect the consumption of electricity needed for the discharge unit volume of water as follows:

$$C = \frac{UI_{t_1 t_2}(t_2 - t_1)}{V_{t_2} - V_{t_1}}. \tag{1}$$

where U is the power voltage/V; $I_{t_1 t_2}$ is the average current from t1 to t2/A; V_{t_1} and V_{t_2} are the cumulative volume of water at t_1 and t_2, respectively; t_1 and t_2 is the energization time.

As indicated in Fig. 10, the energy consumption coefficient change of the F2, F7, F15, and F16 test groups increases slowly before 19 h of electro-osmosis at lower levels. Then, the energy consumption coefficient increases rapidly [20]. For example, in the F16 experiment test group, 98% of the water is discharged, which consumes 70% of the total energy. It is not economical to continue the test after 19 h.

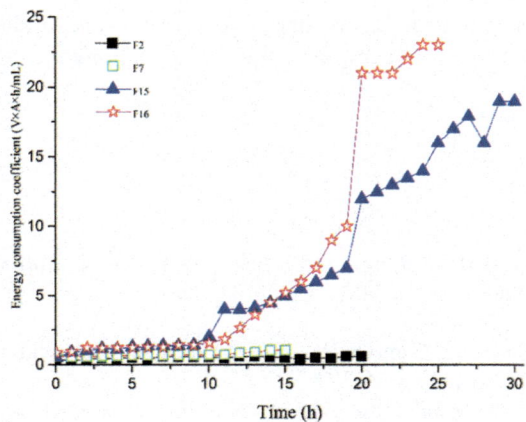

Fig. 10. Chart of energy consumption coefficient change.

4 Conclusions

In this study, the life source polluted soils at various seepage depths (with different pollution degrees) were taken as the research object to perform the experiment tests of 150 samples for the determination of basic physical, electrochemical indicators and microstructure characters. The primary conclusions can be listed as follows:

(1) A critical depth exists at 23 cm. Such parameters as void ratio, clay content, and saturated permeability coefficient of the polluted soils are positively correlated to the seepage depth above the critical depth, while dry density are negatively correlated to seepage depth, with their opposite trends below the critical depth.

(2) With the increase of seepage depth, such plasticity parameters of polluted soils as plastic and liquid limit, plasticity index generally increase, and however such microstructure parameters as pore number, pore size, pore perimeter, apparent void ratio increase initially and decrease afterwards, which is exactly opposite to the variation trend of the fractal dimension negatively correlated to soil permeability;

Using 16 groups of formal tests, this study focused on five factors affecting the effects of electro-osmosis reinforcement. Therefore, this study obtained the order and optimal level of the electro-osmosis effects on life source-contaminated soil, which can be listed as follows: power voltage, power-on time, initial moisture content, CaCl2 supplementation, electrode material. The primary conclusions can be listed as follows:

(1) The range of the power voltage has the highest level in different experiment indexes, which has a critical effect on the test, and its optimal level is approximately 50 V.

(2) The energy consumption coefficient increased by 100% after 20 h which indicating that the power-on time should be designed based on the duration and energy consumption.

(3) The impact of the $CaCl_2$ supplementation is similar to that of the electrode material, and its optimal level is approximately 0.2 g/L.

(4) For the electrode material, the drainage effects of iron are similar to that of copper, which has a good performance compared to the aluminum electrode and graphite electrode.

References

1. Kjeldsen, P., Barlaz, M.A., Rooker, A.P., Baun, A., Ledin, A., Christensen, T.H.: Present and long-term composition of MSW landfill leachate: a review. Crit. Rev. Environ. Sci. Technol. **32**, 297–336 (2002)
2. Gabr, M.A., Valero, S.N.: Geotechnical properties of municipal solid waste. Geotech. Test. J. ASTM **18**, 241–251 (1995)
3. Yeung, A.T., Hsu, C., Menon, R.M.: Physicochemical soil-contaminated interactions during electrokinetic extraction. J. Hazard. Mater. **55**, 221–237 (1997)

4. Kaniraj, S.R., Yee, J.H.S.: Electro-osmotic consolidation experiments on an organic soil. Geotech. Geol. Eng. **29**, 505–518 (2011)
5. Esrig, M.I.: Pore pressure, consolidation and electro-kinetics. J. Soil Mech. Found. Div. ASCE **94**, 899–921 (1968)
6. Tao, Y.L., Zhou, J., Gong, X.N.: Experimetal study on function mechanism of electrode materials upon electro-osmotic process. J. Zhejiang Univ: Engineering Science **48**, 1618–1623 (2014)
7. Li, Y., Gong, X.N., Zhang, X.C.: Experimental research on effect of applied voltage on one-dimensional electro-osmotic drainage. Rock Soil Mech. **32**, 709–714 (2011)
8. Bucheli, T.D., Müller, S.R., Siegrun Heberle, A., Schwarzenbach, R.P.: Occurrence and behavior of pesticides in rainwater, roof runoff, and artificial stormwater infiltration. Environ. Sci. Technol. **32**, 3457–3464 (2015)
9. Li, D., Tan, X.Y., Wu, X.D., Pan, C., Xu, P.: Effects of electrolyte characteristics on soil conductivity and current in electrokinetic remediation of lead-contaminated soil. Sep. Purif. Technol. **135**, 14–21 (2014)
10. Liu, S., Chen, L., Han, L.: Experimental study on electrical resistivity characteristics of contaminated soils. Can. J. Criminol. **34**, 3–4 (2008)
11. Dong, X., Bai, X., Lv, Y.: Experimental study on the electrical resistivity of cemented soil contaminated by vitriol. In: International Conference on Remote Sensing, Environment and Transportation Engineering, pp. 4975–4978. IEEE (2011)
12. Liu, S.Y., Du, Y.J., Chen, L., Liu, Z.B., Jin, F.: Application of electrical resistivity for cement solidified/stabilized heavy metal contaminated soils. In: Advances in Environmental Geotechnics, pp. 259–264 (2010)
13. Burnotte, F., Lefebvre, G., Grondin, G.: A case record of electro-osmotic consolidation of soft clay with improved soil electrode contact. Can. Geotech. J. **41**, 1038–1053 (2004)
14. Zhuang, Y.F., Wang, Z.: Interface electric resistance of electro-osmotic consolidation. J. Geotech. Geoenviron. Eng. **133**, 1617–1621 (2007)
15. Harris, A., Nosrati, A., Addai-Mensah, J.: The effect of clay type and dispersion conditions on electroosmotic consolidation behaviour of model kaolinite and na-exchanged smectite pulps. Chem. Eng. Res. Des. **101**, 56–64 (2015)
16. Fourie, A.B., Johns, D.G., Jones, C.F.: Dewatering of mine tailings using electrokinetic geosynthetics. Can. Geotech. J. **44**, 160–172 (2007)
17. Lefebvre, G., Burnotte, F.: Improvements of electro-osmotic consolidation of soft clays by minimizing power loss at electrodes. Can. Geotech. J. **39**, 399–408 (2002)
18. Alshawabkeh, A.N., Sheahan, T.C., Wu, X.: Coupling of electro-chemical and mechanical processes in soils under DC fields. Mech. Mater. **36**, 453–465 (2004)
19. Wang, J., Ma, J.J., Liu, F.Y., Mi, W., Cai, Y., Fu, H.T., Wang, P.: Experimental study on the improvement of marine clay slurry by electro osmosis-vacuum preloading. Geotext. Geomembr. **44**, 615–622 (2016)
20. Otsuki, N., Yodsudjai, W., Nishida, T.: Feasibility study on soil improvement using electro-chemical technique. Constr. Build. Mater. **21**, 1046–1051 (2007)

Effect of Sorption Induced Swelling on Gas Transport in Coal

Renato Zagořščak[(⊠)] and Hywel Rhys Thomas

Geoenvironmental Research Centre (GRC), School of Engineering,
Cardiff University, The Queen's Buildings, The Parade, Cardiff CF24 3AA, UK
ZagorscakR@cardiff.ac.uk

Abstract. In this study, an investigation of carbon dioxide sorption induced coal swelling and its effects on gas transport in coal is shown. The model presented is based on an existing coupled thermal, hydraulic, chemical and mechanical (THCM) model. A series of numerical simulations dealing with high pressure carbon dioxide injection in coal sample is presented. In particular, the effect of carbon dioxide sorption induced swelling on permeability evolution and gas breakthrough is investigated. Different cases are considered accounting for the difference in coal seam properties and its sorption characteristics. Under the conditions considered, it is demonstrated that the permeability response of coal to gas is affected by the carbon dioxide sorption induced volumetric strain. The results suggest that medium and high porous coals that swell gradually over the range of pressures considered in this work would lose a smaller portion of injectivity during gas injection, compared to low porous coals that swell significantly at low pressures, allowing quick breakthrough of gas through the domain.

Keywords: Carbon sequestration · Coal swelling · Coupled modelling

1 Introduction

Sequestration of carbon dioxide in deep, unmineable coal seams is one of the promising technologies to mitigate the climate change. Storage of carbon dioxide into a coal seam can also enhance the recovery of methane from the seam offsetting the costs of carbon dioxide capture, transport and injection. Numerous studies have shown that coal can hold at least twice the volume of CO_2 as CH_4 [1, 2]. The depth interval for CO_2 storage in coal is between 300 and 1500 m of depth where CO_2 predominantly exists in its supercritical state [2]. It was estimated that the worldwide CO_2 storage in coal seams is large with a potential of storing up to 964 Gt of CO_2 [2, 3].

Although coal seams have a great potential to store CO_2, the presence of a sorptive gas such as CO_2 swells the coal matrix leading to porosity and permeability reduction under in situ conditions [4]. It has been demonstrated both experimentally and in situ that such technical issue represents a major challenge before putting a large-scale CO_2 enhanced coal bed methane (CO_2-ECBM) project into practice. Also, coals exhibit different porosity and affinity to store gases with respect to rank resulting in limited understanding of coal behaviour under different conditions [5, 6]. Hence, further understanding of coal response to carbon dioxide injection is required.

© Springer Nature Singapore Pte Ltd. 2018
A. Farid and H. Chen (Eds.): GSIC 2018, *Proceedings of GeoShanghai 2018*
International Conference: Geoenvironment and Geohazard, pp. 386–394, 2018.
https://doi.org/10.1007/978-981-13-0128-5_43

In the present work, a theoretical model considering coal swelling induced by carbon dioxide sorption is implemented within the existing thermal, hydraulic, chemical and mechanical (THCM) numerical model. By applying the model under in situ conditions, theoretical changes in permeability and gas breakthrough during supercritical carbon dioxide injection over time are assessed, taking into account volumetric expansion of coal induced by gas sorption. The aim of the present research work is to investigate the impact of selected major parameters affecting the gas transport in coal. Different cases are considered representing variations in coal seam porosity and sorption properties.

2 Constitutive Model

A constitutive model employing the laws of mass conservation and stress equilibrium is implemented within an existing thermal, hydraulic, chemical and mechanical (THCM) numerical model COMPASS developed at the Geoenvironmental Research Centre, Cardiff University by Thomas and co-workers [7, 8].

In the model presented, the continuum is considered to be a two-phase system, consisting of a solid skeleton and pore gas. The deformation behaviour is governed by a constitutive relationship previously developed using an elastic model for highly swelling porous medium. Conditions are considered to be isothermal. Details of the developed model are presented elsewhere [9]. The governing equations are expressed in terms of two primary variables, i.e. gas chemical concentration and displacement.

2.1 Governing Equations

In a single porosity medium, the conservation equation can be expressed mathematically as [7]:

$$\frac{\partial(\theta_g c_g \delta V)}{\partial t} = -\delta V \nabla J_g - \delta V R_g \qquad (1)$$

where t is the time, θ_g is the volumetric gas content, c_g is the gas concentration, δV is the incremental volume, ∇ is the gradient operator, J_g is the total gas flux and R_g represents the sink/source for geochemical reactions. In Eq. (1), the sink/source term is expressed using a retardation factor via solid density and the Langmuir equation [10], since it is assumed that the majority of gas is stored as an adsorbed gas in the solid phase.

Langmuir equation, a common approach for calculating the equilibrium adsorbed amount s_g, can be expressed as:

$$s_g = s_{max}\frac{u_g}{P_L + u_g} \qquad (2)$$

where s_{max} and P_L are the Langmuir constants for the maximum sorption capacity and pressure at which half of the maximum sorption is achieved, respectively.

Following that the net stress is defined as the difference between the total stress and gas pressure, as well as that the equilibrium is achieved when the resultant of the forces in any direction is zero, stress equilibrium can be expressed as [7]:

$$\mathbf{P}d\sigma'' + \mathbf{P}\mathbf{m}du_g + d\mathbf{b} = 0 \tag{3}$$

where $\mathbf{b}$ is the vector of body forces, σ'' is the net stress, u_g is the gas pressure and $\mathbf{P}$ is the strain matrix.

In this work, the assumption is made that coal can be considered as an elastic porous material, which during the increment of stress produces only recoverable strains. Hence, the elastic stress-strain relationship is expressed through a generalized Hooke's law [7]. The elastic component of strain due to sorption induced swelling is expressed through Langmuir equation:

$$\varepsilon_{sw} = \varepsilon_{max} \frac{u_g}{P_L + u_g} \tag{4}$$

where ε_{max} is the Langmuir constant for the maximum volumetric strain.

Chemical equilibrium is assumed to exist between the solid phase and the porous system meaning that any amount of sorption induced swelling is based on the gas concentration within the pores.

In this work, appropriate relationships are employed to consider key gas transport properties, i.e. real gas compressibility, viscosity and diffusivity following the models developed by Peng and Robinson [11], Chung et al. [12] and Reid et al. [13], respectively.

The relationship between the porosity and permeability is expressed using a widely used approach as [14]:

$$\frac{K}{K_0} = \left(\frac{n}{n_0}\right)^3 \tag{5}$$

The expanded governing equations for gas transport and stress equilibrium can be expressed in the following general form:

$$C_{c_g c_g} \frac{\partial c_g}{\partial t} + C_{c_g \mathbf{u}} \frac{\partial \mathbf{u}}{\partial t} = \nabla \left[K_{c_g c_g} \nabla c_g\right] + J_{c_g} \tag{6}$$

$$C_{\mathbf{u}c_g} dc_g + C_{\mathbf{u}\mathbf{u}} d\mathbf{u} + d\mathbf{b} = 0 \tag{7}$$

where C and K matrices are the storage and flux terms, respectively. Binary subscripts are assigned to illustrate how each primary variable may be influenced in the coupled system. The term J represents the flux and $\mathbf{u}$ is the vector of displacement.

2.2 Numerical Solution

The finite element method (FEM) is employed to spatially discretise the system of equations, whereas the finite difference method (FDM) is applied to achieve temporal discretisation. Such method has been previously shown to be suitable for coupled flow and deformation equations [15, 16]. Through application of the Galerkin spatial discretisation approach, the system of differential equations is expressed in matrix form as:

$$\mathbf{A}\varnothing + \mathbf{B}\frac{\partial\varnothing}{\partial t} + \mathbf{C} = \{0\} \tag{8}$$

where $\mathbf{A}$, $\mathbf{B}$ and $\mathbf{C}$ are the matrices of coefficients and $\varnothing$ is the vector of variables.

Many of the fundamental aspects of these equations have been described in detail elsewhere [7, 17, 18].

3 Numerical Simulations

Simulations of supercritical carbon dioxide injection in a large coal sample are performed in this section. The outcome of these simulations is to better understand the major mechanisms which control the reactive transport of CO_2 in coal. A sensitivity analysis is conducted to evaluate the potential impact of coal porosity and Langmuir pressure on the permeability evolution and gas breakthrough. For each parameter, a "base case" value is selected, along with reasonable lower and upper limits. Using such approach, consideration was given to represent the potential variability among coals of different ranks.

3.1 Computational Domain and Material Parameters

The system considered is a 1 m long domain with a 0.5 m height, discretised into 100 equally sized 4-noded quadrilateral elements. A variable time step is used which allows the size of the time step to vary depending on the state of convergence.

The domain is initially saturated with CO_2 at atmospheric conditions. In each simulation, a fixed atmospheric pressure is applied at the outlet boundary, while at the inlet boundary a time-dependent gas concentration is imposed. In particular, gas pressure increases monotonously from atmospheric conditions up to 7.5 MPa over the duration of 3600 s and then remains constant until the end of the simulations. The duration of each simulation is six hours.

All boundaries of the column are restrained for deforming vertically to simulate uniaxial conditions, as expected in situ. Also, the outlet boundary of the sample is fully restrained from deforming horizontally. Conditions are isothermal, with a fixed temperature of 308 K. Selected temperature and injection pressure conditions represent approximate conditions at 750 m below the ground level [19].

A summary of the material parameters, equal in each simulation, is provided in Table 1. Also, cases considered in the sensitivity analysis are shown. Base case values of the porosity and Langmuir pressure are 2% and 2.5 MPa, respectively. The low and high Langmuir pressure cases consider Langmuir pressure values of 0.5 MPa and

Table 1. Material parameters

Material parameters	Value	Reference
Initial permeability, K_0 (m^2)	1.0×10^{-15}	[2]
Elastic modulus, E (GPa)	2.0	[6]
Poisson's ratio, υ (-)	0.35	[6]
Coal density, ρ (kg m^{-3})	1380	[20]
Langmuir capacity, s_{max} (mol kg^{-1})	2.0	[6]
Langmuir vol. strain, ε_{max} (%)	2.0	[6]

Sensitivity analysis parameters					
	Low Langmuir pressure case	High Langmuir pressure case	Base case	Low porosity case	High porosity case
Initial porosity, n_0 (%)	2.0	2.0	2.0	1.5	2.5
Langmuir pressure, P_L (MPa)	0.5	5.0	2.5	2.5	2.5

5 MPa, respectively, while having the value of porosity equal to the base case, i.e. 2%. In a similar manner, low and high porosity cases consider porosity values of 1.5% and 2.5%, respectively, while using the base case Langmuir pressure value of 2.5 MPa.

3.2 Results of the Sensitivity Analysis

Figure 1 shows the temporal evolution of permeability throughout the duration of each simulation. Permeability evolution is assessed at an arbitrary chosen point close to the inlet boundary, i.e. 0.1 m away from the injection point. The results revel that using the base case values for porosity and Langmuir pressure, permeability continuously decreases during the first hour of the simulation and then reaches a value of 5×10^{-17} m^2.

The same final permeability value is predicted for the low Langmuir pressure case, however, with a different shape of the curve during the first hour of simulation. In particular, the reduction in permeability occurred quicker achieving the final value of permeability after half an hour into the simulation.

Between the high Langmuir pressure and high porosity cases, a small difference between the slopes of the curves and final permeability values is predicted. In such cases, permeability values are reduced to 1.1×10^{-16} m^2 and 1.4×10^{-16} m^2 after one hour into the simulations, respectively.

For the low porosity case, it is predicted that the permeability continuously drops throughout the duration of the simulation resulting in maximum reduction in the permeability value of 1.5×10^{-17} m^2 at the end of the simulation.

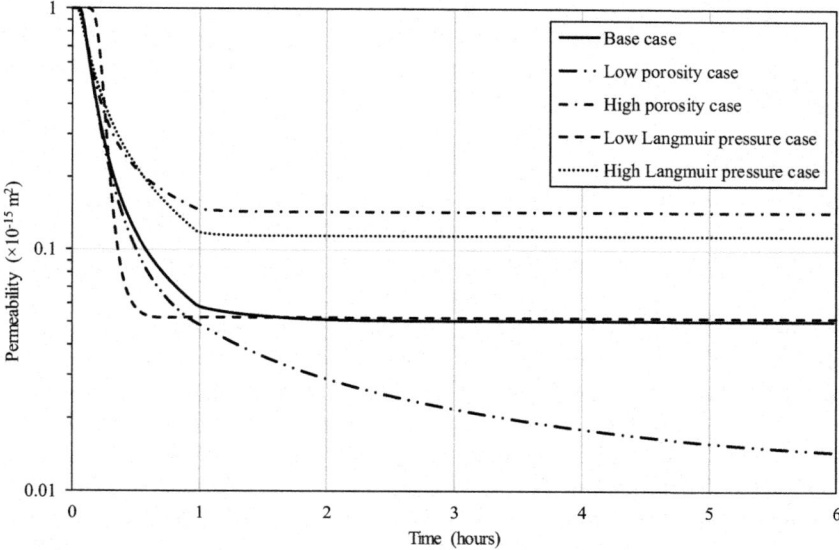

Fig. 1. Predicted permeability evolution for five different combinations of initial porosity and Langmuir pressure, evaluated 0.1 m from the injection point.

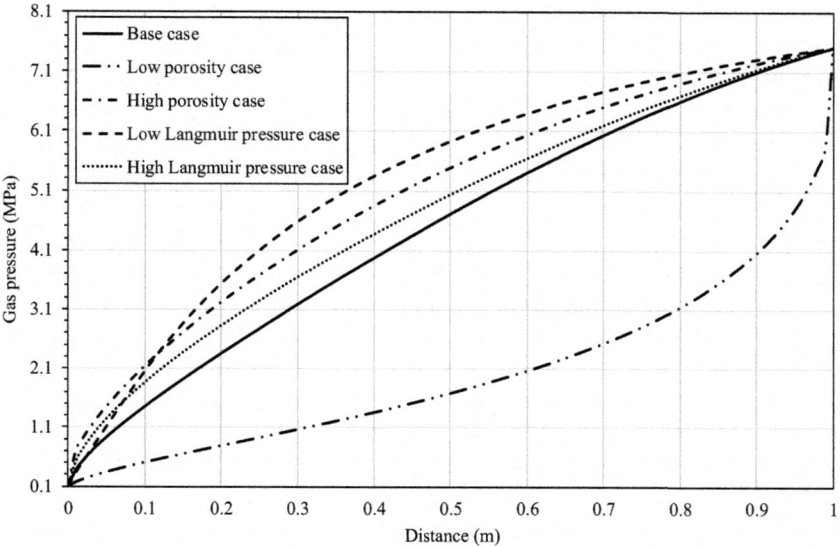

Fig. 2. Profiles of CO_2 in the domain at the end of simulations for five different combinations of initial porosity and Langmuir pressure.

Based on such observations, the Langmuir pressure parameter has a greater effect on the shape of the curve than on the final reduction value. This is related to the fact that for coals with low value of P_L, half of the sorption induced swelling occurs at low pressures reducing the permeability significantly at early stages of injection. Hence, at high pressures when the sorption induced swelling is almost complete, difference in absolute permeability values between low P_L and high P_L cases would be negligible.

Opposite to the Langmuir pressure parameter, the initial porosity has a greater effect on the final reduction value than on the slope of the curve. In other words, for the same amount of the sorption induced swelling to be accommodated, smaller pore volumes are more affected than larger pore volumes.

In order to investigate the effect of permeability reduction on gas breakthrough, pressure profiles across the domain at the end of the simulations for each case considered are assessed and presented in Fig. 2. The results show that all cases, except the low porosity case, exhibit non-linear profiles, typical for highly compressible gases, with more than 4.7 MPa of gas pressure at the middle of the domain. This suggests a near-complete breakthrough of CO_2, i.e. steady-state, throughout the domain at the end of simulation time. The significant reduction in permeability for the low porosity case resulted in limited gas flow through the sample where the gas pressure of 1.67 MPa was observed half way between the injection and the abstraction points.

4 Conclusions

A sensitivity analysis was conducted to investigate the response of coal subject to supercritical carbon dioxide injection through a series of numerical simulations. The base case as well as lower and upper values of the coal porosity and Langmuir pressure were selected to examine the influence of such parameters on the permeability evolution and gas breakthrough in coal. The numerical modelling results suggest that coals with low Langmuir pressure experience strong reduction in permeability in the early stages of gas injection. Also, the results demonstrate that low porosity is a parameter having the strongest influence on the final permeability and having the largest impact on the gas flow throughout coal. Hence, based on the results of this study, coals with high value of the Langmuir pressure and large volume of pores available for flow are expected to be the least affected by the coal swelling and would offer a stable injection of CO_2.

This study offers useful information on the importance of various factors on coal response to CO_2 injection providing an enhanced understanding of the coupled processes during carbon sequestration. These factors strongly vary among coals of different ranks, hence, it is recognised that reliable parameter measurement and determination is crucial for assessing their impact. Continued research is required to incorporate additional controlling factors that affect coal behaviour and obtain an optimum combination of such factors which would result in the greatest carbon storage potential.

Acknowledgements. The financial support from WEFO, for the first author, is gratefully acknowledged.

References

1. Jones, N.S., Holloway, S., Smith, N.J.P., Browne, M.A.E., Creedy, D.P., Garner, K., Durucan, S.: UK Coal Resource for New Exploitation Technologies. DTI Report No. COAL R271, DTI/Pub URN 04/1879. DTI Cleaner Coal Technology Transfer Programme, Harwell (2004)
2. White, C.M., Smith, D.H., Jones, K.L., Goodman, A.L., Jikich, S.A., Lacount, R.B., Dubose, S.B., Ozdemir, E., Morsi, B., Schroeder, K.T.: Sequestration of carbon dioxide in coal with enhanced coalbed methane recovery – a review. Energy Fuels **19**(3), 659–724 (2005)
3. Kuuskraa, V.A., Boyer, C.M., Kelafant, J.A.: Coalbed gas: hunt for quality basins goes abroad. Oil Gas J. **90**(40), 49–54 (1992)
4. Larsen, J.W.: The effects of dissolved CO_2 on coal structure and properties. Int. J. Coal Geol. **57**(1), 63–70 (2004)
5. Rodrigues, C.F., Lemos de Sousa, M.J.: The measurement of coal porosity with different gases. Int. J. Coal Geol. **48**(3–4), 245–251 (2002)
6. Durucan, S., Ahsan, M., Shi, J.Q.: Matrix shrinkage and swelling characteristics of European coals. Energy Procedia **1**(1), 3055–3062 (2009)
7. Thomas, H.R., He, Y.: Modelling the behaviour of unsaturated soil using an elasto-plastic constitutive relationship. Géotechnique **48**, 589–603 (1998)
8. Thomas, H.R., Sedighi, M., Vardon, P.J.: Diffusive reactive transport of multicomponent chemical under coupled thermal, hydraulic, chemical and mechanical conditions. Geotech. Geol. Eng. **30**(4), 841–857 (2012)
9. Zagorščak, R.: An investigation of coupled processes in coal in response to high pressure gas injection. Ph.D. thesis, Cardiff University, Wales, UK (2017)
10. Langmuir, I.: The adsorption of gases on plane surfaces of glass, mica and platinum. J. Am. Chem. Soc. **40**(9), 1361–1403 (1918)
11. Peng, D.Y., Robinson, D.B.: A new two-constant equation of state. Ind. Eng. Chem. Fundam. **15**(1), 59–64 (1976)
12. Chung, T.-H., Ajlan, M., Lee, L.L., Starling, K.E.: Generalized multiparameter correlation for nonpolar and polar fluid transport properties. Ind. Eng. Chem. Res. **27**, 671–679 (1988)
13. Reid, R.C., Prausnitz, J.M., Sherwood, T.K.: The Properties of Gases and Liquids, 3rd edn. McGraw-Hill, New York (1977)
14. Somerton, W.H., Söylemezoğlu, I.M., Dudley, R.C.: Effect of stress on permeability of coal. Int. J. Rock Mech. Min. Sci. Geomech. Abstr. **12**(5), 129–145 (1975)
15. Thomas, H.R., He, Y.: Analysis of coupled heat, moisture and air transfer in a deformable unsaturated soil. Geotechnique **45**(4), 677–689 (1995)
16. Thomas, H.R., He, Y., Onofrei, C.: An examination of the validation of a model of the hydro/thermo/mechanical behaviour of engineered clay barriers. Int. J. Numer. Anal. Meth. Geomech. **22**, 49–71 (1998)
17. Thomas, H.R., He, Y., Sansom, M.R., Li, C.L.W.: On the development of a model of the thermo-mechanical-hydraulic behaviour of unsaturated soils. Eng. Geol. **41**, 197–218 (1996)

18. Thomas, H.R., Cleall, P.J.: Inclusion of expansive clay behaviour in coupled thermo hydraulic mechanical models. Eng. Geol. **54**, 93–108 (1999)
19. Gensterblum, Y.: CBM and CO_2-ECBM related sorption processes in coal. Ph.D. thesis, RWTH Aachen University, Germany (2013)
20. Connell, L.D., Mazumder, S., Sander, R., Camilleri, M., Pan, Z., Heryanto, D.: Laboratory characterisation of coal matrix shrinkage, cleat compressibility and the geomechanical properties determining reservoir permeability. Fuel **165**, 499–512 (2016)

Experimental Study on Sodium Sulfate Exciting Steel Slag Fine in Muddy Soil

Y. K. Wu[1], T. Han[2(✉)], J. L. Yu[2], and X. S. Hu[2]

[1] Shandong Provincial Key Laboratory of Civil Engineering Disaster
Prevention and Mitigation, Shandong University of Science
and Technology, Qingdao, China
wuyankai2000@163.com
[2] School of Civil Engineering and Architecture, Shandong University of Science
and Technology, Qingdao, China
15764240023@163.com

Abstract. Steel slag is a kind of solid waste generated in iron and steel smelting. It has the potential cementitious characteristics when the steel slag has been milled to the steel slag fine. The steel slag fine (SSF) is mainly composed of silicate phase. It is very useful to improve muddy properties by mixing with cement and SSF. So by this way, the steel slag not only can be fully utilized and avoid polluting the environment but also can save resources by replacing part of the cement. However, the low activity levels of SSF make it difficult to increase its early strength. In this paper, the chemical excitations method was used. As a stimulant, Na_2SO_4 is mixed into SSF to stimulate its activity, making the early strength to meet the engineering requirements. Unconfined compressive strength (UCS) test and x-ray diffraction (XRD) test have been used to analyze the effects. The results show that the strength of cement-soil is greatly improved after the addition of Na_2SO_4 reagent. When cured for 7 days, the strength of SSF-cement soil with Na_2SO_4 has reached basically at the strength of that without Na_2SO_4 in 28 days. It indicates that the incorporation of Na_2SO_4 has an obvious improvement on the early strength. The materials are analyzed by XRD test, and it is found that the addition of activator has the function to promote the hydration process of SSF - cement and to promote the formation of hydration products such as $Ca(OH)_2$ and $CaSO_4$.

Keywords: Sodium sulfate · Steel slag fine · Unconfined compressive strength
X-ray diffraction test

1 Introduction

The Cement-soil is a kind of hydraulic materials with high strength, low compression, low permeability and some other superior characteristics. It has been widely used in the area of treating the soft soil, like industrial, civil construction, highways, railways and airports. However, the problem that the strength of cement-soil is not high enough and the late deformation of the composite foundation is too large in the engineering practice which limits the further popularization and development of the reinforcement technology to cement- soil [1]. It is an important method to add admixture to improve the

© Springer Nature Singapore Pte Ltd. 2018
A. Farid and H. Chen (Eds.): GSIC 2018, *Proceedings of GeoShanghai 2018*
International Conference: Geoenvironment and Geohazard, pp. 395–404, 2018.
https://doi.org/10.1007/978-981-13-0128-5_44

performance of cement-soil. The admixture can accelerate the chemical reaction rate, catalyze the degree of reaction, and regulate the composition in the cement hydration process. And it can also improve the physical and chemical properties of the material and then improve the performance of cement-soil [2].

Steel slag is a kind of waste residue from smelting industry. The main way to use the steel slag as an effective resource is to make it a kind of admixture of cement and concrete after being grinded into the fine. The chemical composition and mineral composition of steel slag fine(SSF) is similar to that of Portland cement clinker, which has hydration activity. Thus the SSF has great potential for its using in cement concrete. However, compared with cement, the SSF has the properties of low activity, low hydration rate and slow development of strength. Therefore, it is used as an auxiliary cementing material in the same way as fly ash and slags. Moreover, it is also much difficult to use it because of its complicated composition and poor cementing property [3].

The gelling activity of SSF originates from the silicates and aluminates and other minerals, which produce a large amount of vitreous body with FeO, MgO and other impurities in the high temperature and quenching production process, resulting in a significant reduction in activity. Therefore, there must be appropriate ways to stimulate the activity of SSF when it is used in the field of cement and concrete [4] Guo et al. [5] found that the humus acid of muddy soil can react with the hydration product of cement, which hinders the formation of cementations material during the curing process of cement- soil and greatly weakens the process of pozzolanic reaction. Therefore, an appropriate alkaline activator will have a dual advantage about neutralizing humus acid and exciting SSF. Shi et al. [6] used the methods of pre-excitation, physical excitation, chemical excitation and thermodynamic to stimulate the activity of SSF. The results show that chemical excitation can achieve satisfactory effects. Sajedi [7], Li et al. [8] used a variety of activators alone to improve the mechanical properties of SSF, and achieved different effects of improvement. Wu et al. [9] believe that the activity of SSF can be excited in the alkaline environment. Han et al. [10] consider that the effect of sodium and sulfate on cement-soil strength plays a super-important role on the theory and practice., although the various methods of excitation are involved and effective in the researchers mentioned above, there are still many unknowns that need to be further studied.

In this paper, the sodium sulfate has been used as the activator of SSF to explore its effect on the strength of muddy SSF-cement soil. The main methods used in the test are the X-ray diffraction (XRD) test, to observe the microscopic reaction product and analyze the excitation mechanism, and the unconfined compressive strength (UCS) test, which is used to analyze the macroscopic features.

2 Experimental Methods

2.1 Experimental Materials

The marine muddy soil is obtained approximately 8.3 m below ground surface at the construction site of Qingdao Economic and Technological Development Zone, Shandong, China. The soft clay is grayish black with slight smell. The properties of the soil are listed in Table 1. The chemical composition of SSF changes greatly with the difference of raw materials and making process. The SSF used in this test is processed by the waste steel slag from Rizhao (Shandong, China) Steel Plant. It is a kind of gray-black fine, with the specific surface area of 402 m^2/kg. The cement is 32.5 # ordinary Portland Cement. The chemical composition and parameters of the SSF and cement are given in Table 2. Activator-sodium sulfate is produced by a reagent factory in Tianjin. The purity is analytically pure and the content is more than 96.0%.

Table 1. The properties of muddy soil

Properties	Values	Properties	Values
Proportion/ds	2.72	Liquid limit W_L/%	32.1
Natural water content ω/%	41.1	Plasticity index I_P/%	16.6
Wet density ρ/(g/cm^3)	1.85	Liquidity index I_L/%	1.522
Dry density ρ_d/(g/cm^3)	1.29	Compression factor α_{2-1}/(MPa^{-1})	0.62
Saturation S_r/%	97.7	Internal friction angle φ/°	2.6
Porosity ratio e	1.18	Cohesion C/kPa	16.9
Plastic limit W_P/%	15.5	Compression modulus E_S/MPa	1.522

Table 2. Chemical composition of SSF and cement

Chemical composition	CaO	Al$_2$O$_3$	SiO$_2$	MgO	Fe$_2$O$_3$	MnO$_2$	SO$_3$	Na$_2$O	P$_2$O$_5$
SSF	45.99	2.55	14.07	4.26	24.15	4.36	—	—	2.6
Cement	65.14	5.03	22.17	4.30	0.51	—	2.70	0.15	—

In the study of chemical excitation, it is found that chemical activators mainly include alkaline and acid excitation. They all activate the SSF by changing the process of mineral formation. The cement and SSF are both alkaline. The pH value of the muddy soil from Qingdao Economic and Technological Development Zone is 8.2. It shows weakly alkaline. For the alkaline excitation, the effect of strongly alkaline excitation has been confirmed, but that of weakly alkaline activator still needs to be further studied. Therefore, the authors select an alkaline activator - sodium sulfate (Sodium sulfate is a neutral salt, but in the case of hydrolysis will produce sulfite and hydroxide ions, so that it shows weakly alkaline) for testing, to study its feasibility as a SSF activator and the effect to improve the muddy soil.

2.2 Specimen Preparation

To dry, grind and sieve the muddy soil collected from the scene. According to the research results of Wu [11], the optimal ratio of cement and SSF in the curing agent is 1:1. On this basis, the mixing ratio of the activator is slightly adjusted. The sum of content of the SSF and the activator is consistent with the cement content. The content of Na$_2$SO$_4$ is controlled at 0.3%, 0.6%, 0.9%, 1.2% and 1.5% of the mass of SSF. The cement content (the ratio of cement quality to muddy quality) is 10%. The test water is tap water, whose weight is the sun of 30% of the muddy soil and 40% of the cementitious material. The mixed cement soil was placed in a standard mold of 70.7 mm × 70.7 mm × 70.7 mm. And the surface was covered with a plastic film after 1 min vibration on the vibrating table. After 48 h, to take the test block out of the mold and place them at a standard curing box for conservation where preset temperature is 20 °C and the relative humidity is 95%. Curing age is set to 7d, 28d and 90d.

2.3   Unconfined Compressive Strength(UCS) Test

The loading equipment is a10 tons of WAW-1000B electrohydraulic servo hydraulic universal testing machine. The loading speed is controlled at 1 mm/min, every group has three test specimens, and the average intensity of the three test specimens is the UCS.

2.4   X - Ray Diffraction (XRD) Test

The XRD test is performed on the experimental materials with the concentration of Na_2SO_4 in the range of 0.3%, 0.9% and 1.5%, age in 7d, 28d and 90d. The material samples are ground into powder and passed through a 0.075 mm square hole sieve. To take some powder to fill in a grooved flat glass plate, and apply a certain pressure to keep the sample sticky and the distribution of particles random. The continuous scanning method is used to test the phase of the material. The starting angle is 5°, the termination angle is 80° and the scanning speed is 0.8 °/s.

3   Test Results and Analysis

3.1   Unconfined Compressive Strength(UCS) Test

The results of the UCS of cement soils cured for 7 days, 28 days and 90 days are shown in Table 3.

Table 3. Unconfined compressive strength measured values

Maintenance age	Control trial		Content of Na_2SO_4				
	Cement (10%)	Cement (10%) + SSF (10%)	0.3%	0.6%	0.9%	1.2%	1.5%
7d	0.306	0.245	0.5022	0.53	0.563	0.5846	0.6076
28d	0.459	0.522	0.8964	0.8104	1.0112	0.9333	1.1105
90d	0.576	1.425	1.2547	1.1661	1.2597	1.3313	1.7175

The data show that the strength of cement soil at early stages declines after mixing 10% of the SSF with 10% cement. Especially the strength of 7 day has even decreased by 20%. But with the curing time increasing, the strength increases gradually, and even beyond the strength of cement-soil without-mixing the SSF. At the curing age of 28 day, it has reached 1.14 times of the strength of cement-soil without-mixing the SSF, and it has a faster growth rate as time goes on. When the curing age comes to 90 day, the strength has increased by as much as 147%. It shows that mixing the SSF to the muddy soil is very beneficial to the soil strength, and also fully demonstrates that SSF's shortcomings of low early activity and the slow development rate of strength exist.

Na_2SO_4 is used as the activator in the experiment. According to the data, the early strength of the cement-soil has obviously improved after the addition of Na_2SO_4, which compensates the low activity of the SSF. At the curing age of 7 day, the strength of the specimens mixing the Na_2SO_4 is the same as the strength of that without Na_2SO_4 at 28d. When there is Na_2SO_4, the strength of the specimens at 28 day has been developed to about 65% of that at 90 day. It has been greatly improved compared to the 37% of the SSF-cement-soil without activator. It can be obtained from Fig. 1 that the strength of the specimens at the curing age of 90 day is gradually increasing with the increase of the content of Na_2SO_4 from 0.6% to 1.5%. When the content of the Na_2SO_4 is 1.5%, the strength is significantly enhanced. It indicates that the Na_2SO_4 as the activator can also be positive to the strength in the late stages. Especially in the content of 1.5%, the early strength and late strength have both improved significantly, which also confirms the success of the test.

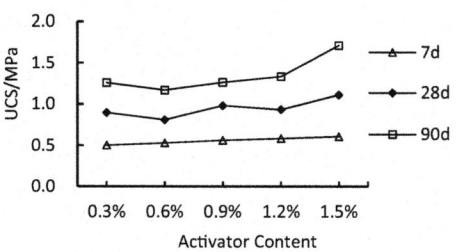

Fig. 1. The relationship between strength and content

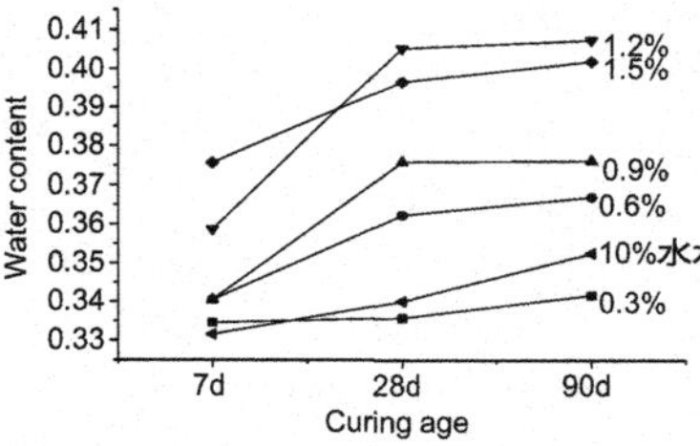

Fig. 2. Changes of water content with curing age

The early strength of the muddy-SSF-cement-soil increases faster than the samples without-mixing the activator. The relations between the specimens' strength and the activator content are shown in Fig. 1. The specimen strength of each curing age is increasing by the increase of the activator. When the content of the activator is constant, the specimen strength increases with the curing age. The strength's changing shows that the Na_2SO_4 can excite the SSF activity effectively.

The water content of cement soil has been also observed while tracking the UCS value of the test pieces. The results show that the water content of the cement soil is increasing with the change of the curing age. The relations between the water content and the curing age are shown in Fig. 2. The water of the cement and SSF is converted from the free state to the combined state in the hydration process. And then it exists in the form of bound water in the hydrated products. The water in this process has been supplemented in the standard curing box after being consumed. Many hydration products, such as C-S-H and AFt, are hydrophilic substances. They absorb a large amount of water in the curing period. So the water content of the specimens gradually increases. In particular, the determination of the moisture content is carried out at a temperature of 105 °C. The AFt will decompose and the bound water will separate and dissipate, so the moisture content will be relatively high. It means that with the increase of hydration products in cement-soil, the water adsorption capacity of the specimen gradually increases.

3.2 Damage Laws

The failure of the SSF-cement-soil specimen is similar to that of the concrete specimen. The trend can be divided into the ascending stage and the descending stage. The ascending phase is composed of elastic deformation and elastic-plastic deformation. The descending stage includes the destruction stage and residual strength stage. Cement-soil is a non-ideal elastomer, so its stress-strain relationship is a nonlinear trend. After loading, there will be an elastic deformation stage where stress and strain have a linear relationship. The range of stress value is about $\sigma \leq 0.8q_u$. And then the curve enters into the elastoplastic stage, where appears irreversible deformation. Later, the specimens gradually yield and reach the peak of the intensity, which is the ultimate compressive strength. The range of stress value is roughly $0.8q_u \leq \sigma \leq 1.0q_u$. After the peak intensity, the specimens start to break. The internal cracks gradually develop and eventually become the penetrating cracks, results that the specimens' strength gradually decreases. But there is no phenomenon of steeply drop in strength, which indicates that the specimens have good ductility. Finally, in the stage of the residual strength, the stress no longer changes as the strain increases, and it keeps at about $0.47q_u$.

The deformation modulus of cement-soil, also known as secant modulus, refers to the ratio of stress to the corresponding strain when the stress and strain have non-linear relationship in the uniaxial compression test [12]. The ultimate deformation modulus E_{su} is the ratio of the stress to the strain at the peak of the curve. The value of E_{su} grows gradually with the increase of the UCS, and it can be regarded as linear growth. The scatter plot is shown in Fig. 4.

The failure strain (ε_f) is the strain value which corresponds to the ultimate compressive strength q_u in the stress-strain relationship of cement-soil [13]. The value of the damage strain reflects the degree of brittleness and toughness of the cement-soil. Figure 3 shows the change trend of the failure strain in the process of UCS's growth.

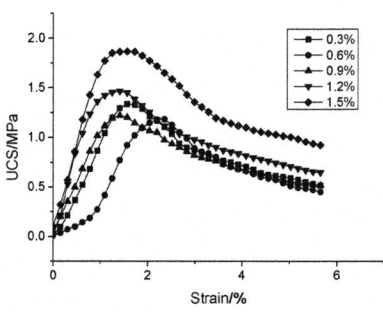

Fig. 3. Stress-strain curves for 90 days

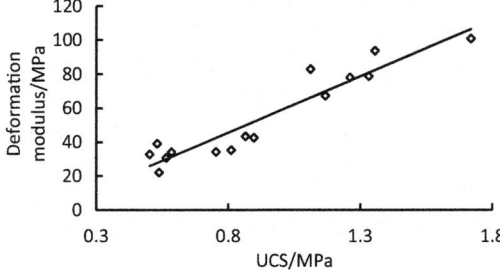

Fig. 4. Change trend of deformation modulus with intensity

The values of failure strain mostly distribute between 2.2% to 3.4%, and they gradually reduce with the UCS. It means that cement-soil will lose some toughness on the basis of increased strength, and the specimen will develop from plastic to brittle.

3.3 X-Ray Diffraction (XRD) Test

The phase changes of the cementitious material before and after mixing the activator have been analyzed by the XRD test, and the relevant reaction mechanism has been judged. The diffraction pattern for curing age of 28d is shown in Fig. 5. The addition of Na_2SO_4 as the activator makes the phase composition of the SSF-cement-soil different. With the increase of the activator, the diffraction peak of the calcium hydroxide (Ca $(OH)_2$ is referred to as CH) in the polymer is increasing. The increase of CH content means that the hydration progress of the SSF-cement has been accelerated. So there presenting more hydration products in the polymer than before at the same curing age. The cementation between the particles is more prominent. From the macroscopic aspect, there are more compact structure and more high intensity.

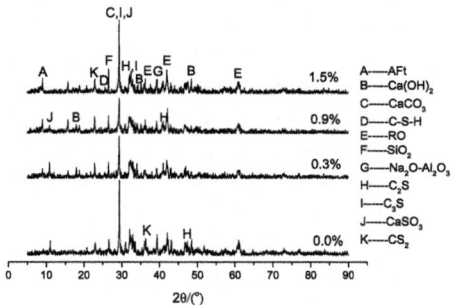

Fig. 5. XRD diffraction patterns of the polymer for 28 days

There is an obvious diffraction peak at 2θ of about 9°, which is ettringite (AFt), increasing with the increase of the Na_2SO_4. AFt and single sulfur-type hydrated calcium sulphoaluminate (AFm) are two kinds of substances that can convert into each other under certain conditions. Their formation is inseparable from the participation of $CaSO_4$. Due to the sufficient sulfur provided by Na_2SO_4, the possibility of producing AFm is reduced. On the contrary, the AFt is an important component of the products. The role of AFt in cement-soil is to bond and expand. Compared to the method of using cement to improve muddy soil, the addition of SSF improves the micro-particle gradation of the mixture. And the large particle diameter of the SSF also increases the pore size between the particles, which enhances the ability of the structure to bear the swelling effect of AFt. So there is no phenomenon of structural damage and reduced strength because of the expansion of the AFt when adding more Na_2SO_4. Contrarily, the specimens have more compact structure and higher strength.

The addition of Na_2SO_4 has introduced a large amount of sodium ions. The chemical reaction produces $Na_2O \cdot Al_2O_3$. It reacts not only with free SiO_2 to form albite ($Na_2O \cdot Al_2O_3 \cdot 6SiO_2$), but also with CH and water to produce C-S-H and NaOH, to further promote the formation of AFt. Consumption of $CaSO_4$ also promotes the positive reaction of Na_2SO_4 and CH, to improve the PH value of the soil.

4   Mechanism Analysis

The particles of the SSF are much larger than that of the cement, which makes the surface energy of the cement larger than that of the SSF. So the cement has the stronger adsorption performance. Because of the low activity of the SSF and the SSF's obstacle effect to the contact of the cement and water, when there is no accelerator, the adsorbed moisture not only fails to hydrate the cement sufficiently, but also leads to the decrease of the specimen strength. After the addition of the activator, the activity of the SSF is stimulated. The SSF can fully hydrate because of the more moisture brought by the cement. And then the SSF particles become smaller after hydration, which makes the surface energy larger and the adsorption of water stronger. It in turn promotes the SSF full hydration. Moreover, there is no weakening to the full hydration of cement. In the whole process, the SSF and cement play a mutual promotion role to promote the hydration process, so the intensity has been greatly improved.

The key of chemical excitation is to create an alkaline environment, where the glass bodies can fully depolymerize and hydrate [6]. The Na_2SO_4 in the water ionizes Na^+ and SO_4^{2+}, and further more SO_4^{2+} hydrolyzes to produce HSO_4^+ and OH^-, which provides a certain alkaline conditions for hydration reaction. It can not only stimulate the hydration speed of the SSF, but also can further promote the process of pozzolanic reaction and other reactions. As we all know, cement hydration can produce C-S-H and CH. So the reaction environment is in alkaline state, which provides the basic conditions for further chemical reactions, lays the foundation for the improvement of the strength of cement-soil. Since the solubility of $CaSO_4$ is less than that of CH, the addition of Na_2SO_4 causes the CH to have more reaction pathways to produce $CaSO_4$ and NaOH. It not only promotes the formation of AFt due to the increase of $CaSO_4$, but increases the alkalinity of the reaction because of the combination of hydroxide and sodium ions. They both further promote the hydration reaction, and constitute a virtuous circle to promote the specimen strength together.

The formation and stabilization conditions of AFt are related to the concentration of $[SO_3]$ in the liquid phase [9]. In the absence of activator, ettringite will be produced in the early hydration. But the sulfate ions can be provided only by the cement, and the SSF contains excess calcium aluminate. So the gypsum will be consumed quickly. And then AFt will convert into a single sulfur-type hydrated calcium sulphoaluminate, which corresponds to the XRD analysis of the AFt's change. Compared with the gypsum as a stimulant, the Na_2SO_4 can provide sulfate ions faster because of the high solubility. So the AFt can be generated directly, avoids the process of mutual conversion between substances. The structure can be stabilized faster, and the specimen strength can be improved steadily. A large number of AFt in the structure plays a role of condensation to form a whole by the various large particles of the polymer together, and to improve the overall properties of the material and the compressive strength of the specimens.

The SSF contains a large amount of RO phases, such as FeO, which exhibit low activity in the alkaline environment in cement-soil. But in the soil, the Fe^{3+} has the best exchange capacity in the common cations. The application value of the SSF will be greatly improved if the RO phases can be rationally exploited and used.

5   Conclusion

In this paper, sodium sulfate is used as the activator. And the effect of the SSF on the strength improvement of the muddy cement-soil has been observed through the contrast between the different specimens with different dosage of Na_2SO_4. The following conclusions are drawn:

1. Sodium sulfate can stimulate the activity of SSF and promote the progress of its hydration process, so the early strength of the SSF-cement-soil can be improved.
2. The sulfate ion ionized from sodium sulfate partially hydrolyzes to produce sulfite and hydroxide ions, which improves the pH value of the mixture and provides an alkaline environment to bring favorable conditions for the activity of the SSF.
3. The formation of AFt in the structure makes the gelled particles in the specimens well connect together, and makes the integrity stronger and the UCS greater. At the same time, the optimum dosage of sodium sulfate is improved due to the specimens' stronger ability to bear the expansion of AFt.

References

1. Wang, W., Zhu, X.: Study on strength property of nanometer silica fume reinforced cemented soil and reinforcement mechanism. Rock Soil Mech. **25**(6), 922–926 (2004)
2. Wang, W.: Study on Reinforcement Mechanism and Damage Performance of Cemented Soil Stabilized with Nanometer Material. Zhejiang University, Zhejiang (2004)
3. Li, Y., Liu, F., Zhou, Z., et al.: Study of improving the activity of steel slag powder using compound activator. J. Wuhan Univ. Technol. **31**(4), 11–13 (2009)
4. Zhang, T., Liu, F., Wang, J.: Recent development of steel slag stability and activating activity. Bull. Chin. Ceram. Soc. **26**(5), 980–984 (2007)
5. Guo, Y.: Study on Stabilization of Muddy Soil and Mechanical Properties of Stabilized Soil. Zhejiang University, Zhejiang (2007)
6. Shi, H., Huang, K., Wu, K., et al.: Research advance on activation and mechanism of steel slag activity. Fly Ash Compr. Utili. **1**, 48–53 (2011)
7. Sajedi, F., Razak, H.A.: The effect of chemical activators on early strength of oxlinand Portland cement-slag mortars. Constr. Build. Mater. **24**, 1944–1951 (2010)
8. Li, Y., Wang, Z., Peng, M., et al.: Study on the effect of different activators on activation of steel slag. Bull. Chin. Ceram. Soc. **31**(2), 280–284 (2012)
9. Wu, Y., Hu, X., Hu, R., et al.: Experimental study on caustic soda activating steel slag powder in muddy soil. Chin. J. Geotech. Eng. **39**, 2187–2193 (2017)
10. Han, P., Liu, X., Bai, X.: Effect of sodium sulfate on strength and micropores of cemented soil. Rock Soil Mech. **35**(9), 2555–2561 (2014)

11. Wu, Y., Hu, R., Zhao, W., et al.: Study on strength characteristic of muddy cement soil stabilized by steel slag powder. Sci. Technol. Eng. **17**(15), 306–311 (2017)
12. Xue, H., Shen, X., Zou, C., et al.: Analysis of the factors affecting the early mechanical properties of cement soil. Bull. Chin. Ceram. Soc. **33**(8), 2056–2062 (2014)
13. Zhu, X., Wang, L., Ding, T.: Study on engineering properties of nanometer silica and cement –stabilized soil. Geotech. Eng. Tech. **4**, 187–192 (2003)

Chemical and Geotechnical Properties of Red Mud at Liulin, China

Jin Li[1], Shi-Jin Feng[1], Hong-Xin Chen[1(✉)], and Hongtao Wang[2]

[1] Department of Geotechnical Engineering, Tongji University,
Shanghai 200092, China
chenhongxin@tongji.edu.cn
[2] College of Environmental Science and Engineering, Tongji University,
Shanghai 200092, China

Abstract. Disposal and storage of red mud are essential concerns due to its huge volume and high alkalinity, and understanding its properties is the basis to mitigate the risk. In this study, the chemical and geotechnical properties of red mud at Liulin, Shanxi Province, China were comprehensively investigated. The major mineral compositions were katoite, andradite, hematite, anorthite, periclase, sodalite, paravauxite, and magnesite. SiO_2 (20.76%), Al_2O_3 (20.40%), CaO (17.56%), Na_2O (17.33%), CO_3 (12.87%), Fe_2O_3 (6.44%) accounted for 95.36% of red mud. The content of Fe_2O_3 (6.44%) was substantially lower than that in other areas. Diameter of the red mud particles was in the range of 0.008–300 µm and 97% of the particles was silt and clay. The specific gravity, density, water content, hydraulic conductivity of the red mud were 2.77, 1.70 g/cm^3, 50.15% and 4.5×10^{-6} cm/s, respectively. The liquid limit, plastic limit and plasticity index were 64, 42, and 22, respectively. The ranges of friction angle and cohesion were 28.7–33.1° and 7.8–35 kPa, respectively, and both of them decreased as water content increasing from 50% to 65%. The results in this study are very useful for the design and operation of red mud stack and utilization of red mud in China.

Keywords: Bauxite · Slope stability · Grain size distribution · Direct shear test
Shear strength

1 Introduction

Red mud is a kind of industrial waste, which is generated when producing aluminum from bauxite ore. Over the past few decades, the production of red mud has substantially increased with the increasing demand of metals. The global red mud inventory has reached about 2.7 billion tons in 2011, annually increasing by approximately 120 million tons [1]. In China, red mud output has reached 47.8 million tons in 2014, accounting for about one-third of the world output. In addition, it is a remarkable fact that red mud is of high alkalinity. It contains heavy metals and even radioactive elements. Once it is accidently released, the waste can pose great danger to the people and environment. Therefore, disposal and storage of red mud become essential concerns, and understanding the properties of red mud is the basis to mitigate the risk.

© Springer Nature Singapore Pte Ltd. 2018
A. Farid and H. Chen (Eds.): GSIC 2018, *Proceedings of GeoShanghai 2018*
International Conference: Geoenvironment and Geohazard, pp. 405–414, 2018.
https://doi.org/10.1007/978-981-13-0128-5_45

Mineral and chemical compositions of red mud are vital for assessing the risk to environment and are also helpful to understand the geotechnical properties. Red mud is mainly composed of Al_2O_3, Fe_2O_3, SiO_2, TiO_2, CaO, and Na_2O [2]. However, the composition of red mud differs substantially from region to region mainly due to different ore origins and production technologies, which may highly influence the geotechnical properties. And geotechnical properties are essential for seepage, deformation and stability of red mud stack. Hence, great efforts have also been made to

Fig. 1. Location of sampling site and fresh red mud.

investigate the geotechnical properties of red mud [3–5]. Similarly, the geotechnical properties can differ significantly from one region to another due to different ore origins and production technologies, and the reported values vary over wide ranges. For example, the liquid limit and plasticity index can vary from 25–66 and 4–32, respectively [4–6].

The chemical and geotechnical properties of red mud are significant for the design of its disposal site. The objective of this study is to conduct a comprehensive investigation of the characteristics of red mud at Liulin, Shanxi Province, China. The tested red mud was sampled from Zuozhu red mud disposal site of Senze Coal Aluminium Limited in Shanxi Province (100°38′ E, 37°60′ N), as shown in Fig. 1. The generated red mud was treated using filter press, turning into red mud cake, and then it was transported to the site. The red mud was compacted by machine to increase the density at the final disposal stage. Bulk sample was obtained and sealed in plastic boxes to maintain the original features. The mineral and chemical compositions, grain characteristics, soil properties (e.g., density, water content, permeability, plasticity index) and shear behavior were studied. The tested results were also compared with the published data to investigate the difference in various regions.

2   Composition Analysis

2.1   Test Methods

X-ray diffractometer (XRD) and X-ray fluorescence spectrometer (XRF) were conducted to investigate the mineral composition and chemical composition of the red mud, respectively. The information of the tests are summarized in Table 1.

XRD analysis was conducted using Bruker's D8 Discover. Sample was dried at 105 °C and pulverized to yield fine powders. The powders were then tested using the device with a Cu Kα radiation source and a single crystal graphite monochromator. The angular range is between 10° and 90°. Additionally, sample was pressed to be pellets to test using Bruker's S4 Explorer in order to determine the chemical composition. The instrument can measure elements from Al to U.

2.2   Mineral and Chemical Compositions

The XRD pattern of the red mud is shown in Fig. 2. The major mineral compositions were katoite [$Ca_3Al_2(SiO_4)(OH)_8$], andradite ($Ca_3Fe_2Si_3O_{12}$), hematite (Fe_2O_3), anorthite ($CaAl_2SiO_8$), periclase (MgO), sodalite [$Na_8(Al_6Si_6O_{24})Cl_{12}$], paravauxite [$FeAl_2(PO_4)_2(OH)_2(H_2O)_8$], and magnesite ($MgCO_3$), some calcite ($CaCO_3$) and diaspore [$\alpha$-AlO(OH)] also existed. The results of XRF analysis are summarized in Table 2. The red mud was mainly composed of SiO_2 (20.76%), Al_2O_3 (20.40%), CaO (17.56%), Na_2O (17.33%), CO_3 (12.87%), Fe_2O_3 (6.44%), accounting for 95.36% of the red mud. SiO_2 and Al_2O_3 were associated with katoite, anorthite and sodalite. The abundance of CaO and Na_2O were caused by the addition of lime (CaO) and NaOH during the Bayer process, respectively, which can also explain the relatively high PH value (13.01) and electrical conductivity (21.1 ms/cm) in the red mud slurry. Fe_2O_3

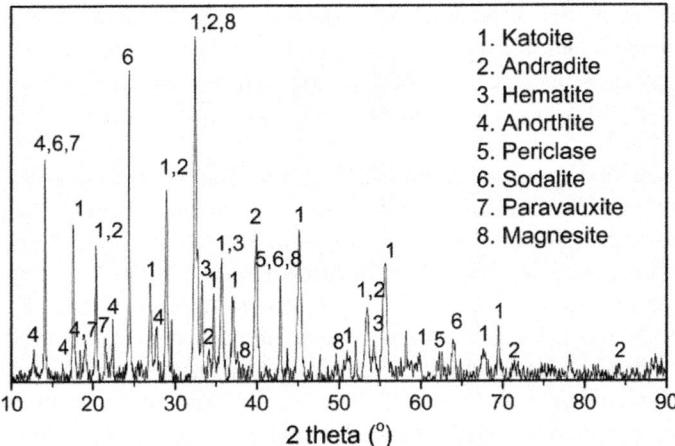

Fig. 2. XRD pattern of the red mud.

Table 1. Summary of tests in this study.

Soil properties	Test	Standard
Mineral composition	XRD	Bruker Guidelines
Chemical composition	XRF	ASTM E1621
Grain size distribution	Laser scattering particle size analysis	HORIBA Guidelines
Specific gravity	Pycnometer	ASTM D854
Density	Density test	ASTM D7263
Water content	Direct heating	ASTM D4959
Plasticity	Atterberg's Limits test	ASTM D4318
Permeability	Flexible wall permeameter	ASTM D5084
Shear behavior	Direct shear test	GB/T 50123

was formed after oxidation and hydration of Fe ion in the bauxite. In addition to the major elements, the red mud also contained some hazardous trace heavy metals, such as V, Cr, Mn, Ni, Cu, Zr, As, Nb, La, Ce, Nd.

3 Soil Property Analysis

3.1 Test Methods

Standard geotechnical tests were conducted to assess the soil properties of red mud. Information of the tests are summarized in Table 1. The grain size distribution of red mud was measured using a laser scattering particle size analyzer (HORIBA LA-960). Specific gravity of the particles was determined using pycnometer. Water content was measured by direct heating method. Atterberg's Limits test was conducted to determine the liquid limit, plastic limit and plasticity index. Specimens for determining density

Table 2. Chemical composition of the red mud at Liulin, China.

Composition	wt (%)	Composition	ppm	Composition	ppm
SiO_2	20.76	V	366.8	Y	140.1
Al_2O_3	20.40	Cr	322.6	Zr	1159.1
CaO	17.56	Mn	61	Nb	81.6
Na_2O	17.33	Co	13.3	Mo	1.6
CO_3	12.87	Ni	68.5	Sn	26.6
Fe_2O_3	6.44	Cu	39.2	Ba	83.4
TiO_2	2.08	Zn	16.2	La	327
K_2O	0.88	Ga	41.9	Ce	608
MgO	0.37	As	13.6	Nd	338
S	0.52	Br	8.3	W	0.8
P	0.21	Rb	23.3	Pb	120
Cl	0.06	Sr	1281.4	Th	107
				U	0.8

were carefully prepared to keep the original characteristics, a cutting ring was cut directly into the bulk red mud in the boxes to get the specimens. Hydraulic conductivity test was performed using flexible wall permeameter. For this test, the specimen was prepared in the same way as described above and then saturated under a vacuum. The specimens were subjected to a confining pressure (50 kPa), and hydraulic conductivity was determined by measuring flow rate under constant gradient conditions.

3.2   Physical Properties

Grain size distribution of the red mud is shown in Fig. 3 (red line). The diameter of the red mud particles was in the range of 0.008–300 µm and a majority of the particles were silt and clay, accounting for 97% of the red mud. The particles mainly concentrated in two regions (0.08–0.5 µm and 2–20 µm). The soil was a little bit gap-graded between 0.5 and 2 µm. d_{10}, d_{30}, d_{50}, d_{60} of the red mud were 0.3, 3, 4.5, 5.7 µm, respectively. The coefficient of uniformity (C_u) and the coefficient of curvature (C_c) were 19 and 5.26, respectively. Therefore, the red mud was poorly graded.

In order to further understand the grain size characteristics of red mud, the grain size distribution of red mud is compared with that of other tailings, as shown in Fig. 3. The results indicate that red mud particles are 1 to 2 orders of magnitude smaller than other common tailings, and hence great attention should be paid to the seepage, deformation and stability of red mud stack.

The basic physical properties of red mud in this study are summarized in Table 3, the properties are also compared to those in other areas. The specific gravity of the red mud at Liulin was 2.77, slightly smaller than the reported values in other studies (3.05–3.7). The main reason is that the content of Fe_2O_3 in the red mud at Liulin was very low (6.44%) compared to other studies. The density (1.7 g/cm³) was enclosed by the reported values in other studies (1.39–2.04 g/cm³), which depend on production technology and consolidation in disposal site. The water content was 50.15% in this study. But it is noteworthy

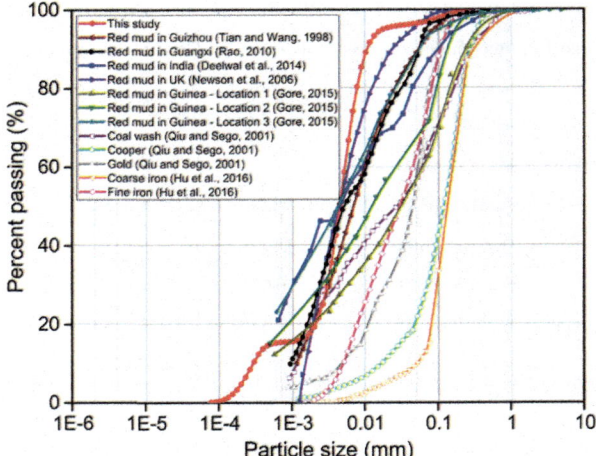

Fig. 3. Comparison of grain size distribution between red mud and other tailings.

Table 3. Comparison of physical properties of red mud in different areas.

	This study	Guangxi [7]	Guizhou [8]	UK [4]	Jamaica [5]
Specific gravity	2.77	-	3.11	3.05	3.6–3.7
Density (g/cm³)	1.70	2.04	1.39	1.75	-
Water content (%)	50.15	27–37	39.4	57	113–140
Permeability (cm/s)	4.5×10^{-6}	4.5×10^{-6}	-	-	-
Liquid limit (%)	64	43	44	54	62
Plastic limit (%)	42	35	38	40	36
Plasticity index	22	8	6	14	26
d_{10} (μm)	0.3	1	1	1.5	76
d_{30} (μm)	3	2.5	4	2.5	400
d_{50} (μm)	4.5	5	7	4.5	1000
d_{60} (μm)	5.7	10	10	6	1490
Uniformity coefficient	19	10	10	4	19.6
Curvature coefficient	5.26	0.63	1.6	0.69	1.41

that the reported values vary over a wide range (33–140%), and that in Jamaica is extremely high (113–140%). The variation of data results from different production processes. For example, the bauxite residue at Liulin was filtered and pressed, and hence the water content was significantly reduced; but the slurried bauxite residue in Jamaica is directly piped to a lagoon so the sampled red mud has extremely high water content. The permeability of the red mud at Liulin was 4.5×10^{-6} cm/s, which was within the reported range in other studies (10^{-5}–10^{-7} cm/s). With the increase of overburden pressure and decrease of water content, the permeability of red mud can be further reduced [9].

The liquid limit, plastic limit and plasticity index of the red mud at Liulin were 64, 42, 22, respectively. So the red mud can be classified as elastic silt (MH) according to ASTM D2487 (Fig. 4). Comparison of plasticity index of red mud in different areas is also illustrated in Fig. 4. The liquid limit and plasticity index of the red mud at Liulin were both relatively high. Most red muds in Fig. 4 belong to ML or MH. CL and CH are quite limited.

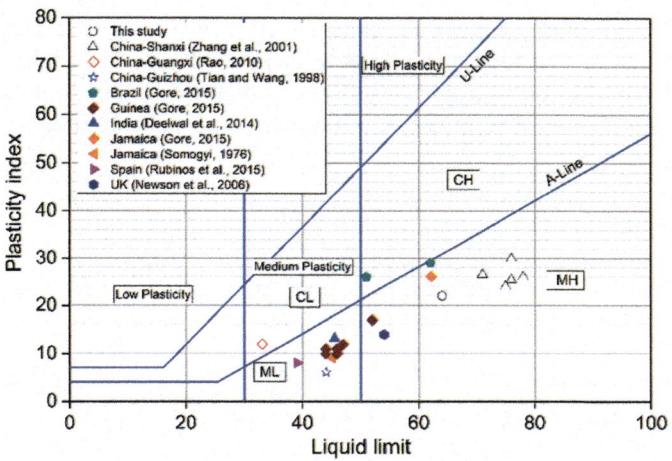

Fig. 4. Comparison of plasticity index of red mud in different areas

4 Direct Shear Test

4.1 Test Method

Direct shear tests were performed to evaluate the shear behavior of red mud according to the method described in GB/T 50123. Specimens were prepared at water contents of 50%, 55%, 60%, 65%. The shear box had dimensions of 61.8 mm in diameter and 20 mm in thickness. Vertical load was applied for one day at various normal stresses respectively for consolidation before shearing. Normal stresses of 100, 200, 300 and 400 kPa were applied to specimen with 50% and 55% water content, and normal stresses of 50, 100, 150, 200 kPa were applied to that with 60% and 65% water content. The reason why smaller normal stress was adopted for 60% and 65% water content was that the red mud became much softer and could not sustain normal stress of 400 kPa. Because the direct shear apparatus cannot strictly control drainage conditions, the shear rates are set to simulate various drainage conditions. Quick shear rate (0.02 mm/min) and slow shear rate (0.8 mm/min) were adopted for the direct shear test. The maximum shear displacement was 4 mm, which was taken as the shear strength.

4.2 Shear Behavior

Quick direct shear test can be seen as a test conducted approximately under undrained condition, while slow direct shear test referred to drained condition [10]. The friction angle and cohesion were determined based on Mohr-Coulomb criterion in term of total stress, as shown in Fig. 5. The friction angle of fast shear case was larger than that of slow shear case while the variation of cohesion was contrary. For both fast shear and slow shear cases, the friction angle and cohesion both overall decreased with increasing initial water content. The reduction of friction angle for fast shear and slow shear cases were 7.9% and 3.4%, respectively, with the initial water content increasing from 50% to 65%; and that of cohesion were 23.1% and 58.6%, respectively. Hence, the influence of change in initial water content on cohesion was more significant, which indicate that it is necessary to reduce the initial water content of red mud before depositional consolidation.

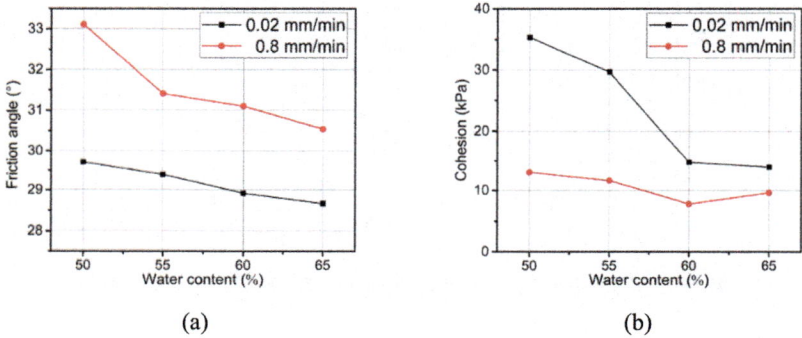

Fig. 5. Variations of friction angle (a) and cohesion (b) with water content and shear rate.

The shear strength parameters for direct shear tests and triaxial shear tests in different areas are summarized in Fig. 6, which are in term of total stress. The shear strength parameters may have some differences due to various drainage conditions and test apparatuses, but we can investigate the regional difference of red mud according to the results under similar test conditions. The friction angle (28.7–33.1°) and cohesion (7.8–35 kPa) in this study were similar to those (29–31°, 5–40 kPa) of fresh red mud reported by Zhang et al. [3], the reason may be that the tested material was also sampled in Shanxi. Extreme high cohesion values were observed by Zhang et al. (75 kPa), which is caused by the recrystallization and cementation of red mud with time [3]. The friction angles obtained by direct shear test are overall smaller than those obtained by triaxial test, probably because the flaw of the direct shear method. It is noteworthy that the friction angles obtained by direct shear test (excluding the Jamaica data) well match the typical range of silt and clay; while those obtained by triaxial test are overall larger than the typical range of silt and clay, but are close to the typical range of sand and gravel.

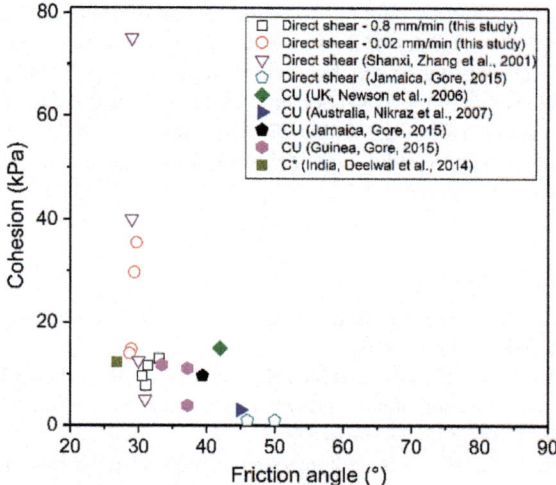

Fig. 6. Comparison of shear strength parameters of red mud in different areas. C*: the drainage condition in the triaxial test was not reported

5 Summary and Conclusions

The chemical and geotechnical properties of red mud at Liulin, Shanxi Province, China were comprehensively investigated. The mineral and chemical compositions, grain characteristics, soil properties (e.g., density, water content, permeability, plasticity index), and shear behavior were studied. Some major conclusions can be drawn based on the test results:

(1) The major mineral compositions were katoite, andradite, hematite, anorthite, periclase, sodalite, paravauxite, and magnesite, some calcite and diaspore also existed. As for chemical composition, the red mud was mainly composed of SiO_2 (20.76%), Al_2O_3 (20.40%), CaO (17.56%), Na_2O (17.33%), CO_3 (12.87%), Fe_2O_3 (6.44%), accounting for 95.36% of the red mud.

(2) The diameter of the red mud particles was in the range of 0.008–300 μm and a majority of the particles were silt and clay, accounting for 97% of the red mud. The particles mainly concentrated in two regions (0.08–0.5 μm and 2–20 μm). The coefficient of uniformity (C_u) and the coefficient of curvature (C_c) were 19 and 5.26, respectively. The red mud was hence poorly graded. The specific gravity, density, water content, permeability of the red mud were 2.77, 1.70 g/cm^3, 50.15% and 4.5 × 10^{-6} cm/s, respectively. The liquid limit, plastic limit and plasticity index were 64, 42, and 22, respectively. So the red mud can be classified as elastic silt (MH).

(3) The ranges of friction angle and cohesion were 28.7–33.1° and 7.8–35 kPa, respectively, and both of them decreased as water content increasing from 50% to 65%. The influence of change in water content on cohesion was more significant than that on friction angle.

Acknowledgments. The authors would like to thank the staff of Senze Red Mud Stack for their assistance in sampling. The majority of the work described in this paper was supported by the National Natural Science Foundation of China under Grant Nos. 41602288 and 41572265, the Newton Advanced Fellowship of the Royal Society, and the Fundamental Research Funds for the Central Universities under Grant No. 2016KJ027. The authors would like to acknowledge all these sources of financial support and express the most sincere gratitude.

References

1. Klauber, C., Gräfe, M., Power, G.: Bauxite residue issues: II. Options for residue utilization. Hydrometallurgy **108**(1), 11–32 (2011)
2. Liu, Y., Naidu, R., Ming, H., Dharmarajan, R., Du, J.: Effects of thermal treatments on the characterisation and utilisation of red mud with sawdust additive. Waste Manag. Res. (2016). https://doi.org/10.1177/0734242X16634197
3. Zhang, Y.S., Qu, Y.X., Wu, S.R.: Engineering geological properties and comprehensive utilization of the solid waste (red mud) in aluminium industry. Environ. Geol. **41**(3–4), 249–256 (2001)
4. Newson, T., Dyer, T., Adam, C., Sharp, S.: Effect of structure on the geotechnical properties of bauxite residue. J. Geotech. Geoenviron. Eng. **132**(2), 143–151 (2006)
5. Gore, M.S.: Geotechnical characterization of bauxite residue (red mud). Ph.D. Thesis, The University of Texas at Austin, Austin, USA (2015)
6. Somogyi, F.: Dewatering and drainage of red mud tailings. Ph.D. Dissertation, Department of Civil Engineering, University of Michigan, USA (1976)
7. Rao, P.P.: Analysis on basic characteristics of Bayer's dry red mud and the operation feature of the yard. J. Eng. Geol. **18**(3), 340–344 (2010). (in Chinese)
8. Tian, Y., Wang, F.X.: The mechanical properties of red mud storage. Light Met. **2**, 32–34 (1998). (in Chinese)
9. Rubinos, D., Spagnoli, G., Barral, M.T.: Assessment of bauxite refining residue (red mud) as a liner for waste disposal facilities. Int. J. Min. Reclam. Environ. **29**(6), 433–452 (2015)
10. Wu, M., Fu, X.D., Xia, T.D., Xu, D.X., Wang, J., Liu, Y.M.: Comparison between unconsolidated undrained simple and direct shear tests on compacted soil. Chin. J. Rock Mech. Eng. **25**(2), 4148–4152 (2006). (in Chinese)

Physical and Mechanical Characteristics of MBT Waste in China

Zhenying Zhang[✉], Yuxiang Zhang, Yingfeng Wang, Hui Xu,
Wenqiang Guo, Dazhi Wu, and Yuehua Fang

Zhejiang Sci-Tech University, Hangzhou 310018, China
zhangzhenyinga@163.com

Abstract. The mechanical-biological treatment (MBT) is a new technique used to treat the municipal solid waste (MSW) and has the advantages of significantly reducing the quantity of MSW and greatly saving the capacity of a landfill. The MBT waste has already become one of the hot global research topics in geo-environmental engineering. In the present study, the basic geotechnical parameters of MBT waste were obtained by performing several physical and mechanical tests, including the composition analysis, sieve analysis, water content test, specific gravity test, compression test, and direct shear test. The test results show that: (1) The main components of MBT waste are plastic, textile, and glass, which collectively account for 55% of the total dry mass. (2) The uniformity coefficient of MBT waste is greater than 21 and the curvature coefficient is 2.1. (3) The relationship between the void ratio and the logarithmic pressure is fitted into a linear equation and the compression index is calculated to be 0.726. (4) The strength characteristics of MBT waste conform to the Coulomb's law, and the cohesion is 15.76 kPa, while the internal friction angle is 39.8°. The conclusions provide a reference for the prediction of reservoir capacity and the stability analysis of an MBT landfill.

Keywords: MBT · Uniformity coefficient · Compression index
Shear strength parameters · Coulomb's law

1 Introduction

Since European Union (EU) passed new laws and regulations that made certain restrictions to the landfills of high organic content waste, the technology of mechanical-biological treatment (MBT) developed rapidly in European countries due to its efficient organic matter reduction capacity. The common steps of MBT are: (1) to separate non-biodegradable, large-volume, or recyclable substances mechanically or manually, (2) to accelerate degradation through biotechnology, such as anaerobic digestion and leaching hydrolysis, (3) derive the final product (referred to as MBT waste in this study) from biological drying under certain conditions. The entire process also contains the steps of collection, purification, and utilization of gas and leachate. The mechanical-biological treatment changes the particle size, composition ratio, moisture content, density, and other physical properties of wastes. The shear strength, compression, consolidation, and

© Springer Nature Singapore Pte Ltd. 2018
A. Farid and H. Chen (Eds.): GSIC 2018, *Proceedings of GeoShanghai 2018*
International Conference: Geoenvironment and Geohazard, pp. 415–423, 2018.
https://doi.org/10.1007/978-981-13-0128-5_46

other mechanical properties change as well, which greatly impact the strength, stiffness, and stability of the waste pile [1].

Several studies on MBT waste were conducted. Comparing the differences between British and German MBT wastes, Siddiqui et al. [2] tested samples collected in both countries to determine the extent of degradation and stability. Bhandari and Powrie [3] carried out numerous tri-axial experiments to study the stress-strain characteristics of MBT wastes. As the technology is developing, there is still no clear standard procedure or classification system for MBT wastes. Even in Europe, the sorting processes and biodegradation durations are quite different between Germany and the U.K.

In 2015, over 19 million tons of municipal solid waste was generated in China. China has been actively exploring and developing the use of the MBT technology to reduce the amount of waste disposal. However, there has been no research demonstrating the physical and mechanical properties of Chinese MBT wastes. Considering that the composition of fresh MSW in China is quite different from that in European countries [4], it is worth investigating the physical and mechanical properties of MBT wastes in China.

In this study, the geotechnical parameters of MBT wastes were obtained by performing basic physical and mechanical tests, such as the composition analysis, sieving test, water content test, specific gravity test, compression test, and direct shear test. The waste sample was obtained from a pilot MBT project at the TianZiling landfill located in Hangzhou, China. The tests were performed in accordance with the Technical Specification for Soil Test of Landfilled Municipal Waste CJJ/T 204-2013 [5].

2 Test Materials

The waste samples collected from TianZiling were subjected to the following processes. The fresh MSW was dumped into a discharge hopper and transported into a drum sieve of 120 mm aperture by a conveyor belt along with an electromagnet working to separate the metals. In the drum sieve, plastic bags were punctured and torn for a better separation of inner objects. The objects that remained on the sieve were large pieces of cardboard, plastic, and textile, which could be recycled manually, while the objects that passed the sieve were sent into the bioreactor to accelerate the degradation of organic components. After extrusion and dehydration, the product became a loose material with medium moisture content.

3 Physical Properties

3.1 Composition Analysis

In order to reduce the non-uniformity of the materials from magnification, representative samples of 15 kg were prepared from bulk samples using the quartering method. They were then sorted manually into various material categories and weighed respectively as wet mass before being sent to the drying oven. After being dried at 70 °C for 24 h, each category was weighed again as dry mass. The moisture content of each component was calculated and shown in Table 1. The "unidentified" category represents a mixture of different components that could not be identified or further separated.

Table 1. Moisture content, dry and wet mass of each component in percent (%)

Component (i)	Wet mass	Dry mass	Moisture content
Paper	0.00	0.00	0.00
Plastic and rubber	30.09	31.48	15.76
Textile	10.72	8.28	56.92
Wood	6.40	4.70	64.88
Stones, ceramics	7.11	8.23	4.72
Glass	14.19	17.10	0.45
Metal	2.31	2.80	0.09
Unidentified > 5 mm	16.36	16.55	19.74
Fines < 5 mm	12.81	10.86	42.80
Total	100.00	100.00	21.11

As shown in Table 1, the percentages of plastic and rubber, glass, fines < 5 mm are the largest three among recognizable substances. The percentage of the paper component is zero because paper is fully degraded and transformed into tiny fibers. Higher than the other components, the moisture contents of textile and wood are 57% and 65%, respectively, indicating that textile and wood are good water retention materials. The composition of the "unidentified" is quite complex, mainly including fibers and adhesive paste-based materials, and its moisture content is also high. Siddiqui et al. [2] explored the landfill characteristics by conducting comparisons of the compositions between the MBT wastes from Germany and the U.K. Few differences were found between the samples from two countries, because the processing technologies are similar.

$$B(i) = |C(i)/Avg(i) - 1| \tag{1}$$

Avg(i) is the average percentage of Germany and the U.K. components and C(i) is the average percentage of China. Using B(i) to represent the deviation of C(i) from Avg (i), we obtain the results presented in Table 2.

The results indicate that the percentages of rubber and plastics, textile, wood, and metal are much higher than the average values. The higher percentage of metals might be caused by the different collection and sorting processes. The wood component, most of which are chopsticks, popsicles, and pits, reflects the typical eating habits of oriental people. The "rubber and plastics" and textile components are mainly plastic bags and clothes, respectively, and the results demonstrate that the recycling capacity of renewable materials in China should be strengthened and the classification system should be improved. The excessive plastic and textile structure of the wastes decreases the efficiency of biological treatment in the MBT process to a certain extent. The public awareness about waste classification should be increased and advocated.

Table 2. Dry mass of each component compared with Germany and the U.K. data (%).

Component	China	UK	Germany	Bi(i)
Paper	0.00	0.00	0.18	-
Plastics and rubber	31.48	11.02	8.56	2.216
Textile	8.28	1.57	3.22	2.457
Wood	4.70	1.33	0.63	3.796
Stones, ceramics	8.23	4.29	7.79	0.363
Glass	17.10	22.77	24.36	0.274
Metal	2.80	0.49	1.49	1.828
Unidentified > 5 mm	16.55	28.95	26.75	0.610
Fines < 5 mm	10.86	29.15	27.02	0.411
Total	100.00	100.00	100.00	0.000

3.2   Moisture Content

A sample of 6 kg was prepared from the bulk samples using the quartering method. The wet mass was weighed as m_w before being sent to the drying oven. After being dried at the temperature of 70 °C for 24 h, the dry mass was weighed as m_s. The moisture content was then calculated using Eq. 2.

$$\omega(\%) = (m_w/m_s - 1) \times 100 \tag{2}$$

Four samples passed the test and their average moisture content was 22.66%, which is close to the calculated value of 21.11% in Table 1, but different from the European samples. Fucale [6] explored the mechanical behavior of MBT wastes and the moisture content of the samples used in the experiments was 36.5%, 35.0%, and 34.92%. The moisture content of the original MBT waste samples used by Kuehle-Weidemeier [7] to compare them with the German boundary values was 35.5%, and the moisture content of the sieved waste was even higher. The difference between the samples from Europe and Hangzhou in moisture content is approximately 12%. The European MBT waste is smaller both in particle size and plastics proportion than the Chinese MBT waste. The difference in the capacity of carrying capillary water leads to different moisture content.

3.3   Natural Density

Barrel weighing method was used to measure the density of the MBT waste. A 120-L barrel and a bench scale of 1 g accuracy were used in the measurement. The sample was tested immediately after being loaded into the container and smoothed with a wire saw. No additional vertical load was applied during the test. The natural density was calculated using Eq. 3 and the result is 0.227 g/cm^3. In Eq. 3, m_{tl} and m_b are the total mass and the barrel mass, respectively. V_0 is the volume of the barrel (120-L)

$$\rho = (m_{tl} - m_b)/V_0. \tag{3}$$

3.4   Particle Analysis

The stability of the MBT waste pile is a significant factor in the process of landfilling and stacking. The grain gradation plays a major role through the collapse of the waste pile and the formation of a sliding surface of the landslide. After being air-dried for 48 h, 3 kg sample was selected using the quartering method and placed on the top of the round-hole sieve. The sieves were stacked according to the apertures, which are 60, 40, 20, 10, 5, 2, and 1 mm. The sample was sieved for 15 to 20 min and the total mass of the material that remained in each layer was weighed. The grain size distribution curve is shown in Fig. 1. Based on this curve, the grain size parameters of d10, d30, and d60 were 1.07, 6.87, and 22.89, respectively.

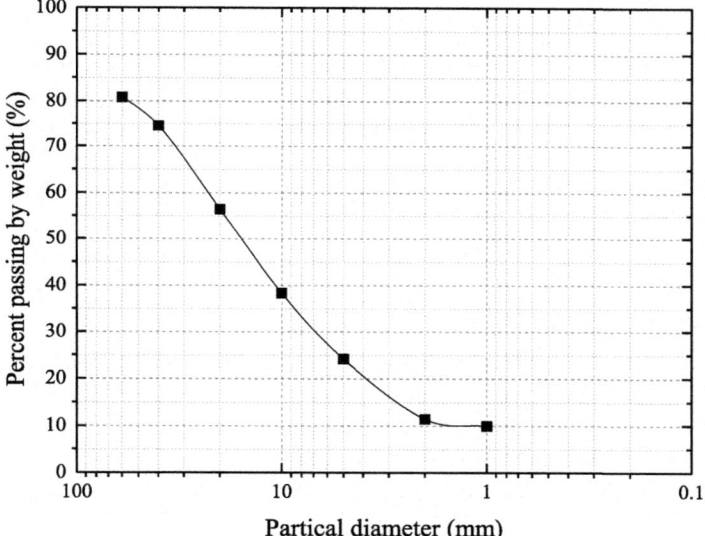

Fig. 1. Grain size distribution curve

A comparison of the grain size parameters between the samples from Schaumburg, Germany [7] and Hangzhou (the present work) is shown in Table 3. The uniformity coefficient C_u and the curvature coefficient C_c were calculated as 21.6 and 1.93, respectively, according to the soil filter stability equation. However, an assumption of the equation is that the soil particles are elliptical; however, most particles in MBT wastes exist in one, two, and three-dimensional forms (materials in strips hanging on the sieve during the sieving process). Therefore, the C_u and C_c values are for reference use only.

In Table 3, the particle size parameters of Hangzhou 1–100 mm are larger than those of LG 1–100 mm at a given mass percentage and their difference increased with the percentage. The particle size of MBT wastes in Hangzhou is generally larger, but the particle density is lower. The high proportion of plastics may be the key factor in this result.

Table 3. Particle analysis results comparation [7]

Material	H Lahe 0–30 mm	SHG 0–20 mm	SHG 0–40 mm	SHG 0–60 mm	SHG 0–150 mm	LG 0–100 mm	HangZhou 0–100 mm
d10	0.4	0.03	0.043	0.052	0.061	0.45	1.07
d15	0.7	0.053	0.082	0.10	0.12	0.8	2.74
d50	2.7	0.73	2.1	4.1	7.0	11	15.70
d60	4.0	1.5	5.3	10	20	12	22.89
d85	10	7.3	24	35	44	31	-
C_u	10	50	123	192	328	27	21.6

3.5 Specific Gravity

The specific gravity of MBT wastes was measured using the vacuum pumping method. By measuring the mass ratio of kerosene and MBT wastes in equal volumes, the specific gravity of the sample was calculated to be 1.575 at the temperature of 36 °C.

4 Mechanical Properties

4.1 Direct Shear Test

The equipment used for the direct shear test was a large-scale solid waste compression-shearing instrument with a diameter of 200 mm and a maximum vertical load of 30 kN. Various stages of vertical loads were applied to the MBT samples before shearing, i.e., 12.5, 25, 50, 100, and 200 kPa.

Figure 2 shows the relationship between shear stress and lateral displacement. The points marked by arrows are the peak shear stresses. The shear stress continues to

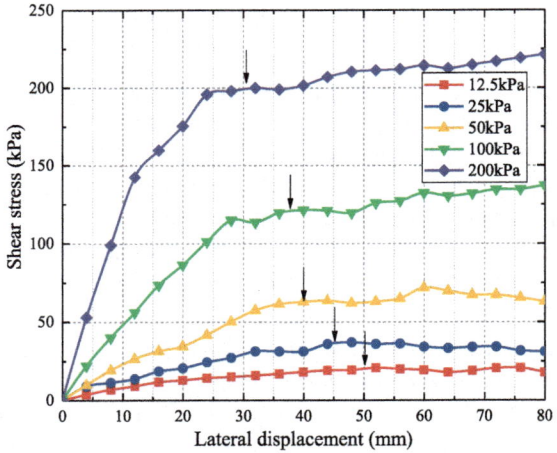

Fig. 2. Shear stress versus lateral displacement

increase following the peak values because of the tensile stress caused by the inner band component. In Fig. 2, the sample has already lost its shear capacity. Figure 2 also demonstrates that the displacement corresponding to the peak shear stress decreases with the increase in vertical load. The peak value of the shear stress increases with increasing vertical load, indicating that the vertical load enhanced the shear strength. Figure 3 shows the relationship between the shear strength and normal stress: the shearing characteristics of the MBT waste conform to the Coulomb's law (Eq. 4). The cohesion c is 15.76 kPa and the internal friction angle φ is 39.8°. The value of cohesion is smaller than the values reported by Fucale [6]: Φmax $\leq$ 40 mm, c = 16–63.1 kPa, φ = 40.1°–48.1°. The internal friction angle is close to the results reported by De Lamare Neto [8]: Φmax $\leq$ 60 mm, c = 29.34–54.18 kPa, φ = 39.04°–43.35°; Φmax $\leq$ 100 mm, c = 32.86–45.04 kPa, and φ = 38.55°–40.33°.

$$\tau_f = c + \sigma \tan\varphi. \tag{4}$$

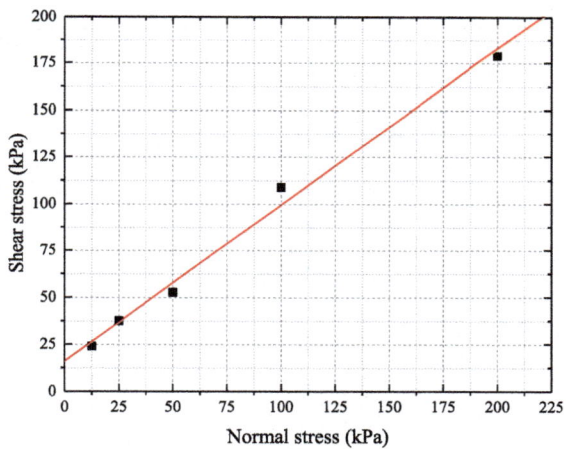

Fig. 3. Shear stress-normal stress relationship

4.2 Compression Test

A large-scale high-pressure solid waste compression apparatus (maximum vertical load = 50 kN, maximum vertical displacement = 300 mm) was used in the compression test. Various stages of axial stresses were applied to each prepared sample, i.e., 12.5, 25, 50, 100, 200, and 400 kPa. The vertical displacement was recorded within 24 h after the application of each load. The time-deformation curves are shown in Fig. 4. Due to the low moisture content, the MBT sample had no leachate outflow even after consolidation under 400 kPa for 24 h.

The relationship between the void ratio and the axial effective stress (plotted in logarithmic coordinates) is shown in Fig. 5. The compression index is 0.726, derived from the slope of the curve, which is almost twice the index of highly compressed soil (0.4).

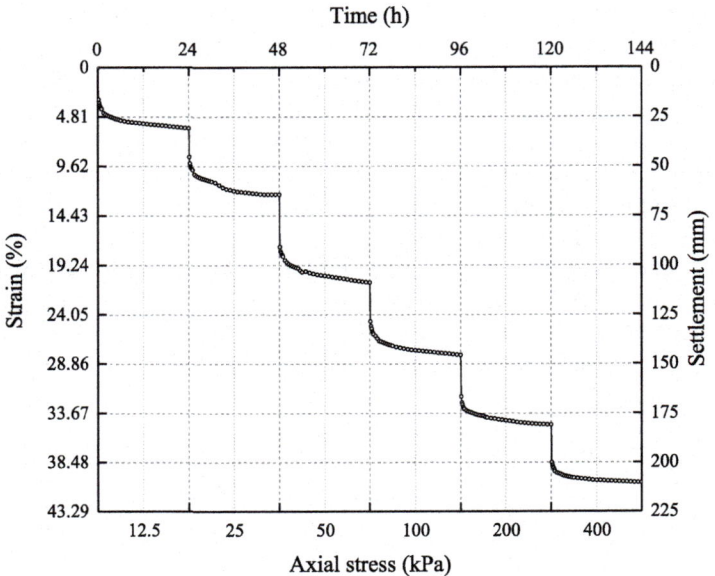

Fig. 4. Time-deformation curve

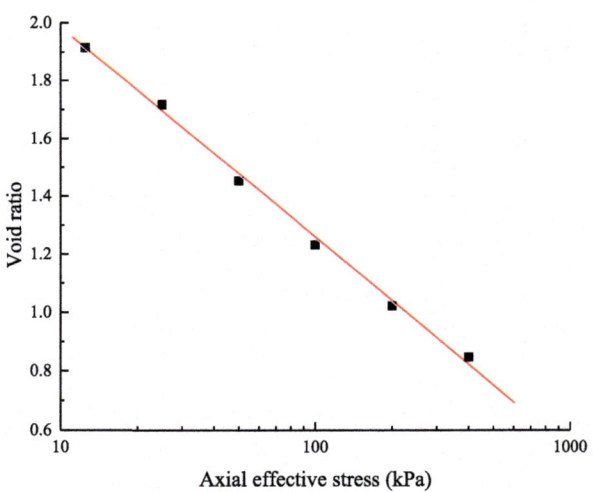

Fig. 5. Void ratio versus axial effective stress

5   Conclusions

1. The MBT waste from the TianZiling Landfill in Hangzhou, China is a material with low water content (22.66%) and low density (0.227 g/cm^3), but has a high percentage of plastic, textile, and glass (collectively account for 55%).

2. The MBT waste is of good gradation. The uniform coefficient is 21.6 and the curvature coefficient is 2.1.
3. The void ratio of the MBT waste linearly decreases with the increase in logarithmic axial effective stress. The compression index is 0.726 and the compressibility is excellent.
4. The shear strength characteristics of the MBT waste conform to the Coulomb's law; the internal friction angle is 39.8°, and the cohesion is 15.76 kPa.

Acknowledgements. The authors thank the reviewers for their valuable comments and suggestions. The work is funded by the National Natural Science Foundation of China (Contract Nos. 51478436, 51678532, and 51708508), which is greatly appreciated. The authors thank the Hangzhou Environmental Group for providing experimental materials.

References

1. Jones, D.R.V., Dixon, N.: Landfill lining stability and integrity: the role of waste settlement. Geotext. Geomembr. **23**, 27–53 (2015)
2. Siddiqui, A.A., Richards, D.J., Powrie, W.: Investigations into the landfill behavior of pretreated wastes. Waste Manag. **32**, 1420–1426 (2012)
3. Bhandari, A.R., Powrie, W.: Behavior of an MBT waste in monotonic triaxial shear tests. Waste Manag. **33**, 881–891 (2013)
4. Zhan, L.T., Xu, H., Chen, Y.M., Lv, F., Lan, J.W., Shao, L.M., Lin, W.A., He, P.J.: Biochemical, hydrological and mechanical behaviors of high food waste content MSW landfill: preliminary findings from a large-scale experiment. Waste Manag. **63**, 27–40 (2017)
5. Technical specification for soil test of landfilled municipal waste CJJ/T 204-2013. China Architecture and Building Press, Beijing (2013)
6. Fucale, S.: The mechanical behavior of MBT waste. Electron. J. Geotech. Eng. (EJGE) Bundle **13**, 5927–5931 (2015)
7. Kuehle-Weidemeier, M.: Landfilling of mechanically-biologically pretreated municipal solid waste. In: International Symposium Waste Management Zagreb, VIII (2004)
8. De Lamare Neto, A.: Resistência ao Cisalhamento de Resíduos Sólidos Urbanos e de materiais Granulares com Fibras, 190p. Tese de Doutorado em Engenharia Civil. Universidade Federal do Rio de Janeir COPPE/UFRJ (2004)

Effect of Microbial Activities on Permeability of Sand

Saswati Ghatak$^{(\boxtimes)}$, Debasis Roy, and G. Vinoth

Indian Institute of Technology, Kharagpur, West Bengal, India
saswatighatak@gmail.com

Abstract. Accumulation of microbial metabolic products namely extracellular polymeric substances (EPS), inorganic minerals and gases often clogs soil pores and reduces its permeability. Reduction in permeability of soil has both advantages and disadvantages. Changes in soil permeability, therefore, need to investigate for successful use as a bioengineering solution or to predict future problems. Permeability reduction of a soil mass can be attributed to both biogenic deposits and biogas related unsaturation. Although several studies were conducted to see the efficacy of biofilm and biomineral in reducing soil permeability, the effect of horizontally flowing growth medium, resembling groundwater in both mineral salt composition and circulation velocity, has not been studied yet. Present study mainly focuses on the influence of horizontal flow of growth media on the amount of metabolic products and saturated and unsaturated permeability of loose quartz sand (relative density of about 40%) inoculated with an aerobic microbial species *Lysinibacillus sp.* DRG3. In this study, microbes were instigated to produce metabolic products through three naturally occurring bioprocesses namely, non-ureolytic calcifying process, ureolytic calcifying process and non-calcifying process. Sand samples with fluvial activities yielded higher amount of biogenic precipitates, unsaturation and permeability reduction. Calcifying process resulted in lesser development of unsaturation than that for non-calcifying process probably because of higher utilization of evolved CO_2 for biocalcification. Unsaturated permeability was generally lesser than the saturated one. After treatment, the unsaturated permeability decreased up to 16% of the original permeability, while the saturated permeability recorded was up to 41% of original permeability.

Keywords: Permeability · *Lysinibacillus sp.* DRG3 · Unsaturation

1 Introduction

Microorganisms are known to produce extracellular polymeric substances (EPS), inorganic minerals and a variety of gases through the metabolic processes (Ehrlich and Newman 2009). In situ production and accumulation of these metabolic products of soil-residing microbes clogs the pores in soil and reduces its permeability (Seki et al. 1998; Ivanov and Chu 2008). Reduction in permeability has both advantages and disadvantages (Seki et al. 1998). Bioclogging can reduce seepage in irrigation channel and reservoir, piping of earthen dam or dikes, increase oil recovery from reservoir and form a barrier to reduce the migration of heavy metals and organic pollutants

© Springer Nature Singapore Pte Ltd. 2018
A. Farid and H. Chen (Eds.): GSIC 2018, *Proceedings of GeoShanghai 2018*
International Conference: Geoenvironment and Geohazard, pp. 424–433, 2018.
https://doi.org/10.1007/978-981-13-0128-5_47

(Khachatoorian et al. 2003; Nemati et al. 2005; Ivanov and Chu 2008). However, reduced permeability due to microbial activities can also create significant geotechnical problems such as impaired sand filtration and difficulty in aquifer recharge (Bonala and Reddi 1998; Dupin and McCarty 2000). Therefore it is necessary to analyze the changes in permeability of soil due to microbial activities, either to use as a bioengineering solution or in order to predict the problems which may arise in future.

Efficacy of biofilm in reducing saturated permeability of sand and silt was investigated by several researchers and it was observed that permeability can be reduced up to 100% after only two to three weeks of treatment (Cunningham et al. 1991; Karimi 1998; Proto et al. 2016). Along with biofilm, biomineral can also be used as bioclogging agent and capable of reducing permeability up to 65% (Nemati et al. 2005; Whiffin et al. 2007). Reduction in permeability of a soil mass can be attributed to both biogenic deposits and biogas related unsaturation (Vandivivere and Baveye 1992). While previous studies have reported a decrease in the permeability of soil because of the biogenic precipitates, the effect of biologically induced unsaturation on the permeability of soil has not been studied so far.

Groundwater circulation is one of the factors often influence the generation of metabolic products. Several studies have attempted to investigate the effect of continuous supply of nutrients on metabolic product formation and/or permeability reduction by vertical flow of growth medium though a sand column (Whiffin et al. 2007; Al Qabany et al. 2012; Martinez et al. 2013). But the hydraulic effect of the horizontally flowing growth medium, resembling groundwater in both mineral composition and velocity, has not been studied yet. The present study intends to investigate the effect of horizontal flow of growth medium on metabolic product formation by an aerobic microbial species *Lysinibacillus sp.* DRG3 through three different bioprocesses—non-calcifying, non-ureolytic calcifying and ureolytic calcifying and related changes in saturated and unsaturated permeability of sand.

2 Materials and Methods

2.1 Sand

Silica sand, having subangular grains, with about 97% quartz grains was used in the experiments. The sand had a co-efficient of uniformity (C_u) of 1.99, Co-efficient of curvature of (C_c) of 0.66 and a median particle size (D_{50}) of 0.55 mm. According to Unified Soil Classification System the sand was classified as poorly graded sand (SP). The specific gravity of the sand grains was 2.65 and the maximum and minimum void ratios were 0.83 and 0.54 respectively (ASTM 2006). The sand was thoroughly washed with 0.25 N hydrochloric acid and then with 0.25 N sodium hydroxide, followed by deionized water until the pH becomes neutralized. Before use, it was autoclaved at 103 kPa and 120 °C for 15 min.

2.2 Microorganism

Lysinibacillus sp. DRG3, an aerobic soil-residing bacterium (Ghatak et al. 2015) iso-lated from a naturally cemented tailings sand deposit on the east coast of India, was used in this study. It is capable of producing EPS with or without precipitating calcite in both ureolytic and non-ureolytic pathways depending on nutrient availability.

2.3 Nutrient Media

A mineral salt medium was prepared with essential minerals roughly resembling the compositions of groundwater. Nutrient medium, for non-calcifying process, was pre-pared by dissolving 20 g glucose, 3 g NH_4Cl, 0.14 g KH_2PO_4, 2.2 g K_2HPO_4, 0.14 g NaCl, 0.6 g $MgSO_4$, 2.3 mg $ZnSO_4$, 17 mg $MnSO_4$, 10 mg $CuSO_4$, 4 mg Na_2MoO_4, 10 mg EDTA, 0.4 mg $NiCl_2$, and 6.6 mg NaI in 1 L of distilled water. For non-ureolytic calcifying process, 0.1 g of $CaCl_2$ was added to the media prepared for non-calcifying process. For ureolytic calcifying process, 0.1 g of $CaCl_2$, and 3 g of urea was added to the medium instead of NH_4Cl. Before use, the media was sterilized by autoclaving (Ghatak et al. 2015).

2.4 Sample Preparation

Loose sand samples – 70 mm in length, 40 mm in width and 55 mm in height – were prepared about 40% relative densities by pluviating autoclaved sand grains under a mixture of sterilized nutrient medium and centrifuged bacterial biomass inside a ster-ilized polypropylene container designed to promote the flow of medium under gravity. The sand is confined between two vertical high density polyethylene (HDPE) meshes supported by autoclaved marbles through which the growth medium was passed. Bacteria-free samples were also prepared in a similar way but not adding bacterial biomass in the nutrient media. A layer of marbles was kept on top of the sand samples with live DRG3 population not to allow the sand to bulge due to evolving gases. Perforated inlet pipes were sterilized and connected to the container through which the medium was allowed to flow. The growth medium was carried from the reservoirs to the container by sterile tubes at a constant rate. The samples were prepared in a laminar air flow chamber to avoid contamination. After preparation, the samples were placed on a level platform inside an incubator and kept undisturbed for 12 h at 30 °C to allow the bacteria to attach to the sand particles. Subsequently the sterilized biomass-free growth medium was allowed to flow through the samples hosting live DRG3 population kept inside the incubator at a constant rate of 0.012 mm/s – a typical groundwater velocity. Another set of samples, similarly prepared, were kept undisturbed in the incubator without any flow of medium to assess the desired parameters at zero velocity. Microbial growth, changes in pH, amounts of biogenic EPS and calcite, and saturated and unsaturated water content and permeability of soil matrix with or without live bacterial population were monitored periodically at an interval of 24 h up to 48 h.

2.5   Microbial Growth and pH

Microbial growth within sand samples with live DRG3 population was measured by collecting the effluent from the sample and estimating the number of colony forming unit (CFU) using serial dilution technique (Ghatak et al. 2015). For samples with no flow of medium, CFU was measured from the effluent collected during permeability measurement. The pH of effluent was measured by a digital pH meter (Oakton, Eutech Instruments).

2.6   EPS and Calcite

For EPS estimation, a portion of sand was collected from bacteria-inoculated samples, oven dried at 70 °C and vortexed with 10 ml sterilized distilled water for 30 min. The supernatant was then centrifuged at 5000 rpm for 10 min at 4 °C to remove bacterial cells to extract EPS. Subsequently, the cell-free supernatant was mixed with double volume cold acetone and kept for 24 h at 4 °C. The mixture was further centrifuged at 12000 rpm for 20 min at 4 °C. The pellet was collected and lyophilized to get dry EPS powder.

Biogenic calcite was measured by a simple gravimetric method (Bauer et al. 1972) for which a portion of sand was collected from bacteria-inoculated samples and oven dried at 70 °C. 5 g of oven dried sand was mixed with 7 ml of 5 N HCl in a beaker and the solution was stirred from time to time for 30 min. The initial and final weights of the beaker containing sand sample and HCl were taken and the difference in weight is the measure of CO_2 evolved through the reaction: $CaCO_3 + 2HCl \rightarrow CaCl_2 + CO_2 \uparrow + H_2O$. Amount of biogenic calcite was estimated stoichiometrically from the measured amount of evolved CO_2.

2.7   Saturated and Unsaturated Water Content

For measuring saturated and unsaturated water content, cylindrical tin cans of negligible wall thickness were pushed into both saturated and unsaturated samples with live DRG3 population and the whole setup was allowed to freeze. Undisturbed samples were then extracted and oven dried at 70 °C. The initial and final weights of the undisturbed sample were taken and the weight difference gives the weight of water for saturated and unsaturated samples. Saturated and unsaturated water contents were then estimated from the ratio of weight of water and weight of dry soil of the respective soil samples. The samples were saturated fully by allowing the sterilized distilled water to flow at the rate of approximately 50 ml/min through the samples and checking the flow rate at constant head from time to time. This was continued until the increasing trend of flow rate diminishes and the flow rate becomes constant.

2.8   Degree of Saturation

Degree of saturation of unsaturated samples was estimated using the following expression assuming the degree of saturation of the saturated samples as 100%.

$$S = \frac{w_{unsat}}{w_{sat}} \qquad (1)$$

Since, $S \times e = w_{unsat} \times G_S$ and $e = w_{sat} \times G_S$ where, S is the degree of saturation of DRG3 treated samples, e is the void ratio of these samples (same for saturated and unsaturated condition), G_S is the specific gravity of soil solids, w_{sat} and w_{unsat} are the saturated and unsaturated water content of treated samples respectively.

2.9   Saturated and Unsaturated Permeability

The permeability of loose bacteria-free sample (k), was estimated from standard permeability test. Then the flow rate of loose bacteria-free saturated sample q was assessed by placing similar sample on a level platform and allowing sterilized distilled water through the sample under a constant head. The flow rates for the saturated and unsaturated DRG3-inoculated samples, q_{sat} and q_{unsat} respectively, were similarly measured for the same constant head. Finally, the permeability of loose saturated and unsaturated DRG3-inoculated samples, k_{sat} and k_{unsat} respectively, were computed by the expressions, $\frac{k_{sat}}{k} = \frac{q_{sat}}{q}$ and $\frac{k_{unsat}}{k} = \frac{q_{unsat}}{q}$.

3   Results and Discussions

3.1   Microbial Growth, pH and Metabolic Products

With increasing flow velocity CFU was found to increase (Fig. 1) indicating enhanced microbial population growth possibly resulting from greater nutrients availability. Under zero circulation, soil pH decreased with incubation duration (Fig. 1) because of production of biogenic acidic intermediates and/or onset of death phase resulting from nutrient depletion, while soil pH increased marginally with increasing flow velocity (Fig. 1) possibly because of continued nutrient availability and washing out of dissolved acidic metabolites. Increased EPS and calcite production with flow velocity (Fig. 1) resulted from increased nutrient availability and microbial population growth.

3.2   Water Content and Degree of Saturation

Saturated and unsaturated water content for treated samples was all lesser than corresponding water content of the untreated sample (Tables 1 and 2). Degree of saturation was monotonically decreasing with incubation duration for all processes (Fig. 2). For all three processes, development of unsaturation was more for the samples exposed to the flow of medium, probably because of enhanced growth of bacteria which resulted in more evolving of CO_2. Figure 2 further revealed that non-calcifying process developed maximum unsaturation whereas non-ureolytic calcifying process developed least unsaturation. Absence of mineralization could be the possible reason for higher unsaturation observed in non-calcifying process since a part of evolved CO_2 was utilized for biocalcification.

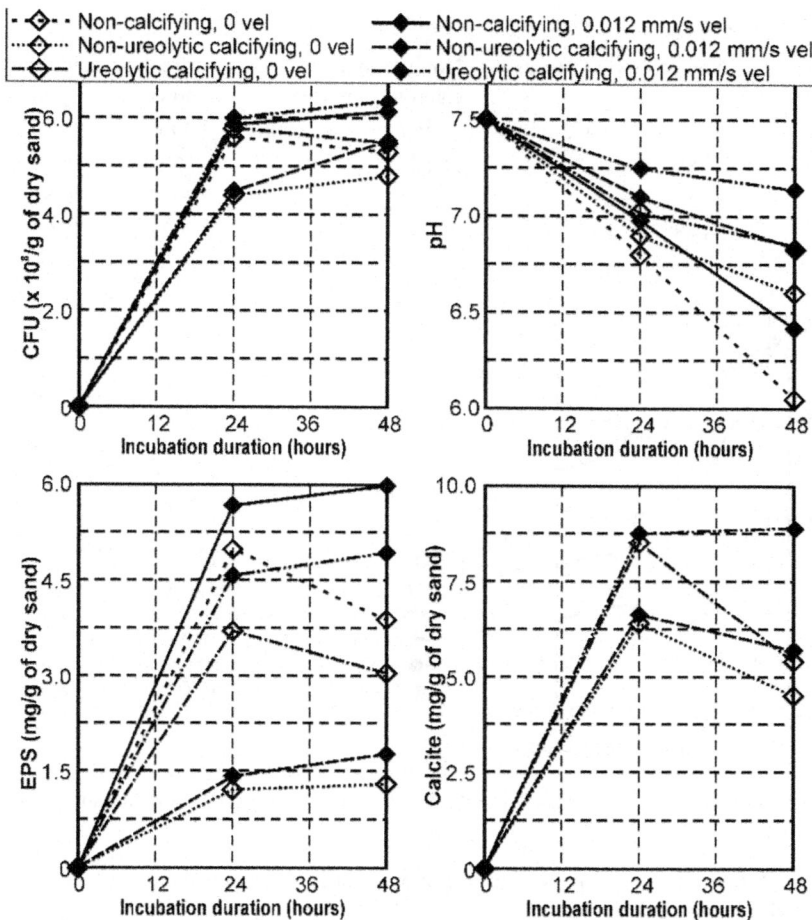

Fig. 1. Influence of incubation duration and flow velocity on CFU, pH, EPS and calcite

Table 1. Influence of incubation duration and flow velocity on unsaturated water content (%).

Incubation duration (hours)	Non-calcifying process		Non-ureolytic calcifying process		Ureolytic calcifying process	
	0 mm/s	0.012 mm/s	0 mm/s	0.012 mm/s	0 mm/s	0.012 mm/s
0	26.89	26.89	26.89	26.89	26.89	26.89
24	23.71	22.53	25.09	24.97	24.32	23.41
48	23.56	22.08	24.81	24.36	24.12	23.16

3.3 Permeability

The unsaturated and saturated permeability was monotonically decreasing with incubation duration, for all the processes and in the presence and absence of flow (Tables 3 and 4). Evidently for all the samples, the saturated permeability was higher than the

Table 2. Influence of incubation duration and flow velocity on saturated water content (%).

Incubation duration (hours)	Non-calcifying process		Non-ureolytic calcifying process		Ureolytic calcifying process	
	0 mm/s	0.012 mm/s	0 mm/s	0.012 mm/s	0 mm/s	0.012 mm/s
0	26.89	26.89	26.89	26.89	26.89	26.89
24	26.34	26.32	26.69	26.68	26.57	26.54
48	26.48	26.36	26.74	26.69	26.67	26.63

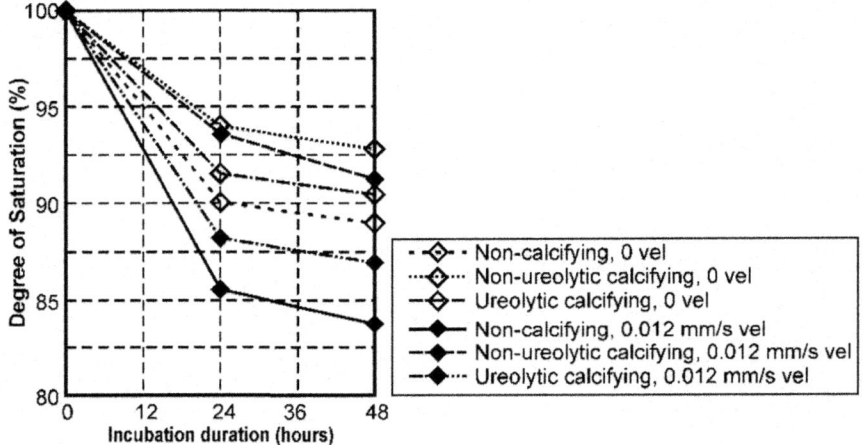

Fig. 2. Influence of incubation duration and flow velocity on degree of saturation

unsaturated one, due to removal of air from voids. The unsaturated and saturated permeability measured in the samples with flow of medium was lesser than the ones without flow of medium, probably because the enhanced microbial growth due to continuous supply of nutrients contributed to higher amount of EPS production (Tables 3 and 4).

Table 3. Influence of incubation duration and flow velocity on unsaturated permeability $(\times 10^{-2}$ cm/s).

Incubation duration (hours)	Non-calcifying process		Non-ureolytic calcifying process		Ureolytic calcifying process	
	0 mm/s	0.012 mm/s	0 mm/s	0.012 mm/s	0 mm/s	0.012 mm/s
0	4.94	4.94	4.94	4.94	4.94	4.94
24	2.08	1.17	2.65	2.35	2.40	1.36
48	1.71	0.99	2.52	2.04	2.16	1.27

The non-calcifying process was observed to induce highest reduction in unsaturated permeability than the other two processes as this process yielded largest EPS

Table 4. Influence of incubation duration and flow velocity on saturated permeability ($\times 10^{-2}$ cm/s).

Incubation duration (hours)	Non-calcifying process		Non-ureolytic calcifying process		Ureolytic calcifying process	
	0 mm/s	0.012 mm/s	0 mm/s	0.012 mm/s	0 mm/s	0.012 mm/s
0	5.52	5.52	5.52	5.52	5.52	5.52
24	3.29	2.73	4.95	4.68	3.54	2.47
48	3.77	2.56	4.77	4.35	4.02	2.20

production and unsaturation (Fig. 3). Reduction in saturated permeability was also found to be highest for non-calcifying process in the presence and absence of flow probably due to highest EPS production. With or without media circulation non-ureolytic calcifying process yielded minimum reduction in permeability since it had least amount of both calcite and EPS precipitation (Fig. 3).

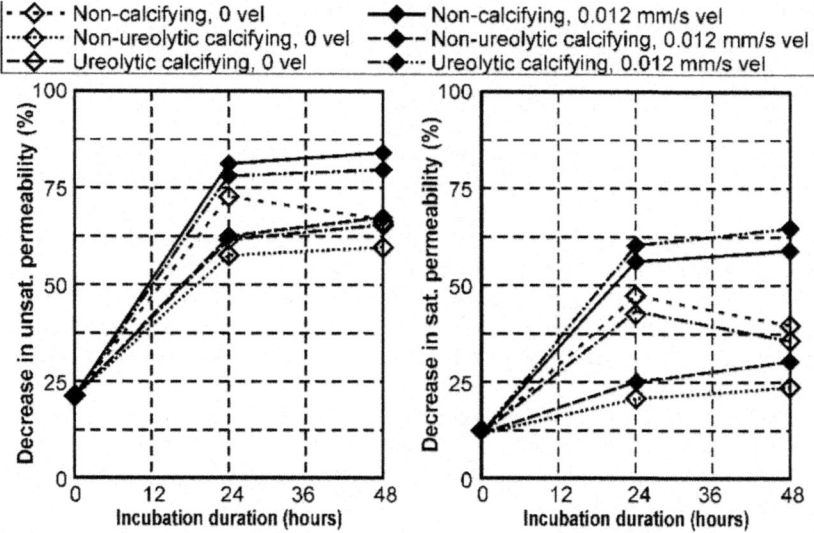

Fig. 3. Influence of incubation duration and flow velocity on saturated and unsaturated permeability

4 Conclusions

In the absence of media circulation, EPS and calcite precipitation was found to increase initially and then decrease with increasing incubation duration due to subsequent nutrient depletion. Whereas with increasing flow velocity microbial population growth as well as EPS and calcite production was found to increase due to increased nutrient availability. pH for all the processes noted to decrease with incubation duration, whereas it increased marginally if the growth medium was continuously supplied.

Degree of saturation had lowered down to 84% for DRG3-inoculated samples with flow of medium and to 89% for samples without flow of medium due to biogenic unsaturation. Higher unsaturation in samples with continuous flow of medium was developed due to the enhanced growth of bacteria and biogas generation. Data obtained in this study further exhibited that reduced permeability for bacteria treated samples was 58% to 84% of the original permeability in unsaturated samples and 21% to 59% of the original permeability in saturated samples. It indicates that biogenic unsaturation does have a significant effect on soil permeability. The samples with flow of medium ended up with higher reduction in permeability because of increased quantity of biogenic precipitation and higher unsaturation.

References

Al Qabany, A., Soga, K., Santamarina, C.: Factors affecting efficiency of microbially induced calcite precipitation. J. Geotech. Geoenviron. Eng. ASCE **138**(8), 992–1001 (2012)

ASTM: Standard Test Methods for Minimum Index Density and Unit Weight of Soils and Calculation of Relative Density. ASTM, West Conshohocken (2006)

Bauer, H.P., Beckett, P.H.T., Bie, S.W.: A rapid gravimetric method for estimating calcium carbonate in soils. Plant Soil **37**, 689–690 (1972)

Bonala, M.V.S., Reddi, L.N.: Physicochemical and biological mechanisms of soil clogging: an overview. ASCE GSP **78**, 43–68 (1998)

Cunningham, A.B., Characklis, W.G., Abedeen, F., Crawford, D.: Influence of biofilm accumulation on porous media hydrodynamics. Environ. Sci. Technol. **25**(7), 1305–1311 (1991)

Dupin, H.J., McCarty, P.L.: Impact of colony morphologies and disinfection on biological clogging in porous media. Environ. Sci. Technol. **34**, 1513–1520 (2000)

Elrich, H.L., Newman, D.K.: Geomicrobiology, 5th edn. Taylor and Francis, Boca Raton (2009)

Ghatak, S., Manna, S., Roy, D.: First-order assessment of the influence of three EPS and calcite producing microbes isolated from a cemented sand site on soil shear strength. Geomicrobiol. J. **32**, 761–770 (2015)

Ivanov, V., Chu, J.: Applications of microorganisms to geotechnical engineering for bioclogging and biocementation of soil in situ. Rev. Environ. Sci. Biotechnol. **7**, 139–153 (2008)

Karimi, S.: A study of geotechnical applications of biopolymer treated soils with an emphasis on silt. Ph.D. Dissertation, University of Southern California (1998)

Khachatoorian, R., Petrisor, I.B., Kwan, C.-C., Yen, T.F.: Biopolymer plugging effect: laboratory-pressurized pumping flow studies. J. Pet. Sci. Eng. **381**(2), 13–21 (2003)

Martinez, B.C., DeJong, J.T., Ginn, T.R., Montoya, B.M., Barkouki, T.H., Hunt, C., Tanya, B., Major, D.: Experimental optimization of microbial induced carbonate precipitation for soil improvement. J. Geotech. Geoenviron. Eng. ASCE **139**(4), 587–598 (2013)

Nemati, M., Greene, E.A., Voordouw, G.: Permeability profile modification using bacterially formed calcium carbonate: comparison with enzymic option. Process Biochem. **40**, 925–933 (2005)

Proto, C.J., DeJong, J.T., Nelson, D.C.: Biomediated permeability reduction of saturated sands. J. Geotech. Geoenviron. Eng. ASCE **142**(12), 1179–1189 (2016)

Seki, K., Miyazaki, T., Nakano, M.: Effects of microorganisms on hydraulic conductivity decrease in infiltration. Eur. J. Soil Sci. **49**, 231–236 (1998)

Vandevivere, P., Baveye, P.: Relationship between transport of bacteria and their clogging efficiency in sand columns. Appl. Environ. Microbiol. **58**(8), 2523–2530 (1992)

Whiffin, V.S., van Paassen, L.A., Harkes, M.P.: Microbial carbonate precipitation a soil improvement technique. Geomicrobiol. J. **24**, 1–7 (2007)

Adsorption of Cr(VI) onto SHMP-Amended Ca-Bentonite Backfills for Slurry-Trench Cutoff Walls

Yu-Ling Yang[1], Yan-Jun Du[1(✉)], Krishna R. Reddy[2], and Ri-Dong Fan[1]

[1] Jiangsu Key Laboratory of Urban Underground Engineering and Environmental Safety, Institute of Geotechnical Engineering, Southeast University, Nanjing 210096, China
duyanjun@seu.edu.cn
[2] Department of Civil and Materials Engineering, University of Illinois at Chicago, Chicago, IL 60607, USA

Abstract. Several series of batch adsorption tests are conducted using slurry-trench cutoff wall backfills composed of sand and 20% Ca-bentonite with and without sodium hexametaphosphate (SHMP) amendment and also using Ca-bentonite alone with and without SHMP amendment. First, amended/un-amended backfills and Ca-bentonites are prepared. Next, aqueous solutions with 0 to 1000 mg/L hexavalent chromium [Cr(VI)] concentration are prepared. Batch adsorption tests are conducted at a liquid to solid ratio of 10:1 to investigate the adsorption of Cr(VI) onto the backfills and Ca-bentonites. The results show that phosphate dispersant causes significant increase in Cr(VI) adsorption of both the soil/Ca-bentonite backfill and the Ca-bentonite. The relationship between the amount of adsorption and the equilibrium liquid concentration is assessed using Langmuir, Freundlich, and Dubinin-Radushkevich isotherm models. The adsorption of Cr(VI) on the backfills or Ca-bentonites is attributed to chemical adsorption mechanism.

Keywords: Calcium bentonite · Slurry walls · Adsorption · Chromium

1 Introduction

Soil-bentonite (SB) slurry trench cutoff walls, composed of backfill prepared using the excavated soil (generally sand with or without clay fraction) and high quality sodium bentonite (Na-bentonite), are commonly used as engineered barriers for containment of the contaminated groundwater (D'Appolonia 1980). The hydraulic conductivity and the contaminant adsorption capacity are two of the key parameters that control the containment efficiency of the walls. Recently, significant attention has been paid on improving contaminant adsorption of the SB walls to passively delay the transport of contaminants through the slurry walls (Malusis et al. 2010; Fan et al. 2014; Hong et al. 2016).

In countries such as China, there is a great potential to use slurry walls at many contaminated sites for the containment of contaminated groundwater. However, high

© Springer Nature Singapore Pte Ltd. 2018
A. Farid and H. Chen (Eds.): GSIC 2018, *Proceedings of GeoShanghai 2018*
International Conference: Geoenvironment and Geohazard, pp. 434–441, 2018.
https://doi.org/10.1007/978-981-13-0128-5_48

quality Na-bentonite is scarce and the commonly available calcium bentonite (Ca-bentonite) is considered as an alternate material for use (Du et al. 2015a, 2015b, 2015c). Due to lower swelling and adsorption capacity as compared to Na-bentonite, the Ca-bentonite may not be considered suitable to use for soil/Ca-bentonite backfills to provide low hydraulic conductivity and high retardation of the contaminants, thus modification of the Ca-bentonite is found necessary before its use.

The authors have conducted extensive studies and demonstrated that sodium hexametaphosphate (SHMP, dispersant) is an effective amendment to enhance workability of the Ca-bentonite/water slurry and reduce hydraulic conductivity of the soil/Ca-bentonite backfills (Yang et al. 2017a, 2017b, 2017c). The optimum SHMP dosage for the Ca-bentonite amendment is found to be 2% (dry weight ratio of SHMP to Ca-bentonite), and optimum SHMP-amended Ca-bentonite content for the soil/Ca-bentonite backfill is found to be 20% (dry weight ratio of the amended Ca-bentonite to the backfill). However, adsorption characteristics of such soil/Ca-bentonite backfill to contaminants such as hexavalent chromium [Cr(VI)], which is ubiquitous at industrial contaminated sites in China, are still unknown.

The main objective of this study are to: (1) investigate the adsorption of Cr(VI) to amended/unamended sand/Ca-bentonite backfill as well as amended/unamended Ca-bentonite alone via batch adsorption tests; (2) assess the adsorption mechanism of the Cr(VI) onto the backfills and bentonites using Langmuir, Freundlich, and Dubinin-Radushkevich (D-R) isotherm models; and (3) compare the Cr(VI) adsorption capacity of the backfills and bentonites before and after SHMP amendment.

2 Materials and Methods

2.1 Materials

Powdered unamended Ca-bentonite (CaB) and SHMP-amended Ca-bentonite (SHMP-CaB), clean sand, and granular SHMP are used as the source materials in this study. The CaB (Panther Creek® 200) is provided by the Colloid Environmental Technologies Company (CETCO, Hoffman Estates, IL, USA). The SHMP-CaB is prepared in the laboratory by the authors by mixing the CaB with SHMP-distilled water solution with SHMP concentration of 10 g/L. The soil-to-water ratio of the resulting slurry is approximately 1:2 (g/mL), and the optimum SHMP-to-bentonite ratio is 2% (dry weight basis). The slurry is then cured, oven dried, milled and screened (using 0.075 mm standard sieve) following procedures described by Yang et al. (2017a, 2017b).

The sand known as Three River Sand is obtained from a quarry in Beloit, WI, USA. The SHMP used is obtained from Humboldt Manufacturing Co., Elgin, IL, USA. Four types of materials are tested in this study: (1) amended bentonite (SHMP-CaB), (2) unamended bentonite (CaB), (3) amended backfill containing 20% of SHMP-CaB (SHMP-20CaB), and (4) unamended backfill mixture containing 20% of CaB (20CaB). The bentonite content of the sand/Ca-bentonite backfill is 20%.

A stock solution of Cr(VI) with concentration of 1000 mg/L is prepared by dissolving a predetermined amount of potassium dichromate ($K_2Cr_2O_7$) in deionized water. Then, Cr(VI) solutions with different initial concentrations (C_0) ranging from 0.1 to 1000 mg/L are prepared by diluting the stock solution with deionized water.

2.2   Batch Testing Procedure

The batch adsorption tests are performed as per the guidelines of ASTM D4646, except that a soil-to-solution ratio of 1:10 is used as recommended by Reddy et al. (2014). 10 g of selected backfill or bentonite and 100 mL of Cr(V) solution are placed in a 200 mL plastic bottle and they are mixed for 24 h using a mechanical tumbler at room temperature. Then, the solids and supernatant are separated by centrifugation at a speed of 7500 rmp. The equilibrium Cr(VI) concentration in the supernatant (C_e) is measured using flame atomic absorption spectrophotometer (Perkin Elmer, CT). Weng et al. (2008) reported an contact time of 1.5 h for Cr(VI) equilibrium adsorption onto clay. Thus, it is reasonable to believe that 24 h is sufficient for establishing equilibrium between the Cr(VI) solution and the tested backfill or bentonite during adsorption process.

3   Results and Analysis

The amount of Cr(VI) adsorbed to the backfill or bentonite (q_e) is calculated using below equation:

$$q_e = \frac{(C_0 - C_e) M_s}{V_L} \tag{1}$$

where M_s is the mass of the dry backfill or bentonite used (kg); and V_L is the volume of Cr(VI) solution added (L).

Figure 1 shows the measured adsorption data (q_e versus C_e) for the tested backfills and Ca-bentonite with and without SHMP amendment. These results show that all of the backfills and bentonites exhibit nonlinear Cr(VI) adsorption behavior, with steep increase in adsorbed concentration (q_e) initially until the equilibrium concentration (C_e) of approximately 420 mg/L, and thereafter it increased gradually until C_e reached 820 mg/L. It should be known that there are other 6 data points for each backfill mixture which displayed overlapped near the origin of the coordinate in Fig. 1. The amended backfill (SHMP-20CaB) and amended bentonite (SHMP-CaB) display greater Cr(VI) adsorption capacity than their respective control or unamended backfill or bentonite, i.e., 20CaB or CaB.

Three nonlinear isotherm adsorption models, including Langmuir model, Freundlich model, and Dubinin-Radushkevich (D-R) model, are applied for fitting adsorption data as presented in Fig. 1. The Langmuir model is given by following equation (Langmuir 1916; Mckay et al. 1982):

$$q_e = \frac{K_L q_{m,L} C_e}{1 + K_L C_e} \tag{2}$$

where $q_{m,L}$ is the maximum adsorption capacity; K_L is the Langmuir isotherm constant; C_e is the equilibrium adsorbate (Cr(VI)) concentration. The Freundlich model can be given as follows (Freundlich 1906; Mckay et al. 1982):

$$q_e = K_F C_e^{1/n_F} \tag{3}$$

where the K_F is the Freundlich isotherm constant; n_F is the adsorption intensity and favorable adsorption is referred when n_F between 1 to 10. The D-R model is given by (Dubinin and Radushkevich 1947; Foo and Hameed 2010):

$$q_e = q_{m,DR} \exp\left(-K_{DR}\varepsilon^2\right) \tag{4}$$

where q_e, and C_e are the same as defined above but with unit of mol/kg, and mol/L, respectively; $q_{m,DR}$ is the saturation adsorption capacity; K_{DR} is D-R isotherm constant; and ε is the Polanyi potential:

$$\varepsilon = RT\ln\left(1 + \frac{1}{C_e}\right) \tag{5}$$

in which R is the ideal gas constant; T is the absolute temperature; E is the mean free energy of adsorption which can be given by following expression (Foo and Hameed 2010; Fan et al. 2014)

$$E = \frac{1}{\sqrt{2K_{DR}}} \tag{6}$$

If E ranges from 8 to 16 kJ/mol, the chemical adsorption is considered to be the predominant mechanism (Vijayaraghavan et al. 2006).

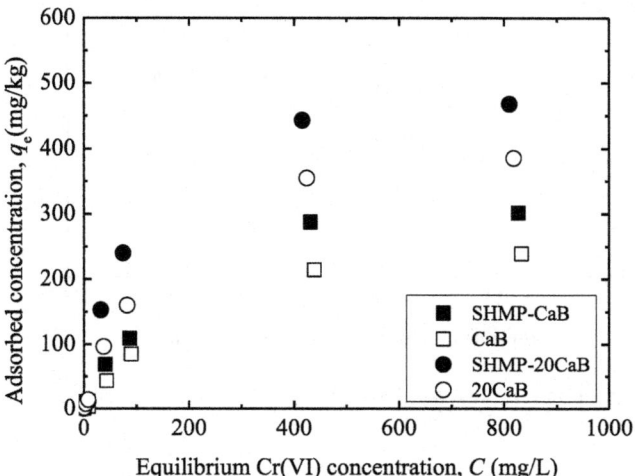

Fig. 1. Measured Cr(VI) adsorption data for backfills and Ca-bentonite with and without SHMP amendment

The measured data are fitted by Langmuir, Freundlich, and D-R models as shown in Figs. 2, 3 and 4, respectively. The relationship between equilibrium Cr(VI) concentration and adsorbed amount are well fitted by all of the three models with high coefficient of determination (r^2) values to the experimental data, e.g., $0.939 \leq r^2 \leq 0.996$.

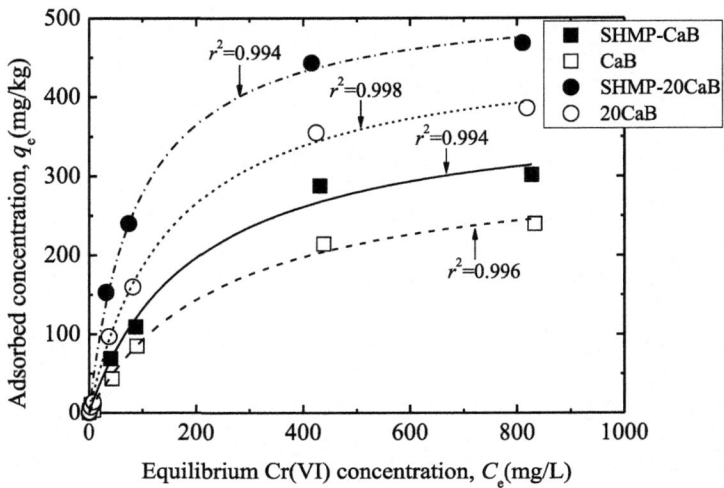

Fig. 2. Langmuir fitting results for backfills and Ca-bentonite with and without SHMP amendment

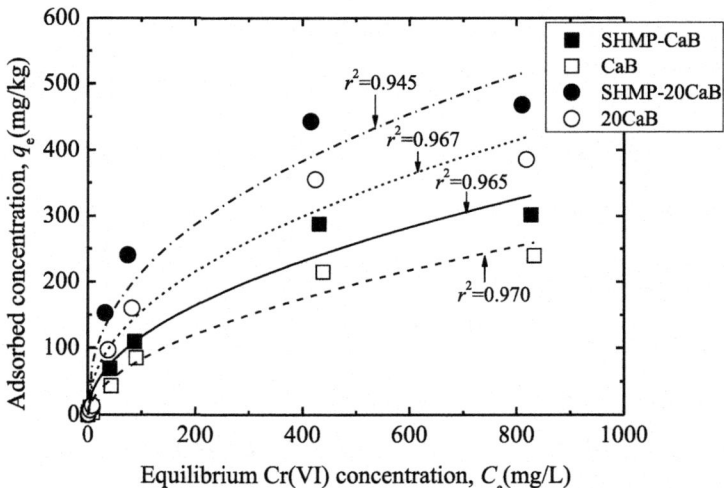

Fig. 3. Freundlich fitting results for backfills and Ca-bentonite with and without SHMP amendment

The best-fit parameter values for the Langmuir, Freundlich, and D-R isotherm models are summarized in Table 1. Based on results of both Langmuir and D-R fitting, the SHMP-CaB possesses 49 to 73 mg/kg higher adsorption capacity than that of the CaB, and the SHMP-20CaB exhibits 62 to 88 mg/kg higher adsorption capacity than the 20CaB. The results indicate that the Cr(VI) adsorption capacities of both the Ca-bentonite and sand/Ca-bentonite backfill are significantly increased by SHMP amendment. This can be attributed to lower pH values of the SHMP amended bentonite and backfill than those of the unamended bentonite and backfill. This finding is consistent with Reddy et al. (1997) who found lower pH conditions favor greater Cr(VI) adsorption, while high pH conditions lead to lower Cr(VI) adsorption. It is noted that the adsorption capacity of each backfill or bentonite calculated from the Langmuir model is lower than that obtained from the D-R model. This is attributed to different assumption on adsorption behavior of the adsorbate onto the surface of the adsorbent in these two models. For example, the Langmuir model assumes that adsorption occurs on monolayer coverage sites within the adsorbent, while the D-R model considers the total specific micropore volume of the adsorbent.

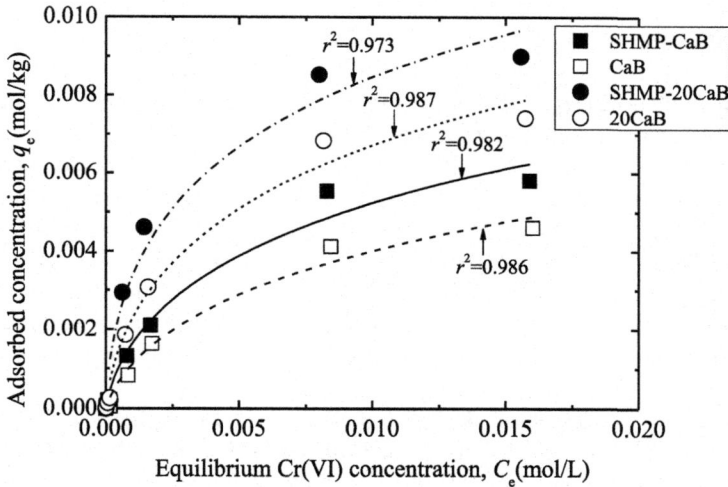

Fig. 4. D-R fitting results for backfills and Ca-bentonite with and without SHMP amendment

The K_L value of the backfills, i.e., 20CaB and SHMP-20CaB, is greater than that of the bentonite samples, i.e., CaB and SHMP-CaB, implying more favorable adsorption of Cr(VI) to the backfills as compared to the bentonite samples (Hong et al. 2016). The adsorption intensity (n_F) value of the tested samples ranges from 1.86 to 2.38, indicating favorable adsorption. The value of the free energy obtained from D-R model ranges from 7.84 to 9.24 kJ/mol, indicating the dominant adsorption mechanism is chemical adsorption.

Table 1. Summary of best-fit Langmuir, Freundlich, and D-R model parameters for Cr(VI) adsorption onto backfill mixtures.

Fitting parameters	CaB	SHMP-CaB	20CaB	SHMP-20CaB
Langmuir model				
$q_{m,L}$ (mg/kg)	318	367	467	529
K_L (L/mg)	0.00412	0.00481	0.00659	0.0111
r^2	0.978	0.987	0.996	0.985
Freundlich model				
K_F (L/kg)	7.0	10.2	17.5	30.8
n_F	1.86	1.97	2.11	2.38
r^2	0.953	0.982	0.967	0.939
D-R model				
$q_{m,DR}$ (mg/kg)	582	655	832	920
K_{DR} (mol^2/kJ^2)	0.00814	0.00751	0.00686	0.00585
E (kJ/mol)	7.84	8.16	8.54	9.24
r^2	0.968	0.989	0.985	0.965

4   Conclusions

The present study investigates the effects of SHMP amendment on adsorption of Cr(VI) onto sand/Ca-bentonite backfills containing 20% bentonite as well as bentonite alone via a series of batch adsorption tests. The results reveal that all of the amended/unamneded backfill mixtures and the amended/unamended bentonites exhibit nonlinear adsorption behavior to the Cr(VI). The Langmuir, Freundlich, and Dubinin-Radushkevich isotherm model are capable of describing the experiment data well. The Cr(VI) adsorption capacity of both sand/Ca-bentonite backfill and Ca-bentonite is significantly increased by the SHMP amendment. The chemical adsorption is considered to be the dominant adsorption mechanism for Cr(VI) adsorption onto the tested backfills and bentonites.

Acknowledgements. Financial support for this project is provided by National Natural Science Foundation of China (Grant No. 41330641, 51278100 and 41472258) and Primary Research & Development Plan of Jiangsu Province (Grant No. BE2017715). The authors thank the Colloid Environmental Technologies Co. (CETCO) for providing the bentonites used in this study.

References

D'Appolonia, D.J.: Soil-bentonite slurry trench cutoffs. J. Geotech. Eng. Div., ASCE **106**(4), 399–417 (1980)

Du, Y.J., Fan, R.D., Liu, S.Y., Reddy, K.R., Jin, F.: Workability, compressibility and hydraulic conductivity of zeolite-amended clayey soil/calcium-bentonite backfills for slurry-trench cutoff walls. Eng. Geol. **195**, 258–268 (2015a)

Du, Y.J., Fan, R.D., Reddy, K.R., Liu, S.Y., Yang, Y.L.: Impacts of presence of lead contamination in clayey soil–calcium bentonite cutoff wall backfills. Appl. Clay Sci. **108**, 111–122 (2015b)

Du, Y.J., Yang, Y.L., Fan, R.D., Wang, F.: Effects of phosphate dispersants on the liquid limit, sediment volume and apparent viscosity of clayey soil/calcium-bentonite slurry wall backfills. KSCE J. Civ. Eng. **20**(2), 670–678 (2015c)

Dubinin, M.M., Radushkevich, L.V.: Equation of the characteristic curve of activated charcoal. Chem. Zentralbl. **55**, 331–337 (1947)

Fan, R.D., Du, Y.J., Liu, S.Y., Yang, Y.L.: Sorption of Pb(II) from aqueous solution to clayey soil/calcium-bentonite backfills for slurry-trench walls. In: 7th International Congress on Environmental Geotechnics, pp. 1566–1573. Engineers Australia, Melbourne (2014)

Freundlich, H.M.F.: Over the adsorption in solution. J. Chem. Phys. **57**, 385–471 (1906)

Foo, K.Y., Hameed, B.H.: Insights into the modeling of adsorption isotherm systems. Chem. Eng. J. **156**(1), 2–10 (2010)

Hong, C.S., Shackelford, C.D., Malusis, M.A.: Adsorptive behavior of zeolite-amended backfills for enhanced metals containment. J. Geotech. Geoenviron. Eng. **142**(7), 04016021 (2016)

Langmuir, I.: The constitution and fundamental properties of solids and liquids. Part I. Solids. J. Am. Chem. Soc. **38**(11), 2221–2295 (1916)

Malusis, M.A., Maneval, J.E., Barben, E.J., Shackelford, C.D., Daniels, E.R.: Influence of adsorption on phenol transport through soil–bentonite vertical barriers amended with activated carbon. J. Contam. Hydrol. **116**(1), 58–72 (2010)

Mckay, G., Blair, H.S., Gardner, J.R.: Adsorption of dyes on chitin. I. equilibrium studies. J. Appl. Polym. Sci. **27**(8), 3043–3057 (1982)

Reddy, K.R., Parupudi, U.S., Devulapalli, S.N., Xu, C.Y.: Effects of soil composition on the removal of chromium by electrokinetics. J. Hazard. Mater. **55**(1–3), 135–158 (1997)

Reddy, K.R., Xie, T., Dastgheibi, S.: Removal of heavy metals from urban stormwater runoff using different filter materials. J. Environ. Chem. Eng. **2**(1), 282–292 (2014)

Vijayaraghavan, K., Padmesh, T.V.N., Palanivelu, K., Velan, M.: Biosorption of nickel (II) ions onto Sargassum wightii: application of two-parameter and three-parameter isotherm models. J. Hazard. Mater. **133**(1), 304–308 (2006)

Weng, C.H., Sharma, Y.C., Chu, S.H.: Adsorption of Cr(VI) from aqueous solutions by spent activated clay. J. Hazard. Mater. **155**(1), 65–75 (2008)

Yang, Y.L., Du, Y.J., Reddy, K.R., Fan, R.D.: Effect of phosphate dispersant amendment on workability of Ca-bentonite slurry for slurry trench cutoff-wall construction. Indian Geotech. J. **47**(4), 445–452 (2017a)

Yang, Y.L., Du, Y.J., Reddy, K.R., Fan, R.D.: Phosphate-amended sand/Ca-bentonite mixtures as slurry trench wall backfills: assessment of workability, compressibility and hydraulic conductivity. Appl. Clay Sci. **142**, 120–127 (2017b)

Yang, Y.L., Reddy, K.R., Du, Y.J., Fan, R.D.: SHMP Amended calcium bentonite for slurry trench cutoff walls: workability and microstructure characteristics. Can. Geotech. J. **55**(4), 528–537 (2017c). https://doi.org/10.1139/cgj-2017-0291

Engineering Properties of Heavy Metal Contaminated Soil Solidified/Stabilized with High Calcium Fly Ash and Soda Residue

Jingjing Liu[1], Fusheng Zha[1(✉)], Long Xu[1], Yongfeng Deng[2], and Chengfu Chu[1]

[1] Hefei University of Technology, Hefei 230009, China
geozha@hfut.edu.cn
[2] Institute of Geotechnical Engineering, School of Transportation, Southeast University, Nanjing 210096, China

Abstract. Both high-calcium fly ash and soda residue are the industrial by-products generated during the combustion of coal for energy production and sodium carbonate production, respectively. Previous studies suggested that both high-calcium fly ash and soda residue can be used to treat heavy metal contaminated soils. Therefore, a series of laboratory experiments were performed to investigate the strength and leaching properties of Pb-contaminated soils solidified/stabilized by high calcium fly ash and soda residue. The results presented that adding high-calcium fly ash to the contaminated soils can significantly improve the soil strength and reduce the Pb^{2+} amount released out. The specimens treated with soda residue got a higher strength, but a relatively lower chemical stability than that of fly ash. SEM results confirmed that $Ca(OH)_2$ and CSH was the main products that improved the engineering properties of the stabilized contaminated soils.

Keywords: Fly ash · Soda residue · Heavy metal contaminated soil
Solidification/stabilization · Unconfined compressive strength
Leaching characteristics

1 Introduction

Solidification/stabilization has emerged as one of the most effective remediation technologies of heavy metal contaminated soils utilization of additives as cement, fly ash and lime.

High calcium fly ash (defined as class C fly ash), which is the by-product of the electric power thermal plants using lignite as the main combustion material. The main compositions of high calcium fly ash includes SiO_2, Al_2O_3, CaO and residual carbon that doesn't burn. Therefore, this material is endowed with capacities of strong adsorption and hydration activity [1]. Researchers have done lots of work on stabilizing heavy metal contaminated soils by utilizing fly ash. The results proved that fly ash was an available binder to treat contaminated soils, which can increase the soil strength and immobilize heavy metals [2–4]. In China, systematic research on utilizing fly ash to solidify/stabilize heavy metal contaminated soils are rarely reported.

© Springer Nature Singapore Pte Ltd. 2018
A. Farid and H. Chen (Eds.): GSIC 2018, *Proceedings of GeoShanghai 2018*
International Conference: Geoenvironment and Geohazard, pp. 442–449, 2018.
https://doi.org/10.1007/978-981-13-0128-5_49

Soda residue is a by-product generated during the ammonia-soda process. Its main compositions concluded $CaCO_3$, $CaSO_4$ and some oxides of aluminum, iron and silicon. Soda residue was extensively used in bolus and embankment engineering as an engineering soil [5, 6]. Besides that, soda residue was composed of the very fine particles, most of which size was less than 0.074 mm. The specific surface area was extremely high, which attributed to the high adsorptive capacity of soda residue. In recent year, researches have been performed on the soda residue to study the adsorptive capacity no matter for the organic or inorganic matters [7]. And the mechanical properties of soda residue or blended with the other materials have been investigated as well [8]. However, few reports on the results of heavy metal contaminated soils solidified/stabilized with soda residue.

The purpose of this study is to investigate the effectiveness of using high-calcium fly ash and soda residue as additives to solidified/stabilized (s/s) heavy metal contaminated soil. Unconfined compressive strength (UCS) test was performed to investigate the mechanical properties and the chemical stabilities was assessed by toxicity characteristic leaching procedure (TCLP). Furthermore, the microstructure characteristic was analyzed with scanning electron microscope (SEM) technology.

2 Materials and Methods

2.1 Materials

The soil used in this investigation was sampled from a foundation pit in Hefei, China, which excavation depth was 4–5 m. Basic physic-mechanical indexes of the soil sample are shown in Table 1.

Table 1. Basic physic-mechanical properties of tested soils

Water content/%	Specific gravity	Density/g/cm³	Liquid limit/%	Plastic limit/%	Plastic index	Optimum water content/%	Maximum dry density/g/cm³
22.54	2.65	1.950	39.9	22.2	17.7	22.34	1.615

The main compositions of the two materials included SiO_2, CaO, Al_2O_3 and MgO, which was determined by X-ray fluorescence (XRF) and listed in Table 2. Otherwise, both fly ash and soda residue consisted of extremely fine particles, about 80% of which size was smaller than 0.075 mm. The two materials both possessed really high specific surface area (529 m^2/kg for fly ash and 793 m^2/kg for soda residue, respectively), which leaded to a strong capability of adsorption.

According to the previous investigation, Pb, Cd, Hg and As contaminated soils were widely distributed in China [9]. The paper choose Pb-contaminated soil as the research object, and $Pb(NO_3)_2$ was selected as the heavy metal contaminant for nitrate had negligible interference in hydration reactions.

Table 2. Compositions of fly ash and soda residue (% wt.)

Chemical compositions	SiO$_2$	Al$_2$O$_3$	CaO	Fe$_2$O$_3$	TiO$_2$	MgO	SO$_3$	K$_2$O	Na$_2$O
Fly ash	37.1	13.9	30.3	0.4	1.8	12.5	2.1	0.6	1.0
Soda residue	10.2	9.0	62.8	1.3	0.2	12.5	0.3	0.2	0.2

2.2 Methods

Soil Sample Preparation. In China, the common Pb^{2+} concentration of the industrial contaminated sites is approximately 5000 mg/kg [10]. And in order to study the effects of Pb^{2+} concentration variations on the stabilized soils properties, the Pb^{2+} concentration of the soils was designed as 1000 and 5000 mg/kg (denoted as Pb0.1 and Pb0.5, respectively).

The oven-dried soils and binders were initially pulverized and sieved with 2 mm and 0.5 mm sieves, respectively. The quantity of distilled water was calculated based on the optimum water content, which was divided into two equal parts. Pb(NO$_3$)$_2$ solutions were prepared by dissolving Pb(NO$_3$)$_2$ AR into one part of the distilled water. Mixing soil with the Pb(NO$_3$)$_2$ solutions thoroughly, and curing for 24 h. The other part of distilled water was used to make binder slurries by mixing with the predetermined mass of fly ash and soda residue, respectively. The mass ratio of fly ash to dry soil was set as 0.05, 0.1 and 0.15 (denoted as FA5, FA10 and FA15), and that of soda residue was 0.1, 0.2 and 0.3 (denoted as SR10, SR20 and SR30). Subsequently, pouring the binder slurries into the prepared Zn-contaminated soils and fully stirred. After that, the mixtures were filled into a cylindrical mould with dimensions of Φ5 cm × H5 cm, statically compacted. Demoulded until soil sample in the mould reached 95% of the maximum dry density. Finally, the specimens were wrapped with plastic bags and stored in the curing room under standard curing conditions (humidity 95 ± 5%, temperature 20 °C) until tested at 0, 3, 7 and 28 days.

Unconfined Compressive Strength. Unconfined compression strength test was conducted on the specimens based on the Road geo-technical testing procedures (JTG E40-2007). The test procedure is almost the same as ASTM D5102-09 (2017). The specimens had a length: diameter ratio of 1:1. The test was performed at a loading rate of 0.4 mm/min.

Toxicity Characteristic Leaching Procedure. In order to investigate the leaching characteristics of specimens, Toxicity Characteristic Leaching test was performed [11]. The diluted acetic acid (mixing 5.7 mL acetic acid with distilled water until the volume of solution reach to 1000 mL) was selected as the leachant, which pH value was 2.88 ± 0.05. The specimens after unconfined compression strength test were pulverized to pass through 2 mm sieve. Weighing 50 g of samples into polypropylene bottles. The bottles were horizontal vibrated at 80 rpm in a water-bathing constant temperature vibrator at 25 °C for 18 h. At the end of the extraction, the liquid was separated from the solids by filtration through a 0.45 μm glass fiber filter. The leachate pH value was measured using a Rex PHS-25 pH meter. The measurement range was 0.00–14.00 and the accuracy was 0.01. The leached Pb^{2+} concentration was determined by WYS2200

Atomic Adsorption Spectrometer. The electric current of the Pb lamp was 3 mA, the wave length was 283.0 nm, the slit width was 0.4 nm and the measurement range was 0.08–20 mg/L.

Scanning Electron Microscope. The samples used for SEM analysis were freeze-dried and broken slightly to pieces. Samples with a dimension approximated to 5 × 5 × 5 mm were selected, and their surface was cleaned using a hairbrush. To achieve a sufficient electrical conductivity, the sample's surface was pretreated by the vacuum metal spraying technology.

3   Results and Analysis

3.1   Unconfined Compression Strength

Curing Time. Figure 1 presented the development of unconfined compression strength in relation to the curing time of the stabilized soils contaminated with Pb0.1 and Pb0.5, respectively.

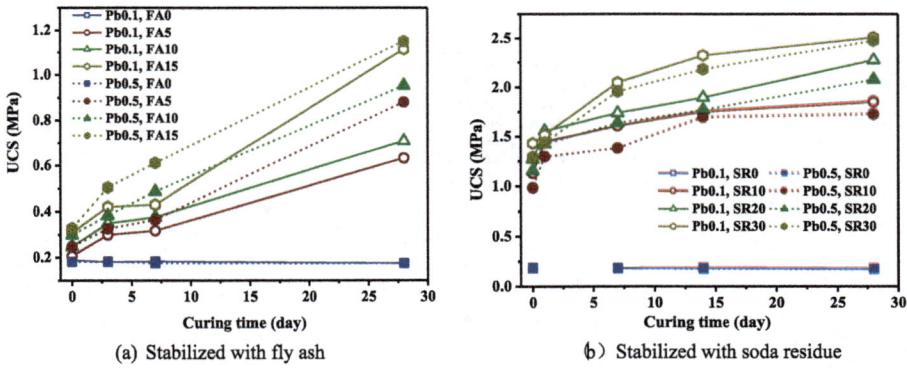

Fig. 1. Relationships of the stabilized Pb-contaminated soils between UCS and curing time

As shown in Fig. 1, adding fly ash and soda residue into the contaminated soils can improve the soil strength obviously, which increased with the increase of curing time. Compared with the specimens treated by soda residue, there was a linear correlation between the UCS and curing time for the fly ash treated specimens.

For the specimens stabilized with fly ash, the increased early strength was attributed to some calcium aluminate hydration products, such as CAH, Aft and AFm that were the early hydration products [12]. Furthermore, Ca^{2+} substituted from hydrated products by Pb^{2+} can accelerate the hardening of soil particles, and then increase the soil strength as well. As curing time increased, the activity of fly ash can be motivated. The spherical particles in fly ash mainly composed by silicon dioxide dissolved gradually, and the concentration of free silicon dioxide in the pore solutions increased. These increased free silicon dioxide participated in the pozzolanic reactions and promoted the formation of calcium silicate hydrates. Thereby, soil strength was extremely enhanced.

It can be observed from Fig. 1(b) that UCS of the soda residue stabilized specimens increased faster at the first day of curing, and then increased continuously with a little slower rate with curing time increased.

The main components of soda residue contains CaO, $CaCO_3$ and $CaSO_4$. In the initial curing period, CaO interacted with water and hydrated immediately to form Ca $(OH)_2$, which can increase the early strength of the specimens. And $CaSO_4$ can reacted with $Ca(OH)_2$ and Al_2O_3 to form the ettringite that was also the early hydrated product and increased the early strength as well. As shown in Table 2, there was a little amount of pozzolanic components in the soda residue. With the curing time increased continuously, the quantity of pozzolanic components contained in the soda residue reduced gradually, which retarded the development of pozzolanic reactions. Therefore, the growth rate of the UCS decreased.

Binders Type and Content. Figure 2 showed the effects of binders type and content on the UCS of specimens contaminated with Pb0.1 and Pb0.5 curing at 0, 7 and 28 days.

It can be seen that increasing both fly ash and soda residue contents resulted in an obvious increase of the soil strength. And the strength growth rate of the soda residue stabilized specimens was faster than that of fly ash. The increased UCS of fly ash

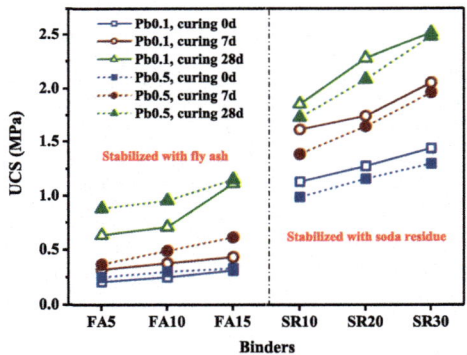

Fig. 2. Effects of binders type and content on the UCS of the specimens

stabilized specimens was mainly attributed to the pozzolanic reactions. Increasing fly ash contents would accelerate the development of pozzolanic reactions and some more CSH, CAH and Aft were formed to enhance the soil strength. For the soda residue stabilized specimens, there was a certain quantity of $CaCO_3$ contained in the soda residue, which can increase the UCS. And the $CaCO_3$ can also incorporated with Ca $(OH)_2$ to form $CaCO_3 \cdot Ca(OH)_2$ that can continuously reacted with SiO_2 to form a new complex $CaSiO_3 \cdot CaCO_3 \cdot Ca(OH)_2 \cdot nH_2O$ [13]. The new complex can fill in the soil pores and enhance the soil structure.

3.2   Leaching Characteristics

The TCLP was performed on the specimens to determine the effectiveness of the fly ash and soda residue treated Pb-contaminated soils, as shown in Fig. 3.

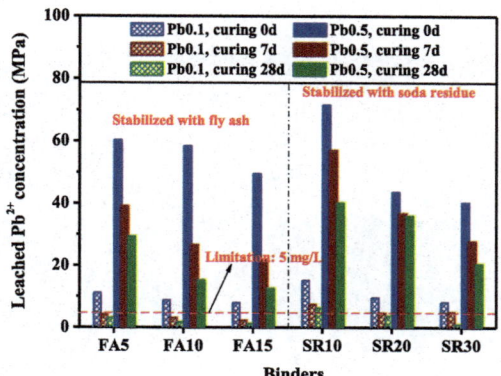

Fig. 3. Leaching characteristics of the stabilized Pb-contaminated soils with curing time

Figure 3 indicated that the quantity of leached Pb^{2+} observably decreased with curing time increased. As the curing time increase to 28 days, the disparity of leached Pb^{2+} concentrations between the Pb0.1 and Pb0.5 decreased gradually. As shown in Fig. 3, the leached Pb^{2+} concentration of the stabilized specimens decreased rapidly during the first 7 days of curing and then the reduction rate slowed down in the follow curing days. During the initial curing period, most of Pb^{2+} dissociated in the pore solutions can effectively adsorb onto the surface of clay particles, the non-burned carbon in the fly ash as well as the soda residue particles [4, 14]. As the curing time increased, a certain quantity of hydrated cementitious materials was produced, which can neutralize the acid leachant and decreased the leached Pb^{2+} concentration [15]. And Pb^{2+} can be immobilized by a more stabilized and solidified crystalline structure performed by combination and substitution with the hydrated products [16, 17].

It can be observed that increasing binders content can reduce the Pb^{2+} mobility effectively. In general, fly ash appeared a better immobilization behavior than that of soda residue. For the specimens with 1000 mg/kg Pb^{2+} after curing for 7 days, the leached Pb^{2+} concentration was lower than the regulatory limit for Pb^{2+} (Pb < 5 mg/L). While for the specimens with 5000 mg/kg Pb^{2+}, all the leached Pb^{2+} concentration exceeded the limitation. Therefore, it can be suggested that fly ash and soda residue can be successfully applied to treat the contaminated soils with lower Pb^{2+} concentration.

3.3   SEM Analysis

The microstructural measurement was performed on the Pb0.1-specimens stabilized with F15 and SR30 at 28 days of curing, respectively. As shown in Fig. 4(a), the lath-shaped $Ca(OH)_2$ can be found, which induced the growth of soil strength. Besides that, a large quantity of fly ash microsphere distributed in the specimens and increased the specific surface area of the specimens. Therefore, some of the Pb^{2+} can be adsorbed. For the sample stabilized with SR30, lots of CSH can be observed from Fig. 4(b). Soil particles aggregated tightly together bonded with CSH, and got a denser structure than that of sample F15. It can be seen that the SEM results was approximately consistent with the aforementioned conclusions.

(a) Sample F15   (b) Sample SR30

Fig. 4. SEM images of the stabilized samples.

4   Conclusions

The paper investigated the strength and leaching properties of high calcium fly ash and soda residue stabilized lead contaminated soils via a systematic laboratory tests. The following conclusions can be drawn:

(1) Both high calcium fly ash and soda residue can improve the strength and lower the leached Pb^{2+} quantities. Strength of soda residue stabilized Pb^{2+} contaminated soils was a little higher than that of fly ash.
(2) The designed binders proportions was available to treat the contaminated soil with a low Pb^{2+} concentration.
(3) In general, the mobility of Pb^{2+} in the fly ash stabilized specimens can be better decreased than that of soda residue.
(4) SEM results confirmed that $Ca(OH)_2$ and CSH was the main products that improved the engineering properties of the stabilized contaminated soils.

Acknowledgment. This research is financially supported by the National Natural Science Foundation of China (grant nos. 41672306, 41372281, 41172273, and 41572282).

References

1. Shi, H.S.: Research on intrinsic characteristics and hydration properties of fly ash with high calcium oxide. J. Tongji Univ. **31**(12), 1440–1443 (2003). (in Chinese)
2. Antiohos, S., Tsimas, S.: Investigating the role of reactive silica in the hydration mechanisms of high-calcium fly ash/cement systems. Cem. Concr. Compos. **27**, 171–181 (2005)
3. Tishmack, J.K., Olek, J., Diamond, S.: Characterization of high-calcium fly ash ashes and their potential influence on ettringite formation in cementitious systems. Cem. concr. aggreg. **21**, 82–92 (1999)
4. Ahmaruzzaman, M.: A review on the utilization of fly ash. Energy Combust. Sci. **36**, 327–363 (2010)
5. Liu, X.W., Liu, X.B., Liu, Z.G., Li, G.F.: Leaching toxicity experiment and treatment of alkaline slag. J. Salt Chem. Ind. **40**(5), 51–54 (2011). (in Chinese)
6. Li, Y.Y., Yan, S.W., Zhang, J.Y., Yin, X.T.: Engineering properties and microstructural features of the soda residue. Chin. J. Geotech. Eng. **21**(1), 100–103 (1999)
7. Cao, X., Sun, J.C., Jin, C.J., Gao, Y., Liu, Y., You, X.H.: The competitive absorption effect of heavy metals on alkaline sludge. Shandong Sci. **22**(6), 17–20 (2009). (in Chinese)
8. Yan, C., Song, X.K., Zhu, P., Sun, H.Y., Li, Y.P., Zhang, J.F.: Experimental study on strength characteristics of soda residue with high water content. Chin. J. Geotech. Eng. **29**(11), 1683–1688 (2007). (in Chinese)
9. Chen, C.: Heavy metal pollution state and evaluation of typical area of Anhui province. Hefei: Hefei University of Technology, thesis (2013). (in Chinese)
10. Liao, X.Y., Chong, Z.Y., Yan, X.L., Zhao, D.: Urban industrial contaminated sites: a new issue in the field of environmental remediation in China. Environ. Sci. **32**(3), 784–794 (2011). (in Chinese)
11. U.S. EPA: Test methods for evaluating solid waste, physical/chemical methods, SW-846. 3rd edn. Method 1311, Washington DC
12. Xu, A., Sarkar, S.L.: Microstructrual study of gypsum activated fly ash hydration in cement paste. Cem. Concr. Res. **21**, 1137–1147 (1991)
13. Sun, S.L., Zheng, Q.H., Tang, J., Zhang, G.Y., Zhou, L.G., Shang, W.T.: Experimental research on expansive soil improved by soda residue. Rock Soil Mech. **33**(06), 1608–1612 (2012). (in Chinese)
14. Bayat, B.: Combined removal of zinc (II) and cadmium (II) from aqueous solutions by adsorption onto high-calcium Turkish fly ash. Water Air Soil Pollut. **136**, 69–92 (2002)
15. Ricou, P., Lecuyer, I., Le Cloirec, P.: Removal of Cu^{2+}, Zn^{2+} and Pb^{2+} by adsorption onto fly ash and fly ash/lime mixing. Water Sci. Technol. **39**, 239–247 (1999)
16. Cocke, D.L.: The binding chemistry and leaching mechanisms of hazardous substances in cementitious solidification/ stabilization systems. J. Hazard. Mater. **24**, 231–253 (1990)
17. Yousuf, M., Mollah, A., Vempati, R.K., Lin, T.C., Cocket, D.L.: The interfacial chemistry of solidification-stabilization of metals in cement and pozzolanic material systems. Waste Manag. **15**(2), 137–148 (1995)

Leaching Properties of Heavy Metals in a Phosphate-Based Binder Stabilized Contaminated Soil

Run Zhang, Bo-Wei Yu, Fei Wang, and Yan-Jun Du[✉]

Southeast University, Nanjing, Jiangsu Province, China
duyanjun@seu.edu.cn

Abstract. Revegetation has emerged as a promising method in the rehabilitation of contaminated sites. As traditional binders (e.g. cement and lime) usually lead to high alkalinity and hardening in soils, it is necessary to develop new binders which are capable of both supporting plant growth and minimizing the leachability leaching of heavy metals. This paper presents an experimental study on the leaching properties of soil with coexisted Pb, Zn, and Cd contaminants sampled at an industrial contaminated site. The soil is solidified/stabilized using a new phosphate-based binder, SS-B. Batch-type leaching tests and a synthetic precipitation leaching procedure are conducted to evaluate the leachability of Pb, Zn, and Cd in SS-B stabilized soil. Frendlich sorption model is used to fit the experimental data and to obtain the sorption parameters of these three heavy metals. The retardation factor (R_d) of heavy metals in the SS-B stabilized soil follows the order of Pb > Zn > Cd. One-dimensional semi-dynamic leaching tests are performed to acquire observed diffusion coefficients (D^{obs}) of these three heavy metals. The results show that the values of D^{obs} of Pb, Zn, and Cd are 2.30×10^{-15} m²/s, 7.47×10^{-15} m²/s, and 2.51×10^{-15} m²/s respectively.

Keywords: Contaminated soil · Heavy metal · Leachability · Diffusion

1 Introduction

As a result of rapid industrialization, a large number of abandoned industrial sites with high levels of heavy metals, typically lead, zinc, and cadmium, can be commonly encountered in urban areas in China, United States and other locations worldwide [1–3]. The presence of these heavy metals has become a serious threat to the environment and the public health, and imposes adverse effects on the redevelopment of contaminated sites. Thus, there is an urgent need in finding effective and economical technologies to immobilize or remove these pollutants.

Stabilization/Solidification (S/S) remediation technique is an effective and economical remediation method that is widely used for heavy metal contaminated sites [1, 4, 5]. According to previous reports, most S/S treated soils are excavated and sent to landfills because of their toxicity, which leads to land consumption [7]. Thus, there is a pursuit in redeveloping contaminated sites [8–10].

Revegetation and site greening is considered as an environmentally friendly and cost-effective means [11]. To return industrial contaminated sites into productive uses,

© Springer Nature Singapore Pte Ltd. 2018
A. Farid and H. Chen (Eds.): GSIC 2018, *Proceedings of GeoShanghai 2018*
International Conference: Geoenvironment and Geohazard, pp. 450–459, 2018.
https://doi.org/10.1007/978-981-13-0128-5_50

a novel binder as a substitute for traditional cement-based materials is required, which is capable of supporting plant growth and preventing the migration of hazardous waste simultaneously. SS-B binder is a self-developed phosphate-based binder, composed of oxalic acid-activated phosphate rock, monopotassium phosphate (KH_2PO_4) and reactive magnesia (MgO) [1]. Previous studies suggest that SS-B could effectively decrease the leachate concentration of Pb and Zn after 7 days of standard curing. The values of which are lower than the Ministry of Environmental Protection (MEP) of People's Republic of China and United States Environmental Protection Agency (USEPA) regulatory limit at 100 mg/L for Zn and 5 mg/L for Pb respectively [12, 13]. The strength and leachability of SS-B stabilized soil could be enhanced when exposed to atmospheric carbon dioxide [9]. Generally, SS-B binder is superior to cement binder in strength, leachability, and long-term durability [1, 6].

The previous study has indicated that after remediated, the pH value, heavy metal concentrations of leachate, and phytotoxicity of SS-B stabilized soil mixed with loam soil decreased remarkably to an acceptable value, which means SS-B stabilized soil could reuse as greening soil [14]. However, heavy metals in stabilized soil may transport when exposed to rainfall or groundwater, which could risk the health of plants and human beings [15]. Studies on leaching behaviors of SS-B stabilized soil used as greening soil are rather limited.

The purpose of this study is to investigate the leaching toxicity of contaminated soils with high levels of Pb, Zn, and Cd after stabilized by SS-B binder. While plants should be supplied with suitable pH, texture, ionic, and necessary nutrients to maintain normal metabolism, the ingredient proportion of SS-B binder is optimized. Moreover, the SS-B stabilized soil is mixed with a commercial loam soil to meet the demand of plants to survival. Batch-type leaching tests and a synthetic precipitation leaching procedure are conducted to evaluate the leachability of Pb, Zn, and Cd in SS-B stabilized soil. One-dimensional semi-dynamic leaching tests are performed to acquire observed diffusion coefficients and retardation factors.

2 Materials and Methods

2.1 Raw Materials

The soil is sampled from an industrial site in Zhuzhou, Hunan province in China, which is contaminated with lead, zinc and cadmium. Samples are passed through a 1 mm sieve after air drying. The soil is classified as low plasticity silt (ML) [16]. The water content of the soil is 1–3%, and the Atterberg limits are tested according to ASTM D4318-10 [17]. Specific surface area of the soil is measured through Brurauer Emmerr Teller (BET) method with a Phys sorption Analyzer ASAP2020. Based on the method recommended by Japan Soil Society, the electric conductivity (*EC*) value of soil is measured at a S/L ratio of 1:5 The pH value of the soil is measured by a pH probe of HORIBA pH/COND METER D-54 maintaining a solid/liquid (S/L) ratio of 1:1 [18]. The concentrations of Pb, Zn and Cd are tested per as HJ/T 299-2007 [19]. The detailed physio-chemical properties and concentrations of heavy metals are listed in Table 1.

Table 1. Basic physical and chemical properties of the sampled soil

Property	Value
Water content, w (%)	1–3
Specific gravity, G_s	2.72
Liquid limit, w_L (%)	36.3
Plastic limit, w_p (%)	24.9
specific surface area (m^2/g)	48.59
EC (S/L = 1:5, ms/cm)	2.83
pH (S/L = 1:1)	7.73
Contaminant concentration (mg/kg)	
Lead	6400
Zinc	14150
Cadmium	310

In order to estimate the leaching behaviors of SS-B stabilized soil at different initial heavy metal concentrations, lead, zinc and cadmium are artificially added into the sampled soil. The nitrate form of source chemical is selected because it is inert to cement hydration and has insignificant effects on engineering characteristics [20]. Stock solutions with different concentrations of heavy metals, which is prepared by dissolving $Pb(NO_3)_2$, $Zn(NO_3)_2 \cdot 6H_2O$ and $Cd(NO_3)_2 \cdot 4H_2O$ (Chemical Analytical Reagent) in distilled deionized water, are mixed thoroughly with the sampled soil, and subsequently sealed in a bucket remaining undisturbed for 7 days under natural condition. The mass of heavy metal nitrates added to sampled soil is shown in Table 2, numbered 1 to 6 in accordance with the decrease of heavy metal concentration.

Table 2. The proportions of heavy metals added to sampled soil

Number	$Pb(NO_3)_2$ (g/kg)	$Zn(NO_3)_2 \cdot 6H_2O$ (g/kg)	$Cd(NO_3)_2 \cdot 4H_2O$ (g/kg)
1	16	68.65	1.35
2	13.6	58.35	1.15
3	11.2	48.05	0.95
4	8.8	37.75	0.75
5	6.4	27.45	0.55
6	4	17.15	0.35

SS-B binder is composed of oxalic acid-activated phosphate rock (PR), monopotassium phosphate (KH_2PO_4) and reactive magnesia (MgO) in a proportion of 1:1:2 by dry weight basis [1]. The phosphate rock powder used as the binder is activated by oxalic acid, oven dried at 60 °C, and is passed a 0.074 mm sieve. The procedure to prepare the SS-B binder and the properties of the components of SS-B is presented in Du et al. [1]. As the previous study demonstrates, 20% is the optimum proportion of MgO, which could not only obtain the most effective stabilization but also maintains a relatively neutral pH and a low EC value. Moreover, the SS-B stabilized soil does not pose repressive influence on planted cabbages [14]. In consequence, the ratio of PR, KH_2PO_4, and MgO in SS-B binder is modified to 2:2:1.

The loam soil used in this study is purchased from the Huai'an Agricultural Science & Technology Development Co., LTD.

2.2 Testing Methods

The SS-B powder with predetermined weight (3%, by dry basis) is added into the contaminated soils, and mixed thoroughly by an electronic mixer for at least 15 min to ensure the binder well-distributed. Subsequently, approximately 127 g of the mixture is compacted evenly by three lifts in a cylindrical stainless mold with 50 mm diameter and 50 mm height, whose inner wall is lubricated with Vaseline beforehand. The density of the compacted soil is controlled at 1.30 g/cm³ with a water content of 25%. Then, the specimens are then extruded form the molds and sealed in polyethylene bags and cured for 28 d under standard curing condition (20 ± 2 °C, relative humidity of 95%). Six types of specimens with predetermined heavy metal concentrations are prepared, numbered 1 to 6 with decreasing contaminant concentrations correspondingly. After 28 d standard curing, the specimens are subjected to the synthetic precipitation leaching procedure (SPLP) per as USEPA Method 1312 afterwards [21]. The leachant used in this study is a compound of sulphuric acid and nitric acid, diluted by distilled deionized water to maintain a pH value at 4.50 to simulate the acid rain. The molar mass ratio of SO_4^{2-} and NO_3^- is 2.68:1.

Frendlich sorption model is used to analyze the SPLP experimental data, and the equation is described as follows [22]:

$$C_S = k_f * C^N \tag{1}$$

where: Cs = the amount of heavy metal adsorbed by per unit mass of stabilized soil (mg/kg), C = the concentration of metal ions in leachate (mg/L), k_f and N = Frendlich constants related to desorption capacity and intensity.

The partition coefficient (Kp) is determined by the following equation, indicating the desorption capacity of heavy metals:

$$K_p = k_f * C^{N-1} \tag{2}$$

where: C_0 = the initial concentration of heavy metals (mg/L).

Then, the retardation factor (Rd) could be obtained as follows:

$$R_d = 1 + (\rho_d * K_p)/\theta \tag{3}$$

where: ρ_d = dry density of stabilized soil, θ = volumetric soil water content.

1-D semi-dynamic leaching test is carried out to evaluate the leaching behavior and long-term environmental security of heavy metals in SS-B stabilized soil based on USEPA 1315 [23] and ASTM 1308-08 [24]. The No. 1 stabilized soil is chosen to simulate the most detrimental condition. After mixed with loam soil, the No. 1 stabilized soil is compacted into a self-developed cylindrical PVC mold with 50 mm diameter and 60 mm height at the optimum water content (25%) and density (1.30 g/cm^3). A glass cylinder with 61 mm diameter and 220 mm height is selected as the container to hold the leachant with a predetermined volume (200 mL) to maintain the ratio of volume of leachant to the surface area of specimen 10:1. The leachant used is consistent with pre-mentioned SPLP procedure, which is poured into the container beforehand. Due to the loose state of stabilized soil, the specimen is carefully put into the container with the mold, and then sealed by polyethylene bags. Duplicate samples are prepared to reduce error. The test duration is 11 days, and the leachant is replaced after time periods of 2 and 7 h during the first day, and then daily for the following 10 days. At each time interval, 50 mL of the leachate is sampled, whose pH, *EC*, and concentrations of Pb, Zn and Cd are measured later on.

The observed diffusion coefficient (D^{obs}) is calculated as follows [23]:

$$D_i^{obs} = \pi \{ M_i / [2\rho_d Q_0 (\sqrt{t_i} - \sqrt{t_{i-1}})] \}^2 \tag{4}$$

where: M_i = cumulative mass of leached heavy metal (mg) at each time interval, ρ_d = dry density of stabilized soil (kg/m3), Q_0 = Initial heavy metal concentration (mg/kg), t_i= the end of each time interval (s), t_{i-1}= the beginning of each time interval (s).

3 Test Results and Discussion

3.1 Leaching Behavior

The results of the batch-type desorption test subjected to a SPLP procedure are shown in Fig. 1. The scatters from right to left represent the stabilized soil from No. 1 to No. 6 respectively, while the Pb concentration of No. 6 stabilized soil is lower than the limit of detection and hence it is not plotted. The sorption isotherm indicates that Pb has the lowest desorption rate among the three heavy metals. The parameters of Frendlich model for desorption of Pb, Zn and Cd is displayed in Table 3. As all of the fitting coefficients are greater than 0.9, Frendlich model fits well with the experimental data.

The K_p and Rd values of Pb, Zn, and Cd are shown in Table 4. The retardation factors (R_d) of Pb, Zn, and Cd are 48283.73, 6558.32, and 1824.65. It can be seen that Cd has the lowest value followed by Zn, whereas Pb reaches the highest, which means Cd has the highest mobility followed by Zn, while Pb presents the opposite behavior.

3.2 Semi-dynamic Leaching Test

The basic physical and chemical properties of specimens used in semi-dynamic leaching test are given in Table 5.

The pH value remains constant from 7.0 to 7.5, and slightly undulates on day 7 and day 9. At the initial stage (0–2 days), *EC* goes up steadily, then gradually falls down until a small rise occurred on day 7, and finally goes down continuously. The *EC* measured on day 11 is in the same range of the initial value (200–250 µS/cm).

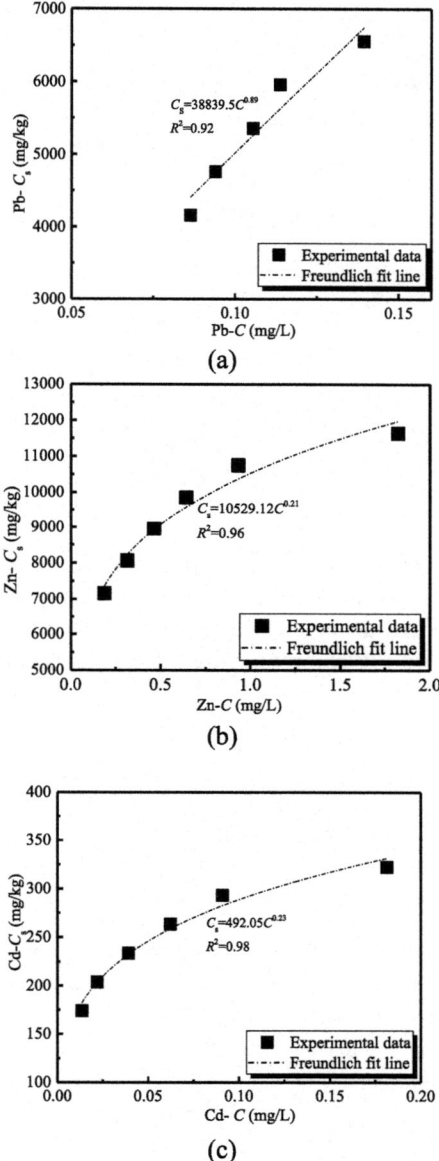

Fig. 1. Desorption behavior of SS-B stabilized soil: (a) Pb; (b) Zn; (c) Cd.

Table 3. Parameters of Frendlich model for desorption of Pb, Zn, and Cd

	k_f	N	R^2
Pb	38839.5	0.1395	0.92
Zn	10529.1	0.2142	0.96
Cd	492.1	0.2316	0.98

Table 4. The retardation factors and partition coefficients of Pb, Zn, and Cd

	C_0 (mg/L)	ρ_d (g/cm^3)	θ	K_p	R_d
Pb	0.1395	0.559	0.560	48369.11	48283.73
Zn	1.8228			6569.05	6558.32
Cd	0.1814			1826.91	1824.65

Table 5. The basic physical and chemical properties of specimens used in semi-dynamic leaching test

Parameters	Value
Initial mass (g)	153
Volume (cm^3)	117.75
Initial volumetric water content θ	0.560
Saturated volumetric water content θ_S	0.8295
Initial saturation degree (%)	67.5
Void ratio e	4.87

The variations of leached Pb, Zn, and Cd concentration with leaching time are displayed in Fig. 2. The scatters in the figure are the average value of three duplicate specimens, and the standard deviation is also given. Generally, the concentrations of Pb, Zn and Cd declines with leaching time. Leached Pb concentration fluctuates at the

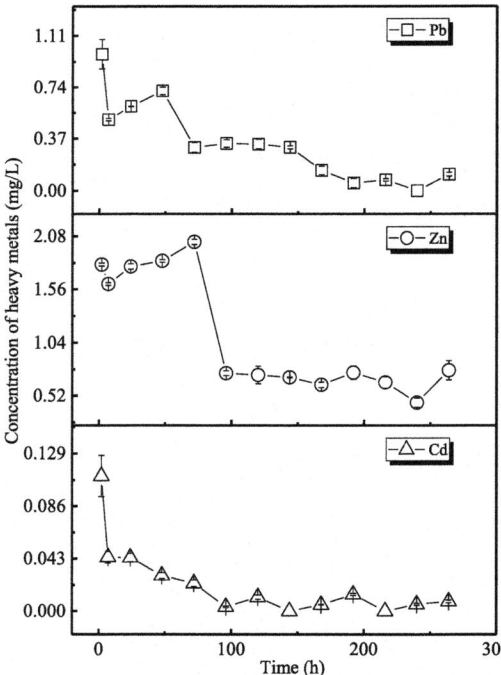

Fig. 2. Variations of Pb, Zn, and Cd concentration with leaching time

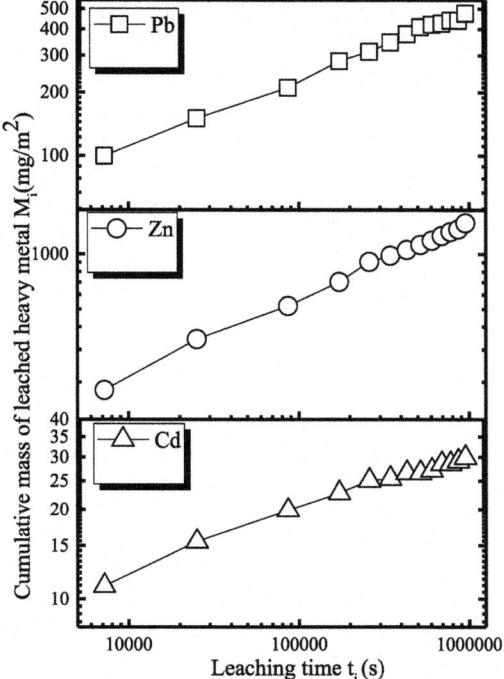

Fig. 3. The log Mi-log ti fitting line

Table 6. The slope of log Mi-log ti line and the related observed diffusion coefficient

No.	Time interval	Pb		Zn		Cd	
		Slope	D^{obs} (m²/s)	Slope	D^{obs} (m²/s)	Slope	D^{obs} (m²/s)
1	2 h	–	–	–	–	–	–
2	7 h	0.33	–	0.51*	1.40e-14	0.37	–
3	1 d	0.27	–	0.34		0.20	–
4	2 d	0.42*	3.20e-15	0.44*	6.72e-15	0.20	–
5	3 d	0.25	–	0.62*	1.38e-14	0.24	–
6	4 d	0.35*	1.69e-15	0.27		0.05	–
7	5 d	0.41*	2.10e-15	0.32		0.20	–
8	6 d	0.43*	2.24e-15	0.35*	3.68e-15	–	–
9	7 d	0.17	–	0.35*	3.49e-15	0.13	–
10	8 d	0.10	–	0.46*	5.72e-15	0.37*	2.51e-15
11	9 d	0.28	–	0.43*	4.91e-15	–	–
12	10 d	–	–	0.32	–	0.20	–
13	11 d	0.81	–	0.81	–	0.29	–
Average (m²/s)		2.30e-15		7.47e-15		2.51e-15	
SD		6.39e-16		4.53e-16		–	

* indicates that pore fluid diffusion dominates.

initial stage, and then declines continuously. The concentration of Zn slightly increases till day 4, suddenly drops to 0.6 mg/L on day 5, and remains constant. The concentration of Cd decreases rapidly and remains unchanged ranging from 0.05–0.10 mg/L.

The log M_i-log t_i fitting line is displayed in Fig. 3. As it is considered that pore fluid diffusion dominates if only the slope of log M_i-log t_i line is greater than 0.35, the slopes of log M_i-log t_i line and the related observed diffusion coefficients at each time interval are given in Table 6. The observed diffusion coefficient (D^{obs}) of Pb, Zn, and Cd are 2.30e–15 m^2/s, 7.47e–15 m^2/s, and 2.51e–15 m^2/s respectively. The result is consistent with the desorption test that Pb has the lowest mobility in SS-B stabilized soil. The relative low D^{obs} shows that SS-B binder could effectively immobilize the heavy metals.

4   Conclusions

Based on the results shown in this study, the following conclusions can be drawn:

(1) The sorption characteristics of SS-B stabilized soil are fitted by Frendlich model. The retardation factors (R_d) of Pb, Zn, and Cd are 48283.73, 6558.32, and 1824.65, which shows Pb has the lowest mobility while Cd presents the opposite.

(2) SS-B stabilized soil experiences a low observed diffusion coefficient of Pb, Zn, and Cd, the he values of D^{obs} of Pb, Zn, and Cd are 2.30×10^{-15} m^2/s, 7.47×10^{-15} m^2/s, and 2.51×10^{-15} m^2/s respectively.

(3) The results of this study suggests that as targeted on revegetation, SS-B stabilized soil could effectively decrease the mobility of Pb, Zn, and Cd on environmental risk.

Acknowledgements. This research is financially supported by the National Natural Science Foundation of China (Grant No. 51278100, 41330641 and 41472258), Natural Science Foundation of Jiangsu Province (Grant No. BK2012022) and National High Technology Research and Development Program of China (Grant No. 2013AA06A206).

References

1. Du, Y.J., Wei, M.L., Reddy, K.R., Jin, F., Wu, H.L., Liu, Z.B.: New phosphate-based binder for stabilization of soils contaminated with heavy metals: Leaching, strength and microstructure characterization. Environ. Manage. **146**, 179–188 (2014)

2. Sharma, H.D., Reddy, K.R.: Geoenvironmental engineering: site remediation, waste containment, and emerging waste management technologies. Wiley, New York (2004)

3. The Word Bank: China Contaminated Site Management Project (2015)

4. Wang, F., Wang, H., Al-Tabbaa, A.: Leachability and heavy metal speciation of 17-year old stabilized/solidified contaminated site soils. Hazard. Mater. **278**, 144–151 (2014)

5. Wang, F., Wang, H., Al-Tabbaa, A.: Time-dependent performance of soil mix technology stabilized/solidified contaminated site soils. Hazard. Mater. **286**, 503–508 (2015)

6. Du, Y.J., Wei, M.L., Reddy, K.R., Wu, H.L.: Effect of carbonation on leachability, strength and microstructural characteristics of KMP binder stabilized Zn and Pb contaminated soils. Chemosphere **144**, 1033–1042 (2016)

7. Van Liedekerke, M., Prokop, F., Rabl-Berger, S., Kibblewhite, M., Louwagie, G.: Progress in the management of contaminated sites in Europe [Report EUR 26,376 EN]. EC Joint Research Centre, Ispra (2014)

8. Commission of the European Communities. Towards a thematic strategy on the urban environment. Technical report, COM 60 Final, Brussels (2004)

9. Burke, H., Hough, E., Morgan, D.J.R., Hughes, L., Lawrence, D.: Approaches to inform redevelopment of brownfield sites: an example from the Leeds area of the West Yorkshire coalfield. UK Land Use Policy **47**, 321–331 (2015)

10. UK Environment Agency: GPLC1-Guiding Principles for Land Contamination (2010)

11. US EPA: Reusing Superfund Sites: Recreational Use of Land Above Hazardous Waste Containment Areas (2006)

12. M.E.P. China GB156180. Environmental quality standard for soils (1995)

13. US EPA: Toxicity Characteristic Leaching Procedure, Appendix 1, Federal Register 51:216 (1986)

14. Yu, B.W., Du, Y.J., Liu, R., Liu, C.Y., Pan, K.Y.: Reuse of stabilized contaminated soils with heavy metals as greening soils: leaching, physicochemical, and phytotoxicity characterization. Geo-Chicago **2016**, 557–571 (2016)

15. Du, Y.J., Wei, M.L., Reddy, K.R., Liu, Z.P., Jin, F.: Effect of acid rain pH on leaching behavior of cement stabilized lead-contaminated soil. J. Hazard. Mater. **271**, 131–140 (2014)

16. A.S.T.M. D2487-10: Standard practice for classification of soils for engineering purposes (Unified soil classification system), ASTM International (2010)

17. A.S.T.M. D4318-10: Standard test methods for liquid limit, plastic limit, and plasticity index of soils, ASTM International (2010)

18. M.E.P. of China: Solid waste-extraction procedure for leaching toxicity-sulphuric acid & nitric acid method. HJ/T 299-2007, Ministry of Environmental Protection of China, Beijing, China (2007)

19. Cuisinier, O., Le Borgne, T., Deneele, D., Masrouri, F.: Quantification of the effects of nitrates, phosphates and chlorides on soil stabilization with lime and cement. Eng. Geol. **117**, 229–335 (2011)

20. Ministry of Construction of China. CJ/T340-2011: Planting soil for greening standard. Ministry of Construction of China, Beijing, China (2011)

21. US EPA. Method 1312: Synthetic precipitation leaching procedure. US (1994)

22. Doung, D.D.: Desorption Analysis: Equilibrium and Kinetics. Imperial College, London (1998)

23. US EPA. Method 1315: Mass transfer rates of constituents in monolithic or compacted granular materials using semi-dynamic tank leaching procedure. US (2013)

24. A.S.T.M. C1308-08: Standard method for accelerated leach test for diffusive release from solidified waste and a computer program to model diffusive, fractional leaching from cylindrical waste forms, ASTM International (2009)

Numerical Simulation of Methane Oxidation in Actively-Aerated Landfill Biocover

Shi-Jin Feng[1,2(✉)] and Tian-Yi Wang[1]

[1] Department of Geotechnical Engineering, Tongji University,
Shanghai 200092, China
fsjgly@tongji.edu.cn
[2] Key Laboratory of Geotechnical and Underground Engineering
of Ministry of Education, Shanghai 200092, China

Abstract. Methane oxidation in landfill biocover is a complex process involving water, gas and heat transport as well as microbial oxidation. Such technology is useful for dealing with landfill gases. With the aim of finding the most suitable design configuration for actively-aerated methane biocover, an air injection system is proposed in this work to ensure that oxygen availability is no longer a limitation for methane degradation. A numerical model is developed that incorporates water-gas-heat coupled transport in layered landfill biocover with the consideration of methane oxidation and air actively-aerated injection. The model is verified and calibrated using published data from a laboratory soil column test. Moreover, parametric studies are carried out to investigate the influences of air injection amount and location. It is found that injecting air into the biocover through several inlets along the biocover bed is a promising approach to enhance methane oxidation in the cover.

Keywords: Landfill biocover · Aeration · Methane oxidation · Gas transport
Numerical simulation

1 Introduction

Landfill is a site stacking the municipal solid waste (MSW) and has wide application worldwide due to its low cost and environmental-friendly characteristics [1]. The cover system, as the essential structure of landfill, is designed to impede the penetration of rainfall and emission of landfill gas (LFG). One main component of LGF is methane (CH_4) which is one of the greenhouse gases and has much higher global warming potential than carbon dioxide (CO_2) [2]. Therefore, it is necessary to mitigate CH_4 emission from landfill effectively and economically.

Landfill biocover is one type of landfill cover, which is effective in impeding the emission of methane from landfill into atmosphere using biotic methane oxidation. It is quite useful in landfills, especially those lacking landfill gas extraction system [3]. The oxidation that occurs in biocover can consume CH_4, generate CO_2 and water, and release heat, which can be described by the following equation:

© Springer Nature Singapore Pte Ltd. 2018
A. Farid and H. Chen (Eds.): GSIC 2018, *Proceedings of GeoShanghai 2018*
International Conference: Geoenvironment and Geohazard, pp. 460–467, 2018.
https://doi.org/10.1007/978-981-13-0128-5_51

$$CH_4 + (2 - x)O_2 \rightarrow (1 - x)CO_2 + (2 - x)H_2O + x \text{ - } CH_2O\text{- } + \text{heat} \qquad (1)$$

where -CH_2O- is organics and x is the reaction coefficient.

The oxidation reaction is affected by numerous factors, such as CH_4 concentration, O_2 concentration, temperature, moisture and other properties. The heat released by reaction can reach 632 kJ/mol [4]. Passive aeration is usually used in traditional bio-cover. O_2 naturally intrudes into the cover by diffusion and provides the required O_2 for oxidation reaction. The diffusion of O_2 depends on the O_2 concentration gradient in cover soils [5]. Recent researches reveal that the maximum reaction depth of CH_4 is approximately 30 cm, which is limited by the O_2 distribution in cover. Increasing O_2 amount in cover artificially is a possible way to enhance the methane oxidation [6].

In this study, a new cover structure is proposed. The new cover structure can adjust gas distribution in cover using actively-aerated injection system, which is beneficial for deepening the O_2 penetration depth and reducing LFG emission. A numerical model is built to simulate water-gas-heat transport in the new landfill cover structure.

2   Conceptual Model of Actively-Aerated Landfill Cover System

In order to better reduce methane emission, a new landfill cover structure is proposed here. As shown in Fig. 1, the landfill cover consists of three layers with different functions. At the bottom, a gravel layer is used to make gas uniformly distributed. The gravel layer is overlain by a clay layer whose function is impeding methane emission. The silt layer is the reaction layer, which is located at the top. As the major place where the methanobacteria grows, the silt layer also provides the environment for the growth of vegetation. Additionally, the silt layer also helps to avoid cracking of the clay layer under dry condition.

In this study, soil is assumed to be a three-phase porous medium containing liquid, gas and solid particles. The liquid phase is water, while the gas phase is a four-component gas mixture (CO_2, O_2, CH_4 and N_2). Gas migration involves advection and diffusion. After the methane migrates from the MSW to the silt layer, methane

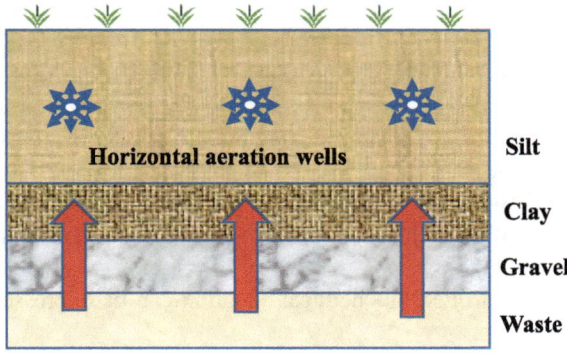

Fig. 1. Actively-aerated landfill cover system

oxidation reaction occurs, generating water, carbon dioxide and organic matter. The heat released by the reaction in turn affects the biochemical reaction rate. Thus it is a water-gas-heat coupled process.

In the silt layer, O_2 is required for the oxidation reaction. Therefore, O_2 concentration has significant impact on the oxidation reaction efficiency. In traditional landfill cover, the available O_2 for oxidation reaction in the landfill cover relies on the diffusion of O_2 from atmosphere to the landfill cover, which makes the O_2 concentration becomes the main restriction for the oxidation reaction. To overcome the limitation, horizontal aeration wells are laid in silt layer injecting air under low pressure. In this way, O_2 concentration in landfill cover can be increased artificially to improve oxidation reaction efficiency.

3   Numerical Model of Actively-Aerated Landfill Cover System

In this section, a multi-field coupled model is developed to describe the water-gas-heat transport processes in the landfill cover.

3.1   Governing Equation

In this model, gas is regarded as the ideal gas. Heat balance is assumed to exist between two phases, so each phase has the same temperature at the same location. Thus only one heat transfer governing equation is required [7].

Based on the conservation of energy and mass, water transfer governing equation can be expressed as

$$\rho_w \frac{\partial \theta_w}{\partial t} = -\nabla(\rho_w v_w) + \rho_{DB} M_{H_2O} r_w \tag{2}$$

where ρ_w is the water density; ρ_{DB} is the dry density of soil; v_w is the water flow velocity; θ_w is the volumetric water content; M_{H2O} is the molar mass of water; r_w is the reaction rate of water.

The gas transfer governing equation can be expressed as

$$\frac{\partial}{\partial t}\left[(1 - S_w)\phi c_g + S_w \phi H_w\right] = -\nabla\left[v g c_g\right] - \nabla\left[v_w H_w\right] - \nabla N_g \pm \rho_{DB} r_g \tag{3}$$

where ϕ is the soil porosity; N_g is the diffusive flux of gas; S_w is the saturation degree; v_g is the gas mixture advective velocity; c_g is the gas molar concentration; H_w is the gas molar concentration dissolved in water; r_g is the gas reaction rate in methane oxidation; "+" indicates gas generation such as CO_2; and "−" denotes gas consumption such as O_2 and CH_4.

The heat transfer controlled biochemical equation can be expressed as

$$\frac{\partial[W(T - T_r)]}{\partial t} = -\nabla(-\lambda_T \nabla T + H_{conv}) + H_{oxi} \tag{4}$$

where W is the soil heat capacity (J m^{-3}°C^{-1}); T_r is the temperature (°C) and its value takes 22 °C in this study; H_{conv} is the heat per unit area (J m^{-2} s^{-1}); H_{oxi} is the heat generation rate of the methane oxidation (J m^{-3} s^{-1}); and λ_T is the thermal conductivity (J m^{-1} s^{-1}°C^{-1}) and the function of water content.

Microbial aerobic methane oxidation (MAMO) is controlled by the concentrations of CH$_4$ and O$_2$, temperature and water content in soil. MAMO can be described by Eq. (1) mentioned. That $x = 0.5$ is adopted in Eq. (1) gives:

$$CH_4 + 1.5O_2 \rightarrow 0.5CO_2 + 1.5H_2O + 0.5 \text{ - } CH_2O\text{- } + \text{heat.} \tag{5}$$

It is necessary to determine the methane oxidation rate r_g^{CH4} (kg m^{-3} s^{-1}) which can help to investigate the effect of methane oxidation on water, gas and heat transfer:

$$r_g^{CH_4} = -\frac{V_{max} y_{CH_4}}{K_m + y_{CH_4}} \tag{6}$$

where V_{max} is the maximum reaction rate; y_{CH_4} is the CH$_4$ concentration; and K_m is the half-saturated constant for CH$_4$.

Methane oxidation efficiency is used to evaluate the performance of the biocover (dimensionless), which can be calculated as follows:

$$\text{Oxidation efficiency}(\%) = \frac{Q_{in}C_{in} - Q_{out}C_{out}}{Q_{in}C_{in}} \times 100\% \tag{7}$$

where C_{in} and C_{out} are CH$_4$ inlet and outlet concentrations, respectively; Q_{in} and Q_{out} are the CH$_4$ flow discharges at the inlet and outlet, respectively.

3.2 Numerical Implementation and Model Assumptions

The model is solved based on ANSYS Fluent platform and selected "mixture" model to solve multiphase conservation equation. The coupled pressure-velocity computation used the "phase-coupled SIMPLE" algorithm, and biochemical and kinetic equations are solved with the first-order implicit algorithm.

Since the main mode of gas transfer is convection and diffusion, the gas in the model is considered as laminar flow instead of turbulent flow, which is also determined by the rate of migration. Soil is considered as an ideal porous medium with uniform pore distribution.

4 Numerical Examples

Since researches about actively-aerated landfill cover system are quite limited. In this part, a soil column test [1] (Fig. 2) is analyzed, which can be viewed as a simple actively-aerated landfill cover.

As shown in Fig. 2, a 2D finite element model with a height of 90 cm and a width of 7.5 cm is used to simulate the soil column test. The holes are used to simulate the aeration probes. CH_4 gas is introduced at the bottom of the model. The boundaries can transfer heat, but not gas and liquid. The top of the column is the exit of gas, which can simulate the real atmosphere environment. The initial soil moisture content is 0.236 and the initial methane and water vapor contents are zero, which are the same as the values in the test. The other needed parameters are summarized in Table 1.

First, the experimental test results [1] are adopted to verify the numerical model. Parametric study is then conducted to investigate the influences of air injection amount and location on CH_4 oxidation efficiency.

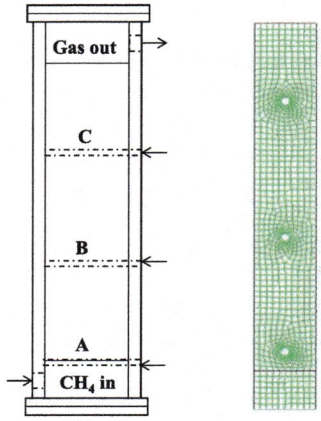

Fig. 2. Experimental setup for soil column test and finite element mesh

Table 1. Parameters used in the simulation

Parameter	Value	Reference
Porosity	0.73	[2]
Soil bulk density (kg m^{-3})	980	[2]
Soil particle density (kg m^{-3})	2280	[2]
Saturated volumetric water content	0.73	[2]
Residual volumetric water content	0.02	[1]
Maximum methane oxidation rate (kg m^{-3} s^{-1})	5.6×10^{-6}	[2]
Half-saturated constant	0.0066	[3]
Intrinsic permeability (m^2)	5.8×10^{-12}	[3]
Van Genuchten's parameter (m)	0.33	[3]
Van Genuchten's parameter, a (m^{-1})	5	[3]

4.1   Model Calibration and Verification

In the test, air aeration is applied through A and B equally (Fig. 2). The daily injection flux of CH_4 was 1235.41 g while the total amount of air aeration was 10 times that of CH_4. The test lasted for 200 days. The comparison of gas concentration profiles on the 28th day is shown in Fig. 3a. It can be found that the calculated results agree with the experimental results reasonably well. O_2 concentration increases firstly and then decreases with depth, indicating that the horizontal aeration wells can effectively increase O_2 in the cover. CH_4 concentration keeps decreasing with depth, which reflects that CH_4 is consumed in the whole cover domain. CO_2, as the product of oxidation reaction, increases from the bottom to the top. N_2 concentration distribution is stable and only affected by the concentration change in the other gases.

Figure 3b shows the comparison of CH_4 oxidation efficiency. After 18 days, the calculated results remain stable. The calculated results are overall enclosed by the observed values. The difference is caused by the change in bacteria quantity during the aeration process, which is not considered in the numerical model.

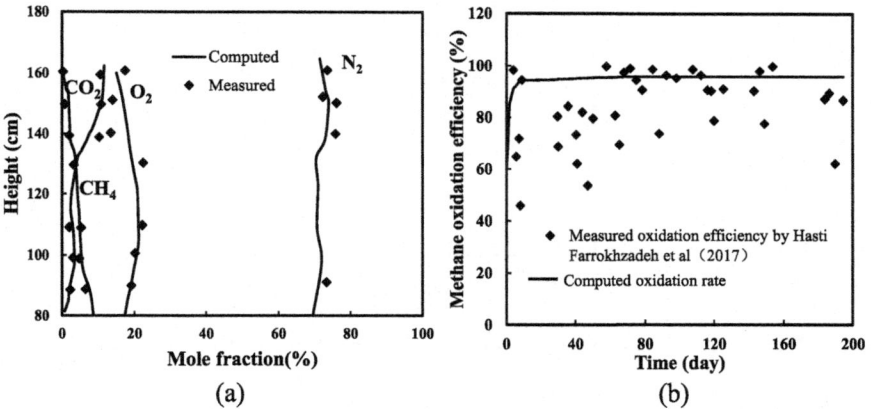

Fig. 3. Comparison between measured and computed results

4.2   Influences of Air Injection

Figure 4 shows the CH_4 oxidation efficiency given different injected air amounts while CH_4 flux keeps unchanged. The daily injection flux of CH_4 is 1235.41 g and air flux varies from 2 to 10 times that of CH_4. Air is injected through A and B equally and lasts for 200 days. The methane oxidation efficiency gradually increases with the aeration process for all the 5 curves and the results remain stable after about 18 days. The methane oxidation efficiency also increases with the air flux level. It is noteworthy that the methane oxidation efficiencies of the 8 times case and 10 times case are very close, indicating that higher air flux level can significantly enhance the oxidation efficiency but it is not economical to further increase the air flux level when it is high enough.

Figure 5 shows the influence of the air injection location on methane oxidation efficiency. Three scenarios are simulated; namely, air injected through A (Scenario 1),

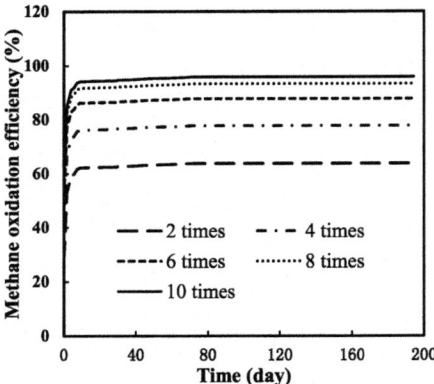

Fig. 4. Influence of injected air amount on methane oxidation efficiency

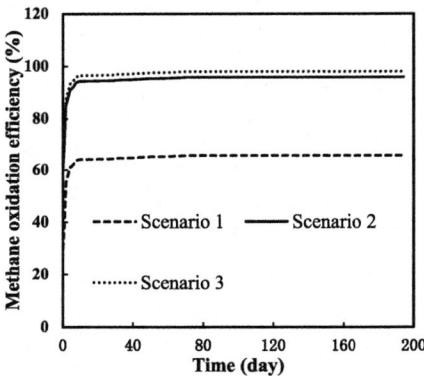

Fig. 5. Influence of air injection location on methane oxidation efficiency

A and B equally (Scenario 2), A, B and C equally (Scenario 3). The gas fluxes for the three scenarios are the same, which is 1235.41 g/d for CH_4 and 5068 g/d for air. The simulation lasts for 200 days. The methane oxidation efficiencies of Scenarios 1 and 2 are 65.8% and 95.8%, respectively, and that for Scenario 3 is slightly higher than that of Scenario 2.

5   Conclusion

A new landfill biocover with horizontal aeration wells is proposed in this study. A numerical model is developed to simulate the water-gas-heat coupled transport in the cover. The numerical model is verified against existing soil column experimental test results. The main conclusions are as follows:

(1) Laying the aeration well at the top of landfill cover system can effectively increase O_2 concentration in the landfill cover, which can enhance the methane oxidation reaction and is beneficial for impeding methane emission to the atmosphere.

(2) Higher air flux level can significantly enhance the oxidation efficiency but it is not economical to further increase the air flux level when it is high enough (e.g., 8 times the methane flux in this study).

(3) Given the same injected air amount, injecting air though more rows of wells gives higher methane oxidation efficiency. In this study, two rows of wells have similar oxidation efficiency as three rows of wells.

References

1. Ng, C.W.W., Feng, S., Liu, H.W.: A fully coupled model for water-gas-heat reactive transport with methane oxidation in landfill covers. Sci. Total Environ. **508**, 307–319 (2014)
2. Farrokhzadeh, H., Hettiaratchi, J.P.A., Jayasinghe, P., Kumar, S.: Aerated biocovers with multiple-level air injection configurations to enhance biological treatment of methane emissions. Bioresour. Technol. **239**, 219–225 (2017)
3. Clapp, R.B., Hornberger, G.M.: Empirical equations for some soil hydraulic properties. Water Resour. Res. **14**(4), 601–604 (1978)
4. Abu-Hamdeh, N.H., Reeder, R.C.: Soil thermal conductivity effects of density, moisture, salt concentration, and organic matter. Soil Sci. Soc. Am. J. **64**(4), 1285–1290 (2000)
5. Poling, B.E., Prausnitz, J.M., O'connell, J.P.: The Properties of Gases and Liquids. Mcgraw-Hill, New York (2001)
6. Gebert, J., Groengroeft, A.: Passive landfill gas emission-influence of atmospheric pressure and implications for the operation of methane-oxidising biocovers. Waste Manag. **26**(3), 245–251 (2006)
7. Huber-Humer, M., Gebert, J., Hilger, H.: Biotic systems to mitigate landfill methane emissions. Waste Manag. Res. **26**, 33–46 (2008)

Determination of Anti-floating Water Level in Guangzhou Based on the Distribution of the Underground Aquifers

Mengxiong Tang[1], Hesong Hu[1], Chenglin Zhang[1], Mingxun Hou[1], and Hang Chen[1,2(✉)]

[1] Guangzhou Institute of Building Science, Guangzhou, China
chenhang.thu@gmail.com
[2] School of Civil Engineering, Guangzhou University, Guangzhou, China

Abstract. Three kinds of underground aquifers are observed in Guangzhou with universal shallow water level. The water content is abundant in the strata along the coastal area of Pearl River and Liuxi River, in Guang-hua alluvial plain, carbonates rock karst area and fault fracture zones. The specific yield of the carbonate rock fissure cavern and paleo-channels reaches 1539–2131 m^3/d and 1054–4680 m^3/d, respectively. The water yield property gradually decreases with the increasing distance from the rivers. On the other hand, the underground water exploitation and running water system have negligible effects on the underground water level, and the construction in the underground space also has no long-term influences on the underground water level due to the abundant recharge resources. It is concluded that the determination of anti-floating water level has strong regional characteristics since the distribution of underground water level is similar to the topography. Some practical advices for determination of the anti-floating water level in Guangzhou are proposed.

Keywords: Aquifers · Water yield property · Water supply and drainage Underground water exploitation · Anti-floating water level

1 Introduction

Due to the insufficient knowledge of the underground water, increasing examples of the buoyancy induced basement floating, structure cracking and even overall damage of the buildings have been reported, which generally lead to serious losses [1]. The damage of the basement induced by groundwater buoyancy can be classified into two major categories, i.e. the breakage of the underground structure backplane after swelling and the floating of integral underground structure. For example, for an ultra-tall building with podium located on Tianhe Rd. in Guangzhou, the floating of the completed basement was found in a rainstorm due to the long-term suspension of the construction of ground structures, which was also encountered by a residential and commercial community in Haikou [2]. The three-floor basement of a tall building with natural raft foundation located on Fazhen Rd. in Guangzhou floated unevenly during a rainstorm, which thereby led to the failure of anti-floating anchor. Arching and cracking were also present in the backplane of the tower and nearby region of a residential and commercial

© Springer Nature Singapore Pte Ltd. 2018
A. Farid and H. Chen (Eds.): GSIC 2018, *Proceedings of GeoShanghai 2018*
International Conference: Geoenvironment and Geohazard, pp. 468–480, 2018.
https://doi.org/10.1007/978-981-13-0128-5_52

building in Jiangmen [3]. The non-uniformly floating of the basement of Sunshine Park of Shenzhen led to synchronous overall oblique of the building [4]. The maximum floating displacement among the different columns of the Anbao Zhonglv Hotel in Shenzhen also reached 160 mm [5].

The anti-floating design of the basement has attracted increasing attentions from scholars all over the world [6–9], where the determination of the anti- floating water level is the knottiest issue. The underground water level, which is a random variable without artificial interference, presents the nature of uncertain relying on environmental factors, such as its natural fluctuation, existence conditions, weather and precipitation. However, due to the urban construction, including dewatering, underground exploitation, as well as impoundment and flood discharge by lakes and reservoirs, the underground water level will be also significantly influenced by artificial activities. The excessive underground water exploitation of Beijing led to the decrease of its elevation from 34–37 m in 1957 to about 19 m in 2005 [10]. The project of South-to-North Water Diversion in China reduces the underground water exploitation of 300 million tons per year, which benefits the restoration of the confined water level up to 4.29 m/year [11]. Therefore, neither the measured water level nor the lowest/highest historical water level is appropriate to be the anti-floating water level. Moreover, the anti-floating water level should be determined by comprehensive analysis based on the environmental situations (such as type, distribution and depth of underground water, the number of aquifers, rock structures, geological structure, recharge and discharge conditions of groundwater [12], long-term dynamical monitoring data, local highest historical water level and the current situation of underground water exploitation).

Through the analysis of the underground aquifers distribution, the water yield property of strata and the measured underground water level in Guangzhou, the principles of determining the anti-floating water level during construction are discussed in this paper.

2   Natural Geographical Conditions

2.1   Geomorphology

Guangzhou is located in the south-central region of Guangdong Province, at the centre-northern boundary of Pearl River Delta, and the interchange of three rivers (i.e. Xijiang River, Beijiang River and Dongjiang River). Its geomorphological characteristics present that it has higher elevation in northeast, lower elevation in southwest, it is backed by mountains and faces the sea, as shown in Fig. 1. The highest peak of the northern hills region is located at the boundary of Conghua districts with the altitude of 1210 m. A national park of China, i.e. Baiyun Mountain, is located in the low mountain area in northeastern Guangzhou. The platform and terraces are generally distributed in central Guangzhou. The altitude of the southern alluvial plain is about 2–10 m.

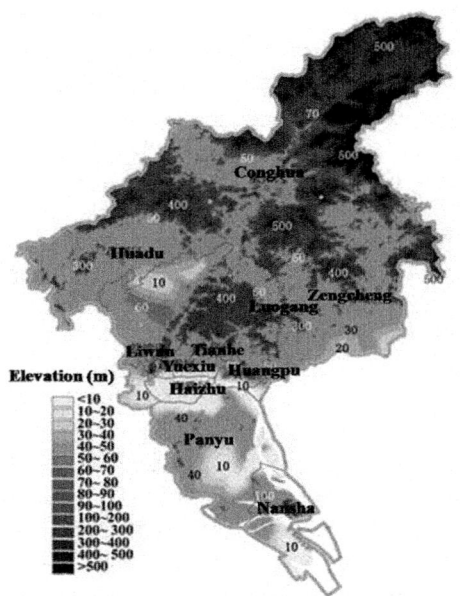

Fig. 1. Ground elevation contours in Guangzhou.

2.2 Distribution of Rainfall

Guangzhou, situated in south subtropical with maritime monsoon climate, has abundant rainfall with the annual precipitation ranging from 1229.6–2491.3 mm (Web-1), 85% of which is accumulated in rainy season (i.e. April to September). As shown in Fig. 2, the annual precipitation is measured as 2000 mm in northern mountain, 1800–2000 mm in centre-northern region, 1800–1900 mm in Nansha district (southernmost area), and less than 1800 mm in Panyu district.

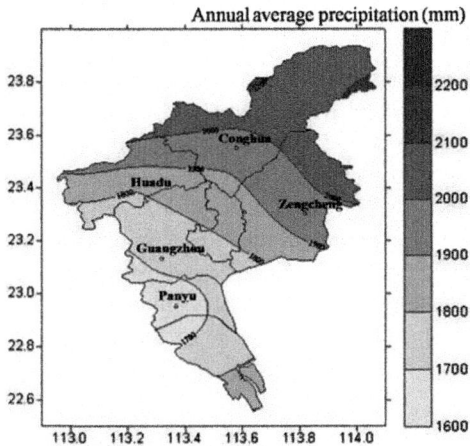

Fig. 2. Distribution of annual average precipitation in Guangzhou.

2.3   Distribution of the Fault, River and Terrace

According to the geological structure, Guangzhou is at the centre of the concave depression in South China Fold System [13]. The geomorphology of Guangzhou is simultaneously controlled by two faults, i.e. Shougouling and Guangcong faults, as shown in Fig. 3 [14]. The secondary level faults, crushing rocks, fissures and developed karst cave are widely distributed and provide passages for recharge, run-off and draining of underground water [15].

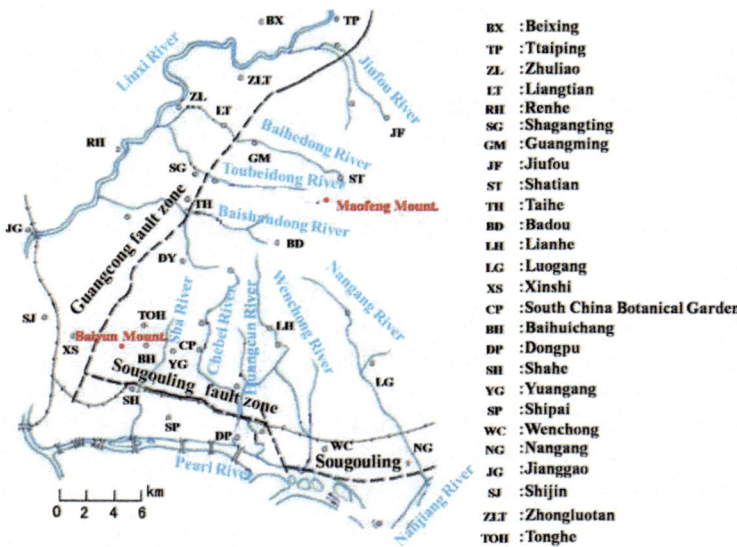

Fig. 3. Schematic diagram of Guangcong and Shougouling faults.

Perpendicular to the Guangcong Fault, several rivers with surrounding terrace of first level flows from east into Liuxi River through western Guanghua Plain. Similarly, perpendicular to the Shougouking Fault, Sha River, Chebei River, Huangcun River, Wenchong River, Nangang River and etc., run southward across the alluvial plain and Pearl River Delta, and finally converge into Zhujiang and Dongjiang, where the riverain terrace of first level is also distributed around their coasts, except that the remaining gravel terrace of second level exists in Longxue Island of Nansha district.

2.4   Recharge, Run-off and Draining of the Underground Water

The recharge, run-off and draining of the underground water present obvious regional characteristics in Guangzhou [16]. Rainfall and ground surface water are the two primary abundant sources for the recharge of underground water, since the precipitation is greater than the evaporation, and the ground surface water is plentiful because of the rivers woven into a network. In the hilly area and low mountains, the fractures and fissures of the exposure bedrocks are well developed so that the rainwater can directly penetrate into the underground space to realize recharge, which is, however, prevented by terraces and plains with thick covering clay. Alluvial sand and gravel can accept the recharge not only from the rivers and lateral fissure of bedrock, but also from the reservoir seepage. In sand stratum, rivers can recharge the underground water during flood or high tide, where the riverbed plays the role of draining passage. On the other hand, the development and utilization of subway and underground space has significant effects on local recharge and draining channels of underground water.

Bedrock fissure water circulates alternately. Ground surface water is recharged by the terrain change from high to low. It flows into fracture zone and is finally drained as spring, which is similar as the insidious water in corrosion cracks and confined water in caves. Along the plains and low hills around the riparian, underground water is excreted into rivers as run-off. The river network in delta is slowly drained into seas in the way of horizontal loop. Artificial draining is commonly adopted in the pumping well, the old mining district, excavation, civil air defense projects and subway tunnels.

3   Types of Underground Water and the Water Yield Property

The pore water in loose sediment, bedrock fissure water and cavern water are the three major types of underground water in Guangzhou. According to causes of formation, the pore water in loose sediment can be divided into alluvial and fluvial sedimentary phases. And the massive jointed, layered jointed and red stratum rocks fissure water are subdivisions of the bedrock fissure water.

The Quaternary unconsolidated sediments are subdivided into three aquifers according to the geological time, hydrodynamic characteristics, causes of formation and lithology: (1) phreatic aquifer mainly composed by the hydraulic filling soil; (2) first confined aquifer and (3) second confined aquifer, of which the lithology consists of oyster shell layer, fine sand, coarse sand and gravel [17]. The aquifuge between two aquifers consists of mud, clay and slit.

According to the statistics of 211 pumping drilled holes with the dissolved aqueous rocks less than 1 g/L in Guangzhou [18], the unit specific yield q, which is defined as the water flow released from the well per unit decline in hydraulic head, has notable discrete, as shown in Table 1, where the ratios of holes with rich ($q > 1.5$ L/(s·m)), medium (0.15 L/(s·m) $< q < 1.5$ L/(s·m)) and poor ($q < 0.15$ L/(s·m)) water yield property are 12% (i.e. 25 holes), 41% (i.e. 86 holes) and 47% (100 holes), respectively.

Table 1. The unit specific yield q of aquifers.

Types of underground water	Number of holes	q_{max} [L/(s·m)]	q_{max} [L/(s·m)]	$q < 0.15$(%)	$0.15 < q < 1.5$ (%)	$q > 1.5$ (%)
Pore water in loose sedi-ment	86	3.62	0.006	26	66	8
Carbonate cavern water	32	16.502	0.003	25	19	56
Bedrock fissure water	93	1.295	0.0001	75	25	0

3.1 The Pore Water in Loose Sediment

Phreatic Layer (as shown in Table 2). The phreatic layer presents lenticular shape around the coastal terraces and river islands in Zhonguotan north, Hengli, Qingtan and the east coast of West River. The phreatic aquifer at in the paleo-channel has abundant water yield property around the Huadu Nitrogen Fertilizer Factory of Huadu, Nianyugang, Qingtan and Bainixu.

Abundant phreatic layers exist in the area of Huadu – Lixi – Renhe, Gangmei, Qingqi, the coasts of Xijiang River and the valley located at higher and middle of Chebei River, where three paleo-channels, including Fuxing–Guigang–Renhe, Donghua–Guanglin–Hengli –Tangbei, and Guanglin–Ancient Louchang–Old Yayao Village, present rich water yield property. The medium water yield property of phreatic layer at the coasts along the west estuary of Beijiang River increases with the distance away from rivers or paleo-channels.

The thickness of the phreatic layer with poor water yield property in the alluvial plain of Tanbu and Chini, piedmont plain, valleys of the hills, Longxue Island and Southern Xinken is about 4–8 m, where the underground water level is about 0.1–4.5 m in depth.

Table 2. Distribution of phreatic layer.

Sites	Lithology	Thickness (m)	Underground water level (m)	Specific yield[a] (m^3/d)
Zhonguotan north, Hengli, Qingtan, the east coast of West River and etc.	Gravel sand	6.0–23.9	0.1–2.6	1014–3079
Huaxian, Lixi, Renhe, Gangmei, Qingqi, the coasts of Xijiang River, the valley located at higher and middle of Chubei River and etc.	Coarse sand, gravel sand, clay gravel	1.7–35.9	0.1–5.7	102–992
Tanbu and Chini, piedmont plain, valleys of the hills	Fine sand with coarse, muddy gravel sand	1.5–17.2	0.2–4.5	8–92
Longxue Island, Southern Xinken and etc.	Fine sand, coarse sand	2.3 (in average)	–	–

[a] The specific yield is defined as the volume of water released from the well per unit time.

Confined Aquifer (as shown in Table 3). At the coasts of Liuxi River from Lixi, Zhuliao, Renhe to Benghu, the top of the confined aquifer is about 3.0–6.6 m lower than the ground, and the specific yield of the paleo-channel reaches 1054–4680 m^3/d, which indicates rich water yield property.

Confined aquifers are scattered in Dali, Shajiao and Zhongcun of Guangfou plain, Zhuliao, Liangtian, Longgui, and Baini-Taiping, where the top of the confined aquifer is 4.3–9.9 m (locally 14.2–16.0 m). The depth of the underground water of these areas is about 0.2–5.3 m with medium water yield property as specific yield 247 m^3/d.

Table 3. Distribution of confined water.

Sites	Lithology	Thickness of aquifer (m)	Underground water level (m)	Specific yield (m³/d)
Coasts of Liuxi River from Lixi, Zhuliao, Renhe to Benghu	Fine gravel, gravel sand, coarse sand	3.0–6.6	0.9–4.5	1054–4680
Dali, Shajiao and Zhongcun of Guangfou plain, Zhuliao, Liangtian, Longgui, Baini-Taiping and etc.	Gravel sand, medium coarse sand, muddy sandy gravel	2.4–9.6	0.2–5.3	109–742
Zhongxitan-Jiahe, Longyandong, Huangnan and Jinli	Fine silty sand, clayey sandy gravel, clayey sand	1.9–14.3	0.4–4.5	5.4–53.5
Lianhua Mount.-Dongchong, Shawan-Lanhe-Lingshan, Xiaohu Island and Nansha-Hengli	Fine silty sand, medium coarse sand, gravel	0.5–45.6		<100

In the piedmont plains and mountain valleys of Zhongxitan-Jiahe, Longyandong, Huangnan and western Jinli, the confined aquifers with the thickness 2–5 m, depth 3.0–8.0 m (locally 47.6 m), and specific yield 5.4–53.5 m³/d present poor water yield property.

The first necessitous confined aquifer with the depth 0.3–25.4 m and specific yield 100 m³/d are composed of paralic sedimentary sand in North Lianhua Mountain, Shawan-Lanhe-Lingshan, northeastern Dongyong, western Huangge and southern Wanqingsha-Nansha. And the lithology of the first confined aquifer consists of fine silty sand followed by medium coarse sand and gravel.

In some local area in Lianhua Mountain, Shilou and Dagang, the first and second confined aquifers are connected by faults.

The second confined aquifer composed of river, marine and deltaic sedimentary sand is deposited below 2.2–47.8 m, the lithology of which is medium coarse sand followed by fine silty sand and gravel. The depth of the aquifer top is commonly larger than 15.0 m with the thickness 8.8 m and necessitous specific yield <100 m³/d in Lianhua Mountain-Dongchong, Huangge-Nansha and Nansha-Hengli. Besides, the depth of the aquifer top is varied with the topography in western Longxue Island.

3.2 Bedrock Fissure Water

Red beds are located in Taiping, Longgui and Chigang, where only those in Chigang are exposed to the air and the rest are covered with Quaternary strata. The lithology in these areas obtained by bored holes with the depth of 32.3–250.7 m presents as consists of mudstone, purple sandstone, conglomerate, siltstone interlayer containing pore fissure water (as shown in Table 4). In the tension, tension-torsion induced faults and fracture interchanges, the underground water is about 0.3–3.1 m below the surface with medium to rich water yield property, which is poor of the Red beds in other sites.

Table 4. The results of the borehole pumping test in red beds.

Sites	Lithology	Underground water level (m)	Specific yield (m³/d)
Epidemic Prevention Station	Sandstone, glutenite	2.7	337.4
Metallurgical Exploration Co.	Sandstone, glutenite	2.1	98.8
Guangzhou Zoo	Sandstone, conglomerate	0.3	59.9
Guangzhou Tower	Silty mudstone	3.1	28.7
Dengfeng Road	Sandstone, conglomerate	2.6	0.7

By borehole pumping tests up to 46.3–138.5 m, the fissure water in massive rock are founded in the migmatite and biotite granite joint fissure buried 0.2–4.9 m underground in the hilly area from Baiyun Mountain to Luogang with medium water yield property (see Table 5). However, the water yield property is poor in the low hills and mounds southern to Luntou-Changzhou area due to the thick destructured strata and undeveloped granite fissure.

Table 5. The results of the borehole pumping test in massive rocks.

Sites	Lithology	Underground water level (m)	Specific yield (m³/d)
The Yuexiu Park	Eroded granite detritus	+0.3	187.7
Huangpu Shipyard	Medium-coarse granite	+0.2	181.8
Nanhu quarry	Granite	0.9	131.2
Troop in Shahe	Granite	+0.3	92.0
Zhuzi Road	Eroded granite detritus	0.7	5.4

The fissure water in stratified rocks, i.e. sandstone, shale, mudstone interbeds, also present poor water yield property (locally medium) with the underground water level 0.3–3.2 m (Table 6), which is obtained from the bored hole up to 44.9–185.4 m in depth.

Table 6. The results of the borehole pumping test in stratified rocks.

Sites	Lithology	Underground water level (m)	Specific yield (m³/d)
Guocun	Calcareous silty siltstone	1.2	173.1
The gate of Yuexiu Park	Limestone, sandstone	+0.3	173.1
Bichong Cement Factory	Fine silty sandstone	+0.3	130.0
Panfu Road	Carbonaceous shale, sandstone	3.2	36.7
Magang of Nanhaizhou	Limestone embedded in sandstone	1.6	0.1

3.3 Carbonate Cavern Water

The carbonate cavern water is mainly distributed in area with boundaries such as Sanyuanli (south), Baini-Fengcun-Lianglong of Huadu (north), Zhuliao-Jiahe (east) and Nanhai Kiln-Tanbu (west), where bedrocks are only discretely exposed and almost covered by Quaternary or red beds. Detected up to 28.7–396.4 m in depth, aquifers are composed of hidden karst, including limestone and dolomitic limestone, which presents drastic changes of the water yield property as tabulated in Table 7. Although the underground water level of these aquifers is shallow as 0.6–10.1 m, the aquifers with developed karst are generally thick and present high water yield property (the specific yield of which reaches 1539–2131 m³/d). At the same time, the cavern water is always connected to the pore water of Quaternary.

The cavern water is abundant in Yagang, Jiangxia and Xiaogang-Sanyuanli, while moderate in Haitou Village, Xiamao-Tan Village and Keziling. However, the cavern fissure water are filled with muddy in southern Guliao Village and Tianxin Village-Luoyong so that the water yield capacity of the strata is poor.

Table 7. The results of the borehole pumping test for carbonate cavern fissure water.

Sites	Lithology	The depth of underground water (m)	Specific yield (m³/d)
Guangzhou Second Coal Industry Co.	P_1q Limestone	5.9	2131
Pengshang Village	P_1q Limestone	3.5	1709.2
Yagang	C_1ds Limestone	0.9	1539
New Stadium	C_1ds Limestone	2.0	46.6
Yuelong Village	$C_{2+3}ht$ Limestone	0.6	1.0

In summary, the phreatic aquifers are widely distributed in Guanghua basin and broad northern area of Guangzhou, while the confined aquifers are mainly concentrated in Guangfou plain and coasts of Xijiang river; the water yield property of loose sediment are is controlled by Quaternary paleogeographic environment so that that the water content of the strata is abundant in paleo-channel and but lacking around the mountains and between hills; phreatic aquifers commonly presents lenticular distribution, and their water yield property is closely related to the groundwater recharge system; high water yield property is found in the faults and their intersection zone, fault fracture zone and spherically weathered granite shell fissure, the area of well-developed pure carbonate karst corrosion fissure, as well as river coasts, paleo-riverbeds and river/marine sedimentary plain with thick sand layers.

3.4 Distribution of the Underground Water Lever of Aquifers

The elevation of the underground water is shallow almost all over Guangzhou. It decreases with the water flows from north to south, passes through Shawan, Dalong, Lianhua Mountain, Dagang and Huangge, then runs into Lion Ocean. In hills, mountains, Shilou, Dalong, Shiqiao, Shawan Street, the underground water level is controlled by topography and is commonly larger than 3 m, while that in plains is smaller than 2 m due to the effects of topography and tide.

Fig. 4. The distribution of the underground water level of Guangzhou.

The distribution of the underground water level of Guangzhou, which is obtained by statistics of 348 engineering survey data as shown in Fig. 4 presents similar shape as topography that the elevation gradually decreases from the northeast hills to the central basin, and until the southern coastal alluvial plain.

4 Effect of Exploitation of Underground Water

The average annual reserves of underground water resource of Guangzhou is 2.175 billion m^3, where 35% of them (i.e. 0.756 billion m^3) is allowable to be exploited. The underground water resource is mainly concentrated on five areas, i.e. Liuxi River terraces, Longdong alluvial plain, Nangang-Xintang alluvial plain, Zengjiang alluvial plain and Guanghua basin, as tabulated in Table 8, where Guanghua basin has the largest reserves and allowable yield as 0.6 billion m^3 and 0.197 billion m^3, respectively.

Until 2011, 111 users have obtained the underground water exploiting permission with the total yield 4.0579 million m^3 per year, as tabulated in Table 9. The permitted exploiter are mainly distributed in Baiyun (in Guanghua basin), Tianhe (in Longdong alluvial plain), Conghua (in Liuxi River terraces), Luogang (Zengjiang alluvial plain) and Huadu (in Guanghua basin) districts, where the underground water is abundant. From the ratio of the annual actual yield to allowable yield in these districts, it is found that the scale of the underground water recharged is far more than that of yielded so that underground water exploitation has little effects on its water level.

Table 8. Reserves and allowable withdrawal of groundwater resources.

Sites	Area (km^2)	Reserves (10^9 m^3)	Allowable yield (10^9 m^3)	Average modulus of recharge (10^4 m^3/a·km^2)	Average modulus of allowable yield (10^4 m^3/a·km^2)
Liuxi River terraces	127	0.46	0.21	51.0	23.37
Longdong alluvial plain	89	0.24	0.07	52.88	15.72
Nangang-Xintang alluvial plain	26	0.05	0.03	31.26	17.83
Zengjiang alluvial plain	300	1.03	0.44	69.57	29.63
Guanghua basin	858	2.6	1.97	44.0	33.36
Total	1400	4.38	2.71		

Table 9. The situation of underground water exploiting in Guangzhou.

District	Number of permitted exploiter	Annual yield (10^5 m^3)	Yield ratio
Conghua	21	165.91	0.72
Baiyun	45	101.65	1.03
Luogang	8	87.18	2.39
Tianhe	23	33.95	3.09
Huadu	12	13.6	0.08
Zengcheng	2	3.5	0.02
Total	111	405.79	0.54

5 Conclusion

The determination of anti-floating water level is related to the safety and construction costs of the building. Guangzhou is located in subtropical southern China coastal area, where heavy seasonal rainfalls yield the quickly rise of underground water level to the ground surface. At the same time, the tide, occurring on the 3rd and 18th days every Lunar month, also has great impact on the underground water level, which results in the maximum evaluation of the Waijiang River. If the peak of the upstream arrives simultaneously with the tide, their superimposition leads to immersion of streets. Therefore, the determination of anti-floating water level in Guangzhou has strong regional characteristics. In this paper, the distribution of underground aquifers, the water yield properties, as well as the effects of underground space construction and water exploitation are analyzed. We also present the distribution of the underground water level elevation in Guangzhou to guide the determination of anti-floating water lever with the combination of specific situation. The main conclusions are summarized below:

(1) The water level measured and provided by geotechnical investigation report lacks representation due to the significant seasonal change. Therefore, it is of high risk

to take the measured water level as the anti-floating level and the most adverse circumstance should be considered.

(2) Under the circumstance that the highest flood level may reach Pearl River embankment, the anti-floating level should be higher than designed floor. The calculation of the intensity of the flood should include the estimation of the maximum flood level in 100 or 50 years for important or common engineering respectively.

(3) Low-lying terrains are commonly distributed around the river coasts, the elevation of construction site may lower than flood control lift. Setting strong emission pump around the lift are employed to prevent flood. If the drainage is not timely in case of heavy rain, the anti-floating water level should be higher than the elevation of designed floor.

(4) Atmospheric precipitation and rivers provide enough recharge for underground water, which is far more than the exploitation yield. The running water for daily life and industry is also mainly from the ground surface. Therefore, both the underground water exploitation and running water system has negligible effects on the underground water level. The drainage during construction leads to obvious local decrease of underground water in tunnels and underground space, especially when the subway station or tunnel passes through fault fracture zone or cave area. But in long term, the underground water level will recover after the backfilling due to the adequate recharge. Therefore, the excavation of foundation should not be involved during the determining of anti-floating water level.

(5) The distribution of underground water level is similar to the topography, which decreases from the northeast hills to the central basin, and until the southern coastal alluvial plain. If the construction site is located in hilly areas, the perennial maximum water level is far higher than the peak flood water level of surrounding rivers so that the anti-floating water level for basements should be discretely set up in stepped-sharp according to the terrain ladder. The underground water level for alluvial plains is generally 0.1–2.0 m so that the anti-floating water level should be set the same as the ground surface. The low-lying terrains are generally distributed in the southern region of the coasts of Pearl River. If the embankment is higher than the elevation of designed floor, the anti-floating water level should be also higher than the elevation of designed floor.

Acknowledgements. The authors acknowledge the financial support from national NSFC (Grants No. 51608139 and 51678171), Guangdong Science and Technology Department (Grant No. 2004B36001028), and China Postdoctoral Science Foundation (Grant No. 2016M592471).

References

1. Tang, M., Hu, H., Zhang, C.: Anti-buoyancy of Underground Structures, pp. 86–98. China Architecture & Building Press, Beijing (2016)
2. Yang, F.: Study on Anti-floating Design of Underground Structure. South China University of Technology, Guangzhou (2013)

3. Lu, Y., Gong, J., Lan, H.: Analysis on local anti-floating failure case of existing building and its strengthening. Guangdong Archit. Civil Eng. **21**(12), 27–28 (2014)
4. Fang, X.: Floating accident treatment in Yangguang garden basement. Eng. Qual. **6**, 43–45 (2003)
5. Li, Y.: An example of the floating treatment of a basement. Geotech. Eng. World **4**(9), 34–35 (2001)
6. Cao, H., Zhu, D., Luo, G., Pan, H.: Study of anti-floating calculation method for underground structures near riverside. Rock Soil Mech. **38**(10), 1–8 (2017)
7. Li, Y., Li, X.: Discussion on anti-floating design of basement in the area with high water level and soft soil. Hans J. Civil Eng. **5**(5), 196–200 (2016)
8. Gan, Q., Yang, B., Liu, W., Li, Y., Yu, F.: Discussion and application of the water-discharging pressure relief technology in the anti-floating design of basement. Build. Struct. **46**(2), 86–90 (2016)
9. Zhang, Z., Sun, B., Xu, H.: Effect of characteristics of ground water distribution and seepage on anti-uplift analysis of building foundations. China Civil Eng. J. **34**(1), 73–78 (2001)
10. Shen, X., Zhou, H., Wang, J., Han, X.: Groundwater and Anti-floating Structure, pp. 148–157. China Architecture & Building Press, Beijing (2013)
11. Li, C.: Reasonable value of water level of building anti-floating design. Geotech. Investig. Surv. **4**, 49–54 (2014)
12. Zhang, K., Qiu, J.: Analysis and discussion on water level for prevention of up-floating and calculation of uplift pressure. Geotech. Eng. Tech. **21**(1), 15–20 (2007)
13. Li, P., Zheng, J., Fang, G.: Quaternary geology in Guangzhou area. South China University of Technology Press, Guangzhou (1989)
14. Zheng, J., Liu, R.: Types and characters of groundwater in Guangzhou area and its affection to City Construction Environment. Trop. Geogr. **3**, 264–268 (2006)
15. Liu, S.: The distribution and characteristics of the river terraces in the Pearl River Delta and its nearby areas. Trop. Geogr. **5**, 400–404 (2008)
16. Pang, Y., Lv, W.: The investigation and evaluation of groundwater resources and management countermeasures of Guangzhou. Pearl River **2**, 27–31 (2014)
17. Wang, Z., Liu, H.: Analysis of the hydrogeological characteristics and aquifer structure in the southern area of Guangzhou. Ground Water **4**, 79–81 (2011)
18. Peng, W., Rong, S.: Analysis of hydrogeology characteristic in Guangzhou. Urban Geotech. Investig. Surv. **3**, 59–63 (2006)

Behavior of Biotreated Geomaterials and Foundations

Investigation on the Environmental Impact of Soil Improvement Techniques: Comparison of Cement Grouting and Biocement

Maryam Naeimi[1] and Abdolhosein Haddad[2(✉)]

[1] Research Institute of Petroleum Industry, 14875-33111 Tehran, Iran
[2] Faculty of Civil Engineering, Semnan University, Semnan, Iran
Haddad@semnan.ac.ir

Abstract. Soil improvement techniques – including both mechanical and chemical stabilization methods such as dynamic compression and grouting – have potential drawbacks such as high cost, high energy consumption and sometimes negative environmental impacts. An alternative approach is to use biocement to improve the engineering properties of soil. Microbially induced calcite precipitation (MICP) has been introduced as a technique for modification of geotechnical properties of sand. Among many studies concerning biocementation of sand, there are few studies considering the comparison of cost and environmental impacts of cement grouting and microbial methods. The environmental concerns in the present study was focused on the produced CO_2 and calcite usage in Portland cement. The primary component of cement is limestone which is a natural resource. Cement manufacturing is highly energy and emissions intensive because of the extreme heat required to produce it. Producing a ton of cement requires 4.7 million BTU of energy and generates nearly a ton of CO_2. Given its high emissions and critical importance to society, cement is an obvious place to work on reducing greenhouse gas emissions. On the other hand, calcium chloride was also used as a crucial reagent in MICP treated samples. Therefore, the present study discussed the environmental aspects of conventional and innovative methods of soil improvement. In the cement grouting method, Portland cement was used as a chemical substance. Portland cement were applied by surface percolation and mechanical mixing to the samples. Then, the results of cement grouting were compared with the results of biocement samples which were gathered from literature. The results for treated samples were discussed and compared based on one cubic meter of soil and final target of 700 kPa. The results show that the amount of calcium usage in the cement grouting was 2.5 times more than bio-treated samples and therefore higher energy and gas emissions.

Keywords: Environmental impacts · Soil improvement · Cement Biocement

© Springer Nature Singapore Pte Ltd. 2018
A. Farid and H. Chen (Eds.): GSIC 2018, *Proceedings of GeoShanghai 2018 International Conference: Geoenvironment and Geohazard*, pp. 483–490, 2018.
https://doi.org/10.1007/978-981-13-0128-5_53

1 Introduction

Increasing the population in sedimentary environments with soft soils shows the necessity of escalating the load capacity of soil. Various soil improvement techniques have been studied such as chemical stabilization methods. The most used stabilizer for sand is cement. Cement is a soft roundness, water absorbent and binder. Neri (2015) reported the effect of cement and the effect of high pressure slurry injection on the hydro-mechanical properties of the soil before and after injection. Results showed that the increment of strength and the elasticity of hardness as well as decrease the permeability over time. However, soil density, particle size distribution and degree of cementation has significant correlation with the value of hardness and strength [2]. However, porosity and moisture content are key parameters to control the strength of the treated soil [3]. For example, the moisture content in samples with water to cement ratio up to 0.4 is not sufficient for hydration of cement particles [4]. Therefore, researches in the application of cement in slope stability, reconstruction of roads, foundation of road construction, refineries, reduction of permeability in reservoirs and protection of rivers was reported [5–7].

On the other hand, cement manufacturing requires a calcium source (usually limestone) and a source of silicon (such as clay or sand). Cement manufactured through the treatment of cooked calcium oxide with silicon oxide and iron oxide. Then, the material turns into almost black colored balls called clinger. In order to adjust the setting time, after cooling, the clinger is mixed and grounded with some gypsum. Therefore, the gray powder is produced. Under this process, lime and carbon dioxide (CO_2) are produced as shown in reaction (1). The process is highly energy and emissions intensive because of the extreme heat requirement. Producing a ton of cement requires 4.7 million BTU of energy and generates CO_2 which is a concern in environmental engineering.

$$CaCO_3 \rightarrow CaO + CO_2 \tag{1}$$

Two sources are known for the release of CO_2 in the cement production process. Firstly, combustion of fossil fuels for the activity of the rotary ovens, which are the largest sources for CO_2 production. Secondly, the chemical process of converting limestone to lime. Thus, a total of 1.3 tons of CO_2 per ton of cement is released into the atmosphere. Hence, it can conclude that soil improvement techniques using cement grouting have potential drawbacks such as high cost, high energy consumption and sometimes negative environmental impacts.

An alternative approach is to use biocement to improve the engineering properties of sand. Microbially induced calcite precipitation (MICP) has been introduced as a technique using reagents for modification of geotechnical properties of sand since 2005.

The process involves two main parts: (1) absorption of urease enzymes or urease cells on sand aggregates; (2) hydrolysis of the urea enzyme and formation of calcium carbonate crystals in the presence of calcium ions.

Biocement has been considered as an appropriate solution for replacement with cement grouting because of low viscosity and being environmentally friendly.

However, among many studies concerning MICP technique, there are few studies considering the comparison of cost and environmental impacts of cement grouting and biocement.

The primary component of cement is limestone while calcium chloride also used as a crucial reagent in bio-cemented samples. Therefore, the present study discussed the comparison of conventional and innovative methods. In the grouting method, Portland cement was used as a chemical substance. Note that, the environmental concerns in the present study was focused on the calcium usage in Portland cement as well as biocement. Sariosseiri and Muhunthan (2009) reported that a compressive strength equal or greater than 345 kPa is required for an effective soil improvement. Therefore, it was chosen as a standard to compare the cement and bio treated methods of the present paper.

2   Materials and Methods

2.1   Materials

In the present study, sand from Garmsar region, Semnan province in Iran was provided. The size distribution curve according to ASTM 2487 has been carried out as shown in Fig. 1. Sand was categorized as SP based on its characteristics with the specific gravity of 1450 kg/m^3. Shahrood Portland cement, Type 2, with a maximum setting time of 4 h (GS = 3.15 g/cm^3) was selected as the stabilizer. Physical characteristic of Portland cement was reported in Table 1 [9].

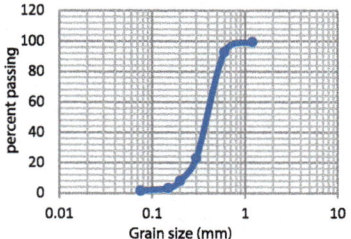

Fig. 1. Size distribution curve

2.2   Methods

Samples was prepared in a PVC mold of 50 * 100 mm. As shown in Table 2, the percentage of Portland cement (C) and water to cement (w/c) ratio has been chosen based on literature [1, 10–12]. Samples were treated with Portland cement using four procedures in room temperature. Procedures was chosen as close to the bio-treatment of sand. Detailed of the procedures has been reported in Table 2. The remarkable point after removing samples from the mold was that the method was fixed only at the points where the injection was performed. Therefore, the sand improvement was not uniform and the sample would be disrupted after the time elapsed. This was seen in samples of

Table 1. Physical characteristics of Portland cement (type II)

Physical characteristics	Average
Fineness (cm^2/gr)	3309
Expansion	0.039
Setting time	
Initial (min)	140
Final (hour)	3:30
Compressive strength (kg/cm^2)	
Day 1	===
Day 3	275
Day 7	372
Day 28	469

procedure II and III. Therefore, the UCS of samples of procedure I and IV was tested. In procedure I, samples were treated using mechanical mixing of sand, cement and water. While, in procedure IV, cement and sand was homogenously mixed and placed in the mold in a dry state. Then, a flow rate of water was injected. Among various setting time reported in literature [12, 13], 7 and 28-days was selected for samples for further investigation. Later on, selected samples were tested using unconfined compressive strength (UCS). Note that the code of selected samples were shown in Table 2.

Table 2. Testing program

Groups	Methods	C%	w/c	Code of samples
I	Mechanical	6-8-10	0.7-1.4	I 6
				I 8
				I 10
II	Grouting with flow from top to bottom	2-4-6-8-10	2-3-4-5-9-19	–
III	Grouting injection with constant head of 2.2 L/h	2-4-6-8-10	2-3-4-5-9-19	–
IV	Water injection with constant head on dry sand mixed with cement of 2.2 L/h	2-4-6-8-10	2-3-4-5-9-19	IV 2-W 19
				IV 8-W 4
				IV 10-W 3
				IV 12-W 2
				IV 6-W 5.6
				IV 4-W 9

3 Results and Discussion

3.1 Comparison of Cement Treated Sand with Procedure I and IV

Energy consumption is one of the most important environmental issues related to cement production. Cement is one of the industries that consumes the most energy which includes direct fuel consumption for the extraction and transportation of raw materials. Therefore, less environmental degradation methods are in a favor.

Comparison of cement treated sand using procedure I and IV based on UCS, amount of Portland cement and calcium are presented in Fig. 2. The results show that the only method to prepare samples with less than 60 (kg/m^3) is the injection method. Calcium is the most important component of cement that supplies it through the use of limestone. Also, it plays a significant role in increasing of compressive strength. In the same calcium content as shown in common part of the graphs of Fig. 2, it can conclude that the injection method produces higher UCS for the same calcium content.

3.2 Comparison of Conventional and Microbial Soil Improvement Techniques

The use of microbial processes has been considered in recent years for in situ soil improvement. Biocementation has some advantages over existing technologies, such as less calcium usage in same UCS. The data obtained from previous studies and present study was gathered in Table 3. It can be stated that the calcium usage in biocement is half of cement treated samples. Therefore, it can surely state that one of the key benefits of bio-stabilization is to significantly reduce the loss of energy and eliminate carbon emission.

Regarding to the cost of biogeochemical stabilization projects, it was reported that the cost was a factor of process and the specific details of each project. Despite the very limited field applications that have been done. The actual cost of various upgrading processes has remained largely unknown. To date, there have been widely differing studies and estimates due to revised and optimal designs that are ongoing to date. For MICP, the cost of materials (urea, calcium) and the total cost of consolidation (materials, equipment, and installation) in saturated soils were estimated. Cheng (2012) studied the economic solutions which led to a reduction in the cost of cultivating bacterial and the cost of chemicals [15]. Production of urease-producing bacteria is a major contributor to costs, including laboratory costs, equipment, implementation, chemicals, sterilization and transfer of the environment from the biotechnology company to the site. In research, they cultivated bacteria in a non-sterile environment, thus reducing the cost of 50% compared with sterilized culture media [15]. Comparison of UCS and cost in biocement (Cheng 2012) and Portland cement treatment (present study) were shown in Fig. 3.

As shown in Fig. 3, the modified biocementation, with non- sterile culture and the use of gravity flow, significantly reduces the costs. However, the results of treated samples with mechanical and injection methods with UCS equal to 700 kPa, are about $ 22 cheaper than the modified biocementation.

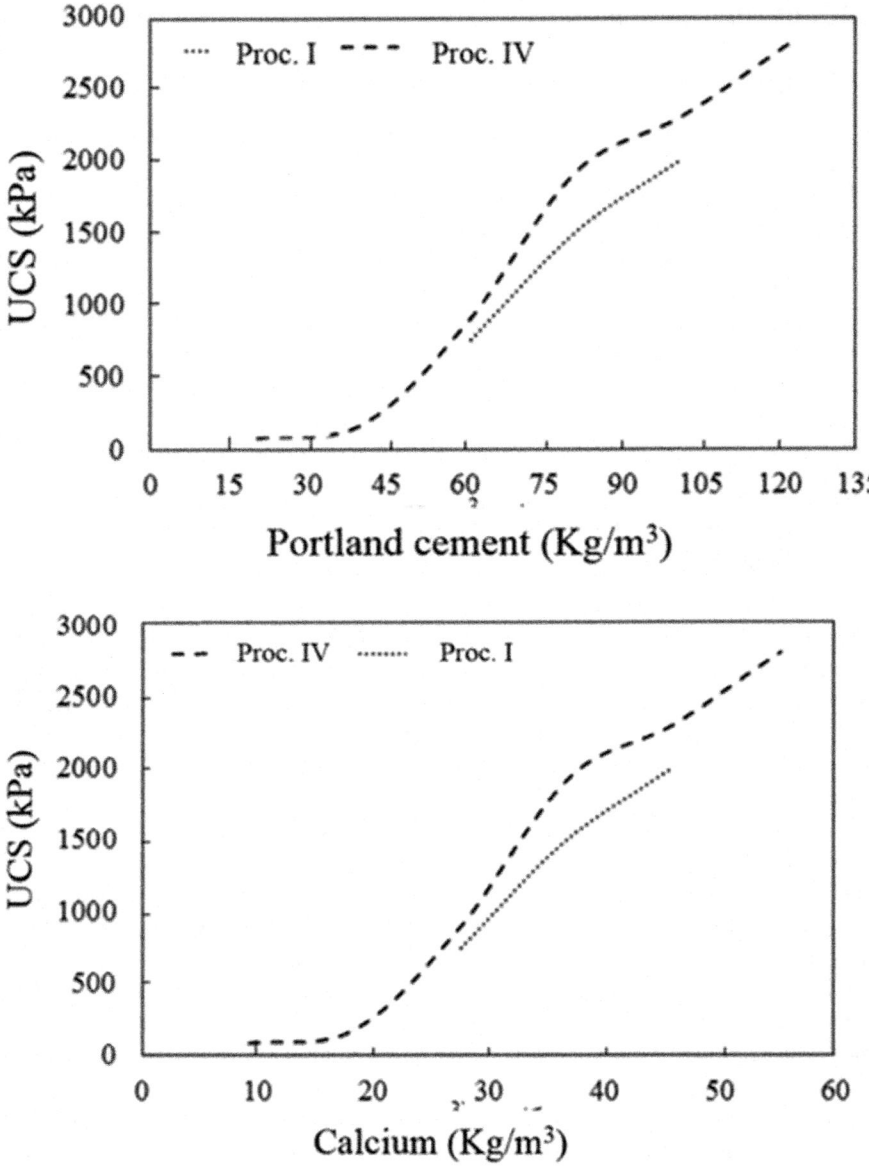

Fig. 2. Compression of UCS and Calcium usage in the Portland cement treatment

Table 3. Comparison of calcium (%) usage in biocement and Portland cement [9]

UCS (KPa)		Ca (Kg/m^3)						
		Biocement					Portland cement (this study)	
—	Whiffin (2007)	Van paasen (2010)	Cheng (2012)	Chu et al. (2014)	Naeimi (2014)		(I)	(IV)
500	8.7	—	8.8	8.74	9.32		26	25
750	—	79.5	12	13.1	13.9		28	27
1000	—	85.4	12.8	17.4	20.3		30	29
1250	—	87.9	16	21.5	26.2		34	32
1500	—	91.1	17	26.5	29.1		36	33

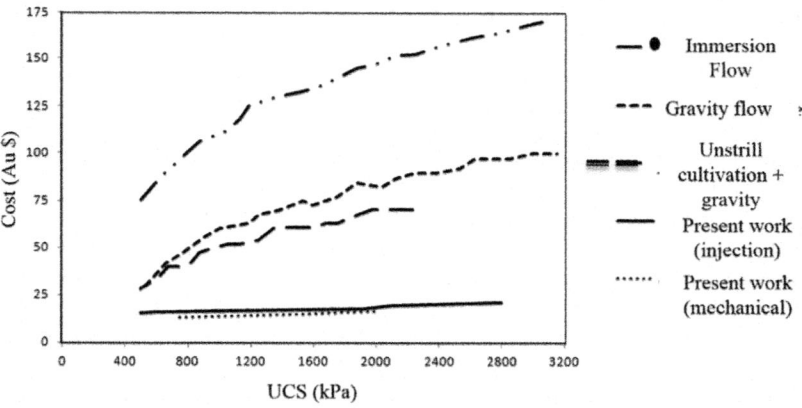

Fig. 3. Comparison of UCS and cost in biocement (Cheng 2012) and Portland cement treatment (present study)

4 Conclusion

Cement is considered as an unhealthy substance because of its high energy consumption, extraction of large amounts of raw materials and land degradation. Therefore, the innovative bio-stabilization has been compared with the results of Portland cement. The results of samples treated by Portland cement indicated the direct correlation between C% and UCS. Samples treated with 6, 8, 10 and 12% after 28-days showed the increment of USC by 70, 108, 25 and 40% respectively in comparison with 7-days.

The results show the only method to prepare samples with cement less than 60 (kg/m^3) is the injection method. In the same calcium content, it can conclude that samples treated with the injection method showed higher UCS. Regarding to the comparison of calcium usage, it can be stated that the calcium usage in biocementation is half of Portland cement grouting. Additionally, the results of samples treated with mechanical and injection methods with UCS of 700 kPa, are cheaper than the modified bio-treatment.

References

1. Néri, R.: Consideration of bonding in the behaviour of a sand-cement mixture simulating jet grouting. Instituto Superior Técnico, Lisbon (2015)
2. Clough, G.W., et al.: Cemented sands under static loading. J. Geotech. Eng. Div. **107**(6), 799–817 (1981)
3. Consoli, N.C., et al.: Key parameters for strength control of artificially cemented soils. J. Geotech. Geoenviron. Eng. **133**(2), 197–205 (2007)
4. Lim, S.K., et al.: Effect of different sand grading on strength properties of cement grout. Constr. Build. Mater. **38**, 348–355 (2013)
5. Ajorloo, A.M., et al.: Experimental investigation of ce-ment treated sand behavior under triaxial test. Geotech. Geol. Eng. **30**, 129–143 (2012)
6. Consoli, N.C., et al.: Key parameters for strength control of rammed sand–cement mixtures: influence of types of portland cement. Constr. Build. Mater. **49**, 591–597 (2013)
7. Szczesniak, M., Rougelot, T., Burlion, N., Shao, J.-F.: Compressive strength of cement-based composites: roles of aggregate diameter and water saturation degree. Cem. Concr. Compos. **37**, 249–258 (2013)
8. Sariosseiri, F., Muhunthan, B.: Effect of cement treatment on geotechnical properties of some Washington State soils. Eng. Geol. **104**(1), 119–125 (2009)
9. Moradi, A.: Study of engineering properties of sand stabilized with cement and lime and comparison with the of biologically improvement method. In: Civil Engineering, Semnan University (2015)
10. Amini, Y., Hamidi, A.: Triaxial shear behavior of a cement-treated sand–gravel mixture. J. Rock Mech. Geotech. Eng. **6**(5), 455–465 (2014)
11. Hashemi, S., et al.: The failure behaviour of poorly cemented sands at a borehole wall using laboratory tests. Int. J. Rock Mech. Min. Sci. **77**, 348–357 (2015)
12. McDowell, G., Bolton, M.: On the micromechanics of crushable aggregates. Geotechnique **48**(5), 667–679 (1998)
13. Beeghly, J.H.: Recent experiences with lime-fly ash stabilization of pavement subgrade soils, base, and recycled asphalt. In: International Ash Utilization Symposium (2003)
14. Naeimi, M.: Biocementation of sand in geotechnical engineering. Nanyang Technological University, Singapore (2014)
15. Cheng, L., Cord-Ruwisch, R.: In situ soil cementation with ureolytic bacteria by surface percolation. Ecol. Eng. **42**, 64–72 (2012)

Development of Innovative Bio-beam Using Microbial Induced Calcite Precipitation Technology

Changming Bu[1], Qian Dong[1], Kejun Wen[2], and Lin Li[2(✉)]

[1] Chongqing University of Science and Technology, Chongqing, China
[2] Jackson State University, Jackson, MS, USA
lin.li@jsums.edu

Abstract. As a new environmental friendly and sustainable technique for soil improvement, microbial induced calcite precipitation (MICP) has been studied widely. Most of previous studies focus on soil modification. This study is to develop innovative bio-beams based on MICP technology. A modified permeable rigid mold has been developed to prepare the bio-beams specimens within the immersing method. The permeable mold provides supports for the bacteria to grow through fully contact cementation media and induce the calcium carbonate precipitated homogeneously within the pores of the sand to form the bio-beam. Bending tests have been conducted to evaluate the flexure behavior of MICP-treated bio-beams. Single and multiple treatment of MICP in the various cementation media concentrations have been conducted. Experimental results show that the vertical load of the MICP-treated bio-beams could reach to 1412 N when the bio-beams were treated for one time. The double treatments of the bio-beams increased to the 163% of bending strength of single-treatment. The peak bending strength of the triple MICP-treated bio-beam specimen increased to 200% of bending strength of single-treatment. The bending strength also increased with cementation media concentrations.

Keywords: Bio-beam · Microbial induced calcite precipitation
Immersing method · Bending strength

1 Introduction

Microbial induced calcite precipitation (MICP) has recently gained much attention from geotechnical engineering researchers for soil improvement worldwide [1, 2]. MICP requires the existence of ureolytic bacteria, urea and calcium-rich solution to drive the MICP biogeochemical reaction [3]. In MICP process, urea is hydrolyzed by microbial urease to form ammonium and carbonate ion. The produced carbonate ions react with calcium ions to precipitate as calcium carbonate crystals. The calcium carbonate crystals can bond the sand partials together and provide an improvement to the geotechnical engineering properties of bio-treated soil. An aerobically cultivated bacterium with highly active urease enzyme (*Sporosarcina pasteurii*) has been widely used to catalyze the MICP process [4]. MICP can be used for a variety of applications such as improvement of strength and stiffness of soil, improvement and remediation of

© Springer Nature Singapore Pte Ltd. 2018
A. Farid and H. Chen (Eds.): GSIC 2018, *Proceedings of GeoShanghai 2018*
International Conference: Geoenvironment and Geohazard, pp. 491–498, 2018.
https://doi.org/10.1007/978-981-13-0128-5_54

concrete, and environmental remediation [5–10]. Achal et al. (2013) used bacteria to heal the simulated cracks of depths up to 27.2 mm in cement mortars and increased 40% compressive strength [6]. Zhao et al. (2014a) developed immersing method with a full contact flexible mold to work with MICP treated standard sand and achieved UCS of 2.13 MPa at 13.4% of calcite precipitation by weight [11]. Bernardi et al. (2014) developed a bio-mediated process for manufacturing bio-bricks [12]. Compression and shear wave were measured on those bio-bricks, the compression stress of bio-brick ranged from 120 kPa to 2200 kPa. Han et al. (2016) found that the MICP grouting can mitigate liquefaction in sandy soil [2]. However, there is no information from the literature reviews about the bio-beam through MICP and its mechanical behavior under compression.

Pumping (or injection) methods have been widely used in MICP studies that typically used PVC plastic columns with a peristaltic pump to inject cementation medium in the influent end. Martinez et al. (2013) used peristaltic pumps to maintain one-dimensional flow conditions in 50-cm-long laboratory sand columns during the MICP process [3]. Their findings indicated that non-uniform calcite was distributed along the column. Near the influent end of the PVC column, most of the calcite precipitated and hindered the cementation reaction in the deeper section of the column. This non-uniformity of MICP-treated soil caused strength variation. Immersing method is another soil preparation method with full contact flexible molds (FCFM) in MICP [13]. Geotextile mold allows the chemicals penetrate into sand pores and maintains a suitable precipitation rate. Their experimental results show that the MICP-treated soil samples are more uniform [13]. However, the fabric structure of this mold is too soft to prepare beam-shape sample, such as beam-shape.

This paper is to develop a new sample preparation mold to form bio-beams based on MICP technique. This new sample preparation mold can retain the efficient penetration advantage of FCFM and increase the stiffness of the mold for bio-beam samples. Four-points bending tests were conducted to test the flexural behavior of MICP-treated bio-beams samples.

2   Methods and Materials

2.1   Sandy Soil

Ottawa silica sand (99.7% quartz) with a median particle size of 0.46 mm was used in this study. According to the USCS soil classification, the Ottawa sand is poorly graded sand.

2.2   Bacteria and Growth Medium

The bacterium used in this study was *Sporosarcina pasteurii* (ATCC 11859). Bacterial cell concentration was controlled by measuring absorbance (optical density) of the suspension using a spectrophotometer at 600-nm wavelength. An Ammonium-Yeast Extract media (ATCC 1376) is used to grow the bacteria cultures to the desired population density. Individual components were autoclaved separately and mixed together post-sterilization. The bacteria were cultivated in culture medium for 40 h at 300C. The bacteria solution was centrifuged at 4000 rpm for 20 min and the supernatant was removed and replaced by fresh culture media. The bacteria solutions were stored in the centrifuge tubes in a fridge under the temperature of 4 °C.

2.3   Cementation Media

Cementation media was used to provide chemicals to induce the calcite precipitation during the treatment. The cementation medium included urea (30 g/L), $CaCl_2 \cdot H_2O$ (73.5 g/L), NH_4Cl (10 g/L), $NaHCO_3$ (2.12 g/L), and nutrient broth (3 g/L). All samples were immersing into 40 L cementation media with a plastic box in an aerobic environment.

2.4   Rigid Full Contact Mold

A rigid full contact mold (RFCM) was fabricated that includes flexible layer and rigid holder. The rigid holder was made with polypropylene perforated sheet with 6.35 mm thickness and 6.35 mm diameter staggered holes. The rigid holder was assembled with different piece of polypropylene sheet by long screw rod and waterproof tape. The size of rigid full contact mold can be varied to meet different needs. Figure 1 shows an example of mold in dimension of 355.6 mm × 101.6 mm × 38.1 mm (L × W × H).

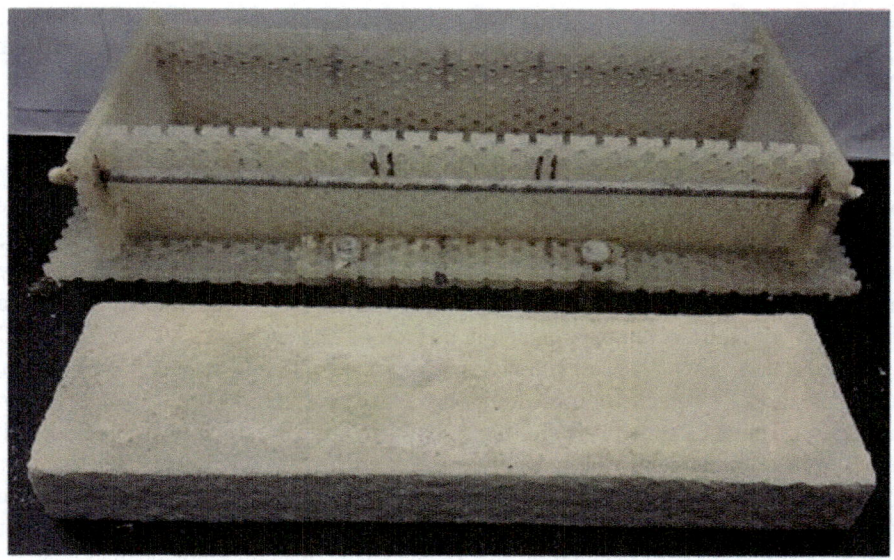

Fig. 1. Rigid full contact mold with image of one MICP-treated bio-beam. The bio-beam size is 355.6 mm × 101.6 mm × 38.1 mm (L × W × H).

To prepare identical three MICP specimens, the rigid holder was revised to include three identical internal cells, as shown in Fig. 2 (a). The size of each divided cell is 177.8 mm × 76.2 mm × 38.1 mm (L × W × H). In these right molds, nonwoven geotextile were used to cover the bottom and side walls of rigid holder. The geotextile was a polypropylene, staple-fiber, needle-punched nonwoven material with grab tensile strength of 1700 kN and apparent opening size of 0.15 mm. Figure 2 (b) shows the holder is covered with geotextile. Then dry sand was poured into the mold. After the bacteria solution was mixed into sand, a layer of geotextile was added on the top of sand to fully cover the soil.

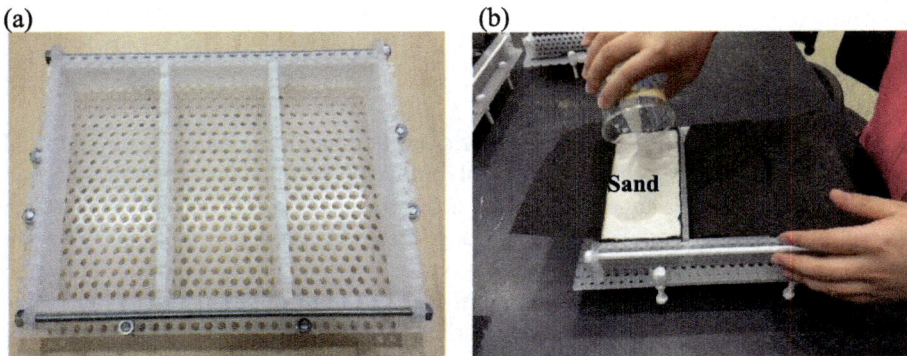

Fig. 2. Rigid full contact mold with geotextile layer: (a) rigid holder, and (b) cover with one layer of geotextile in the bottom & side wall of the rigid holder then pour sand into the mold and cover with geotextile in the top.

2.5   Sample Preparation

For one cell in the rigid full contact mold, 900 g dry sand were compacted into the mold gradually. After the soil was mixed uniformly with bacteria, a cover of geotextile was added on the top to cover the soil. The soil samples in the rigid mold were immersed into a batch reactor with 0.5 molar Ca cementation media and were allowed to react for 7 days, followed the immersing method of Zhao et al. (2004b). During the 7 days of reaction, only air was continuously pumped into the reactor. No bacteria, growth medium, or cementation medium was added to or pumped out of the reactor. The 7-day treatment is one cycle. In this study, a multiple treatment was studied with up to 3 cycles of treatment. During each treatment cycle of MICP, a new bacteria and a new cementation media was used. After 7–21 days of reaction, all samples in the rigid mold were removed from the batch reactor. Then the MICP-treated soil sample was taken out from each cell as shown in Fig. 3. The beam-shaped specimens had a length of 177.8 mm and a width of 76.2 mm and a height of 38.1 mm. All specimens were oven-dried at 50 °C for at least 2 days before any testing.

(a)   (b)

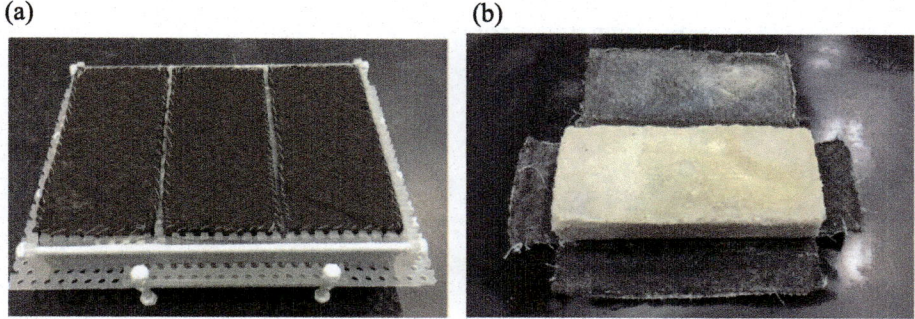

Fig. 3. MICP treated bio-beam with: (a) rigid full contact mold with geotextile, and (b) the MICP specimen (177.8 mm × 76.2 mm × 38.1 mm (L × W × H)).

2.6   Four Point Bending Test

The four point bend tests were conducted on the bio-beam specimens after 7 days' treatment to explore the flexure behavior of MICP treated soil. The MICP-treated specimen was flipped to have 76.2 mm as height. As shown in Fig. 4, the specimens were sit on two supports that were 152.4 mm apart from each other, and the stress applied on top two points at middle of the specimen with 50.8 mm distance. Following the testing method of ASTM D 6272, the vertical load was conducted under strain controlled conditions at a uniform loading rate of 1.5%/min until bio-beam fails.

(a)   (b)

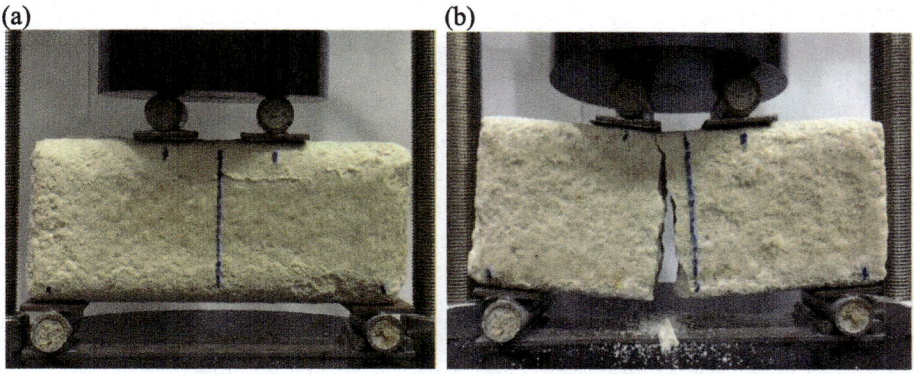

Fig. 4. Four point bending test: (a) the setup of four point bending test; and (b) failure specimens after four point bending test.

3   Results

For the MICP-treated samples prepared by rigid full contact mold, the soil specimen was supported by polypropylene perforated sheet, and was enclosed by the flexible and permeable materials. The rigid holder helped to improve the stiffness of the mold to

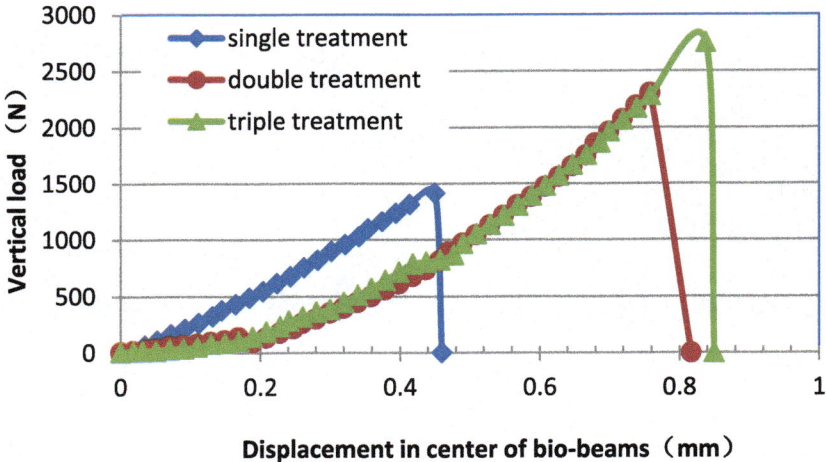

Fig. 5. Load-displacement curves of different treatment bio-beams made by 100% standard sand

develop the beam-shape for large dimensions of soil samples. The geotextile as a flexible film increased the contact area between the specimen and cementation media which can promote the $CaCO_3$ distributed in the bio-beam uniformly.

Figure 5 shows the different load-displacement curves in the center of the bio-beams as a function of treatment times. The vertical peak load of the MICP-treated bio-beam can reach to 1412 N when the bio-beams were treated for one time of MICP. The quickly decline of vertical load after the peak load shows the brittle behavior of bio-beam.

When the treatment period increased from one MICP treatment to two MICP treatments, the peak load reached to 2304 N, with 63% increasing from that of single MICP treatment. The failure strain also increases nearly 80%. The increase of peak load and failure strain indicates that more calcite precipitation form in the double MICP treatment. The more calcite can form strong bonding on the bio-beam. The brittle behavior did not improve after the double MICP-treatment. When the MICP treatment increased to three times, the peak load only increased 19.5% from that of double treatment.

By increasing the contact between the cementation solution and the bacteria, the permeable rigid mold for preparing bio-beams samples not only enhanced the sample uniformity but also significantly improved the engineering properties of the sand. The more times it reacted, the higher strength of the bio-beams became.

4   Conclusion

A rigid full contact mold with geotextile was developed, and it can be used to prepare larger size bio-beam specimens for MICP treatments with immersing method. Beam shape specimens were treated in MICP process. Four point bending test was conducted to evaluate the flexure behavior of MICP-treated specimens. The result indicated that the MICP-treated sample prepared by rigid full contact mold with geotextile can form the regular beam shape and the uniform calcite precipitation inside the sample. The geotextile with fibrous structure increased the contact area of MICP-treated sample, which effectively increased the penetration of cementation media. Furthermore, the large amounts of pores allowed more precipitation occurring inside the mold to improve the strength of samples. The four-point bending results indicates that the vertical peak load of the MICP-treated bio-beams could reach to 1412 N when the bio-beams were treated for one time. The double treatments of the bio-beams increased to the 163% of strength of single-treatment. The peak strength of the triple MICP-treated bio-beam specimen increased to 200% of strength of single-treatment.

Acknowledgements. This paper is based upon work supported by the National Science Foundation Grant No. 1531382 and U.S. Department of Transportation Grant No. (DTRT13-G-UTC50FHWA) through Maritime Transportation Research and Education Center. The first author acknowledges financial support from the China Scholarship Council under Grant No. 201408505113, from Chongqing Research Program of Basic Research and Frontier Technology No. cstc2014jcyjA30005, from Chongqing University of Science and Technology, No. CK2015Z27 and Chongqing BaYu Plan.

References

1. DeJong, J.T., Mortensenb, B.M., Martinez, B.C., Nelson, D.C.: Bio-mediated soil improvement. Ecol. Eng. **36**(2), 197–210 (2010)
2. Han, Z., Cheng, X., Ma, Q.: An experimental study on dynamic response for MICP strengthening liquefiable sands. Earthq. Eng. Eng. Vib. **15**, 673 (2016). https://doi.org/10.1007/s11803-016-0357-6
3. Martinez, B.C., DeJong, J.T., Ginn, T.R., Montoya, B.M., Barkouki, T.H., Hunt, C., Tanyu, B., Major, D.: Experimental optimization of microbially-induced carbonate precipitation for soil improvement. J. Geotech. Geoenviron. Eng. **139**(4), 587–598 (2013)
4. Martin, D., Dodds, K., Ngwenya, B., Butler, I., Elphick, S.: Inhibition of *Sporosarcina pasteurii* under anoxic conditions: implications for subsurface carbonate precipitation and remediation via ureolysis. Environ. Sci. Technol. **46**, 8351–8355 (2012)
5. Whiffin, V., van Paassen, L., Harkes, M.: Microbial carbonate precipitation as a soil improvement technique. Geomicrobiol. J. **24**, 1–7 (2007)
6. Achal, V., Mukerjee, A., Reddy, M.S.: Biogenic treatment improves the durability and remediates the cracks of concrete structures. Constr. Build. Mater. **48**, 1–5 (2013)
7. DeJong, J.T., Soga, K., Kavazanjian, E., Burns, S., van Paassen, L., Al Qabany, A.: Biogeochemical processes and geotechnical applications: progress, opportunities, and challenges. Geotechnique **63**(4), 287–301 (2013)
8. Li, M., Li, L., Ogbonnaya, U., Wen, K., Tian, A., Amini, F.: Influence of fiber addition on mechanical properties of micp-treated sand. J. Mater. Civ. Eng. **28**(4), 04015166 (2016)
9. Wen, K., Li, L., Ogbonnaya, U., Li, Y., Bu, C., Liu, S., Li, C., Amini, F.: Development of an improved immersing method to enhance microbial induced calcite precipitation through multiple treatments in low cementation media concentration. Geotech. Geol. Eng. (in review)
10. Harkes, M.P., Booster, J.L., Paassen, L.A., Loosdrecht, M., Whiffin, V.S.: Microbial induced carbonate precipitation as ground improvement method – bacterial fixation and empirical correlation $CaCO^3$ vs strength. In: International conference on BioGeoCivil Engineering, Delft, The Netherlands (2008)

11. Zhao, Q., Li, L., Li, C., Li, M., Zhang, H., Amini, F.: Factors effecting improvement of engineering properties of MICP-treated soil catalyzed by bacteria and urease. J. Mater. Civ. Eng. (2014a). https://doi.org/10.1061/(ASCE)MT.1943-5533.0001013
12. Bernardi, D., DeJong, J.T., Montoya, B.M., Martinez, B.C.: Bio-bricks: biologically cemented sandstone bricks. Constr. Build. Mater. **55**, 462–469 (2014)
13. Zhao, Q., Li, L., Li, C., Zhang, H., Amini, F.: A full contact flexible mold for preparing samples based on microbial induced calcite precipitation technology. Geotech. Test. J. **37**(5), 1–5 (2014b)

A High-Pressure Plane-Strain Testing System to Evaluate Microbially Induced Calcite Precipitation as a Sand Production Control Method

Ning-Jun Jiang[1]([⊠]), Kenichi Soga[2], and Koji Yamamoto[3]

[1] University of Hawaii at Manoa, Honolulu, HI 96822, USA
`jiangn@hawaii.edu`
[2] University of California-Berkeley, Berkeley, CA 94720, USA
[3] Japan Oil, Gas and Metals National Corporation (JOGMEC),
Chiba 261-0025, Japan

Abstract. Sand production is the phenomenon that solid particles (sand in many cases) from an oil/gas reservoir move into a production well along with oil/gas and water flows. Sand production in weakly consolidated sand reservoirs is a growing concern in petroleum industry. In this study, a high-pressure plane-strain sand production testing system was designed, which was able to simulate the stress and flow conditions in the field. This testing device was used to evaluate the effectiveness of microbially induced calcite precipitation (MICP) as a sand production control method. The erosion rate and hydraulic gradient were measured for both Non-MICP and MICP treated specimens, with results verified by visual observations. It is found that the MICP method was able to substantially reduce sand production rate. In the current experimental set-up, the enlargement of upstream cavity created by sand production was delayed by the MICP treatment. The spatial and temporal variations in hydraulic gradient during sand production experiment were evaluated and the results are discussed.

Keywords: Bio-cementation · Sand production · Testing device

1 Introduction

The first ever field trial of methane gas production from hydrate bearing marine sediments was conducted in the spring of 2013 in the Eastern Nankai Trough off the Pacific Coast of Japan by the Japan Oil, Gas and Metals National Corporation (JOGMEC) [1]. Although this field trial was deemed successful in the sense of stable depressurization and continuous gas production, sand production occurred and terminated the field flow test after six days of gas production.

Sand production is the phenomenon that solid particles (sand in many cases) from a reservoir move into a production well along with oil/gas flows [2]. Sand production is usually initiated due to the redistribution of stress and pore pressure inside sandstone or sand formation by drilling and production operations. The produced sand may corrode the valve and pipe system, leading to leakage problem. It also involves substantial costs

A. Farid and H. Chen (Eds.): GSIC 2018, *Proceedings of GeoShanghai 2018*
International Conference: Geoenvironment and Geohazard, pp. 499–506, 2018.
https://doi.org/10.1007/978-981-13-0128-5_55

to separate sand with hydrocarbon flow on the surface, particularly in an offshore production platform. On the other hand, if the flow rate inside the well is not sufficiently high, produced sands start to gradually deposit inside the production well and eventually the well is clogged, which results in the reduction of production rate and substantial maintenance cost and time to clean up the well.

A credible sand control solution should not only stop sand migration, but also maintain the productivity of the well by preventing the clogging of the near-well formation. This provides a good opportunity for microbially induced calcite precipitation (MICP) to be applied for this problem, considering its highlighted advantages, like enhancing strength and stiffness, retaining soil permeability at small calcium carbonate precipitation content (usually smaller than 5–6%), energy-efficient treatment in the field compared to conventional chemical grouting, and fast bio-geochemical reaction rate [3, 4]. The challenge is whether MICP can work effectively under the high flow velocity condition during depressurization that triggers the dissociation of gas hydrate and the initiation and progression of sand production.

In order to investigate the erosional, geomechanical and hydraulic behaviors of MICP treated sand sediments, a high-pressure plane-strain testing system was designed. The system is capable of simulating the stress and flow conditions in the field. The erosion rate and hydraulic gradient variations were measured for both Non-MICP and MICP treated sand specimens. The measured results were verified by visual observations.

2 High-Pressure Plane-Strain Testing System

The general layout of the testing device developed for this study is schematically shown in Fig. 1. The photograph of the overall layout is shown in Fig. 2. The philosophy of this apparatus was to reproduce a deepwater unconsolidated sand sediment in a laboratory scale and to simulate sand production behavior under the hydraulic condition similar to that applied in the 2013 Nankai Trough field test. More specifically, the plane-strain design was to simulate one horizontal layer of sand formation around the wellbore in a radial section. The high-pressure applied on top of the sand was to reproduce the high in-situ stress of the sand formation. The gravel pack and sand screen placed next to the sand was to simulate the gravel-packing completion used in the 2013 Nankai Trough field test. The applied horizontal hydraulic flow was to reproduce the near-wellbore drawdown due to the depressurization process in the field trial.

As shown in Fig. 1, this apparatus consisted of a pressurized aluminum tank, a vertical loading system, a hydraulic control system, a sand collection system, a MICP implementation system, and an instrumentation system. The pressurized tank (370 mm (L) $\times$ 200 mm (H) $\times$ 200 mm (B)) accommodated sand and gravel pack and withstood vertical loads and horizontal hydraulic flow. The vertical loading system provided a stress similar to the field condition on the bulk of the soil (up to 2 MPa). The hydraulic control system was able to maintain a flow rate up to 864 L/h (0.006 m/s, pore velocity). The sand collection system served to measure the amount of eroded sand from the formation. The MICP implementation system supplied biological and

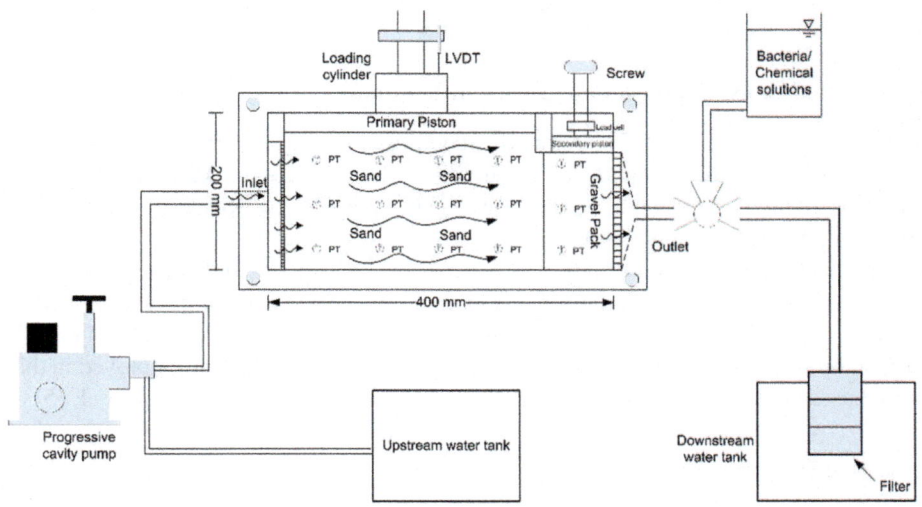

Fig. 1. General layout of the high-pressure plane-strain sand production apparatus

Fig. 2. Photograph of the layout of the high-pressure plane-strain sand production apparatus

chemical agents for MICP reactions and facilitated the injection of these agents into the sand formation. The instrumentation system served as a platform for real-time monitoring of pore pressure, vertical displacement and visualization of sand particles migration.

3 Testing Materials and Procedures

3.1 Tested Soils, Bacteria and Chemical Solutions

In this study, the test sand formation was composed using the British Standard sand (Fraction D, supplied by David Ball Group plc), which has d_{50} of 0.2 mm. The gravel pack was mimicked by natural gravel aggregate (supplied by Ridgeons UK) in this study, which has a d_{50} of 7 mm.

The bacteria used in this study was *S. pasteurii*. Its storage and cultivation procedures were the same as detailed in Jiang and Soga [4]. The cementation solution used in this study comprised equal molar concentration of urea and calcium chloride and 3 g/L nutrient broth. The concentration of urea/calcium chloride (1.0 M) was employed in this study based on the optimization studies by Jiang and Soga [4] and Gomez et al. [5].

3.2 Soil Preparation

The sand and gravel were prepared using the drying pouring method. Before soil pouring, a temporary separation sheet was inserted, forming two individual compartments for sand and gravel respectively. Predetermined weights of dry gravel and sand were loosely poured into each compartment in a consistent manner to achieve constant densities. The initial dry density for sand formation and gravel were 1.746 g/cm^3 and 1.994 g/cm^3, respectively. The corresponding void ratios of sand and gravel were 0.52 and 0.33, respectively. When the sand and gravel finally reached the designed height, the temporary plate was retrieved slightly. Water was then slowly poured into the gravel compartment so that it saturated the sand formation from the bottom, which minimized the entrapment of air bubbles. After the water level rose 0.5 cm higher than the surface of sand formation, a thin layer (<5 mm) of kaolin clay was placed on top of the sand and gravel to seal potential flow leakage at the interface between the piston and the soil. Afterwards, the secondary piston, permanent separation plate and primary piston were carefully laid in sequence on the top of the paved clay. The temporary separation sheet was completely retrieved before the placement of the primary piston. Finally, the top lid was covered and fixed by bolts after O-rings were attached.

3.3 MICP Treatment

The MICP treatment was implemented through gravitational percolation from the downstream side. The percolation direction was selected to simulate the injection from the wellbore to the formation in the field. Firstly, 6 L (approximately 1.2 PV of the sand formation) bacteria solution was filled up in the downstream bucket and was allowed to gradually percolate through the gravel and sand formation by gravity. The hydraulic gradient corresponding to this gravitational percolation was estimated between 2 and 3. The injection took around 1 h to complete. With 6-h retention time, 6 L chemical solution was then filled up in the downstream bucket and was percolated through the gravel and sand formation to trigger MICP reactions in the same way as the bacteria injection. After the chemical percolation, the gravel and sand formation were retained for another 18 h before the start of the hydraulic sand production test.

3.4   Sand Production Test

The saturated sand and gravel, with or without MICP treatment, were subjected to horizontal water flow in the direction from the sand to the gravel (Fig. 1). The flow rate was controlled at 353 L/h. Flushed sands were collected at the pipe exit using meshed plastic bags at 0.5, 1.5, 3, 5, and 7 min after the start of the test. Each sand collection lasted for 15 s. Meanwhile, a high-resolution video recorder kept recording the sand movement through the Perspex wall. The pressure transducers continuously collected pore pressure data during the test.

During the test, it was found that sand migrated too fast, which led to substantial vertical displacement in a short period. This further resulted in the loss of full contact between piston and sand formation after around 50–60 s from the start of the test. The loss of full contact could fundamentally change the sand production pattern from internal erosion to surface erosion. Therefore, in this study, the data interpretation and discussion in this particular test are based on the results from the initial 50–60 s, when only internal erosion occurred.

4   Testing Data Interpretation

4.1   Visual Observations

In the Non-MICP specimen, immediately after the start of pumping, sand quickly migrated into the gravel zone and escaped through the outlet screen. Only after 5 s, most visible pore spaces in the gravel zone were filled up with fluidized sand as shown in Fig. 3(a). When sand kept moving to the exit end, a cavity was formed at the inlet end of the sand formation as shown in Fig. 3(b).

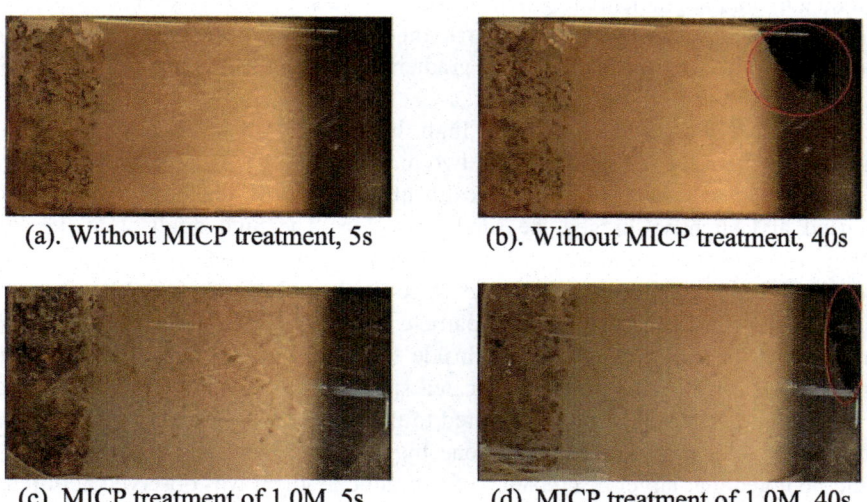

| (a). Without MICP treatment, 5s | (b). Without MICP treatment, 40s |
| (c). MICP treatment of 1.0M, 5s | (d). MICP treatment of 1.0M, 40s |

Fig. 3. Visual observations of the sand production process

In the specimen subjected to the MICP treatment, right after the start of the pump, sand fluidization and migration occurred but in a much slower pattern than what was observed in the Non-MICP specimen, as shown in Fig. 3(c). The gravel zone was fully filled with sand after 15 s. The upstream-side cavity did not form until 30 s after the start of the test. At 40 s, the size of the upstream-side cavity was much smaller than that in the Non-MICP specimen, as shown in Fig. 3(d). Meanwhile, limited sand was observed to come out from the outlet during the initial 40 s, which was much less than in the Non-MICP specimen.

4.2 Sand Production Behavior

Sand production rate was calculated based on the weight of flushed sand at an interval of 15 s through the whole area of the outlet screen (0.036 m^2). Due to the limitation discussed before, only the sand production rate of 30–45 s are reported here. It can be seen that, without the MICP treatment, sand production rate was as high as 70 g/(m^2·s). After the MICP treatment by 1.0 M chemicals, the average sand production rate was reduced to 18 g/(m^2·s). This indicated that, even though single MICP treatment could not fully stop the sand fluidization and migration, it was able to substantially retard the process. This retardation was consistent with the visual observations.

4.3 Hydraulic Behavior

The measured pore pressure data were used to analyze the hydraulic behavior during the sand production process. Horizontal hydraulic gradients were calculated based on the measured pore water pressure by two adjacent pore pressure transducers (PTs) at the same elevation. The spatial and temporal distribution of hydraulic gradient for the Non-MICP and MICP treated specimens are shown in Fig. 4.

For the Non-MICP specimen, a moderate and relatively uniform hydraulic gradient (2.5–7.5 m/m) was established along the entire specimen at 10 s, as shown in Fig. 4(a). A gradual hydraulic gradient drop then occurred at the inlet side due to the formation of cavity. In the gravel zone, a hydraulic gradient increase was observed, which was due to sand filling-up. At 40 s, the hydraulic gradient at the inlet end was almost zero while it was up to 7 m/m in the gravel zone.

For the MICP treated specimen, high hydraulic gradients (5–15 m/m) were established at both the inlet end and the two middle points at 10 s, as shown in Fig. 4 (b). This indicated that the MICP treatment with 1.0 M chemical concentration enhanced and slightly clogged the sand formation. The high gradient in the sand formation was retained until 40 s, which was longer than the untreated case. This was because produced calcite precipitation bridged sand particles, thus make it less vulnerable to skeleton rearrangement and particle fluidization/migration. The aggregated sand particles by MICP also jammed inside the gravel zone to retard further sand migration. Meanwhile, for the MICP treated specimen, the hydraulic gradient within the gravel zone increased slower compared to the Non-MICP one. This was attributed to the slower clogging of the gravel zone by mobilized sand particles. However, it should be noted that larger discrepancy of hydraulic gradient was observed at different locations within the gravel pack zone at 40 s in the MICP treated specimen. This was possibly due to the formation of excessive non-uniform calcite precipitation inside [6].

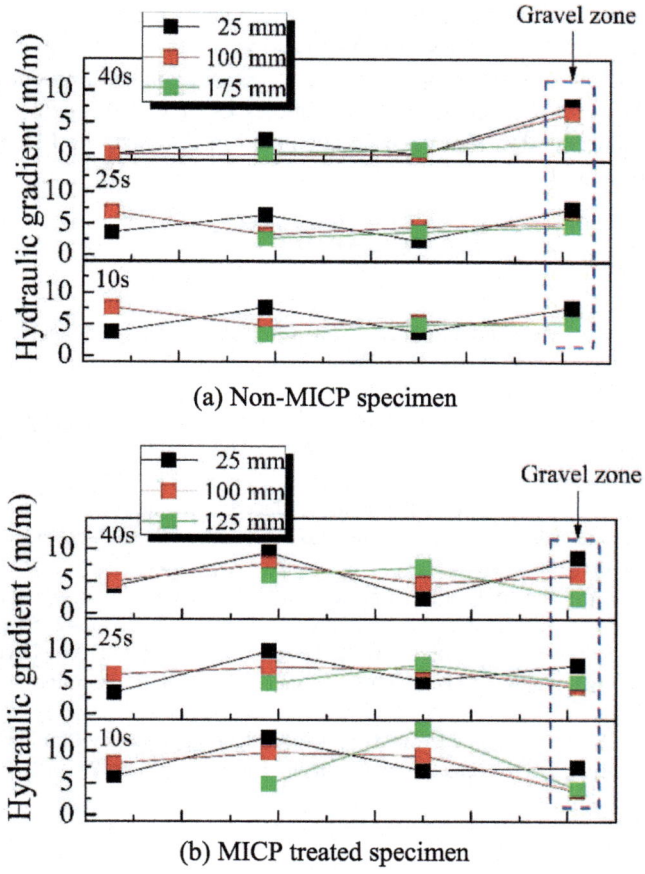

Fig. 4. Spatial and temporal distribution of hydraulic gradient during the sand production process

5   Conclusions

In this study, a high-pressure plane-strain sand production testing system was developed to simulate sand production behavior of a deepwater unconsolidated sand sediment in a laboratory scale. The testing system was used to evaluate the effectiveness of MICP as a sand production control method. It was concluded that the MICP treatment with 1.0 M chemical concentration significantly reduced the sand production rate. The enlargement of upstream cavity generated by sand production was also delayed in the MICP treated specimen case. The hydraulic gradient in the gravel zone gradually accumulated while that in the soil at the inlet side dropped. In the MICP treated specimen case, the hydraulic gradient change was retarded.

It should be noted that this was a preliminary test and substantial future work is needed for better understanding of mechanisms of the sand production process in weak sandy sediments. For example, sand production test can be conducted at a reasonable flow rate range to identify the critical pressure drawdown. Different sand and gravel pack materials, sand screen slots, and vertical loads can be used to investigate the effects of particle size, outlet completion, and stress state on sand production patterns. Moreover, various MICP implementation strategies (e.g., chemical concentration, injection rate, and retention time) can be attempted to optimize the MICP process for sand production control.

References

1. Yamamoto, K.: Overview and introduction: pressure core-sampling and analyses in the 2012–2013 MH21 offshore test of gas production from methane hydrates in the Eastern Nankai Trough. Mar. Pet. Geol. **66**, 296–309 (2015)
2. Ranjith, P.G., Perera, M.S.A., Perera, W.K.G., Choi, S.K., Yasar, E.: Sand production during the extrusion of hydrocarbons from geological formations: a review. J. Pet. Sci. Eng. **124**, 72–82 (2014)
3. Jiang, N., Soga, K., Kuo, M.: Microbially induced carbonate precipitation (MICP) for seepage-induced internal erosion control in sand-clay mixtures. J. Geotech. Geoenvironmental Eng. **143**(3), 04016100 (2017)
4. Jiang, N., Soga, K.: The applicability of microbially induced calcite precipitation (MICP) for internal erosion control in gravel-sand mixtures. Géotechnique **67**(1), 42–55 (2017)
5. Gomez, M.G., Hunt, C.E., Martinez, B.C., DeJong, J.T., Major, D.W., Dworatzek, S.M.: Field-scale bio-cementation tests to improve sands. Proc. ICE - Ground Improv. **168**(3), 206–216 (2015)
6. Al Qabany, A., Soga, K., Santamarina, C.: Factors affecting efficiency of microbially induced calcite precipitation. J. Geotech. Geoenvironmental Eng. **138**(8), 992–1001 (2012)

Long-Term Behavior of Colloidal Nano-silica Reinforced Sand in Different Environments

Mayao Cheng[1(✉)] and Nadia Saiyouri[2]

[1] Foshan University, Foshan, China
568492181@qq.com
[2] Université Bordeaux 1, Talence, France

Abstract. Colloidal nano-silica was firstly used for the injection of soil in the 1990s, which is relatively new compared with other materials. Nano-silica can be used into ground with low porosity. The aim is primarily to prevent or mitigate internal erosion. Since there are not any published data about the efficiency and the durability of nano-silica treated soil in different engineering environments, this paper focus on evaluating long-term improvement effect, including compressive strength and capacity to resist erosion. Nano-silica used in this paper is Levasil 300/30% hardened by NaCl of different concentration. Cylindrical specimens were prepared by mixing fine sand with nano-silica. The environments discussed in this paper include dry air, pure water, salt solution and acid solution. The test was carried out on specimens stored for 1, 7, 15, 30, 90, 180 and 365 days. The results show that, the colloidal nano-silica reinforced sand keeps stable in saturated environments (pure water, salt and acid) and becomes fragile in dry environment.

Keywords: Colloidal nano-silica · Internal erosion · Long-term behavior
Compressive strength · Capacity of resisting erosion

1 Introduction

Internal erosion (IE) is a complex phenomenon which causes major problems of levees, dikes and embankment dam stability. Several examples of failure of embankment dams have been reported in literature (Fry et al. 1997; Foster et al. 2000b) and most of them are related to this phenomenon.

It is very important to improve the erosion resistance of erodible soils using appropriate cost - effective techniques. Chemical stabilization is an effective technique for controlling erosion (Karol 2003). A great deal of research has been carried out on the engineering behavior of stabilized erodible and dispersive soils with traditional admixtures such as lime, cement, gypsum, slag, alum, and fly ash (Perry 1977; Rosewel 1977; Machan et al. 1997; Indraratna 1996; Indraratna et al. 2008a). However, traditional chemical stabilizers are not always readily acceptable due to stringent occupational health and safety issues. They also pose a threat to the environment by changing the soil pH, thus limiting the scope of vegetation and also affecting the quality of the ground water. Nanometric colloidal silica (also known as nano-silica), has shown promise at stabilizing some problematic soils. Nanosilica was firstly used in the 1990s in civil engineering, as

© Springer Nature Singapore Pte Ltd. 2018
A. Farid and H. Chen (Eds.): GSIC 2018, *Proceedings of GeoShanghai 2018*
International Conference: Geoenvironment and Geohazard, pp. 507–514, 2018.
https://doi.org/10.1007/978-981-13-0128-5_56

mineral addition in concrete; soil porosity reducer by the gel formation in the oil deposits to facilitate pumping through the holes; grout in fine soils to form waterproof barriers against the migration of pollutants (Noll 1992; Persoff et al. 1999) and grout for mitigating the risk of liquefaction of saturated soil (Gallagher 2000, 2007). Characterized by low (near water) viscosity and a nano-scale particle distribution, colloidal nano-silica can be considered as a grout injection for fine sand. These colloids are composed of SiO_2 and dispersed in aqueous solution by electrostatic repulsion. This suspension turns into hard frost in the presence of electrolyte (cation, polymer, etc.) for taking the product in the porosity of the soil and improving its properties. This chemical broth freeze time is perfectly adjustable depending on conditions of injection from a few minutes to a few hours (Guefrech 2010). Unlike other chemical grout such as sodium silicate, colloidal silica in gel form is physically stable and exhibits no shrinkage or withdrawal in a saturated environment. This silica is chemically composed of pure silicon dioxide SiO_2 which gives the product an "environmentally friendly character".

The objective of this paper is to develop an experimental setup for the study of sustainability and durability of soil treated with nano-silica.

In addition, it is known that the reinforced structures are typically exposed to natural environments. They are permanently exposed to wind, rain, temperature changes, etc. They are also immersed in fresh water or sea water for a very long period. In short, they are affected by a variety of mechanical and chemical stresses. Therefore, the sustainability of soil chemical grout injected into its environment attracts a major concern of engineers.

In this paper, we focus on the durability of soil treated with nano-silica, depending on the time, the environment, etc.

2   Materials

Fontainebleau sand (FBS) (a fine material silica ($SiO_2 > 98\%$)) was adopted. The grain size is between 50 μm and 400 μm, it's the particle and bulk densities are respectively 2600 kg/m^3 and 1450 kg/m^3, the cohesion is 70 Pa and the internal friction angle is 40°–45°.

The solution of colloidal silica is a kind of liquid composed of dispersed colloidal amorphous silica (SiO_2) diluted and mixed with a reagent. This reagent is generally a salt solution (NaCl solution used in this test) that causes irreversible solidification of solution. In our study, the nano-silica used is Levasil 300/30%. Density is 1300 kg/m^3, particle size is 11 nm–40 nm, and pH is 10. (Iler 1975) proposed the hypothesis about the formation of bonds between silica particles through cations. In an aqueous solution, water molecules are adsorbed against sodium by their oxygen atoms (see Fig. 1).

A constant volume ratio of 1/4 between saline solution and nano-silica was kept. Two saline solutions with different concentrations were prepared to control the gelling time: 75 g/L and 95 g/L which were diluted by water. The choice of the two saline solutions with different concentrations was depending on the mechanic properties and the setting time. The setting time was 60 min (C_{NaCl} = 75 g/L) and 30 min (C_{NaCl} = 95 g/L) at 20° C, respectively. The LN (75 g/L) was noted for formula1: nano-silica + NaCl (75 g/L), while LN (95 g/L) for formula2: nano-silica + NaCl (95 g/L).

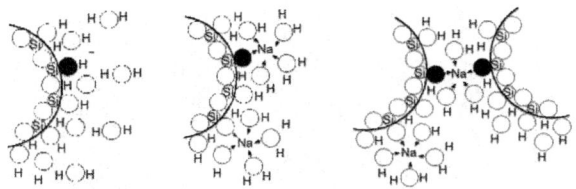

Fig. 1. Coagulation of silica particles in the presence of cations (large circles are oxygen atoms).

3   Laboratory Procedure

3.1   Environments Conditions

Four kinds of conservation environments were controlled (favorable or aggressive to the binders): for samples conserved in air, the relative humidity is 50%, and actually the dry atmosphere might be unfavorable to the stability of silica and silicate gel because of evaporation of water; for samples conserved in water, the samples were immersed in water after gelling, and this environment could be favorable due to humidity retention or unfavorable due to gel dissolves; for samples conserved in saline solution, NaCl solution concentration was 1 g/L, and this environment was studied in case that ground soil reinforced by chemical binders perhaps immersed in seawater, chloride ion could be aggressive to binders; for samples conserved in acid solution, pH was adjusted to 4 using sulfuric acid, and this case was for studying acid environments in contrast with the proprieties of alkaline gel. During these tests, the temperature of all the environments remains 20 °C, while the distilled water was adopted for water used in all solutions.

3.2   Unconfined Compressive Strength Test

Simple compression tests were performed on a T5K LLOYD, a kind of electric press with a capacity ranged from 500 N to 5 kN. An overview of the experimental device is shown in Fig. 2.

Compressive stress was measured through a force sensor placed on the upper base of the test piece and fixed to the press frame. It was considered that the deformation of the cylindrical samples corresponds to the displacement of the upper press plate.

The displacement and the force were recorded every 5 s. The test was controlled in movement at the velocity of 0.3 mm/min. The test was stopped when the force measured by the sensor decreased by at least 10% compared with the largest measured force. Throughout this test, the measured displacement was one of the high plateaus, so we would get an overall measurement of displacement.

Columns (5.2 cm in diameter and 10 cm in height) were filled with a mixture of FBS consolidated by Levasil 300/30%. The binders occupied 25% of the pore volume of the matrix of granular sand. In order to get reasonable time of IE and approach to the reality of consolidated soils, the columns were prepared by successive layers of 2 cm after compaction that was achieved by dropping a circular (256.5 g) so as to obtain a homogeneous density of 1.6 g.cm^{-3} at each level. Then the columns were stored in

Fig. 2. Coagulation of silica particles in the presence of cations (large circles are oxygen atoms).

four different environments: dry atmosphere (50% humidity), pure water, saline solution and acid solution. Preservation time was 1 day, 7 days, 15 days, 30 days, 90 days, 180 days and 365days.

3.3   Internal Erosion Test

The sample preparation was the same as compression test. In this study, erosion under constant head water was used to finish the erosion test (Figure 3).

In the system of constant head, the water level was set from which an injection pressure was constant. The maximum water flow rate entering the cell mortar was regulated by the valve at the top. A tank was used to collect and lead water discharged to the water discharge system in the lab. The effluent samples (40 ml) were collected regularly to monitor the permeability distribution of nano-scale particles, the mass eroded, conductivity and pH after erosion over time.

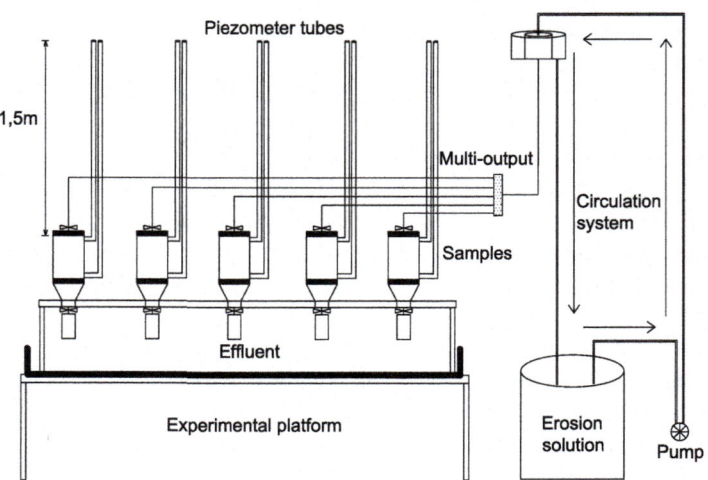

Fig. 3.   Schematic diagram of erosion test under constant head

Before test, it was important to decide the experimental conditions that serve the objective. Preliminary tests were carried on to define the imposed pressure, the maximum flowing entering the device and the preservation time. An average pressure of no more than 0.15 bar was obtained to avoid liquefaction. The suitable entering flow of 30 ml/min was proved to make the mortar erodible and non-self-locking. The preservation time was set as 1 day to 365 days. The duration of erosion was tested for one week from the beginning to the collapse of mortar.

4 Results and Discussion

4.1 Unconfined Compressive Strength

Figure 4 shows unconfined compressive strength of two formulas in different environments.

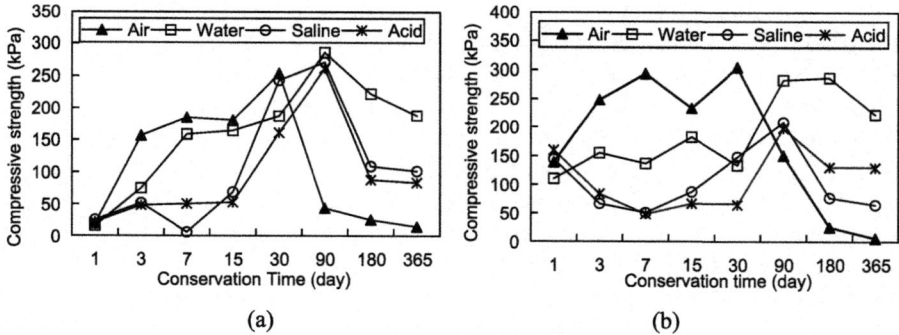

Fig. 4. Compressive strength of LN (75 g/L) and LN (95 g/L)

The maximum compressive strength of LN was 300 kPa for one year. This nano-silica reinforced has no strong resistance, thus it is suitable for filling the pores of sand but not suitable for enhancing mechanical property.

With the increase of NaCl concentration, the resistance does not change a lot. After 1 day of storage, the mortars of LN (95 g/L) have a greater resistance than those of LN (75 g/L). After 1 year of conservation, two formulations have almost the same resistance.

For both formulations stored in air, the resistance increases enormously during the first 3 days, then shows a slight increase from 3 days to 30 days. After 30 days, two formulations show a drop in mechanical strength.

For both formulations stored in water, the resistance increases gradually during the first 90 days, and it decreases gradually after that.

In salty and acid environments, the resistance of LN (75 g/L) increases gradually during the first 90 days, then a drop appears in the mechanical performance. The resistance of LN (95 g/L) decreases in both environments during the first 7 days, then increases from 7 days to 90 days, and a drop in mechanical strength is observed after 90 days.

By summarizing the above, the conservation condition seems the most important parameter for the mechanical strength of mortars made in our experimental study, the air is the most aggressive environmental conservation, the salt and acid environments have less effect on this type of nano-silica, and the environment of pure water has the least effect on this type of nano-silica.

4.2   Internal Erosion Test Under Constant Head Water

Erosion time was measured from the beginning of water injection till the collapse of mortar, which is varied from a few minutes for more fragile samples to a few weeks for more consolidated samples. The result of erosion time of LN (75 g/L) and LN (95 g/L) is shown in Fig. 5.

From Fig. 5, it can be seen that the longest erosion time of LN (75 g/L) was 400 h for 1 year's conservation, while he longest erosion time of LN (95 g/L) was 510 h for 1 year's conservation. With the increase of NaCl concentration, the samples have a better capacity to resist erosion.

For both formulations stored in water, the erosion time increases gradually during the first 90 days, and then decreases gradually after that.

For samples conserved in pure water, the erosion time of LN (75 g/L) increases gradually with the increase of conservation time from 1 to 90 days. After 90 days, the erosion time of LN (75 g/L) tends to be stable. From 1 to 30 days, the erosion time of LN (95 g/L) increases gradually with the increase of conservation time. After 30 days, the erosion time of LN (95 g/L) tends to be stable, which indicates that both formulations have a good ability to resist erosion after 1 year of conservation,; and the erosion time tends to be stable earlier for LN (95 g/L).

For samples conserved in saline solution, the erosion time of LN (75 g/L) increases gradually with the increase of conservation time from 1 to 365 days. From 1 to 30 days, the erosion time of LN (95 g/L) increases gradually with the increase of conservation time. After 30 days, the erosion time of LN (95 g/L) tends to be stable, which indicates that both formulations have a good ability to resist erosion after 1 year of conservation; The erosion time tends to be stable earlier for LN (95 g/L), we have got the same conclusion in cases of dry environment and pure - water environment.

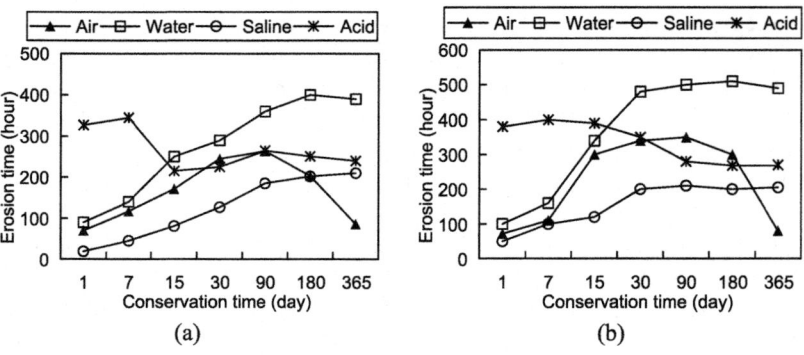

Fig. 5. Erosion time of LN (75 g/L) and LN (95 g/L)

For samples conserved in acid solution, the erosion time decreases with the increase of conservation time. Generally speaking, the capacity of insist erosion of LN in acid environment is good.

Through summarizing the above, the ranking of favorable degree of the four environments is: Pure water > Saline > Acid > Dry air.

5 Conclusions

By analyzing the results of compressive strength and capacity of resist erosion respectively, we have gotten the influence of environments on LN during long time and some same conclusions from two tests: conservation condition seems the most important parameter for the mechanical strength of mortars made in our experimental study. Air is the most aggressive environment, salt and acid environments have less effect on this type of nano-silica, and pure water is the most favorable for this type of nano-silica.

It is worth noting that the influences of long time on compressive strength and capacity of erosion are not completely consistent. In dry air, the compressive strength and capacity of erosion are basically consistent over time, and the values increase first then decrease. In pure water and saline solution, the compressive strength increase first then decrease, while the capacity of erosion is better over time. In acid environment, the compressive strength increases first then decreases, while the capacity of erosion becomes worse over time.

References

Fry, J.J., Degoutte, G., Goubet, A.: L'érosion interne: typologie, détection et réparation. Barrages Réservoirs 6, 126 (1997)

Foster, M., Fell, R., Spannagle, M.: Method for assessing the relative likelihood of failure of embankment dams by piping. Can. Geotech. 37, 1025–1061 (2000b)

Gallagher, P.M., Pamuk, A., Abdoun, T.: Stabilization of liquefiable soils using colloidal silica grout. J. Mater. Civ. Eng. 19, 33–40 (2007)

Gallagher, P.M.: Pasive remediation for mitigation of liquefaction risk. Thèse de Doctorat, Virginia polytechnic Institute (2000)

Guefrech, A.: Injection et pérennité des coulis à base de silices colloïdales nanométrique. Thèse de doctorat à l'Ecole Centrale de nantes, Nantes, France, pp. 23–64 (2010)

Iler, R.K.: Coagulation silica by calcium ions mechanism and effect of particles size. J. Colloid Interface Sci. 53(3), 476–488 (1975)

Indraratna, B.: Utilization of lime, slag and fly ash for improvement of a colluvial soil in New South Wales, Australia. Geotech. Geol. Eng. 14, 169–191 (1996)

Indraratna, B., Muttuvel, T., Khabbaz, H.: Investigating erosional behaviour of chemically stabilised erodible soils. In: Geocongress 2008, New Orleans, Louisiana, vol. 178, pp. 670–677. Geotechnical Special Publication (2008a)

Karol, R.H.: Chemical Grouting and Soil Stabilization: Engineering & Technology, 3rd edn, pp. 122–148. CRC Press, New York (2003)

Machan, G., Diamond, S., Leo, E.: Laboratory study of the effectiveness of cement and lime stabilization for erosion control. Transportation Research Record, pp. 24–28 (1997)

Noll, M.R.: In situ permeability reduction and chemical fixation using colloidal silica. In: Proceedings of 6th National Outdoor Action Conference, Les Vegas, pp. 443–457 (1992)

Perry, J.P.: Lime treatment of dams constructed with dispersive clay soils. Trans. Am. Soc. Agric. Eng. **20**, 1093–1099 (1977)

Persoff, P., Apps, J., Moridis, G., Whang, J.M.: Effect of dilution and contaminants on sand grouted with colloidal silica. J. Geotech. Geoenvironmental Eng. **125**, 61–469 (1999)

Rosewel, C.J.: Identification of susceptible soils and control of tunnelling failure in small earth dams. In: Sherard, J.L., Decker, R.S. (eds.) Dispersive Clays, Related Piping and Erosion in Geotechnical Projects, ASTMSTP 623, pp. 362–369. American society for testing and materials (1977)

Application Research of Bio-grouting in Hydraulic Fill Fine Sands for Reclamation Projects

Jijian Lian, Hongyin Xu, Xiaoqing He, Yue Yan[(⊠)],
and Dengfeng Fu

School of Civil Engineering, Tianjin University, Tianjin 30072, China
yueyan_geo@126.com

Abstract. A kind of fine black sands is always hydraulic filled onto the back-filled silt to support the equipment in reclamation projects in Tianjin, China. Bio-grouting, as a new approach of soil improvement, was adopted to cement black sands on the surface of silt to rapidly increase the carrying capacity. The results showed that the black sands were well cemented by calcium carbonate crystals produced in the process of bio-mineralization. A static load test was carried out on the cemented sand layer, which resulted of an increase of the loading-bearing capacity by about sixty times. The calcium carbonate contents of different positions in the cemented sand layer were detected, which showed an uneven distribution. The micro-structures of the produced calcium crystals were observed by using an electron microscopy. Different forms of crystals were observed, illustrating that the concentrations of bacteria in the sand layer were not uniform. From discussion, the bio-grouting method can cement the black sand layer and the solidification effect can be improved by some reasonable measures, which indicates the feasibility of this method in reclamation projects. Moreover, the bio-grouting method can bring a lot of time saving and material saving compared with traditional methods.

Keywords: Bio-grouting · Black fine sand · Static load test
Calcium carbonate contents

1 Introduction

Reclamation is an effective way to alleviate the shortage of land resources in coastal cities [1]. In Tianjin, China, more than 400 km^2 of land is going to be constructed through reclamation projects on the coast of Binhai New Area [2]. In these projects, silt from channel dredging is the main material for the land foundation, which has high water content, low strength and high deformability [3, 4]. Vacuum preloading technique is often adopted to improve the silt foundation after about two years of natural drying, thus a hard shell forms on the surface of silt to support equipment [5]. To reduce time consumed, Yan et al. [2] proposed a new method to hydraulic fill some black fine sands on the surface of silt to meet the requirements of carrying equipment according to hard shell theory. This method can greatly save time in the projects but bring a lot of consumption of fine sands because 0.5–2 m thickness of sands are needed [6], especially in Tianjin, where sand resources were relatively short.

© Springer Nature Singapore Pte Ltd. 2018
A. Farid and H. Chen (Eds.): GSIC 2018, *Proceedings of GeoShanghai 2018*
International Conference: Geoenvironment and Geohazard, pp. 515–524, 2018.
https://doi.org/10.1007/978-981-13-0128-5_57

Bio-grouting is a new method for soil improvement [7], which can form calcium carbonate crystals in the interspaces of sands to improve the strength and reduce the permeability of sand [8, 9], and provide a new method to cement the black fine sands on the silt in reclamation projects. Some basic experiments have been carried out to study the influencing factors and mechanical behaviors of the bio-grouted black fine sands [10]. The results showed that 10 cm-high black sand column can reach a compressive strength of 1.94 MPa, which is far higher than the required carrying capacity (about 80 kPa). To explore the feasibility of application of this method, this article conducted a bio-grouting test in hydraulic fill fine sands on the surface of silt. After bio-grouting, a static load test was carried out. Then the calcium carbonate contents of different positions of the cemented black sand layer were detected. In the end, the micro-structure of the solidified sand was observed by using an electron microscopy.

2   Experimental Details

2.1   Materials

The black fine sands used in reclamation projects in Tianjin, as well as being bio-grouted in this experiment, was taken from Luan River, Tangshan, China. The basic properties of the sands are shown in Table 1.

Table 1. The basic properties of the black fine sands

Loose bulk density: g/cm^3	1.415
Tight packing density: g/cm^3	1.701
Uniformity coefficient: C_u	2.09
Permeability coefficient: m/s	6.9×10^{-5}
D_{90}: mm	0.382
D_{50}: mm	0.218
D_{10}: mm	0.115

The urease producing bacteria (number of ATCC11859) used in this experiment were cultivated under sterile conditions in a YE medium [11], 500 mL of fresh medium was inoculated at a volume ratio of 1% and cultivated in a shaker for 24 h until to an optical density of approximately 2.5, which was measured at a wavelength of 600 nm (OD_{600}). Before experimental use, the urease activity of the bacterial suspension was detected by using the same methods utilized by Harkes et al. [12].

The urea and the calcium ions used in the process of bio-mineralization are provided by cementing solution, which was a mixed solution of urea and calcium chloride with an equal molar concentration of 0.75 M.

2.2   Bio-grouting Test

The bio-grouting test of black fine sands were conducted in a box with a size of 400 mm × 300 mm × 400 mm, as shown in Fig. 1a. The inner wall of the plexiglass box was coated with a layer of vaseline to reduce friction. About 30 cm-thickness of silt was poured into the box slowly and placed aside for five days until reaching a thickness of 24 cm and a water content of 71.3%. The water on the silt was poured out and 18.5 kg of black fine sands were sprinkled on the surface of the silt. After one day, the height of the silt was 23 cm and the thickness of the black sand layer was 9.5 cm, as shown in Fig. 1a.

The bacterial suspension and cementing solution were injected into sands by ten grouting piles (diameter of 6 mm and length of 80 mm). Six of them (numbering A) formed a hexagon with 80 mm between two of them and were arranged in the center of the sand layer. The other four (numbering B) distributed outside the hexagon with a distance of 80 mm. Ten outlet pipes (diameter of 10 mm and length of 80 mm) were arranged near the walls of the box, as shown in Fig. 1b. The pipes were all blocked at the bottom and half buried under the sands. For the part buried in the sand, some holes were drilled on the sidewalls of the pipes to make the liquid to pass through and a layer of gauze was wrapped outside the pipes to prevent sand particles being washed into. The grouting pipes and outlet pipes were connected with peristaltic pumps respectively, as shown in Fig. 1c.

Two times of bacterial suspension were injected into the black sand layer and after each injection were five injections of cementing solution. The detailed procedure of bio-grouting was displayed in Table 2. It needed to note that the flow rate was 5 mL/min in the process of bio-grouting.

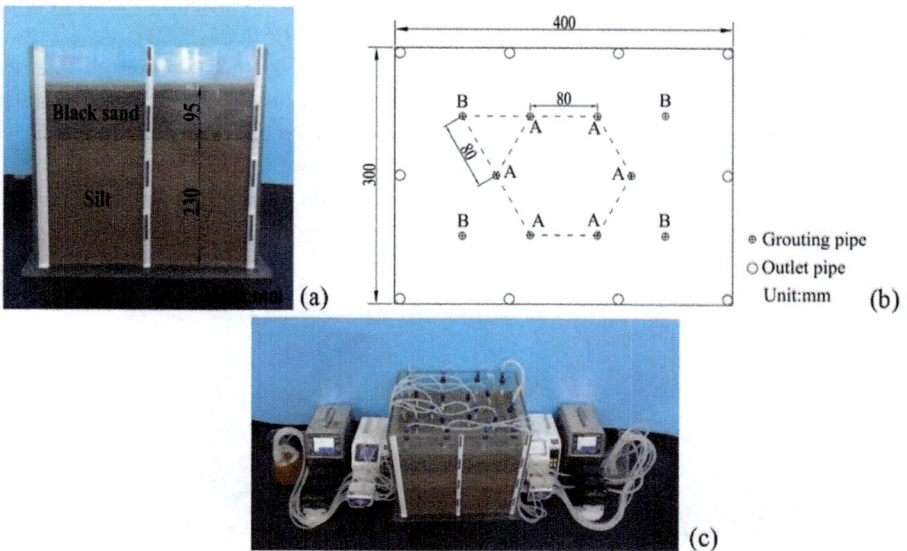

Fig. 1. Experimental set-up: (a) Vertical view; (b) The arrangement of grouting pipes; (c) The overall layout of device

Table 2. The procedure of bio-grouting

Grouting solution	Injection positions	Volume	Concentration	Retention time	Comments
First bacterial suspension	A grouting pipes	2.5 L	$OD_{600} = 2.5$	3 h	Urease activity = 17.76 mM urea/(L·min)
Five times of cementing solution	First three times: A grouting pipes The latter two times: B grouting pipes	Each time was 2.5 L	0.75 M	3 h after each batch	
Second bacterial suspension	B grouting pipes	2.5 L	$OD_{600} = 2.7$	3 h	Urease activity = 18.53 mM urea/(L·min)
Five times of cementing solution	First two times: B grouting pipes The latter three times: A grouting pipes	Each time was 2.5 L	0.75 M	3 h after each batch	

2.3 Static Load Test

After the black sand layer was cemented, a static load test was carried out according to Code for Investigation of Geotechnical Engineering (GB50021-2001). A 10 mm-thickness of plexiglass plate with a size of 100 mm × 100 mm worked as loading board, between which and the cemented sands was a thin layer of sand to make a flat contact surface. Some standard cubic concrete blocks were used to increase the pressure load and two dial indicators were arranged to measure the settlement of sands in the static load test process (as shown in Fig. 2c). The static load was also conducted on the surface of the uncemented black sands as a comparison.

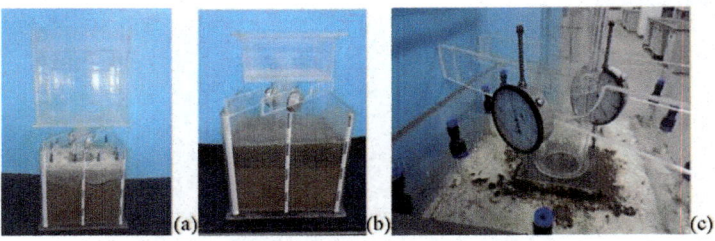

Fig. 2. (a) Static load test on bio-cemented sand; (b) Static load test on uncemented sand; (c) The arrangement of dial indicators

2.4   Calcium Carbonate Content Test

The solidified sand layer was crushed into several parts after static load test. Sand samples were taken at different positions in different parts to detect the calcium carbonate contents (as shown in Fig. 3), adopting a method of hydrochloric acid dissolution which was utilized by Whiffin et al. [7].

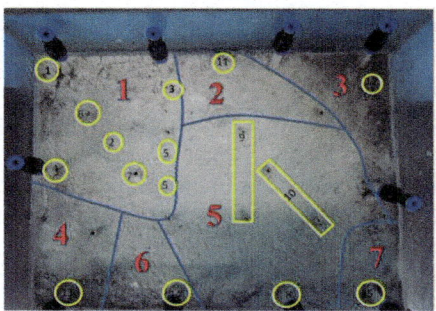

Fig. 3. Sampling positions of calcium carbonate content detection

3   Experimental Results

The black fine sand layer was solidified after bio-grouting experiment with a white calcium carbonate layer on the surface (as shown in Fig. 4a) and the sands in bottom was also cemented (see Fig. 4c), which proved that bio-grouting method was feasible to improve the black sands. A plugging pipe test was carried out to simulate the actual situation of plugging drive pipes in reclamation Projects. No mud was brought out (see Fig. 4d), indicating that it is effective to avoid large amounts of silt being driven out by the drive pipe of the draining boards. It must be noted that the black sand layer was cracked in the process of bio-grouting (as shown in Fig. 4b). The reason may be the uneven settlement of the silt.

Fig. 4. The cemented black sand layer: (a) The top surface; (b) The cemented sand layer was cracked; (c) The cemented sand blocks after static load test; (d) Plugging pipe test

3.1   The Results of Static Load Test

Figure 5 drew the *p-s* curves of the cemented black sand layer and uncemented sand layer from the static load tests. From the figure, the curve of the cured sand layer had obvious proportional limit load p_0 and ultimate load p_u, which divided the curve into three parts. The first part was a straight line from 0 kPa to 18.3 kPa (p_0), which was called elastic deformation stage, in which all the loads were hold by the solidified sand layer. The second part was the partial shear stage from 18.3 kPa to about 30 kPa (p_u), in which plastic deformation occurred in the cemented sand shell and some load began to pass to the silt. The third part was the overall destruction stage with a stress greater than 30 kPa, displaying a steep dropped line, in which the sand layer was broken and the stress on the silt increased greatly. At this time, the foundation was completely destroyed. Thus, the load-bearing capacity of the cemented sand layer reached 18.3 kPa, same with the value of p_0.

The uncured sand layer had no demarcation point in the p-s cure (see Fig. 5). With the increase of load, the settlement of the foundation grew faster and faster. According to Code for Investigation of Geotechnical Engineering (GB50021-2001), The ultimate load of the foundation was 1.2 kPa, which was the load before the settlement reached 6 mm (0.06 times of the loading board's length). Moreover, the load-bearing capacity was estimated to be 0.3 kPa when the settlement reached 1.5 mm (0.015 times of the loading board's length).

According to the results of the static load tests, the loading-bearing capacity of the black sand layer increased by about sixty times from 0.3 kPa to 18.3 kPa and the corresponding settlement decreased from 1.5 mm to 0.36 mm, indicating that both strength and stiffness were improved after bio-grouting. Considering the cemented sand layer was cracked in the process of bio-grouting, it was possible to further improve the mechanical properties of the sand layer.

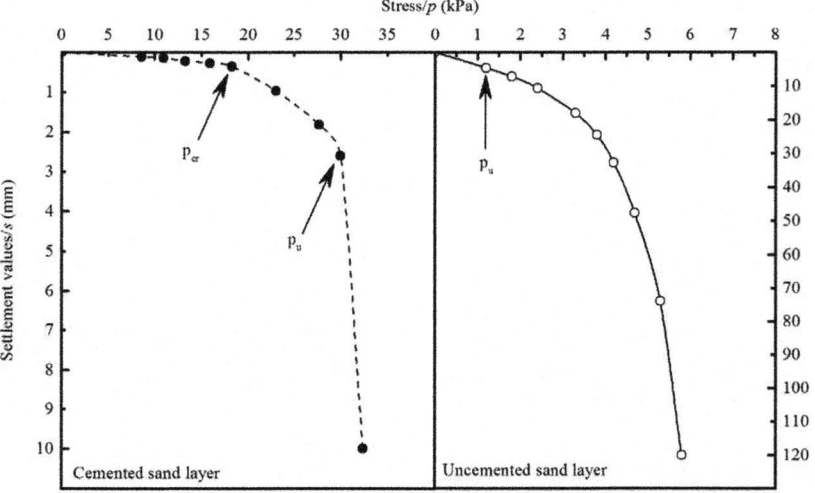

Fig. 5. The p-s curves of the cemented sand layer and uncemented sand layer

3.2   Calcium Carbonate Contents

Figure 6 displays the calcium carbonate contents of the upper cemented sand layer, with a range from 3.9% to 11.5%. In this figure, the whole plane was divided into three sections. Section A was the compression position in the load test; Section B and C were the parts close to grouting pipes and outlets respectively.

The distribution of calcium carbonate in different locations were presented in Fig. 7. In Fig. 7a, the average calcium carbonate contents in Section A, B and C were presented. The crystals had a lowest average content in Section A, indicating that this section had a poor cementing effect. Besides, the crystals distributions in the three sections were all uneven, resulting from the big error bars in the figure.

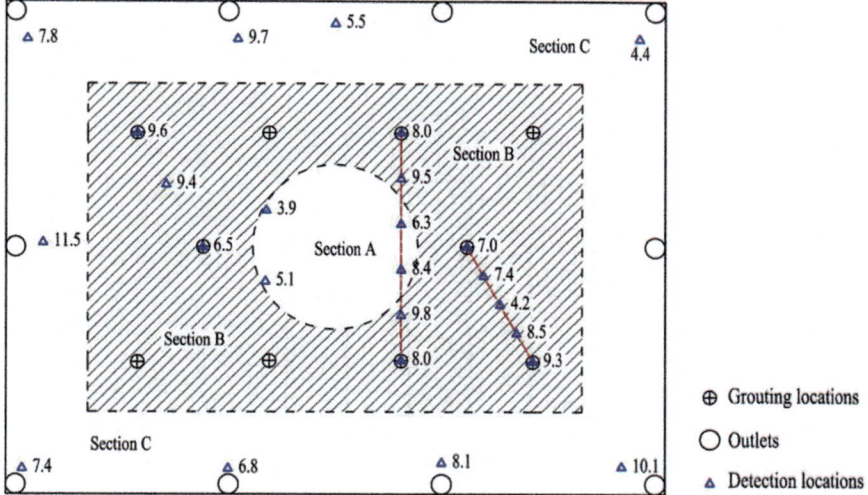

Fig. 6. The horizontal distribution of calcium carbonate contents in the upper sand layer

The crystals contents in the entrances and exits were picked out and compered with the other locations. Clearly, the cured sand near the entrances and the exits had higher crystals contents of about 8.0% than the common positions of about 6.0% (see Fig. 7b). The produced calcium carbonate mainly accumulated near the entrances and exits.

The calcium carbonate contents in different depths of the 6# and 7# measuring points were detected and the results were shown in Fig. 7c. It can be seen that the content decreased first and then increased with the depth increasing and reached the lowest value at 2–4 cm. When bio-grouting, the calcium carbonate produced in the bottom of sand firstly and gradually accumulate upwards, which has been confirmed in our previous studies [13]. After several times of cementing, the grouting solution moved upward to the surface of the sand layer along the grouting pipes, for that the gauze wrapped outside the grouting pipes was jammed and the solution could no longer pass through. Thus, the calcium carbonate produced on the top layer of the sand.

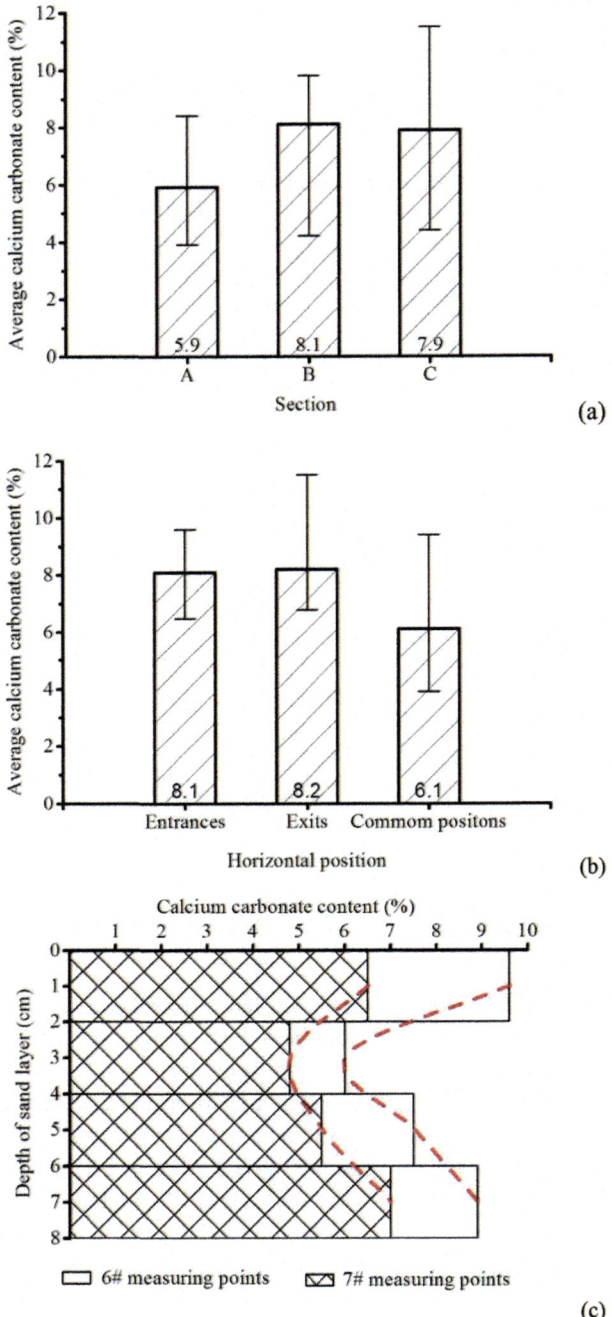

Fig. 7. The calcium carbonate contents in the cemented sand layer: (a) Different sections; (b) Different positions. (c) Different depths

3.3   Micro-structure of the Crystals

The micro-structure of the solidified sand particles was observed by using an electron microscopy. The calcium carbonate crystals attached to the surface of the sand particles or bridged between particles. There were different kinds of crystals being found: spherical crystals, agminated crystals and cuboid crystals (as shown in Fig. 8). That was also a proof of the uneven distribution of bacteria in the bio-grouting process [14], which resulted in the uneven distribution of calcium carbonate in the sand layer.

Fig. 8. The SEM images of the crystals: (a) Spherical crystals; (b) Agminated crystals; (c) Cuboid crystals

4   Discussion

According to the test results, bio-grouting soil improvement method increased the loading-bearing capacity of the black sands layer by about sixty times to be 18.3 kPa, indicating that this method can cement sands effectively. However, it still not reached the requirement of the working layer of equipment (about 80 kPa) in reclamation projects. That was mainly because of the low calcium carbonate contents in Section A in Fig. 6, which was the pressure area in the static load test. Moreover, the uneven crystals distributions also lead to poor mechanical behavior of the cemented sand layer.

Therefore, some improvement measures can be adopted to raise the quality of bio-grouting. From Fig. 7b, the calcium carbonate mainly concentrated in the vicinity of entrances and exits. Thus, reasonable arrangement of grouting pipes and outlets can contribute to cementing quality. In addition, increasing the cementing times and improving the sand thickness would also be helpful.

If the quality of bio-grouting could be improved, it would be a potential method in reclamation projects, for that a huge time saving could be achieved. In this test, a total of two days was consumed before the black sand layer was cemented, while the traditional vacuum preloading method need 2–3 years. Moreover, the consumed sands in the projects would be reduced greatly.

5   Conclusions

A 10 cm thickness of black sand layer was cemented on the surface of silt in the laboratory by bio-grouting method, proving that this technology can improve the mechanical behavior of the hydraulic fill black sands in a short time. The loading-bearing capacity of the sand layer was increased by about sixty times to be 18.3 kPa. The calcium carbonate in the cemented sand distributed unevenly with the contents ranging from 3.9% to 11.5%. The crystals mainly concentrated in the vicinity of the entrances and exits in the horizontal direction, and in the top layer and the bottom layer in the vertical direction. Different forms of calcium carbonate crystals were found in the sand layer, confirming uneven distribution of bacteria.

References

1. Chu, J., Yan, S., and Indraranata, B.: Vacuum preloading techniques—recent developments and applications. In: Geocongress, pp. 586–595 (2008)
2. Yan, S., Guo, B., Sun, L., Wei, L.: Application of crust layer to vacuum preloading dredge fill. Chin. J. Rock Mech. Eng. **32**(7), 1497–1503 (2013)
3. Yan, S., Guo, B., Sun, L., Lei, Z.: Theoretical analysis and model test of rock berm protection. J. Coast. Res. **73**, 747–752 (2005)
4. Li, K., Liu, X., Zhao, X., Guo, W.: Effects of reclamation projects on marine ecological environment in Tianjin Harbor Industrial Zone. Procedia Environ. Sci. **2**(1), 792–799 (2010)
5. Shang, J., Tang, M., Miao, Z.: Vacuum preloading consolidation of reclaimed land: a case study. Can. Geotech. J. **35**(5), 740–749 (1998)
6. Qiu, C., Fei, N., Yan, S., Ji, Y.: Laboratory tests on consolidation of dredger fill under vacuum load. J. Tianjin Univ. **6**, 004 (2014)
7. Whiffin, V.S., Van Paassen, L.A., Harkes, M.P.: Microbial carbonate precipitation as a soil improvement technique. Geomicrobiol J. **24**(5), 417–423 (2007)
8. DeJong, J.T., Mortensen, B.M., Martinez, B.C., Nelson, D.C.: Bio-mediated soil improvement. Ecol. Eng. **36**(2), 197–210 (2010)
9. Achal, V., Mukherjee, A.: A review of microbial precipitation for sustainable construction. Constr. Build. Mater. **93**, 1224–1235 (2015)
10. Lian, J., Xu, H., He, X., Yan, Y., Fu, D., Yan, S., Qi, H.: Bio-grouting of Hydraulic fill fine sands for reclamation projects (Unpublished)
11. Zhang, Y., Guo, H., Cheng, X.: Role of calcium sources in the strength and microstructure of microbial mortar. Constr. Build. Mater. **77**, 160–167 (2015)
12. Harkes, M.P., Van Paassen, L.A., Booster, J.L., Whiffin, V.S., Van Loosdrecht, M.C.M.: Fixation and distribution of bacterial activity in sand to induce carbonate precipitation for ground reinforcement. Ecol. Eng. **36**(2), 112–117 (2010)
13. Lian, J., Xu, H., He, X., Yan, Y., Fu, D., Yan, S., and Qi, H.: Quantification of crystals produced in sand in bio-grouting process under multiple factors (Unpublished)
14. Cheng, L., Qian, C., Wang, R., Wang, J.: Study on the mechanism of calcium carbonate formation induced by carbonate-mineralization microbe. Acta Chim. Sin.-Chin. Edi. **65**(19), 2133 (2007)

Geosynthetics

Investigation of the Prediction Capability of the EM Design Method Under Working Stress Conditions

S. H. Mirmoradi[(⊠)] and M. Ehrlich

Federal University of Rio de Janeiro, Rio de Janeiro, RJ 21945-970, Brazil
shm@ufrj.br, me@coc.ufrj.br

Abstract. A new method for the internal design of reinforced soil walls has recently been developed by Ehrlich and Mirmoradi (2016), EM (2016), based on working stress conditions. This method was based on the Ehrlich and Mitchell (1994) design procedure. There are three key differences between the proposed new method and Ehrlich and Mitchell's (1994) procedure: (1) the effect of the facing inclination is considered in the new method, while the original method was developed for vertical walls; (2) the calculation of the maximum reinforcement load, T_{max}, using the proposed method does not need iteration, which was required by the original method; and (3) the equations are simpler to use. Data from physical and numerical model studies were used to verify the prediction capability of the EM (2016) method for walls with different heights, facing inclinations, reinforcement spacing and stiffness, compaction effort and toe conditions. The results, in general, indicated that the EM (2016) method is capable of properly capturing T_{max} considering different conditions.

Keywords: EM method · Reinforced soil walls · Working stress conditions

1 Introduction

Reinforced soil (RS) walls are widely utilized throughout the world. They have several advantages over other wall types, such as low cost, simple construction, and the ability to accommodate deformation. Prediction of the maximum reinforcement load, T_{max}, is a major objective in the design of reinforced soil structures. Most current methods used in RS wall design are limit equilibrium methods (e.g., FHWA 2009; AASHTO 2014).

Ehrlich and Mirmoradi (2016) proposed an analytical procedure for the calculation of T_{max} under working stress conditions. This method explicitly takes into account the effect of compaction-induced stress, reinforcement and soil stiffness properties and facing inclination. The proposed method was based on Ehrlich and Mitchell's (1994) procedure. There are three key differences between the proposed method and Ehrlich and Mitchell's (1994) procedure: (1) the effect of the facing inclination is considered in the new method, while the original method was developed for vertical walls; (2) the calculation of T_{max} using the EM (2016) method does not need iteration, which was required by the original method; and (3) the equations are simpler to use.

© Springer Nature Singapore Pte Ltd. 2018
A. Farid and H. Chen (Eds.): GSIC 2018, *Proceedings of GeoShanghai 2018*
International Conference: Geoenvironment and Geohazard, pp. 527–536, 2018.
https://doi.org/10.1007/978-981-13-0128-5_58

The method proposed by Ehrlich and Mitchell (1994) has already been validated against data from several full-scale walls (e.g., Ehrlich and Mitchell 1994; Ehrlich et al. 2012; Stuedlein et al. 2012; Ehrlich and Mirmoradi 2013; Riccio et al. 2014; Mirmoradi and Ehrlich 2014b; 2015a, 2015b; Mirmoradi et al. 2016; Mirmoradi and Ehrlich 2016). In this paper, data from two inclined wrapped-face physical model studies were used to assess the prediction capability of the new simplified method. The current paper evaluates the prediction capability of this simplified method by comparing calculated values with data from numerical and physical model studies.

2   Experimental Study

Instrument data and measurements from three large-scale physical model walls constructed at the Geotechnical Laboratory of COPPE/UFRJ and one full-scale reinforced soil wall built at the Royal Military College of Canada (RMC) are used to evaluate the prediction capability of the EM 2016 method under working stress conditions.

2.1   UFRJ Walls

The characteristics of the physical model walls identified as Walls 1, 2 and 3 constructed at the Geotechnical Laboratory of COPPE/UFRJ are as follows.

The physical models were built within a concrete U-shaped wall box 1.5 m high, 3 m long and 2 m wide. The height of each physical model wall was 1.2 m. The length and the vertical spacing of the geogrid were 2.2 m and 0.4 m, respectively. The backfill soil was lightly compacted (8 kPa) and the soil unit weight after compaction was 21 kN/m^3. The soil friction angles, considering the measured unit weight, were determined by triaxial and plane strain tests as 42° and 50°, respectively.

A flexible polyester geogrid was used as reinforcement. Based on the mechanical properties provided by the company, the tensile stiffness modulus of the reinforcement, J_r, is equal to 917 kN/m. Three layers of reinforcement were installed along the height of the wall, placed at 0.2 m, 0.6 m, and 1.0 m above the wall bottom. The walls were constructed with the facing inclination of 6° to the vertical. At the end of construction of all walls, a vertical surcharge loading of up to 100 kPa was applied to the top of the walls.

Wall 1 was constructed using wrapped facing. In Walls 2 and 3 precast concrete blocks were used for the face of the walls. In all walls, a 1-m wide zone at the bottom of the walls was lubricated. The difference between Walls 2 and 3 are related to the toe restriction. For Wall 2 lateral movement of the toe was restricted by a steel beam fixed to the concrete U-shape wall box. In this wall, after the end of construction and application of the surcharge, the load was kept constant on 100 kPa. With the surcharge in place, the toe of Wall 2 was released step by step (0.5 mm horizontal movement allowed in each step). Using this procedure, the toe of the wall was gradually released to the free base condition (Mirmoradi et al. 2016; Mirmoradi and Ehrlich 2017b). In Wall 3, the toe was free to move from the beginning of the test. In Fig. 1 a view of a block-face wall is shown.

Fig. 1. View of a block-face wall at the end of construction.

Figure 2 is a comparison of the measured values of the sum of the maximum reinforcement loads, ΣT_{max}, for Wall 1 with those calculated using the EM (2016) method at the end of construction and during the surcharge. The reinforcement loads were calculated by the EM method using the values of 590, 0.6 and 0.7 obtained in laboratory for k, n and R_f, respectively (Ehrlich et al. 2012; Ehrlich and Mirmoradi 2013). As shown in this figure, the new method was able to capture the measured reinforcement loads properly for different surcharge load values.

Figure 3 compares the measured and calculated values of ΣT_{max} for Walls 2 and 3. Figure 3a shows that for Wall 2, the EM (2016) method overestimated the reinforcement loads during the surcharge application, with average variation of about 2 times. However, this overestimation decreases during toe release (see Fig. 3b). These results are consistent with the expected behavior. For Wall 2, the combined effect of facing stiffness and toe restraint leads to the lower reinforcement loads. This effect is not considered in the EM (2016) method. In Wall 3, as no toe restraint was applied, the influence of the block facing on reinforcement load decrease disappears. Therefore, as shown in Fig. 3a, the reinforcement load values measured in Wall 3 were closely matched by the EM (2016) method. Figure 3b indicates that during the toe release, the accuracy of the proposed method increases. These results call attention to the influence of toe conditions on the reinforcement load. Of note, as stated by Mirmoradi and Ehrlich (2017a) this effect is appreciably dependent on several controlling factors such as wall height, reinforcement stiffness and spacing.

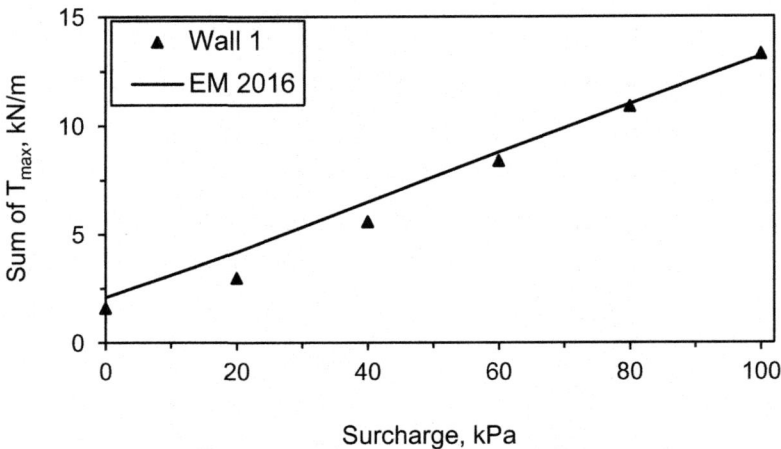

Fig. 2. Comparison of the measured and calculated values of ΣT_{max}.

2.2 RMC Wall

Data from a 3.7 m-high wrapped-face wall was presented by Reeves (2003). Six layers of welded-wire mesh (WWM) steel reinforcement with the tensile stiffness modulus, J_r, of 3100 kN/m were used for this wall. The facing inclination was 8° to the vertical. The compaction surcharge magnitude of 16 kPa, the soil unit weight of 16.7 kN/m³ and the soil friction angle of 44° were presented for this wall. For the parameters of k, n and R_f the values of 510, 0.5 and 0.86 were used, respectively. These values were used by Hatami et al. (2008); Liu (2016), who employed the same wall for the validation of the numerical model studies and analysis method, respectively.

Figure 4 compares the measured values of the reinforcement loads versus wall height with those calculated using the EM (2016) method. The results show, in general, a good agreement between the measured and predicted values. As shown in this figure, the proposed method overpredicted T_{max} in the reinforcement layers placed at the bottom of the walls due to the foundation effect on lateral restriction at the bottom of the wall. This is because of the lateral restriction that occurred at the wall bottom due to the influence of the foundation. Note that this effect is not taken into consideration in the proposed method.

3 Numerical Study

Numerical modeling was carried out using the two-dimensional finite-element program PLAXIS (Brinkgreve and Vermeer 2002). The model was first validated against the results of a large-scale wrapped reinforced soil wall at the Geotechnical Laboratory of COPPE/UFRJ. Ehrlich and Mirmoradi (2013) and Mirmoradi and Ehrlich (2014a, 2014b; 2015b) provide details on the validation of the numerical modeling.

Parametric studies were carried out with different combinations of wall height, backfill soil compaction effort, facing inclination and reinforcement stiffness and

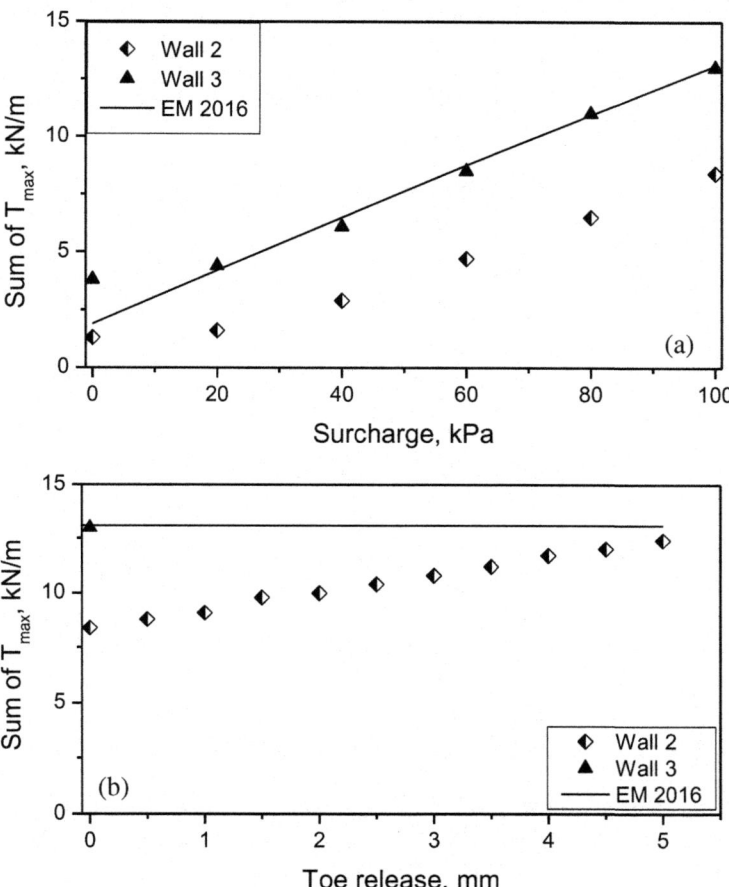

Fig. 3. Comparison of the measured and calculated values of ΣT_{max} during surcharge applications and toe release.

spacing. Two different wall heights were considered: 4 m and 8 m. The values of the reinforcement spacing for 4 m and 8 m high walls were 0.4 m and 0.8 m respectively. For the 4 m high wall, the values of the tensile stiffness modulus of reinforcement were equal to 480 kN/m, 4800 kN/m and 48000 kN/m. For the model with 8 m height, a tensile stiffness modulus of reinforcement of 960 kN/m, 9600 kN/m and 96000 kN/m was used. Therefore, the values of the soil-reinforcement stiffness index, S_i, were the same for these two walls despite the different reinforcement spacing. The parameter S_i is the relative soil-reinforcement stiffness index, which was developed by Ehrlich and Mitchell (1994) and can be calculated as follows:

$$S_i = \frac{J_r}{kP_aS_v} \tag{1}$$

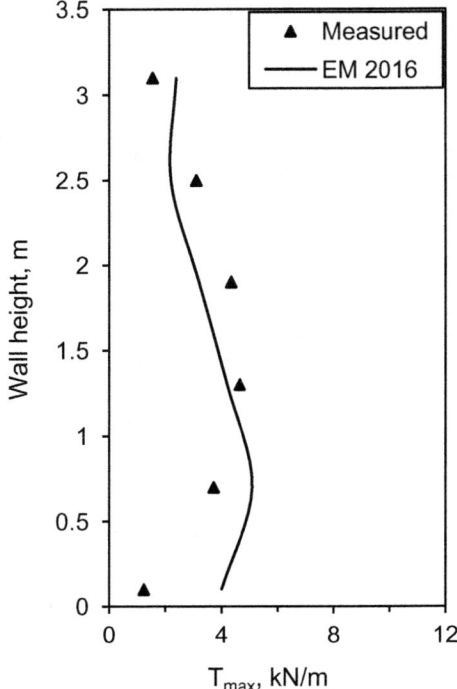

Fig. 4. Comparison of the measured and calculated values of T_{max}.

where J_r is the tensile stiffness modulus of reinforcement, k is the modulus number (hyperbolic stress-strain curve model), P_a is the atmospheric pressure, and S_v is the vertical reinforcement spacing. The facing inclination values of 0, 10° and 20° to the vertical were considered in the analyses. Additionally, the effect of the backfill soil compaction was evaluated considering three different values: 30 kPa, 60 kPa and 120 kPa. The vertical stress induced during backfill compaction, $\sigma'_{zc,i}$ was modelled using two distributed loads at the top and bottom of each backfill soil layer. The details for this procedure can be found in Mirmoradi and Ehrlich (2014a; 2015b). In Table 1, the input parameters of the numerical analysis are presented.

The walls were constructed in stages, in which 0.2 m and 0.4 m-thick soil lifts were placed for 4 m and 8 m high walls, respectively, until the final heights of the walls were reached. A fixed boundary condition in the horizontal direction was applied to the right lateral border. At the bottom of the model, a fixed boundary condition in both the horizontal and vertical directions was employed.

3.1   Results of the Numerical Analysis

Figure 5 compares the normalised reinforcement loads calculated using PLAXIS and the EM (2016) method for different compaction efforts (i.e., no compaction, 30 kPa, 60 kPa and 120 kPa) for 4 m-high walls with $S_i = 0.2$ and facing inclination of 10° to the vertical. The figure shows the acceptable accuracy of the EM (2016) method.

Table 1. Input parameters for the numerical analysis.

Property	Value
Peak plane-strain friction angle, ϕ, ($^\circ$)	40
Cohesion, c, (kPa)	0.1
Dilation angle, ψ, ($^\circ$)	11
Soil unit weight, γ, (kN/m^3)	20
E_{50}^{ref}, (kPa)	42 500
E_{oed}^{ref}, (kPa)	31 800
E_{ur}^{ref}, (kPa)	127 500
Stress dependence exponent, m	0.5
Poisson's ratio, υ	0.2

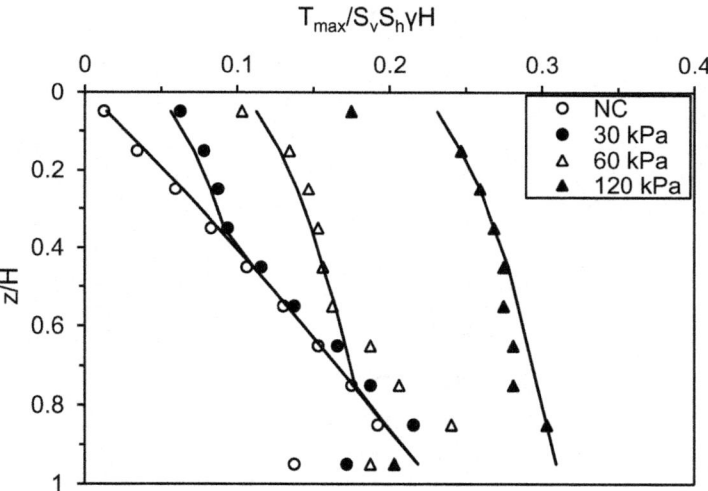

Fig. 5. Comparisons of the normalised T_{max} determined using the PLAXIS (markers) and the EM (2016) method (solid lines).

Figure 6 shows a comparison of the normalised reinforcement load with the normalised wall height calculated using PLAXIS and using the EM (2016) method for different facing inclinations, soil friction angles, and S_i values. In this analysis, the effect of the backfill soil compaction was not considered.

The calculations using the EM (2016) method were carried out for the values of k and n of 593 and 0.6, respectively. Figure 6 shows good agreement between the values calculated using PLAXIS and the proposed method for both geosynthetic ($S_i = 0.02$) and metallic ($S_i = 2$) reinforcements. The most significant discrepancies between the calculated T_{max} are related to the reinforcement layers placed at the bottom of the walls, in which the EM (2016) method overestimated the values determined by PLAXIS. As stated before, this is because of the effect of the foundation on lateral restriction at the bottom of the wall, which is not taken into consideration in the proposed method.

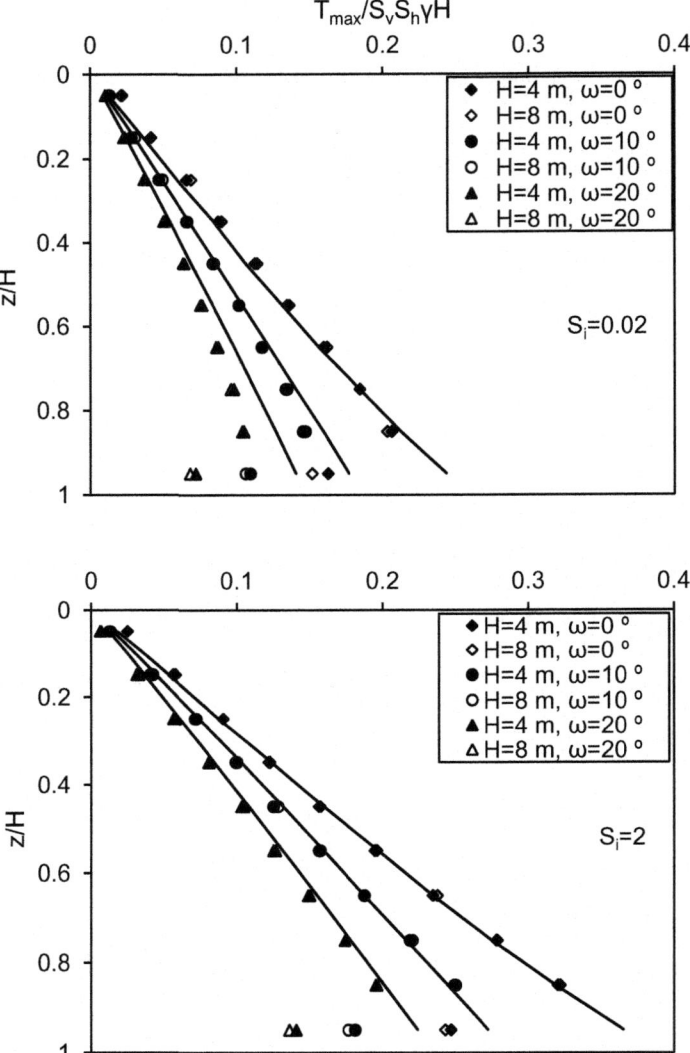

Fig. 6. Comparisons of the normalised T_{max} determined using the PLAXIS and the EM (2016) method (solid lines).

4 Conclusion

The prediction capability of the Ehrlich and Mirmoradi (2016), method was investigated under working stress conditions. The values calculated by this method were compared with the data obtained from both experimental and numerical studies.

The results showed a good predictive capability of the EM (2016) method for different values of the wall height, facing inclination, reinforcement stiffness, compaction efforts and backfill soil parameters. The most significant difference between the results is observed near to the bottom of the walls, for which the reinforcement loads determined by PLAXIS were lower due to the lateral restriction to movement imposed by the foundation. As with the Ehrlich and Mitchell (1994) method, the EM (2016) method does not take this restraint into account.

Moreover, the EM (2016) method may overestimate the T_{max} values for the short wall with structural facing elements (e.g., block facing). Due to the combined effect of facing stiffness and toe restraint, T_{max} may decrease. Nevertheless, even for walls with block facing restricted at the toe, the accuracy of this method significantly increases with the decrease of the facing effect, which could occur with an increase in the wall height and reinforcement stiffness (Mirmoradi and Ehrlich 2017a).

It is important to note that there are two principal differences between the Ehrlich and Mitchell (1994) and EM (2016) methods. The effect of the facing inclination is considered in the EM (2016) method, while the Ehrlich and Mitchell (1994) method was developed for vertical walls. In addition, soil cohesion could be considered in the calculation using the Ehrlich and Mitchell, (1994) method, while the EM (2016) method was developed for cohesionless backfill soil in order to make it simpler. As stated by Ehrlich and Mirmoradi (2016) "from the practical point of view, ignoring the soil cohesion in the calculation is a common design assumption owing to the critical conditions, such as heavy and/or prolonged rain. However, if the soil cohesion is significant and can be guaranteed during the lifespan of the structure, the results calculated using the proposed simplified method are conservative. In this case, for better representation of the actual conditions, the original procedure may be considered for use in the calculations".

Acknowledgments. The authors greatly appreciate the funding of this study by the Brazilian Research Council, CNPq, and the Brazilian Federal Agency for Support and Evaluation of Graduate Education, CAPES. We also thank Flavio Montez and Andre Estevao Ferreira da Silva from the Huesker Company for their support, as well as Cid Almeida Dieguez for his support in performing the experiments.

References

AASHTO (American Association of State Highway and Transportation Officials). LRFD bridge design specifications, 7th edn. Washington, DC, USA (2014)

Berg, R.R., Christopher, B.R., Samtani, N.C.: Design and Construction of Mechanically Stabilized Earth Walls and Reinforced Slopes. No. FHWA-NHI-10-024 Vol. I and NHI-10-025 Vol II. Federal Highway Administration, Washington DC (2009)

Brinkgreve, R.B.J., Vermeer, P.A.: PLAXIS: Finite element code for soil and rock analyses, version 8. CRC Press/Balkema, Leiden (2002)

Ehrlich, M., Mirmoradi, S.H., Saramago, R.P.: Evaluation of the effect of compaction on the behavior of geosynthetic-reinforced soil walls. J. Geotextile Geomembr. **34**, 108–115 (2012)

Ehrlich, M., Mirmoradi, S.H.: Evaluation of the effects of facing stiffness and toe resistance on the behavior of GRS walls. J. Geotextile Geomembr. **40**, 28–36 (2013)

Ehrlich, M., Mitchell, J.K.: Working stress design method for reinforced soil walls. J. Geot. Eng. ASCE **120**(4), 625–645 (1994)

Hatami, K., Witthoeft, A.F., Jenkins, L.M.: Influence of inadequate compaction near the facing on the construction response of wrapped-face MSE walls. Transp. Res. Rec. **2045**, 85–94 (2008)

Ehrlich, M., Mirmoradi, S.H.: A simplified working stress design method for reinforced soil walls. Géotechnique **66**(10), 854–863 (2016)

Liu, H.: Nonlinear elastic analysis of reinforcement loads for vertical reinforced soil composites without facing restriction. ASCE J. Geotech. Geoenviron. Eng. **142**(6), 04016013 (2016)

Mirmoradi, S.H., Ehrlich, M.: Modeling of the compaction induced stresses in numerical analyses of GRS walls. Int. J. Comput. Methods **11**(2), 1342002 (2014a)

Mirmoradi, S.H., Ehrlich, M.: Geosynthetic reinforced soil walls: experimental and numerical evaluation of the combined effects of facing stiffness and toe resistance on performance. In: Proceedings of 10th International Conference on Geosynthetics, International Society of Soil Mechanics and Geotechnical Engineering (ISSMGE), London (2014b)

Mirmoradi, S.H., Ehrlich, M.: Numerical evaluation of the behavior of GRS walls with segmental block facing under working stress conditions. ASCE J. Geotech. Geoenviron. Eng. **141**(3), 04014109 (2015a)

Mirmoradi, S.H., Ehrlich, M.: Modeling of the Compaction-induced Stress on Reinforced Soil Walls. J. Geotextile Geomembr. **43**(1), 82–88 (2015b)

Mirmoradi, S.H., Ehrlich, M., Dieguez, C.: Evaluation of the combined effect of toe resistance and facing inclination on the behavior of GRS walls. J. Geotextile Geomembr. **44**(3), 287–294 (2016)

Mirmoradi, S.H., Ehrlich, M.: Evaluation of the effect of toe restraint on GRS walls. Transp. Geotech. **8**, 35–44 (2016). SI: Geosynthetics in Tpt

Mirmoradi, S.H., Ehrlich, M.: Effects of facing, reinforcement stiffness, toe resistance, and height on reinforced walls. J. Geotextile Geomembr. **45**(1), 67–76 (2017a)

Mirmoradi, S.H., Ehrlich, M.: Experimental evaluation of the effects of surcharge width and location on geosynthetic-reinforced soil walls. Int. J. Phys. Model. Geotech. (2017b). https://doi.org/10.1680/jphmg.16.00074

Reeves, J.W.: Performance of a full-scale wrapped face welded wire mesh reinforced soil retaining wall. M.Sc. thesis, Department of Civil Engineering, Royal Military College of Canada (RMC), Kingston, Ontario, Canada (2003)

Riccio, M., Ehrlich, M., Dias, D.: Field monitoring and analyses of the response of a block-faced geogrid wall using fine-grained tropical soils. J. Geotextile Geomembr. **42**(2), 127–138 (2014)

Stuedlein, A.W., Allen, T.M., Holtz, R.D., Christopher, B.R.: Assessment of reinforcement strains in very tall mechanically stabilized earth walls. ASCE J. Geotech. Geoenviron. Eng. **138**(3), 345–456 (2012)

Geotechnical In-Situ Testing and Monitoring

Settlement Monitoring and Analysis of Changheba Dam with the Application of Terrestrial Laser Scanning

Hai-bo Li[1], Xing Han[2], Xing-guo Yang[1], Shi-ling Zhang[3], and Jia-wen Zhou[1(✉)]

[1] State Key Laboratory of Hydraulics and Mountain River Engineering, Sichuan University, Chengdu 610065, China
jwzhou@scu.edu.cn
[2] Sinohydro Bureau 5 Co. Ltd., Power Construction Corporation of China, Chengdu 610066, People's Republic of China
[3] College of Water Resource and Hydropower, Sichuan University, Chengdu 610065, China

Abstract. Settlement deformation distribution and evolution has an important impact on the stability of the earth-rock dam during dam construction process. In contrast to traditional single-point monitoring, the terrestrial laser scanning (TLS) has an advantage of presenting the changes in shape for the whole object, through scanning and creating the accurate and detailed geometric models. By using TLS technology, the high resolution digital elevation models (DEM) of the Changheba dam is re-constructed, and the settlement deformation between two respective DEMs and their distributions are calculated. Deformation distribution characteristics indicated that, the settlements in the middle are general bigger than the values in both sides near the banks, and the settlements are increase with the increasing of the elevations. Monitoring results also show that, the application of TLS permits a better understanding of distribution and evolution of settlement deformation for the dam during construction process.

Keywords: Earth-rock dam · Terrestrial laser scanning
Digital elevation model · Settlement deformation distribution

1 Introduction

The Changheba dam is one of the highest earth-rock dam in China with a maximum dam height of 240 m (Fig. 1). It is located on the Dadu River about 360 km far away from the Chengdu, Sichuan province. The dam is a core wall earth-rock dam including eight construction zones, and its gross filling capacity is about 3.42×10^7 m^3, which is the largest volume of earth-rock dam have been constructed in China.

The settlement deformation distribution and evolution has an important impact on the stability of the earth-rock dam during construction. In contrast to traditional single-point monitoring, the terrestrial laser scanning (TLS) has an advantage of presenting of changes in shape for the whole object by scanning and creating the accurate and detailed geometric models [1–3]. The principle of extraction deformations with

© Springer Nature Singapore Pte Ltd. 2018
A. Farid and H. Chen (Eds.): GSIC 2018, *Proceedings of GeoShanghai 2018*
International Conference: Geoenvironment and Geohazard, pp. 539–546, 2018.
https://doi.org/10.1007/978-981-13-0128-5_59

TLS is based on scanning, creating and comparing surface models of object in different epochs [4, 5]. Modeling surface gives more accurate representation of the object and can improve the accuracy of the monitoring by taking a group of points to represent the element of surface, rather than a single point for traditional monitoring [6, 7].

This study uses TLS to scan and create high resolution digital elevation models (DEM) of the Changheba dam, the settlement deformations can be determined between two respective DEMs. It permits a better understanding of distribution and evolution of the settlement deformation of the earth-rock dam during the construction.

Fig. 1. Site photo of the Changheba dam.

2   Method

2.1   Terrestrial Laser Scanning

Terrestrial laser scanning (TLS) is know as a Ground based LIDAR (Light Detection and Ranging system) which transmits a laser pulse and emits the reflected signal by the object of reflectorless, and utilizes the time-of-flight (*TOF*) technology to determine the distances between the instrument and a point on a reflective surface [8]. The scanner captures the 3D position of data points in the survey area by a rectangular coordinate (X, Y, Z) setting at the centre of the TLS instrument, where X is in the horizontal plane, Y is vertical to X in the horizontal plane and Z is vertical to horizontal plane. As shown in Fig. 1 the coordinates (x, y, z) of each 3D laser point are calculated from Eqs. (1) and (2) [9].

$$\rho = c \times (TOF/2) \tag{1}$$

$$\begin{cases} x = \rho \cdot \cos\theta \cdot \cos\alpha \\ y = \rho \cdot \cos\theta \cdot \sin\alpha \\ z = \rho \cdot \sin\theta \end{cases} \tag{2}$$

where *TOF* is time of the flight of laser pulse; ρ is the determined distance between the instrument and the reflective surface; α is the horizontal angle; and θ is the vertical scanning angle (Fig 2).

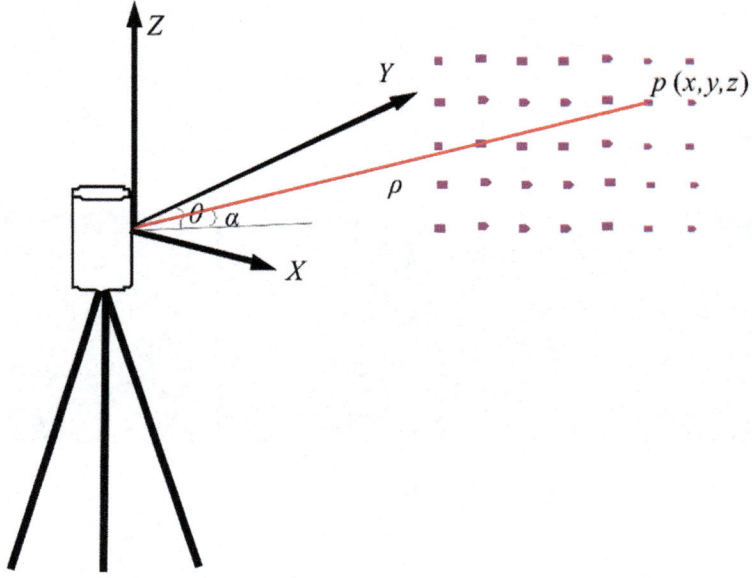

Fig. 2. Measuring principle of the terrestrial laser scanner.

The scanner captures space position of millions points in the survey area to create a 3D geometrically-correct "image" of the survey objects. The "image" of the survey objects which is referred to as the point cloud can then be used to create accurate digital elevation model (DEM) or 3D surface model, for mapping purposes, engineering surveying or further geotechnical analysis.

2.2   Field Data Acquisition

Field surveys were carried out during the construction of the Changheba dam, and the first dataset referred to as reference point cloud was acquired in September 2015 when the dam was filled at the height of 166 m (elevation 1623 m). Because of it was difficult to acquired the whole data using only one scan station, five scan stations were set along the toe of the downstream Changheba dam to acquired the entire point cloud data of the downstream Changheba dam. Overlapping scan of about 30% area between the adjacent scan stations were chosen in order to minimize occluded areas, align and merge the acquired cloud points into a single file and create a final 3-D model. Four circular reflector targets with radius of 0.2 m were set into the scanned scene as tie point for the geodetic referencing process in the post-processing stage. The mean point spacing of the acquired data ranged from 5 to 10 cm. The other two datasets were acquired in March 2016 and October 2016 when the dam construction has finished.

2.3   Data Processing

The acquired dataset of point clouds are in each own coordinate systems relative to the scan stations, alignment and merging is required to combine the different point cloud into one single file. As shown in Fig. 3, the point cloud alignment was a two-step process. First process was done by means of a preliminary rough alignment using manually selected homologous points. The second process is multi station adjustment which was performed using the iterative closest point (ICP) algorithm originally proposed by Besl and McKay to obtain a best fit alignment with an acceptable error of 0.002 m [10, 11].

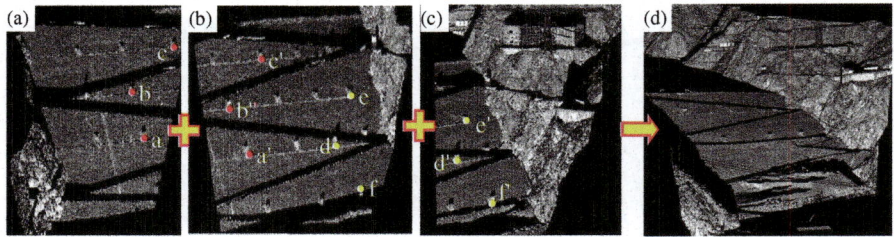

Fig. 3. Point clouds alignment and merging.

After alignment, the merged point clouds must be transformed into the geodetic coordinate system for further calculation process. This was done by using the set of four circular reflector targets as ground control points whose central positions were measured by total station. Usually three control points are sufficient; the left one of the control point can be used as the verification point.

For further study, the resulting point clouds were extracted, and then meshed to derive high resolution digital elevation models (DEM). The resulting high accurate DEM captures the morphology, geometry and elevation changes of downstream of the Changheba dam in great detail [12, 13]. It can be investigated in respect to the settlement deformation of the dam (as shown in Fig. 4).

In order to monitor the settlement deformation distribution and evolution of downstream of the Changheba dam, different DEM models of the dam were created for user defined time periods, which shown the dam construction process and the settlement deformation evolution during the dam construction over the monitored time interval. Settlement deformation between two respective DEMs and their distributions can be calculated, for the DEM can be defined as a function $z = f(x, y)$, and the different between DEMs allows measurement of the deformation in the direction z [14, 15].

The measurement results in a distance between two DEM along the direction z. These distances are not usually in the true direction of displacement, rather the shortest distance between models along the direction z. The shortest distances are useful for the analysis of dam settlement monitoring because the most part of dam displacement during the construction are settlement along the direction z [16].

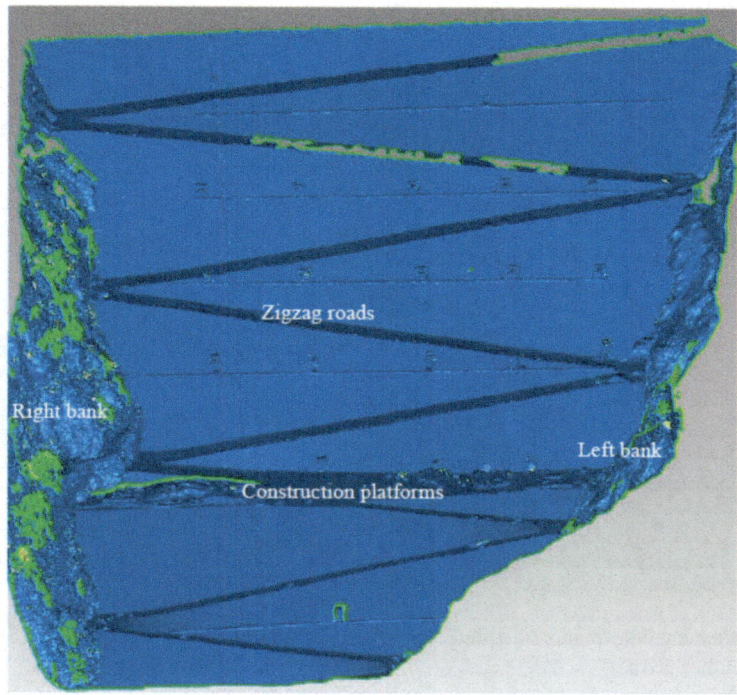

Fig. 4. DEM model for the downstream of the Changheba dam.

3   Results

3.1   Settlement Deformation Monitoring by TLS

The settlement deformation distribution and evolution has an important impact on the stability of the dam during construction. Because deformation of the dam is mostly expressed in surface settlements during construction, a multi-temporal comparison of the dam models provides the most important information on the deformation evolution of the dam.

The settlement deformation distribution and evolution of the dam from September 2015 to October 2016 is shown in Fig. 5. Positive elevation changes representing dam settlement is shown in red color. It can be seen that, since the first survey in September 2015, the dam appeared the situation of settlement deformation because of consolidation of filling materials and the new filling layers, except for some surface bulge in local areas in virtue of construction and human intervention (in blue color). Due to the monitor error, it has limited sensitivity for the extremely small settlement values in the areas near the bank. There are some occlusions existed, especially for the zigzag roads and construction platforms where the incidence angles are highly oblique.

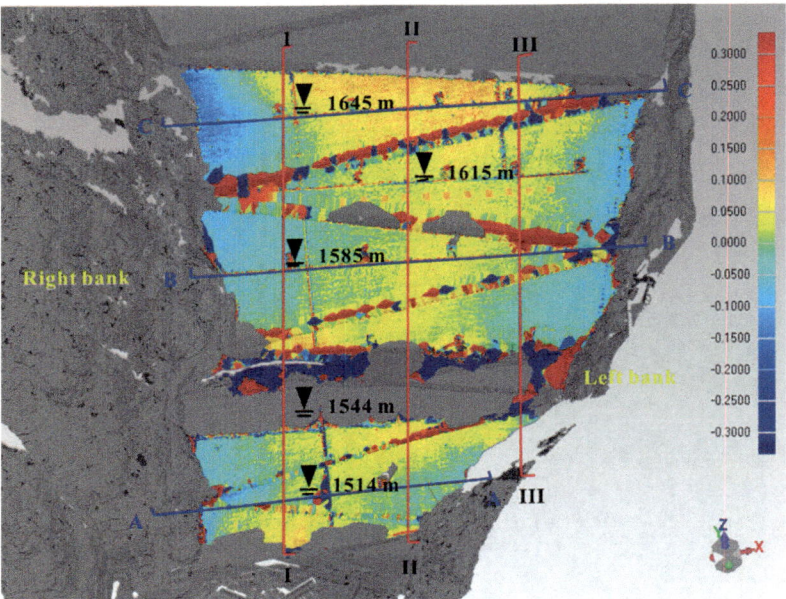

Fig. 5. Settlement deformation distribution of downstream of Changheba dam (from September 2015 to October 2016).

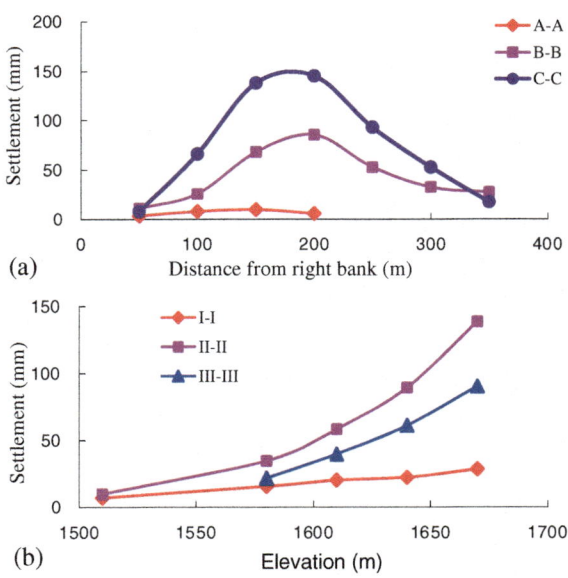

Fig. 6. The settlements of downstream of the Changheba dam along different sections: (a) along transverse sections; (b) along vertical sections.

3.2   Settlement Deformation Distribution and Evolution

Figures 5 and 6(a) show some settlements along the transverse sections of downstream of the Changheba dam in elevation 1514 m, 1585 m, and 1645 m, respectively. It can be seen that the dam had a rapid settlement during the construction, and the largest settlement reaches nearly 150 mm. Along any of the three transverse sections, the settlements in the middle are general bigger than the values in both sides near the banks. The settlements in the elevation 1514 m are much smaller than the values in other two elevations, for the majority settlement at the bottom of the dam had already accomplished in the early construction stage.

Figures 5 and 6(b) show some settlements along the vertical sections I-I, II-II and III-III, respectively. Because of existence of occlusions and data missing, the settlements in the elevation 1540 m did not get. The statistical figures show that the settlements are increase with the increasing of the elevations, which displays that the filling materials in the new filled layers at the top of the dam had a rapid settlement because of consolidation and increasing gravity loadings during the survey.

4   Conclusions

The settlement deformation distribution and evolution has an important impact on the stability of the earth-rock dam during construction. In contrast to traditional single-point monitoring, the deployment of TLS permits a better understanding of distribution and evolution of the settlement deformation of dam during the construction. The following can be concluded:

(a) By means of TLS, the high resolution digital elevation models (DEM) of the dam can be created. Settlement deformation between two respective DEMs and their distributions can be calculated, applying the shortest distance algorithm along the direction z-axes.

(b) Since the first survey in September 2015, the dam appeared the situation of settlement deformation because of consolidation of filling materials and the new filling layers.

(c) The deformation distribution displays the characteristic of that the settlements in the middle are general bigger than the values in both sides near the banks, and the settlements are increase with the increasing of the elevations.

Acknowledgments. We gratefully acknowledge the support of the Key Project of the Power Construction Corporation of China (ZDZX-5) and the Youth Science and Technology Fund of Sichuan Province (2016JQ0011).

References

1. Kuhn, D., Prüfer, S.: Coastal cliff monitoring and analysis of mass wasting processes with the application of terrestrial laser scanning: a case study of Rügen, Germany. Geomorphology **213**(7), 153–165 (2014)
2. Heritage, G., Hetherington, D.: Towards a protocol for laser scanning in fluvial geomorphology. Earth Surf. Process. Land. **32**(1), 66–74 (2010)

3. Bremer, M., Sass, O.: Combining airborne and terrestrial laser scanning for quantifying erosion and deposition by a debris flow event. Geomorphology **138**(1), 49–60 (2012)
4. Monserrat, O., Crosetto, M.: Deformation measurement using terrestrial laser scanning data and least squares 3D surface matching. ISPRS J. Photogramm. Remote Sens. **63**, 142–154 (2008)
5. Pfeifer, N., Lichti, D.: Terrestrial laser scanning. GIM Int. **18**(12), 50–53 (2004)
6. Schäfer, T., Weber, T., Kyrinovic, P., Zámecniková, M.: Deformation measurement using terrestrial laser scanning at the hydropower station of Gabcikovo. In: Proceedings of the INGEO 2004 and FIG Regional Central and Eastern European Conference on Engineering Surveying, Bratislava, Slovakia, 11–13 November 2004
7. Van Gosliga, R., Lindenbergh, R., Pfeifer, N.: Deformation analysis of a bored tunnel by means of terrestrial laser scanning. Int. Arch. Photogramm. Remote Sens. Spat. Inf. **36** (Part 5) (2006). (on CD-ROM)
8. Abellán, A., Vilaplana, J.M., Calvet, J., Garcíasellés, D., Asensio, E.: Rockfall monitoring by terrestrial laser scanning – case study of the, basaltic rock face at Castellfollit de la Roca (Catalonia, Spain). Nat. Hazards Earth Syst. Sci. **11**(3), 829–841 (2011)
9. Yue, D., Wang, J., Zhou, J., Chen, X., Ren, H.: Monitoring slope deformation using a 3-D laser image scanning system: a case study. Min. Sci. Technol. **20**, 898–903 (2010)
10. Besl, P.J., McKay, N.D.: A method for registration of 3D shapes. IEEE Trans. Pattern Anal. Mach. Intell. **14**(2), 239–256 (1992)
11. Zhang, Z.: Iterative point matching for registration of freeform curves and surfaces. Int. J. Comput. Vis. **13**(2), 119–152 (1994)
12. Gruen, A., Akca, D.: Least squares 3D surface and curve matching. ISPRS J. Photogramm. Remote Sens. **59**(3), 151–174 (2005)
13. Lemmens, M.: Product survey 3D lasermapping. GIM Int. **18**(12), 44–47 (2004)
14. Hesse, C., Stramm, H.: Deformation measurements with laser scanners—possibilities and challenges. In: Proceedings of the International Symposium on Modern Technologies, Education and Professional Practise in Geodesy and related Fields, Sofia, 4–5 November 2004, pp. 228–240 (2004)
15. Bitelli, G., Dubbini, M., Zanutta, A.: Terrestrial laser scanning and digital photogrammetry techniques to monitor landslides bodies. Int. Arch. Photogramm. Remote Sens. Spat. Inf. Sci. **35**(Part B5), 246–251 (2004)
16. Oppikofer, T., Jaboyedoff, M., Blikra, L., Derron, M.H., Metzger, R.: Characterization and monitoring of the Åknes rockslide using terrestrial laser scanning. Nat. Hazards Earth Syst. Sci. **9**, 1003–1019 (2009)

Landfills and Contaminated Soil

Evaluation of Hydro-Mechanical Behaviour of Hydraulic Barriers of Landfill Covers

P. V. Divya[1], B. V. S. Viswanadham[2(✉)], and J. P. Gourc[3]

[1] Department of Civil Engineering, IIT Palakkad, Palakkad, Kerala, India
divya.pv.nair@gmail.com
[2] Department of Civil Engineering, IIT Bombay, Mumbai, India
viswam@civil.iitb.ac.in
[3] LTHE, University of Joseph-Fourier, Grenoble, France
jean-pierre.gourc@ujf-grenoble-alpes.fr

Abstract. Bentonite amended local soils are used as hydraulic barriers (CB) in municipal solid waste landfills and for the disposal of low level radioactive waste stored containers. Excessive settlements can cause cracks in the CB which affect the strength and provide preferential flow paths to fluids compromising the sealing efficiency of the barrier. Thus, the integrity of CB with respect to their hydraulic conductivity and flexural strength is very important to prevent the environment hazards. In the present study, a series of hydraulic conductivity tests, direct tensile tests and centrifuge model tests were conducted on model barriers. The influence of waste settlement on the long term performance of the barriers was studied by conducting a series of centrifuge model tests. Digital Image Cross-Correlation technique was used in the present study for obtaining the tensile strain in the barrier during settlement. Since, clay is having low tensile strength and low tensile strain at failure, an attempt was made to improve its tensile strength by using polyester discrete geofibers. A custom designed direct tensile test set-up was used to evaluate the tensile strength-strain characteristics of the unreinforced CB and fiber reinforced soil barriers (FB). A flexible wall permeameter was used in the present study for studying the influence of geofibers on the hydraulic conductivity of the barriers. From the present study, it was observed that for the type of fibers, fiber dosage and length and soil used in the present study, the tensile strain at failure was increased to a maximum of 2.5 times and hydraulic conductivity of the soil has not varied drastically and satisfied target hydraulic conductivity requirements for the hydraulic barriers of the landfills.

Keywords: Landfills · MSW settlements · Tensile tests
Flexible wall permeameter · Centrifuge modelling

1 Introduction

The hydraulic barriers of landfill covers should be designed to have a low hydraulic conductivity (k) of less than 1×10^{-9} m/s to avoid the entry of rainwater into the waste and escape of landfill gas to the atmosphere. There are a wide variety of hydraulic barriers available like geomembranes, geosynthetic clay liners, mineral barriers etc.

© Springer Nature Singapore Pte Ltd. 2018
A. Farid and H. Chen (Eds.): GSIC 2018, *Proceedings of GeoShanghai 2018*
International Conference: Geoenvironment and Geohazard, pp. 549–556, 2018.
https://doi.org/10.1007/978-981-13-0128-5_60

Bentonite amended local soils can be used as hydraulic barriers in municipal solid waste landfills and for the disposal of low level radioactive waste stored containers wherever suitable soils are locally available. Differential settlements are inevitable in landfill covers due to the on-going biodegradation of the waste in MSW landfills and due to the tilting of waste containers in radioactive disposal facilities. Tensile strains will be induced in landfill covers at the zone of maximum curvature due to the differential settlements. Excessive settlements can lead to the formation of cracks in the compacted soil layers of landfill covers. Tensile cracks can affect the strength and provide preferential flow paths to fluids compromising the sealing efficiency of the barriers. Thus, the integrity of these bentonite amended-compacted local soil barriers (CB) with respect to their hydraulic conductivity and tensile strength is very important to prevent the environment hazards. Thus, the motivation behind this study is primarily to evaluate the influence of waste settlement on the performance of landfill cover barriers. In the present study, a series of hydraulic conductivity tests, direct tensile tests and centrifuge model tests were conducted on model barriers.

Soil is generally having low tensile strength and low tensile strain at failure. The tensile strain at failure of compacted soil is typically between 0.1% and 4%, in the case of betnonite [1]. Hence, there is a need for adopting strengthening measures to improve its tensile strength-strain characteristics, while maintaining the hydraulic conductivity of the soil barrier less than the permissible value of 1×10^{-9} m/s. An attempt was made in the present study to improve the tensile strength by using polyester discrete geofibers. Several investigators have used discrete geofibers in various geotechnical applications [2–12]. However, the systematic knowledge pertaining to the use of geofibers to landfill cover barriers, particularly, emphasizing the influence on the hydraulic conductivity, tensile strength and differential settlements of the soil is very limited.

2 Materials

Locally available soil was collected from an excavation site near IIT Bombay, India. The hydraulic conductivity of the soil was found to be 2.1×10^{-7} m/s. Sodium based bentonite was blended with the soil at different percentages of 5–20% in increments of 5% by dry weight to reduce the hydraulic conductivity. Addition of bentonite can be considered as a viable option to reduce the hydraulic conductivity of soil. Hosney and Rowe [13] reported the use of polymer-enhanced bentonite-sand mixture as a cover for gold mine tailings. Local soil mixed with 15% bentonite was selected as a suitable model barrier material. This was done by comparing the properties of the model barrier materials according to the database given in [14]. Soil was classified as CH and hydraulic conductivity (k) was found to be 3.2×10^{-10} m/s. In the present study, polyester fibers with a special triangular cross section having equivalent diameter 40 microns were used and blended with the soil uniformly. A fiber content of 0.5% and length of 90 mm was used in the present study. The model barriers for all the tests in the present study were compacted at a water content 5% on the wet side of optimum (OMC + 5%) and at the corresponding dry unit weight ($\gamma_d = 15.2$ kN/m^3) as obtained from the compaction curve of standard Proctor compaction test.

3   Test Equipment and Methodology

3.1   Direct Tensile Tests

A custom designed tensile test set-up was used to evaluate the tensile strength-strain characteristics of the unreinforced CB and fiber reinforced soil barriers (FB). The test set up was developed in such a way to apply direct tension to the barrier material. The specimen mold having dimensions of 152 mm length, 152 mm breadth and 152 mm height was split in to two equal halves with an open top resting on a horizontal platform. Four triangular wedges were attached inside the box which facilitates proper contact between the soil and mold as tension is developed across the plane of separation. One half of the mold is free to move longitudinally on frictionless bearings that ride on polished guide rails. The other half is held fixed to the horizontal platform. The tensile load was measured by a load cell positioned between the moving part of the mold and the motor which applies the tensile load. The movable box is pulled away at a constant strain rate of 1.25 mm/min in horizontal direction until the soil specimen fails in tension. Two potentiometers were attached to the moving half of the mold for measuring displacements. These measured displacements were used to calculate the average strain along the direction of the movement. Figure 1 shows the view of unreinforced barrier soil specimen (CB) after direct tensile test.

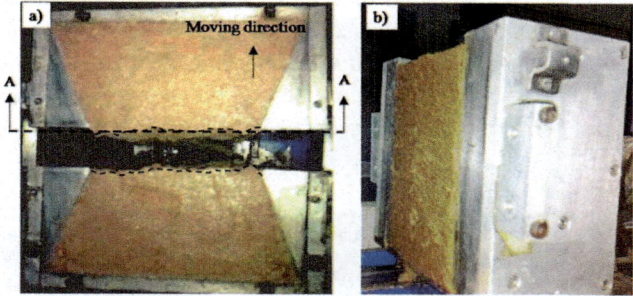

Fig. 1. View of barrier soil specimen (CB) after tensile test (a) top view (b) view along A-A

The entire proceedings were continuously recorded using a charge coupled device (CCD) video camera mounted to view the soil surface from the top. The displacement vectors and strain fields were obtained by performing Digital Image Cross-correlation (DIC) of the photographs captured during various stages of the test by using the strain master 2D package of DaVis 7.2 software [15]. The tensile stress σ_t is obtained by dividing the tensile load by the cross-sectional area at the middle of soil specimen i.e. the cracked zone. Peak tensile stress $\sigma_{pt}(kN/m^2)$ is obtained from the tensile stress-strain plots of the soil corresponding to peak tensile load. The average strain at which visible cracks were initiated on the soil was defined as the tensile strain at crack initiation $\varepsilon_c(\%)$. Figure 2 shows the top view and strain field pattern at crack initiation and at the end of the test. It was observed that the peak tensile strength σ_{pt} and strain at crack initiation ε_c was increased by 2 times and 2.5 times respectively, when reinforced with geofibers. During tensile tests, unreinforced soil showed a brittle behavior compared with the ductile behavior of fiber-reinforced soil.

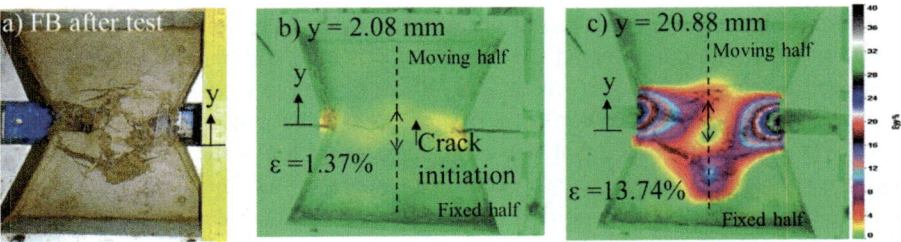

Fig. 2. FB after tensile test (a) top view (b) strain field at crack initiation (c) strain field at the end of the test.

3.2 Hydraulic Conductivity Tests

Considering the advantage of Flexible-wall permeameters to offer complete control over stresses and suitability for minimizing side-wall leakage [16], a flexible wall permeameter was used in the present study for determining the hydraulic conductivity of soils as per ASTM D5084 (2010) [17]. Soil specimens of 100 mm diameter and 100 mm in height were used. Figure 3 shows the view of soil specimen mounted on the pedestal and flexible wall permeameter used in the present study. Pressure and volume change in the cell and the soil specimen was controlled by two separate flow pumps (Geocomp, USA make) with embedded controls and pressure transducers. De-aired tap water was used as permeant according to ASTM D5084 [17] with the help of a

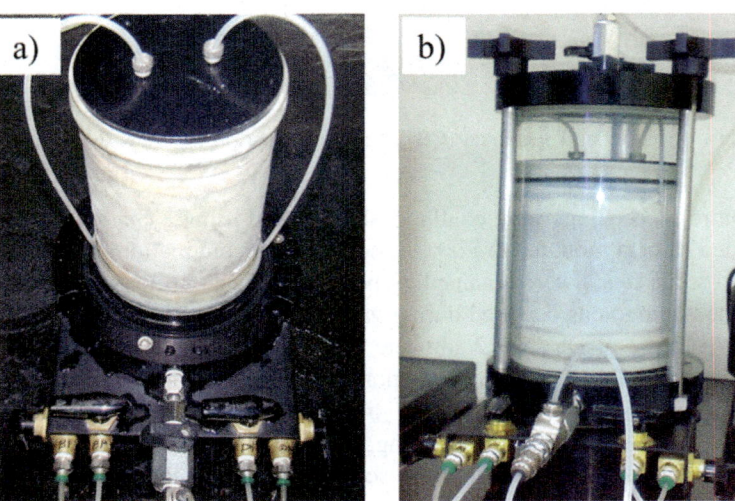

Fig. 3. View of (a) barrier soil specimen mounted on the pedestal (b) flexible wall permeameter used in the present study

deaeration tank, vacuum source and a water trap. Once the test specimen was prepared and mounted; the system was checked for initial inspection of leakages. Soil samples were fully saturated with the help of back pressure. Once the value of pore pressure coefficient B reaches near 0.95, the sample was isotropically consolidated to an effective stress of 25 kN/m^2, since landfill cover barriers are subjected to low confining pressures. After the consolidation stage, the hydraulic conductivity stage was started by adjusting and applying the bottom pressure and top pressure on the sample to get the desired hydraulic gradient of 15. The tests were ended when the measured inflow is equal to measured outflow, or when hydraulic conductivity values are stabilized whichever occurred earlier, according to the termination criteria explained in ASTM D5084 [17]. The hydraulic conductivity of FB was found to be 1.1×10^{-10} m/s and remained within the permissible limit required for use as barrier material.

3.3   Centrifuge Model Tests

Centrifuge tests were conducted at 40 gravities by inducing differential settlements to a maximum of 1 m at a rate of 1 mm/min (in model dimensions) with the help of a motor based settlement simulator. Centrifuge model tests were adopted in the present study considering the advantage of providing stress similarity between the prototype and model which is achieved in a centrifuge by accelerating a model of scale 1/N to N times the Earth's gravity. A 4.5 m radius large beam centrifuge having a capacity of 2500 g-kN available at Indian Institute of Technology Bombay was used for conducing the tests. The description of settlement simulator setup, calibration at 1 g and 40 g has been explained extensively in [18]. It consists of a central support system with trap door settlement unit and two side supports provided on either side. Trapdoor settlement unit can be lowered with the help of a screw jack, and thereby hinged plates resting on the central support undergo an angular rotation and a trough-shaped deformation profile can be achieved gradually. With the help of this system, a landfill area corresponding to 415 m^2 at 40 g (720 mm in length and 360 mm in breadth in model dimensions) can be simulated. All dimensions are given in model scale with those in prototype scale within parenthesis. The thickness of model clay barriers was 25 mm (1 m) prepared on top of a 30 mm (1.2 m) coarse sand layer followed by a 30 mm (1.2 m) fine sand layer. Digital Image Cross-Correlation (DIC) technique was used in the present study for deriving the strain fields during strain localization and fracture when subjected to flexural distress. A settlement rate of 1 mm/min. (36 mm/day at 40 g) was adopted and the distortion level a/l (the ratio of central settlement of clay barrier a over an influence length l) was varied from 0 to 0.125.

Figure 4 shows the flexural distress and cracking pattern along with the strain field in the cracked zone for the bentonite amended-compacted local soil barriers (CB) and fiber reinforced (FB) barriers reinforced with PET fibers.

Strain fields were obtained from the DIC analysis of the photos captured during the centrifuge tests. Crack propagations could be easily located by DIC and it is clear from the figure that very wide cracks of about 15 mm (600 mm) width were observed for unreinforced bentonite amended barriers CB whereas very minute ruptures were only formed for FB. Water-tightness of the barriers was assessed using five pore pressure transducers by measuring the change in height of water which was inundated on the

a) CB
i) Top view of CB   ii) Close view of cracked zone   iii) Strain field of CB

b) FB
i) Top view of FB   ii) Close view of cracked zone   iii) Strain field of FB

Fig. 4. Status of model barriers after subjected differential settlements: (a) CB; (b) FB

barrier surface. Water pressure exerted on clay barrier at different settlement stages can be obtained from the PPT measurements. Numerical integration of the area under the obtained water profile will yield volume of water per unit width. Ratio of numerical difference between initially retained volume of water v_0 and volume of water at any settlement v_a to the initially retained volume of water v_0 is defined as the infiltration ratio, IFR. The strain at water leakage, ε_l for CB was 0.94%. The ε_l was increased as 2.03% for FB. The strain at water leakage for FB barrier was increased by around 2.2 times compared to identical CB. It was also observed that 80% of the water stored on top of the FB was retained without infiltration even after inducing a distortion level of 0.125. Summary of the results of direct tensile tests, hydraulic conductivity tests and centrifuge model tests conducted in the present study are summarized in Table 1.

Table 1. Summary of the results of direct tensile tests, hydraulic conductivity tests and centrifuge model tests

Test legend	Direct tensile tests		Hydraulic conductivity tests	Centrifuge model tests	
	$\sigma_{pt}(kN/m^2)$	$\varepsilon_c(\%)$	k(m/s)	$\varepsilon_l(\%)$	IFR
CB	12.89	0.55	3.2×10^{-10}	0.94	0.4
FB	27.79	1.36	1.1×10^{-10}	2.03	0.2

CB - bentonite amended-compacted local soil barriers; FB - fiber reinforced barriers; σ_{pt} - Peak tensile stress; ε_c - tensile strain at crack initiation; ε_l - strain at water leakage; IFR - Infiltration ratio.

4 Conclusions

Based on the analysis and interpretation of the direct tensile tests, hydraulic conductivity tests and centrifuge model tests the following conclusions can be drawn:

- The tensile strength-strain characteristics of bentonite amended-compacted local soil barriers was found to be significantly improved by the inclusion of discrete geofibers. It was observed that the peak tensile strength σ_{pt} and strain at crack initiation ε_c was increased by 2 times and 2.5 times respectively, when reinforced with geofibers. During tensile tests, unreinforced soil showed a brittle behavior compared with the ductile behavior of fiber-reinforced soil.
- The hydraulic conductivity of FB was found to be 1.1×10^{-10} m/s and remained within the permissible limit required for use as barrier material.
- From the series of centrifuge model tests, it was found that the strain at water leakage, ε_l for CB was 0.94%. The ε_l was increased as 2.03% for FB. The strain at water leakage for FB barrier was increased by around 2.2 times compared to identical CB. It was also observed that 80% of the water stored on top of the FB was retained without infiltration even after inducing a distortion level of 0.125 which indicate the sealing efficiency of the barriers.

References

1. LaGatta, M.D., Boardman, B.T., Cooley, B.H., Daniel, D.E.: Geosynthetic clay liner subjected to differential settlement. J. Geotech. Geoenviron. Eng. ASCE **123**(5), 402–410 (1997)
2. Miller, C.J., Rifai, S.: Fiber reinforcement for waste containment soil liners. J. Environ. Eng. ASCE **130**(8), 891–895 (2004)
3. Tang, C.S., Shi, B., Zhao, L.-Z.: Interfacial shear strength of fiber reinforced soil. Geotext. Geomembr. **28**(1), 54–62 (2010)
4. Sivakumar Babu, G.L., Vasudevan, A.K.: Strength and stiffness response of coir fiber-reinforced tropical soil. J. Mat. Sci. Civil Eng. ASCE **20**(9), 571–577 (2008)
5. Viswanadham, B.V.S., Jha, B.K., Pawar, S.N.: Influence of geofibers on the flexural behaviour of compacted soil beams. Geosynth. Int. **17**(2), 86–99 (2010)
6. Falorca, I.M.C.F.G., Pinto, M.I.M.: Effect of short, randomly distributed, polypropylene microfibres on shear strength behaviour of soils. Geosynth. Int. **18**(1), 2–11 (2011)
7. Divya, P.V., Viswanadham, B.V.S., Gourc, J.P.: Evaluation of tensile strength-strain characteristics of fiber reinforced soil through laboratory tests. J. Mat. Civil Eng. ASCE **26**(1), 14–23 (2014)
8. Divya, P.V., Viswanadham, B.V.S., Gourc, J.P.: Centrifuge model study on the performance of fiber reinforced clay-based landfill covers subjected to flexural distress. Appl. Clay Sci. Elsevier **142**, 173–184 (2017)
9. Divya, P.V., Viswanadham, B.V.S., Gourc, J.P.: Centrifuge modelling of geofiber reinforced clay-based landfill covers subjected to flexural distress. J. Geotech. Geoenviron. Eng. ASCE **143**(1), 04016076(1–11) (2017)
10. Tang, C., Wang, D., Cui, Y., Shi, B., Li, J.: Tensile strength of fiber-reinforced soil. J. Mat. Civil Eng. ASCE **28**, 04016031 (2016). https://doi.org/10.1061/(asce)mt.1943-5533.0001546

11. Oliveira, P.J.V., Correia, A.A.S., Teles, J.M.N.P.C., Custódio, D.G.: Effect of fibre type on the compressive and tensile strength of a soft soil chemically stabilised. Geosynth. Int. **23**(3), 171–182 (2016)
12. Festugato, L., Menger, E., Benezra, F., Kipper, E.A., Consoli, N.C.: Fibre-reinforced cemented soils compressive and tensile strength assessment as a function of filament length. Geotext. Geomembr. **45**(1), 77–82 (2017)
13. Hosney, M.S., Rowe, R.K.: Performance of polymer-enhanced bentonite-sand mixture for covering arsenic-rich gold mine tailings for up to 4 years. Can. Geotech. J. **54**(4), 588–599 (2017)
14. Benson, C.H., Daniel, D.E., Boutwell, G.P.: Field performance of compacted clay liners. J. Geotech. Geoenviron. Eng. ASCE **126**(5), 390–403 (1999)
15. Lavision: DaVis 7.2-strain master software manual, Gottingen, Germany (2009)
16. Daniel, D.E., Trautwein, S.J., Boynton, S.S., Foreman, D.E.: Permeability testing with flexible-wall permeameters. Geotechn. Test. J. ASTM **7**(3), 113–122 (1984)
17. ASTM Standard, D5084: standard test methods for measurement of hydraulic conductivity of saturated porous materials using a flexible wall permeameter. American Society for Testing and Materials, Philadelphia (2010)
18. Rajesh, S., Viswanadham, B.V.S.: Evaluation of geogrid as a reinforcement layer in clay based engineered barriers. Appl. Clay Sci. **42**(3–4), 460–472 (2009)

A Numerical Model for Error Analyses of Static Chamber Method Used at Landfill Site

Xinru Zuo, Siliang Shen, Haijian Xie$^{(\boxtimes)}$, and Yunmin Chen

College of Civil Engineering and Architecture, Zhejiang University,
866 Yuhangtang Road, Hangzhou 310058, China
xiehaijian@zju.edu.cn

Abstract. Static chamber method is widely used to measure emission of landfill gas by deploying the chamber at the surface of landfill cover soil. However, there will be errors between the measured fluxes and the real fluxes due to the increase of gas pressure in the chamber. A numerical model based on dust gas model was developed to investigate the factors affecting the errors. It is found that, height of static chamber is the most sensitive factor. When the height of chamber increases from 0.12 to 0.5 m, the error decreases from 32% to 10%. It will be easier for the chambers of smaller sizes to accumulate higher concentration and lead to greater errors. The proposed numerical model was successfully used in the analysis of the field static chamber tests carried out at the Xi'an landfill site. The evaluated errors for the field tests can be 30%. The model would be very useful for the error assessment of the static chamber tests.

Keywords: Static chamber · Multi-component landfill gas · Cover soil

1 Introduction

Methane (CH_4) and carbon dioxide (CO_2) are the major components of landfill gas (LFG) generated from degradation of municipal solid waste (MSW). Landfill gas will transport from the cover layer to atmosphere. Before taking further measurements to control LFG or analyzing the influence of it on the environment, efforts should be made to determine actual gas emission flux. The static chamber method is widely used to measure landfill gas emission flux from the MSWs [1]. It is easy to operate, economic, and is available for point measurement. The static chamber is placed on the surface of landfill covers and gas concentration in it is measured at certain time intervals to obtain the gas emission flux. Many factors such as improper operation, varying gas pressure and temperature will result in errors of measured flux. Many works is have been focused on optimal chamber design, including using a fan to mix the gas, using a vent tube to minimize pressure change in the chamber, selecting a proper size of chamber and decreasing time interval [2–4]. However, it is assumed that gas concentration at the soil surface keeps constant after deployment static chamber. In fact, gas concentration will increase when gas migrates into the chamber. This will lead to a decrease of vertical concentration gradient and gas transport rate. This is also called "chamber effect". Therefore, the error between measured flux and actual flux should be investigated.

© Springer Nature Singapore Pte Ltd. 2018
A. Farid and H. Chen (Eds.): GSIC 2018, *Proceedings of GeoShanghai 2018*
International Conference: Geoenvironment and Geohazard, pp. 557–566, 2018.
https://doi.org/10.1007/978-981-13-0128-5_61

Senevirathna demonstrated that the error was related to the height of the chamber, real flux and deployment time [5]. Livingston developed the non-linear relationship between gas concentration and flux based on Fick's law and Laplace transform [6]. However, the disadvantage of the relationship is that advection was not considered. Senevirathna developed a two-dimensional model to simulate carbon dioxide transport from a two-layer soil cover to static chamber [7], which also considered methane oxidation Sahoo developed a two-dimensional model considering gas transport in chamber space [8]. However, the effect of advection was also not considered. In addition, most of the proposed models did not consider interaction of different component gas diffusion, i.e., the multi-component gas diffusion. Dust Gas Model (DGM) can describe gas transport and gas concentration in the chamber when the gas consists of many components (e.g., LFG) [9]. The aim of the paper is to develop a numerical model for simulating multi-component gas diffusion and advection in the soil and the static chamber. The proposed model based on DGM can also be used to get the relationship between real flux and measured flux at field.

2 Model Description

2.1 Governing Equations

After static chamber was installed at the soil surface, gas concentration in the chamber was measured at specific time. The flux of component i, $N_{i,measure}$ can be obtained from the gradient of C_i [5]:

$$N_{i,measure} = \left(\frac{V_{chamber}}{A_{chamber}}\right) \frac{dC_{i,chamber}}{dt} \tag{1}$$

where $V_{chamber}$ and $A_{chamber}$ are chamber volume (m^3) and based area (m^2); and $C_{i,chamber}$ (mol/m^3) represents mole concentration of component i in the chamber.

Because of "chamber effect", gas will accumulate in the chamber and result in the decrease of the vertical concentration gradient. The error induced by this effect is [5]:

$$Error = \frac{N - N_{measure}}{N} \times 100\% \tag{2}$$

where $Error$ is the error rate, N (mol/m^2/s) is actual flux without static chamber; and $N_{measure}$ (mol/m^2/s) is flux calculated by concentration-time curve.

It is assumed that gas is mixed well in the chamber and gas concentration in chamber is equal to that of the soil surface. Because the model developed is one-dimensional, lateral gas transport is neglected. This will lead to an error when test time is long. The other basic assumptions are as follows: (1) the soil was homogeneous; (2) gas flux distributed uniformly at the bottom of cover soil and kept constant; and (3) effects of atmospheric pressure and temperature fluctuation were neglected. Based on these assumptions, adopting mass balance equation and equation of continuity, the transient model was developed to describe the relationship between gas concentration and gas flux [10]:

$$\theta_g \times \frac{\partial C_i}{\partial t} + \frac{\partial N_i}{\partial z} = 0 \qquad (3)$$

where θ_g is air volume ratio; z is the depth. C_i (mol/m^3) is mole concentration of component i; R (m^3/Pa $\cdot$ K/mol) is ideal gas constant; T (K) is the absolute temperature; and N_i (mol/m^2/s) is flux of component i consisting of advection flux N_i^V and diffusion flux N_i^D [11]:

$$N_i = N_i^V + N_i^D \qquad (4)$$

Advection flux can be described by [12]

$$N_i^V = -x_i \frac{k_{rg}k_i P}{RT\mu} \times \frac{dP}{dz} \qquad (5)$$

where k_{rg} and k_i (m^2) are relative and intrinsic permeability, respectively; μ (Pa s) is the gas viscosity; and x_i is the mole fraction of component i:

$$x_i = P_i / \sum_{i=1}^{n} P_i \qquad (6)$$

The relative gas permeability depends on the effective degree of saturation [13]:

$$k_{rg} = (1 - S_e)^{\frac{1}{2}} \left(1 - S_e^{1/\alpha}\right)^{2\alpha} \qquad (7)$$

where α can be obtained from soil water characteristic curve test and S_e is effective degree of saturation:

$$S_e = (S - S_r)(1 - S_r)^{-1} \qquad (8)$$

where S is degree of saturation, S_r is residual degree of saturation.

Diffusive flux can be described by the DGM model [11]:

$$\sum_{j=1, j\neq i}^{n} \frac{x_i N_j^D - x_j N_i^D}{\tau D_{ij}} - \frac{N_i^D}{D_{iM}^e} = \frac{\partial C_i}{\partial x} \qquad i = 1, 2, \cdots, n \qquad (9)$$

where D_{ij}(m^2/s) is binary diffusion coefficient between component i and j [14]; D_{iM}^e (m^2/s) is Knudsen diffusion coefficient of component i and τ is tortuosity coefficient [15]:

$$\tau = \theta_g^{2.5} / n \qquad (10)$$

where n is the porosity of soil.

Knudsen diffusion coefficient can be calculated as a function of Klinkenberg parameter b_i, permeability, viscosity and molecular mass M_i [14]:

$$D_{iM} = \frac{k_{rg}k}{\mu_i}b_i \tag{11}$$

where b_i is Klinkenberg parameter of gas component i:

$$b_i = 5.57(k_{rg}k)^{-0.24}\frac{\mu_i}{\mu_{air}}\sqrt{\frac{M_{air}}{M_i}} \tag{12}$$

and M_{air} and M_i(g/mol) are molecular mass of air and component i, respectively.

For idea gas, the partial pressure is related to mole concentration [16]:

$$P_i = RT \cdot C_i \tag{13}$$

where P_i is partial pressure of component i.

2.2 Boundary and Initial Conditions

The boundary condition at soil cover bottom is assumed to be a constant flux boundary:

$$N_i = N_{Bottom} \quad z = L, t \geq 0 \tag{14}$$

The bottom flux of LFG N_{bottom} is equal to gas production from waste layer, while flux of O_2 and N_2 is 0.

The top boundary for the case without static chamber can be constant concentration:

$$C_i = x_i\left(\frac{P_{atm}}{RT}\right) \quad z = 0, t \geq 0 \tag{15}$$

where P_{atm} is atmospheric pressure.

The top boundary for the case with static chamber needs to be considered when gas enters the chamber and gas concentration increases with time:

$$C_{i,chamber} = x_i\left(\frac{P_{atm}}{RT}\right) + \frac{A_{chamber}\int N_i dt}{V_{chamber}} \quad z = 0, t \geq 0 \tag{16}$$

where the first item in Eq. (16) represents atmospheric concentration; and the second item which expressed in integral form represents mole concentration of species i entering the chamber during time t.

This model was separated by two steps. Firstly, the steady condition for the case without static chamber is calculated. The calculated gas concentration was then used as an initial condition of second step for the case with static chamber. It is assumed that the soil cover was initially filled by air. The initial condition for first step calculation is the same as the top boundary in the case without chamber. The initial conditions of the second step are based on the results of the first step calculation.

3 Model Verification

The proposed numerical model was verified by the experimental data from a laboratory static chamber [17]. Details on the experiment were given in this research [17]. CO_2 was used for gas transport in the compacted soil specimen and the static chamber. The porosity, degree of saturation, gas permeability of the soil are $n = 0.38$, $S = 25\%$, $kg = 1 \times 10^{-10}$ m^2, respectively. The test was firstly carried out without the static chamber. The gas concentration at the top boundary was atmospheric concentration, i.e., $P = 101$ Kpa, $x_{CO_2}:x_{N_2}:x_{O_2} = 0.03:78.5:21.5$. The bottom flux of CO_2 is 5.2×10^{-5} $mol/m^2/s$. When the gas concentration in the soil column remained constant, steady state was achieved. The CO_2 concentration profile was obtained at the steady state. The calculated CO_2 concentration profile agrees relatively well with the experimental data (see Fig. 1).

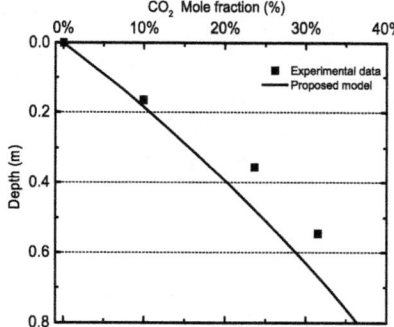

Fig. 1. CO_2 concentration profiles from experimental and model results

Secondly, different sizes of static chambers were deployed at the soil surface. Gas concentration in the chamber was recorded at a certain time interval. For different sizes of chambers, the test data and the calculated results by the proposed model were shown in Fig. 2a, b and c. It can be found that when the chamber is the smallest (h: 0.05 m; id: 0.1 m), the results of the proposed model consists well with the experimental data in the first 500 s. At 800 s, the difference of those two is about 3%. When the chamber is the biggest one (h: 0.16 m; id: 0.25 m), with a height of 0.165 m, result of model calculation and experiment consists well until 20 min. For medium size chamber, with a height of 0.12 m, results of model calculation and experiment consist well in first 600 s. The difference of them is 0.5% at 800 s.

4 Application of Proposed Model in Field

The field test was conducted at a landfill located at Xi'an (Northwest China). The static chamber was installed into the final cover of the landfill. A layer of 0.9 m compacted loess was the main material of final cover of the landfill. Gas permeability of loess at field is 2.86×10^{-13} m^2 [18]. Two points in the field were selected randomly to conduct static chamber test.

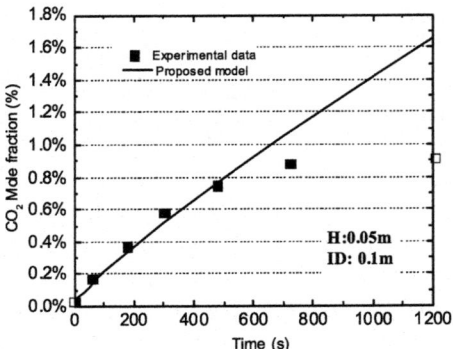

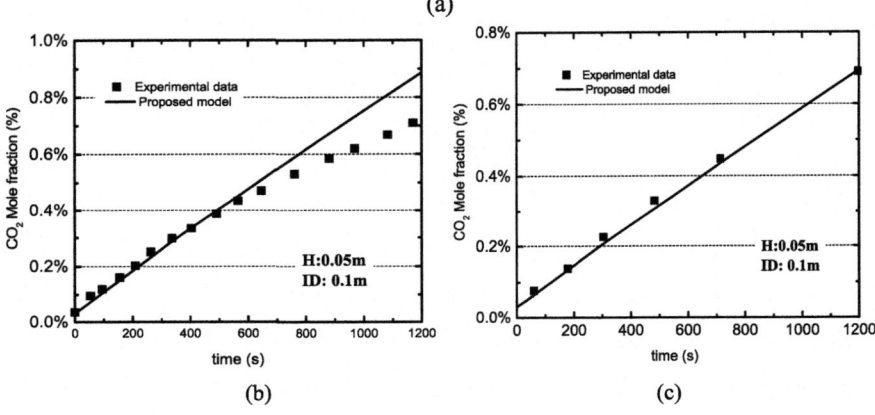

Fig. 2. CO_2 concentration over time in chambers with different sizes, (a) small-size chamber (h:0.05 m; id:0.1 m); (b) medium-size chamber (h:0.12 m; id:0.2 m) and (c) big-size chamber (h:0.16 m; id:0.25 m)

To analyze error of measured flux at field, the proposed model was used on the basis of the parameters obtained in the field site. The parameter values were obtained from the field site test at a landfill in west China (see Table 1). Because LFG mainly consists of CH_4 and CO_2, the 4 components, i.e., CH_4, CO_2, O_2 and N_2 were considered. The Gas properties were shown in Table 2.

Table 1. Parameter values used for analysis

k_g (m^2)	n	S	τ	N	h_s
2.86×10^{-13}	0.52	30%	0.15	5×10^{-5} mol/m^2/s	0.12–0.5 m

Figure 3 shows the errors of methane and carbon dioxide flux for chambers of different heights. The errors decreased with the increase of the height of the static

Table 2. Gas properties for simulations [14]

	Ordinary diffusion coefficient D_{ij} (m²/s)		
	O_2	CO_2	CH_4
N_2	2.083×10^{-5}	1.649×10^{-5}	2.137×10^{-5}
O_2		1.635×10^{-5}	2.263×10^{-5}
CO_2			1.705×10^{-5}

chamber. When the height of chamber is 0.12 m, the error for the flux of CH_4 and CO_2 are both higher than 20% at 1200 s. When the height of the chamber is 0.5 m, the error is 9% at 1200 s. These results demonstrated that chamber effect cannot be neglected. Even the chamber is relatively high (e.g., 0.5 m). The measured flux still needs to be corrected. The reason for the higher errors for the cases with relatively small heights is that the concentration in the static chamber increases more quickly due to the lower volumes. For the same chamber, the error for CO_2 was smaller than that of CH_4, especially when the chamber is small. When the height of chamber is 0.12 m, difference between the errors of those two gases is 3% at 1200 s. This is due to the fact that the binary diffusion coefficients for CO_2 are smaller than that of CH_4.

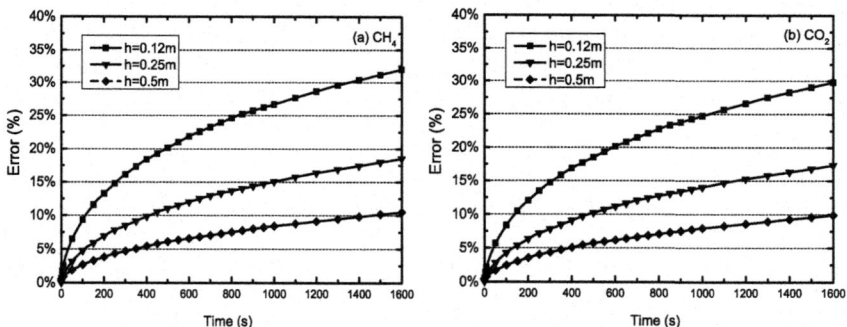

Fig. 3. Flux error for different height of chamber in terms of (a) CH_4 and (b) CO_2

The static chamber, with a height of 55 cm and diameter of 50 cm, consisted of a glass cylinder and equipped with a small fan for internal air recirculation. An external and flexible pedestal was used to seal the static chamber to the ground, thus preventing gases exchange between the chamber and the atmosphere when the groove in the steel pedestal was filled with water (see Fig. 4). Shut off all the valves, waited for accumulation of the gas in the chamber, sampled the gas into Tedlar bags every 10 min for three times (0 min, 10 min, 20 min) and then methane fluxes were determined (the slope of the gas concentration increase versus time). Gas concentration in those gas sampling bags was analyzed by gas chromatograph (GC9800) in laboratory.

By inputting gas concentrations data of different time into Eq. (1), $N_{measure}$, the measured emission flux can be obtained. The flux can be corrected by assuming a flux error. The gas concentrations corresponding to the corrected flux can then be obtained

Fig. 4. Photograph of the static chamber used at a landfill site

by inputting the corrected flux to the model. The calculated gas concentrations were compared to the measured data to evaluate the ranges of the error. In Fig. 5a, when assuming that the errors are 20% and 30%, the corrected concentration predicted by the proposed model is within the range of measured concentrations of methane and carbon dioxide. For the test point 2, the errors of the measured CO_2 and CH_4 data was 0-20% (see Fig. 5b). These results indicate that the errors with respect to field static chamber tests can be quite high and it should be modified by the numerical model.

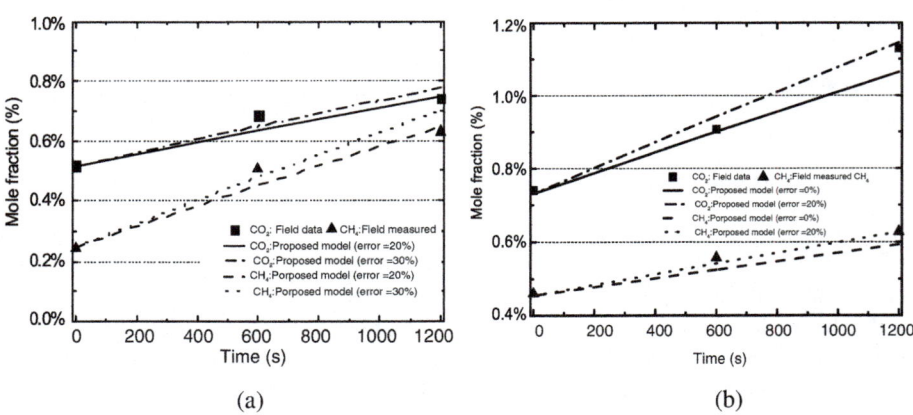

Fig. 5. Comparisons of observed gas concentration curve with results of proposed model, (a) field test point 1 and (b) field test point 2

5   Conclusions

A one-dimensional model for gas transport from the soils to the static chamber was developed. The proposed model is verified by the experimental data. It can be used to analyze effect of chamber height, real flux and degree of saturation on the errors induced by the chamber effect. The main conclusions are as follows:

1. Chamber size is the main factor affecting the measurement error. When the height of chamber ranges from 0.12 to 0.5 m, the error ranges from 32% to 10%. It will be easier for the chambers of smaller sizes to accumulate higher concentration and lead to greater errors. For chamber height of 0.12 m, the error can be 30% at 1200 s. However, the deployment time for smaller chamber will be shorter. The optimal deployment time for small chambers needs further research.
2. The proposed numerical model was successfully used in the analysis of the field static chamber tests carried out at the Xi'an landfill site. The evaluated errors for the field tests can be 30%. The model would be very useful for the error assessment of the static chamber tests.

Acknowledgements. The financial supports from the National Natural Science Foundation of China (Grants Nos. 41672288, 51478427, 51625805, 51278452, and 51008274), and the Fundamental Research Funds for the Central Universities (Grant No. 2017QNA4028) are gratefully acknowledged.

References

1. Pihlatie, M.K., Christiansen, J.R., Aaltonen, H., Korhonen, J.F.J., Nordbo, A., Rasilo, T., Benanti, G., Giebels, M., Helmy, M., Sheehy, J., Jones, S., Juszczak, R., Klefoth, R., Lobo-do-Vale, R., Rosa, A.P., Schreiber, P., Serça, D., Vicca, S., Wolf, B., Pumpanen, J.: Comparison of static chambers to measure CH4 emissions from soils. Agric. For. Meteorol. **171–172**, 124–136 (2013)
2. Christiansen, J.R., Korhonen, J.F., Juszczak, R., Giebels, M., Pihlatie, M.: Assessing the effects of chamber placement, manual sampling and headspace mixing on CH4 flux in a laboratory experiment. Plant Soil **343**(1–2), 171–185 (2011)
3. Hutchinson, G.L., Livingston, G.P.: Soil-atmosphere gas exchange. In: Dane, J.H., Topp, G. C. (eds.) Methods of soil analysis. Part 4. SSSA Book Series 5, pp. 1159–1182. SSSA, Madison (2002)
4. Rochette, P., Eriksenhamel, N.S.: Chamber measurements of soil nitrous oxide flux: are absolute values reliable? Soil Sci. Soc. Am. J. **72**(2), 331–342 (2008)
5. Senevirathna, D.G.M., Achari, G., Hettiaratchi, J.P.A.: A laboratory evaluation of errors associated with the determination of landfill gas emissions. Can. J. Civ. Eng. **33**(3), 240–244 (2006)
6. Livingston, G.P., Hutchinson, G.L., Spartalian, K.: Trace gas emission in chambers: a non-steady-state diffusion model. Soil Sci. Soc. Am. J. **70**(5), 1459–1469 (2006)
7. Senevirathna, D.G.M., Achari, G., Hettiaratchi, J.P.A.: A mathematical model to estimate errors associated with closed flux chambers. Environ. Model. Assess. **12**(1), 1–11 (2007)
8. Sahoo, B.K., Mayya, Y.S.: Two dimensional diffusion theory of trace gas emission into soil chambers for flux measurements. Agric. For. Meteorol. **150**(9), 1211–1224 (2010)

9. Molins, S., Mayer, K.U.: Coupling between geochemical reactions and multicomponent gas and solute transport in unsaturated media: a reactive transport modeling study. Water Resour. Res. **43**(5), 687–696 (2007)

10. Binning, P.J., Postma, D., Russell, T.F., Wesselingh, J.A., Boulin, P.F.: Advective and diffusive contributions to reactive gas transport during pyrite oxidation in the unsaturated zone. Water Resour. Res. **43**(2), 329–335 (2007)

11. Mason, E.A., Malinauskas, A.P.: Gas Transport in Porous Media: The Dusty-Gas Model. Elsevier, Amsterdam (1983)

12. Clifford, K.H., Webb, S.W.: Gas transport in porous media. Encycl. Ecol. **14**(8–9), 3576–3582 (2006)

13. Parker, J.C.: Multiphase flow and transport in porous media. Rev. Geophys. **27**(3), 311–328 (1989)

14. Thorstenson, D.C., Pollock, D.W.: Gas transport in unsaturated zones: Multicomponent systems and the adequacy of Ficks Law. Water Resour. Res. **25**(3), 477–507 (1989)

15. Moldrup, P., Olesen, T., Gamst, J., Schjønning, P., Yamaguchi, T., Rolston, D.E.: Predicting the gas diffusion coefficient in repacked soil: water-induced linear reduction model. Soil Sci. Soc. Am. J. **64**(1), 1588–1594 (2000)

16. Reid, R.C., Prausnitz, J.M., Sherwood, T.K.: The Properties of Gases and Liquids. McGrawHill, New York (1977)

17. Perera, M.D.N., Hettiaratchi, J.P.A., Achari, G.: A mathematical modeling approach to improve the point estimation of landfill gas surface emissions using the flux chamber technique. J. Environ. Eng. Sci. **1**(1), 451–463 (2002)

18. Zhan, L.T., Qiu, Q.W., Xu, W.J., et al.: Field measurement of gas permeability of compacted loess used as an earthen final cover for a municipal solid waste landfill. J. Zhejiang Univ.-Sci. A **17**(7), 541–552 (2016)

Contaminated High-Plasticity Clay by Hydraulic Fracturing Fluids

Zhenning Yang[✉] and Carlton L. Ho

University of Massachusetts, Amherst, MA, USA
zhenning@umass.edu

Abstract. This study aims to examine the changes of geotechnical properties of high-plasticity clay by hydraulic fracturing fluid and to predict the contaminated fluid-clay behavior based on Hattab-Chang model. Similar to electrical double layer van der Waals forces, repulsive and attractive forces derived from energy potentials are used to describe soil behavior under different pore fluid concentrations. The designed contaminated samples are composed of remolded saturated high-plasticity clays and hydraulic fracturing fluids ranging from 0 to 100% industry-supplied concentration, designated as C_0, $C_{0.1}$, $C_{0.5}$, and C_1. The relationship between local model parameters and pore fluid concentration is obtained using the executed geotechnical experiments including Atterberg limits, direct shear, and one-dimensional load increment consolidation. The geotechnical experiments provide the results including soil consistency, hydraulic properties, and shear strength with respect to different pore fluid concentrations. The Hattab-Chang model is supplemented with a relationship between the surface potential characteristic value of $\widetilde{A_N}/\tilde{B}$ and different pore fluid concentrations.

Keywords: High-plasticity clay · Hydraulic fracturing fluid
Contaminated soil

1 Introduction

When considering the success of hydraulic fracturing, multiple issues remain regarding to environmental safety and engineering properties. Due to the contamination of produced water, geotechnical properties of the contaminated soil can change, which could have a detrimental impact on the soil mechanics or adjacent structures. Plenty of researchers studied the effect on geotechnical properties by leaked gas oil on soils. However, the contamination caused by of fracturing fluids has been omitted.

The studies on contaminated fine-grained soil has been many decades. The influence of petroleum-derived oil contamination on the geotechnical properties of different soils has been widely studied [1–9]. Among the factors influencing the engineering properties, the effect of electrostatic force, change in concentration of pore fluid, chemicals in pore fluids, aging effects, and dielectric constant could be found in previous studies [10–14]. However, to authors' knowledge, no previous investigation is done on contaminated high plasticity clays (CH) with industry-supplied fracturing fluid. The specific effect of contaminated fluid concentration is unknown.

© Springer Nature Singapore Pte Ltd. 2018
A. Farid and H. Chen (Eds.): GSIC 2018, *Proceedings of GeoShanghai 2018*
International Conference: Geoenvironment and Geohazard, pp. 567–574, 2018.
https://doi.org/10.1007/978-981-13-0128-5_62

This study aims to examine the changes of geotechnical properties of CH clay by fracturing fluid and to predict the contaminated fluid-clay behavior based on Hattab-Chang model, which consider clay clusters, inter-aggregate forces, and energy potential effects [15]. Like electrical double layer van der Waals forces, repulsive and attractive forces derived from energy potentials are used to describe soil behavior under different pore fluid concentrations. The designed samples are composed of remolded saturated high-plasticity clays mixed with hydraulic fracturing fluids ranging from 0 to 100% industry-supplied concentrations, designated as C_0, $C_{0.1}$, $C_{0.5}$, and C_1. The relationship between local model parameters and pore fluid concentrations is obtained using the executed experiments including Atterberg limits, direct shear, one-dimensional consolidation. The modelling will be simulated to investigate the relationship between characteristic value of A_N/B and different concentrations of pore fluid.

2   Materials and Methods

2.1   Sample Preparation

A Heiden clay sample was obtained from Norman, OK with a dark grey/black color. All the soil index properties could be found in previous study [18]. Produced water/fracturing fluid sample was provided from an industry site in Houston, TX. The industry-supplied fracturing fluid samples were of negligible viscosity and colored urine yellow. To prepare different concentrations of produced water, the stock solution concentration C_1 was diluted with Mili-Q water to 2 times and 10 times, designated as $C_{0.5}$ and $C_{0.1}$. C_0 was defined as the solution concentration of Mili-Q water. Pulverized clay samples were mixed thoroughly with different fracturing fluid samples at concentrations of C_0, $C_{0.1}$, $C_{0.5}$, and C_1 reaching their liquid limit and left in a moisture room for a week to allow to reach equilibrium.

2.2   Methods

To obtain a fundamental comprehension on the change of shear behavior of clayey soil contaminated by fracking fluid, a series of geotechnical laboratory tests were implemented including Atterberg limits, 1-D incremental consolidation test (IL), and direct shear tests (DS).

To further investigation of the proposition that clay can be regarded as an assembly of clusters or aggregates, the workflow by Hattab and Chang [15] is adopted to provide a reasonable method to consider inter-aggregate forces and energy potential effect on clay deformation [15]. Different concentrations of contamination in pore fluid are expected to have reasonable correlation with local parameters in the model.

It was assumed that the change of intra-cluster pores is negligible. In the saturated medium, the two clusters in the model have the radius of R, with the length connecting the two centers designated as l_c. Two types of potentials govern the neighboring clusters: repulsive type ψ_R, and attractive type ψ_A. The resulting potential ψ can be described as:

$$\psi = \psi_R + \psi_A \tag{1}$$

$$\psi_R = \tilde{B}Re^{-d_{min}^{-1}(l_c - 2R)} \tag{2}$$

$$\psi_A = -\tilde{A}\left[\frac{2R^2}{l_c^2 - 4R^2} + \frac{2R^2}{l_c^2} + ln\left(\frac{l_c^2 - 4R^2}{l_c^2}\right)\right] \tag{3}$$

where $\tilde{B}$ and $\tilde{A}$ are parameters related to surface potential of clusters depending on the mineralogy of the clay and the pore fluid chemistry; $2d_{min}$ is the minimum distance between two clusters in the model. The inter-cluster force f is introduced in Eq. (4) which represents the sum of the repulsive and attractive forces

$$f = -\tilde{B}Rd_{min}^{-1}e^{-d_{min}^{-1}(l_c - 2R)} + \tilde{A}R^2\left[\frac{l_c}{(l_c^2 - 4R^2)^2} + \frac{1}{l_c^3} - \frac{2}{l_c(l_c^2 - 4R^2)}\right] \tag{4}$$

where $\tilde{B}$ and $\tilde{A}$ can also be normalized as $\tilde{A}_N = \frac{\tilde{A}}{d_{min}}$; $R_N = \frac{R}{d_{min}}$; $l_N = \frac{l_c}{d_{min}}$. From the experimental data, the formula derived by Hattab and Chang between branch length l_c and void ratio e can be described by the following [15]:

$$l_c = 2R\sqrt[3]{(1+e)/(1+0.35)} \tag{5}$$

where assuming the closely packed assembly has a void ratio 0.35 as hexagonal packing. The coordination number of the assembly is assumed to be 12. The relation between force f, displacement l_c, and stress p' are then given as

$$f = \frac{p'\left(\frac{\pi}{3}\right)(2R)^3(1+e)}{l_c} \frac{}{12} \tag{6}$$

where the stress p' in the model was isotropic effective stress through experiments. In macro-scale, the experimental relation w-p' is presented in terms of water content w with respect to p'. In micro-scale, the calculated relation l_c-f can be expressed between branch length l_c and local force f between the two neighboring clusters. Then, the three parameters $\tilde{B}$, $\tilde{A}$, and d_{min} can be determined based on the three known points in the l_c-f curve.

3   Results and Discussion

3.1   Geotechnical Experiments

Atterberg limits are determined as the preliminary assessment of the soil's mechanical properties for different mixture samples (Fig. 1 (a)). The results indicated that the liquid limit (LL) and plastic limit (PL) decreased with increasing fracturing fluid concentration percent, which is the contrary to the effect exhibited by the role of gas oil. The decrease can be explained by the discussion of Arasan [16], who summarized the effect of chemicals on geotechnical properties of different types of clay [16]. For CH clays, the salt solution tends to reduce the thickness of the Diffuse Double Layer (DDL) and flocculate the CH clay particles, resulting in a decline of LL of CH clay. Figure 1 (b) – (d) present results of 4 series of IL consolidation tests for different mixtures of fracturing fluids. The consolidation curves are shown and compared either in the form of end of increment (EOI) or end of primary consolidation of (EOP) for each loading stage. Casagrande method was used to calculate the Coefficient of Consolidation (C_v) values. Results of C_v show a clear increase for contaminated CH samples ranging from $C_{0.1}$ to C_1. Yet, it also reveals that different contaminated concentrations show minor effect on variation of permeability properties of the contaminated CH soil. In general, the C_v values show a similar decreasing tendency for both the clean and contaminated CH clay as the consolidation stress increases. The Compression index (C_c) values were obtained from the virgin compression curves. The overall decrease trend shows similarity with the gas oil contamination study on Kaolinite [17].

Results of both undrained and drained DS tests are shown in Fig. 2. The quick DS tests were carried out at a consolidation period of 24 h prior to a rate of shear

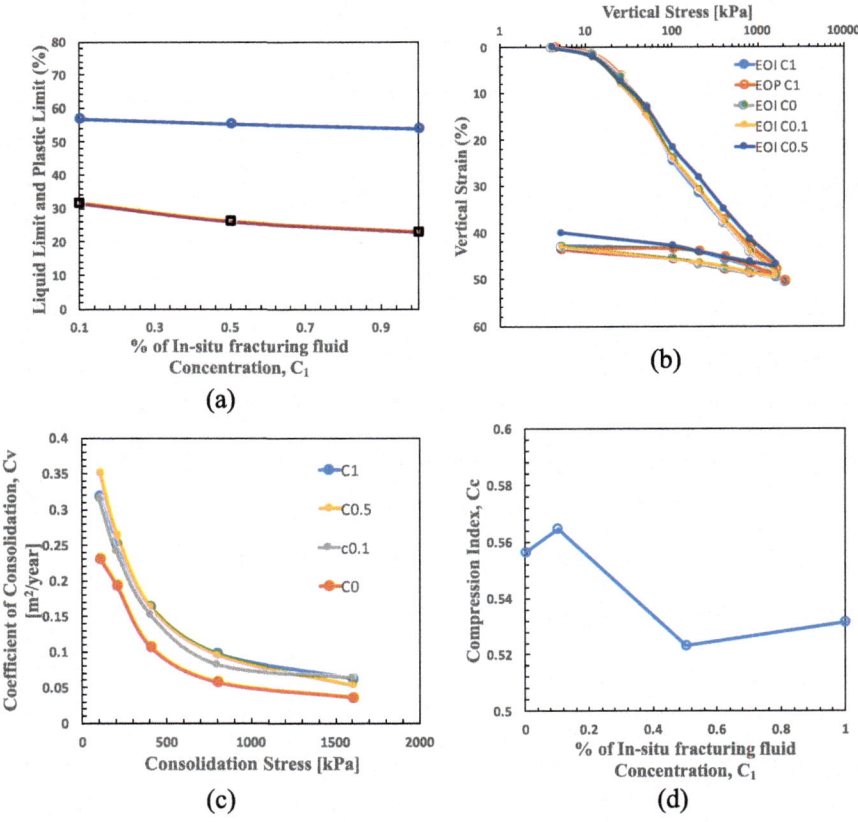

Fig. 1. (a) Atterberg Limits Results; (b) ε_v-p curve of IL consolidation test; (c) C_v-p curve of IL consolidation test; (d) C_c variation with respect to different fracturing fluid concentrations.

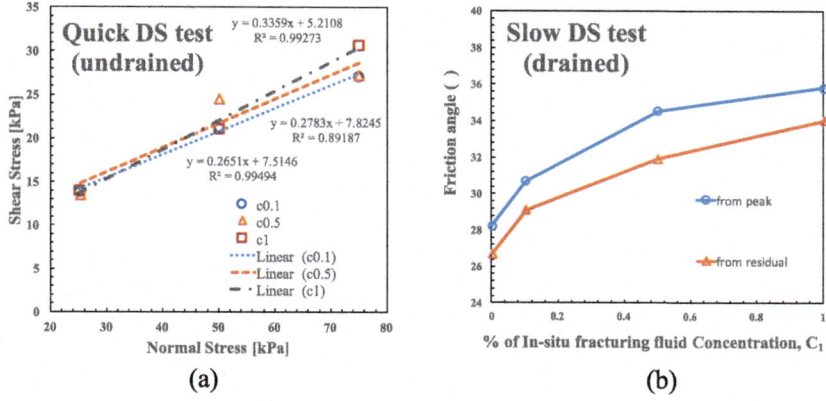

Fig. 2. (a) Quick DS test (undrained); (b) Slow DS test (drained).

deformation equal to 0.5 mm/min at normal stress of 25, 50, and 75 kPa. The slow DS tests were implemented at a consolidation period of 24 h following a shear deformation of 0.00694 mm/min (24 h). To verify the fully drained condition, a shear deformation of 0.00347 mm/min (48 h) was performed, which was proved to have a nearly identical curve. It can be seen from Fig. 2 (a) that the apparent cohesion is generated by building up the pore pressure when the specimen is sheared fast. The friction angle increases with increasing contaminated concentration of fracturing fluids, which is found to the opposite for contamination by gas oil [9]. For drained DS test results on remolded mixture samples (cohesion is 0), both friction angle from peak and residual value increase with increasing contaminated concentrations (Fig. 2 (b)).

3.2   Modelling and Prediction

As in the previous modelling for kaolinite or mixture with montmorillonite, a radius of 4 μm was assumed for single clay cluster in this modelling. Figure 3 (a) shows the macro-scale w-p' relationship for different fluid-soil mixtures ranging from C_0 to C_1. Due to the inconsistency and man-made uncertainty of preparing the specimen at initial condition, initial water content of sample with $C_{0.1}$ shows a bit higher than its LL. This is the main reason of the deviation of the curve from others. After the calculation using Eqs. (4), (5), and (6), micro-scale l_c-f as shown in Fig. 3 (b) could be obtained. Based on three known points on each curve, three characteristic parameters $\tilde{B}$, $\tilde{A}$, and d_{min} can be determined for each fluid-soil mixture. The obtained model parameters were plotted in Fig. 4 (a) as well as mixtures of kaolinite and montmorillonite from Chang's data [15]. The fluid-clay mixtures fell in the range between kaolinite and montmorillonite in the validity domain of the model. In addition, Fig. 4 (b) presents a relationship between characteristic values and different pore fluid concentrations.

The simulation curve can be obtained when three characteristic parameters $\tilde{B}$, $\tilde{A}$, and d_{min} were put into Eq. (4). The simulated local behavior l_c-f was found to be approximately identical to the experimental results. Even though the Eq. (4) has the

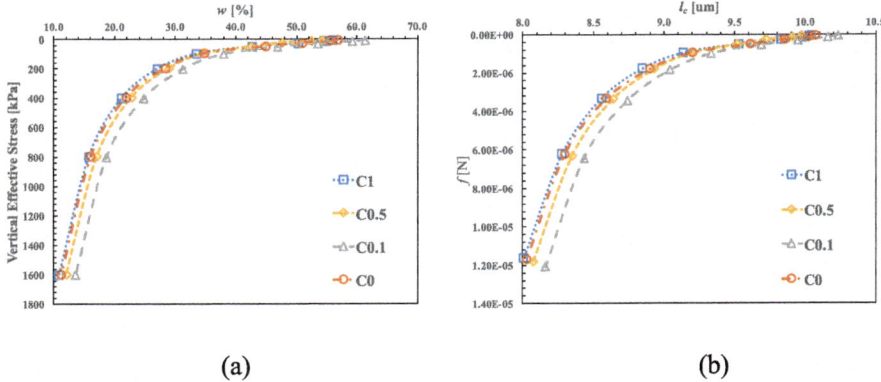

Fig. 3. (a) Experimental data from IL consolidation; (b) Calculated local parameters.

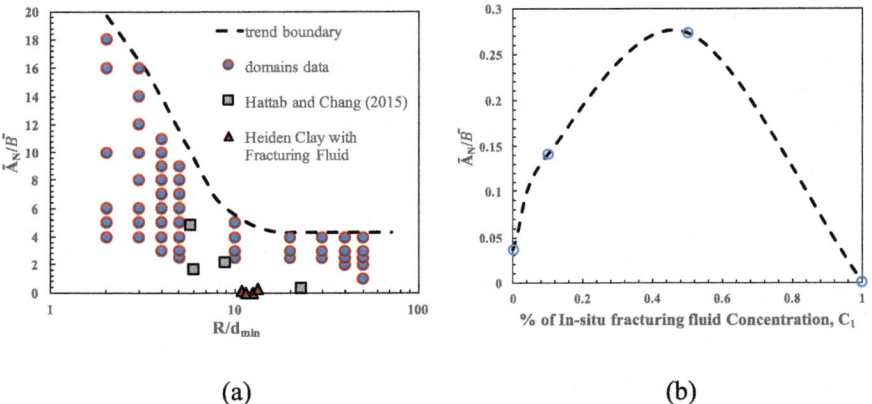

Fig. 4. (a) Model's valid domain for the Heiden Clay data and literature; (b) Effect of different concentrations of fracturing fluid on characteristic values of $\widetilde{A_N}/\tilde{B}$.

realistic physical significance to describe micro-scale behavior, a simple exponential equation is a better fit than Eq. (4). Their relationship cannot be verified until further investigation is accomplished.

4   Conclusions

The study presents a series of experimental investigations and modelling on the geotechnical properties of CH clay when contaminated with different concentrations of fracturing fluids. The following conclusions are made, based on the results and discussions:

(1) LL and PL decrease with increasing contaminated concentration for CH clay.
(2) The value of C_v increases to a nearly uniform range with different contaminated concentrations from $C_{0.1}$ to C_1.

(3) Shear strength of CH clay increases with fracturing fluid contamination.
(4) Neither positive nor negative correlation is found for the C_c of fracturing fluid contaminated CH clay. The compressibility of high-plasticity clay shows a decrease and an increase afterward as the pore fluid concentration increases. Unless further studies on the variation of C_c with pore fluid concentration are accomplished, a general conclusion cannot be given.
(5) DS test with a shear deformation of 0.00694 mm/min is adequate to ensure no pore pressure is generated after verification with 48 h of shearing. Performing DS test at a faster shear rate cannot provide accurate strength index for the contaminated mixture.
(6) The Hattab-Chang model is applicable for Heiden clay mixing with fracturing fluid. The effect of different pore fluid concentrations is provided with respect to the characteristic value of $\widetilde{A_N}/\widetilde{B}$.
(7) The assumed K_0 condition in this modelling is easier to fabricate in 1-D consolidation test compared to the triaxial isotropic loading method.

References

1. Al-Sanad, H.A., Eid, W.K., Ismael, N.F.: Geotechnical properties of oil-contaminated Kuwaiti sand. J. Geotech. Eng. **121**, 407–412 (1995)
2. Aiban, S.A.: The effect of temperature on the engineering properties of oil-contaminated sands. Environ. Int. **24**, 153–161 (1998). https://doi.org/10.1016/S0160-4120(97)00131-1
3. Puri, V.K.: Geotechnical aspects of oil-contaminated sands geotechnical aspects of oil-contaminated sands. Soil. Sediment. Contam. **9**, 359–374 (2000)
4. Shin, E.C., Das, B.M.: Bearing capacity of unsaturated oil-contaminated sand. Int. J. Offshore Polar Eng. **11**, 220–227 (2001)
5. Khamehchiyan, M., Hossein Charkhabi, A., Tajik, M.: Effects of crude oil contamination on geotechnical properties of clayey and sandy soils. Eng. Geol. **89**, 220–229 (2007). https://doi.org/10.1016/j.enggeo.2006.10.009
6. Rahman, Z.A., Hamzah, U., Taha, M.R., et al.: Influence of oil contamination on geotechnical properties of basaltic residual soil. Am. J. Appl. Sci. **7**, 954–961 (2010). https://doi.org/10.3844/ajassp.2010.954.961
7. Jia, Y.G., Wu, Q., Shang, H., et al.: The influence of oil contamination on the geotechnical properties of coastal sediments in the Yellow River Delta, China. Bull. Eng. Geol. Environ. **70**, 517–525 (2011). https://doi.org/10.1007/s10064-011-0349-8
8. Nazir, A.K.: Effect of motor oil contamination on geotechnical properties of over consolidated clay. Alexandria Eng. J. **50**, 331–335 (2011). https://doi.org/10.1016/j.aej.2011.05.002
9. Nasehi, S.A., Uromeihy, A., Nikudel, M.R., Morsali, A.: Influence of gas oil contamination on geotechnical properties of fine and coarse-grained soils. Geotech. Geol. Eng. **34**, 333–345 (2016). https://doi.org/10.1007/s10706-015-9948-7
10. Moore, C.A., Mitchell, J.K.: Electromagnetic forces and soil strength. Géotechnique **24**, 627–640 (1974)
11. Barbour, S.L., Yang, N.: A review of the influence of clay–brine interactions on the geotechnical properties of Ca-montmorillonitic clayey soils from western Canada. Can. Geotech. J. **30**, 920–934 (1993). https://doi.org/10.1139/t93-090

12. Meegoda, N.J., Ratnaweera, P.: Compressibility of contaminated fine-grained soils. Geotech. Test. J. **17**, 101–112 (1994). https://doi.org/10.1520/GTJ10078J
13. Al-Sanad, H.A., Ismael, N.F.: Aging effects on oil-contaminated Kuwaiti sand. J. Geotech. Geoenvironmental. Eng. **123**, 290–293 (1997)
14. Olchawa, A., Kumor, M.: Compressibility of organic soils polluted with diesel oil. Arch. Hydroengineering Environ. Mech. **54**, 299–307 (2007)
15. Hattab, M., Chang, C.S.: Interaggregate forces and energy potential effect on clay deformation. J. Eng. Mech. **141**, 4015014 (2015). https://doi.org/10.1061/(ASCE)EM.1943-7889.0000898
16. Arasan, S.: Effect of chemicals on geotechnical properties of clay liners: a review. Res. J. Appl. Sci. Eng. Technol. **2**, 765–775 (2010)
17. Khosravi, E., Ghasemzadeh, H., Sabour, M.R., Yazdani, H.: Geotechnical properties of gas oil-contaminated kaolinite. Eng. Geol. **166**, 11–16 (2013). https://doi.org/10.1016/j.enggeo.2013.08.004
18. Lutenegger, A.J., Rubin, A.: Tensile strength of some compacted fine-grained soils. Unsaturated Soils: Adv. Geo-Eng. **608** (2008). http://scholarworks.umass.edu/cee_faculty_pubs/608

Low-Temperature Leachability and Strength Properties of Contaminated Soil with High Moisture Content Stabilized by Novel Phosphate-Based Binder

Yasong Feng, Yanjun Du$^{(\boxtimes)}$, Weiwei Ren, and Weiyi Xia

Institute of Geotechnical Engineering, Southeast University,
Nanjing 210096, China
fengyasongys@126.com, duyanjun@seu.edu.cn

Abstract. Although cement-based materials are the most extensively used binders in the solidification/stabilization (S/S) technique, the substantial retardation in the cement hydration by high concentration of heavy metals and low temperature curing condition deteriorate the performances of stabilized soils. A novel binder, KMP, composed of oxalic acid-activated phosphate rock, monopotassium phosphate and reactive magnesia, shows lower leachability for heavy metal and higher strength for the stabilized soils with optimum moisture content. However, the study on low-temperature properties of contaminated soil with high moisture content stabilized by KMP is very limited. The present study validates the low-temperature effectiveness of KMP in S/S techniques by evaluating the strength and heavy metals leached properties of stabilized smelting industrial contaminated soil. A series of tests, including moisture content, dry density, soil pH, leachate electrical conductivity, toxicity characteristic leaching procedure and unconfined compressive strength test are undertaken. Portland cement is selected as a control binder for comparison purposes. The results show that the effect of low-temperature curing condition to the UCS of KMP stabilized soil specimens is lower relative to PC stabilized soil specimens. The KMP stabilized soil specimens with initial moisture content of 25% and cured at −10 °C can fulfill the requirement in USEPA Guidance (≥ 350 kPa). The effect of low-temperature curing condition to the heavy metals leached concentrations of KMP stabilized soil specimens are lower relative to PC stabilized soil specimens. The KMP stabilized soil specimens with initial moisture content of 35% and cured at −10 °C can fulfill the requirement in China MEP threshold limit. The UCS of stabilized soil decreases with increasing initial moisture content. The UCS of KMP stabilized soil is lower than the PC stabilized soil specimen with high initial moisture content of 35%, while the UCS of KMP stabilized soil is higher than the PC stabilized soil with initial moisture content of 25%.

Keywords: Solidification/stabilization · Low-temperature
High moisture content · Phosphate-based binder · Leachability
Strength

© Springer Nature Singapore Pte Ltd. 2018
A. Farid and H. Chen (Eds.): GSIC 2018, *Proceedings of GeoShanghai 2018 International Conference: Geoenvironment and Geohazard*, pp. 575–594, 2018.
https://doi.org/10.1007/978-981-13-0128-5_63

1   Introduction

With rapid urbanization, the heavy metals contamination is becoming more serious resulting from the toxic chemical emissions of anthropogenic activities [1, 2]. The heavy metals pose potential threat to human health and surrounding environment due to their highly persistence and unapparent in soil. Solidification/stabilization technique is widely used for treatment of soils contaminated with heavy metals [3, 4]. It involves the addition of cementitious binder to contaminated soils to cause physical encapsulation and chemical fixation of contaminants and enhance the strength of the soils.

Previous studies and applications of binders have focused on high-alkaline cementitious materials, such as Portland cement (PC) and blend of cement and other cementitious materials like fly ash and lime [4, 5]. It is reported that the production of PC is associated with intensive energy use and is reported to account for 5–8% of anthropogenic carbon emission [6]. In addition, the presence of certain heavy metal in the soils, such as Zn, has a significant negative effect on the hydration of the cement-based binders and strength development of the stabilized soils. Because the product of calcium zincate dihydrate ($CaZn_2(OH)_6 \cdot 2H_2O$) would wrap around the cement grains and hinder the subsequent reaction of binder with clay mineral and contaminants [4]. Furthermore, when exposed to external conditions, such as acid rain, freeze-thaw cycling and wet-dry cycling, heavy metals in the high-alkaline binders stabilized soils would leach easily [7–9]. Therefore, the utilization of PC is limited in above special conditions. Under the drive of searching more sustainable and endurable binders, some low-cost phosphorous-based materials are drawing researchers' attention recently [10]. A novel binder, KMP, consists of oxalic acid-activated phosphate rock (APR), monopotassium phosphate (KH_2PO_4) and reactive magnesia (MgO) in 1:1:2 proportion by dry weight basis [9–11]. Extensive investigations show that the KMP stabilized soil has many advantages compared to PC stabilized soils, such as high immobilization efficiency for Zn and Pb, high strength and relatively lower pH of stabilized soil [10], and better durability in external conditions [9, 11]). Hence, KMP is a promising alternative to cement-based binders due to its economic and environmental merits.

In addition, the cold regions are distributed widely in the Earth, and the area of cold regions in China is about 4.17 million sq. km, accounts for nearly 43.5% of the country's land area [12]. A typical western city of Baiyin is taken as an example, according to the climate monitoring data recorded by the local meteorological department, the daily mean temperature and the minimum daily mean temperature in winter (December, January and February) are 5 °C and −10 °C, respectively. The engineering construction in cold regions will definitely be affected by the low temperature [13, 14], and the in-situ solidification/stabilization practices for heavy metal contaminated soils remediation are no exception. Temperature affects the mechanics properties of cement-based materials through its effect on the hydration of cement and its associated microstructural development [15–18]. In addition, the low temperature properties of the traditional magnesium phosphate cement (MPC) has been investigated by many researchers [19, 20], and the results shows that MPC can set and harden at temperatures as low as −20 °C due to the rapid formation of chemical bonding by violent acid-base reaction between magnesia and phosphate [21]. However, to date, studies on the impacts of low temperature on the leaching properties of heavy metals from PC or KMP stabilized soil are very limited. Moreover, if the contaminated soils are in high natural moisture content (for example, drainage ditch sediment), it is time-consuming for the contaminated soils to be air-dried and to achieve the optimum moisture content, especially when the remediation is conducted in winter. Hence, it is necessary to investigate the low-temperature properties of stabilized soils with high moisture content.

The objectives of this study are to (1) investigate the influence of low-temperature to the binder hydration and further reactions in the binder-contaminant-soil system; (2) compare the strength and heavy metal leaching properties of KMP and PC stabilized soils with high moisture content (more than liquid limit). The extent of reactions in the binder-contaminant-soil system is evaluated by moisture content and dry density of the stabilized soils. The strength and leaching performance are evaluated by unconfined compression strength (UCS) and toxic characteristics leaching procedures (TCLP) test. Furthermore, the soil pH test is conducted as an evaluation parameter for the land reuse.

2   Materials and Methods

2.1   Soils and Binders

A large number of lead-zinc smelters were founded in Baiyin City of Gansu Province since the 1970s, and the local land is heavily polluted by the untreated smelting sewage disposal, especially, in the upper reaches of Dongdagou ditch stream. The contaminated soil used in this study were screened and collected from this contaminated site (36°33′ N, 104°12′ E). Based on previous site investigation, twenty typical soil samples were collected within a depth of 0.5 m below the surface using a plastic grab, stored in polyethylene bags and kept in a seal box for return to the laboratory. Soil samples were then air dried, mixed thoroughly, ground, passed through a 10-mesh sieve with opening size of 2 mm and stored in seal plastic containers until analysis.

The basic physicochemical properties of the contaminated soil are shown in Table 1. The Atterberg limits are measured as per ASTM D4318. The grain size distribution is evaluated using a laser particle size analyzer (Mastersizer 2000). The optimum water content and maximum dry density of the soil are determined using electric compaction device as per ASTM D698, and the compaction curves shown in Fig. 1. Soil pH is measured at a 1:1 (w/w) soil: water suspension using a glass pH meter (HORIBA D-54) as per ASTM D4972. The electrical conductivity test is performed according to the procedures presented by Cai et al. [22]. The X-ray fluorescence (XRF) analysis of the original soil is performed by using an ARL 9800 XP+XRF spectrometry, and the result listed in Table 2. Soil total heavy metal concentration is measured by digesting soil with HNO_3-$HClO_4$-HF, followed by the concentration analysis using ICP-AES. It can be found that the Pb, Zn and Cd concentration are 9710 mg/kg, 17300 mg/kg and 2425 mg/kg, respectively. The leachability of heavy metals from the original soils is determined using the toxicity characteristic leaching procedure (TCLP) in accordance with the USEPA Method 1311, and the leached concentration of Pb, Zn and Cd are 17.96 mg/L, 1593.29 mg/L and 27.73 mg/L, respectively.

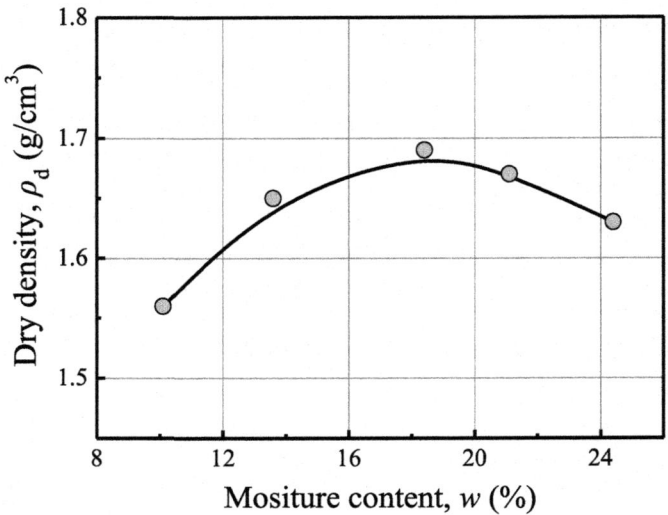

Fig. 1. Compaction curve for contaminated soil.

Table 1. Physicochemical properties of soil.

Property	Value
Natural moisture content, w_n (%)	25.0–35.0
Optimum moisture content, w_{opt} (%)	18.4
Maximum dry density, ρ_d (g/cm^3)	1.69
Plastic limit, w_p (%)	33.3
Liquid limit, w_L (%)	17.2
Plasticity index, I_p (%)	16.1
Specific gravity, G_s	2.73
pH	5.67
Electrical conductivity, EC (ms/cm)	6.21
Grain size distribution (%)	
Clay (<0.005 mm)	6.54
Silt (0.005–0.075 mm)	43.95
Sand (0.075–2 mm)	49.51

The KMP powder is a mixture of the APR, KH$_2$PO$_4$ and reactive MgO in 1:1:2 proportion by dry weight, as this ratio is found to yield relatively low leachability and high unconfined compressive strength for the stabilized soils in previous studies [10]. The chemical analytical reagent, KH$_2$PO$_4$, is obtained from Sinopharm Chemical

Reagent Co. Ltd., China, and industrial reagent reactive MgO is obtained from Fange Chemical Co. Ltd., China. Both the KH_2PO_4 and MgO are used without further treatment. Commercial phosphate rock, obtained from Rongguan Chemical Reagent Co. Ltd., China, is crushed and ground to pass through a 100-mesh sieve with opening size of 0.075 mm. The APR powder is prepared by the oxalic acid method as detailed in Du et al. [10]. The chemical analytical sodium borate ($Na_2B_4O_7 \cdot 10H_2O$) obtained from Sinopharm Chemical Reagent Co. Ltd. is used as retarder. The Portland cement produced by the Nanjing Hailuo Cement Co. Ltd. is used as a conventional binder. The chemical compositions of the APR and PC are measured using an X-ray fluorescence spectrometer (ARL 9800XP + XRF) and are shown in Table 1. The reactivity of the MgO is measured as 90 to 100 s using the acetic acid testing method proposed by Shand [23], and the reactive MgO used in this study is classified as of moderate reactivity [24]. All of these tests are conducted in replicate and the mean values are calculated. The coefficients of variation (COV) of the results are less than 3%, indicating good repeatability of the test results.

Table 2. Oxide compositions of APR, PC, weak-alkaline cement and original soil.

Oxide chemistry	APR	PC	Original soil
Calcium oxide (CaO)	45.93	44.7	2.33
Silicon oxide (SiO_2)	6.14	27.4	58.21
Aluminum oxide (Al_2O_3)	1.23	13.1	12.79
Phosphorus oxide (P_2O_5)	25.10	0.13	–
Sulphate oxide (SO_3)	–	3.96	0.56
Ferric oxide (Fe_2O_3)	–	3.34	6.78
Magnesium oxide (MgO)	–	1.19	2.65
Potassium oxide (K_2O)	–	1.14	2.21
Fluorine (F)	2.35	–	–
Loss on ignition (950 °C)	13.12	2.53	9.01

2.2   Sample Preparation

According to the climate monitoring data recorded by the local meteorological department of Baiyin city, the daily mean temperature and the minimum daily mean temperature in winter (December, January and February) are 5 °C and −10 °C, respectively. Hence, the curing temperature of 5 °C and −10 °C are determined in this study. The standard curing temperature of 20 °C is also used as a temperature control. To simulate an actual environment, the binders and contaminated soil are mixed at room temperature, and then the stabilized soils are cured at low temperature condition of −10 °C and 5 °C to keep consistent with the low temperature at night in field environment.

Preliminary results suggest that the treatment design of a dosage of 6% (on dry weight of soil basis) and a moisture content of 25% (on dry weight of soil basis) shows relatively low leachability of heavy metals and high unconfined compressive strength of the stabilized soils with moisture content of 20–35%. Hence, this design is used as a control to investigate the effects of curing temperature on the leachability and strength properties of the stabilized soils. In addition, the maximum nature moisture content of the field contaminated soil is approximate 35%, and it is slightly higher than the w_L of the contaminated soil (33.3%), as shown in the Table 1. The moisture content of 35% is defined as high moisture content and further selected as target moisture content in this study. However, the dosage of 6% shows an unsatisfactory strength performance for the contaminated soil with high moisture content of 35% in the preliminary study. Therefore, an increased dosage of 12% is determined for the soil with moisture content of 35% in the subsequent test.

The binders with predetermined weights (6% or 12%, on dry weight of soil basis) are poured into the contaminated soil with initial moisture content (IMC) of 25% or 35%, respectively. The mixtures are immediately introduced into an electronic mixer, and the mixtures are agitated at a low speed for approximate 1 min and then at a high speed for approximate 4 min to achieve homogeneity. Then, different sampling methods are used for the stabilized soil with moisture content of 25% (in solid-state) and with moisture content of 35% (in slurry-state). A certain quantity of the solid-state mixture is compacted into a 50 mm diameter by 100 mm height stainless steel cylindrical mold under the conditions of maximum dry density (1.69 g/cm^3). The specimens are then carefully extruded from the mold using a hydraulic jack, sealed in black polyethylene bags and cured for the predetermined 7 or 28 days under controlled ambient conditions (temperature of -10 °C, 5 °C or 20 °C and relative humidity of 95%). Later the slurry-state mixtures are also poured into a 50 mm diameter by 100 mm height PVC cylindrical mold. Due care is taken to remove trapped air bubbles from the mixture during the placement into the molds. The molds are wrapped with polythene caps to minimize moisture loss, and the specimens are cured under the same condition to solid-state specimens. For comparison, the original soil is also prepared and cured under the same conditions. The time consumed for a sample to be prepared is approximate 15 min. Furthermore, at the end of the curing period the specimens cured at temperature of -10 °C are transferred to the standard curing condition (20 °C and relative humidity of 95%) to thaw for 1 day before subsequent tests. The specimens cured in plastic bags or sealed in PVC cylindrical mold, remained sealed during the 1-day equilibration period to prevent moisture loss. Soil samples are designated as PCi (k d) or KMPi (k d) to denote specimens with binder PC or KMP dosage of i% and curing k days, respectively. Blank specimens denote the contaminated soil samples without any binder addition.

2.3   Testing Methods

The unconfined compression strength (UCS) test is conducted as per ASTM D4219. A certain quantity of fragments from the broken UCS test specimens is immediately collected for the subsequent tests. The dry density test is conducted as per ASTM D7263. The moisture content test is conducted as per ASTM D2216. What needs to illustrate is that the stabilized soil is soaked in anhydrous ethanol for 24 h to cease the further hydration of binders before heated in oven. The stabilized soil pH is conducted as per ASTM D4972. Some other soaked fragments from the broken UCS test specimens are air-dried, crushed, and passed through a 2 mm sieve to ensure the homogeneity of the samples. The pH values are obtained by placing the portable HORIBA D-54 pH meter into the supernatant. The electrical conductivity (*EC*) measurements, which are simple, inexpensive, and rapid, can be used as an index of the chemical composition of electrolyte solutions [25]. In this study, the *EC* data are used as indirectly criterion for mobility of heavy metals in the stabilized soils. The *EC* test is performed according to the procedures presented by Cai et al. [22]. The *EC* values of the supernatant are measured by employing the portable DDS-22C conductivity meter into the supernatant. The leachability of heavy metals from the stabilized soils is determined using the toxicity characteristic leaching procedure (TCLP) in accordance with the USEPA Method 1311. Fragments from the broken UCS test specimens are air-dried, crushed, and passed through a 9.5 mm sieve. Measurements are made in triplicate for the concentrations of Pb, Zn and Cd, and the average values are calculated. The results show that the values of the standard deviation are less than 5% for the triplicates, demonstrating the reproducibility of the test results. Triplicate specimens are tested for UCS, w, ρ_d, pH, *EC* and TCLP and the average of these values are calculated, presented and discussed in this study.

3   Results and Analysis

3.1   Moisture Content and Dry Density

Figure 2 shows the moisture content variations with different curing temperature and IMC. The IMC for Fig. 2(a) and (b) are 25% and 35%, respectively. It can be observed that the moisture content of the stabilized soil drops remarkably with increasing curing time and curing temperature. In this study, the IMC and binder dosage are almost same for each kinds of soil specimens, so the higher moisture content of stabilized soil with same curing time reflects that a lesser amount of water is involved in the reactions in the binder-contaminant-soil system. In other words, retardation degrees of low temperature to the binder hydration and further reactions are greater. The lower reduction of moisture content at lower curing temperature can mainly be attributed to that the lower curing temperature has hindered hydration of binders and further reactions in the binder-contaminant-soil system in greater extent [15, 16, 19, 20]. It can be observed from Fig. 2(a) that the effect of the curing temperature (−10 °C) to the PC stabilized soil is more significant compared to KMP stabilized soil, and it reveals that KMP is very suitable for application to the low temperature remediation because it doesn't need any heat-curing or use of antifreezing admixture.

As shown in Fig. 2(a), it is evident that smaller change of moisture content is observed for PC stabilized soil, as compared to KMP stabilized soil with the same IMC of 25% and same curing time, and the difference is mainly determined by the chemical properties of binders. However, the changes of the moisture contents for the stabilized soil specimens with IMC of 35% are different at different curing temperature, as shown in Fig. 2(b). They are in the order of KMP > PC, PC > KMP and PC > KMP at the curing temperature of −10 °C, 5 °C and 20 °C, respectively. This suggests that the IMC and binder dosage, as well as curing temperature, are important factors influencing the reactions of the binder-contaminant-soil system. It can be seen that PC

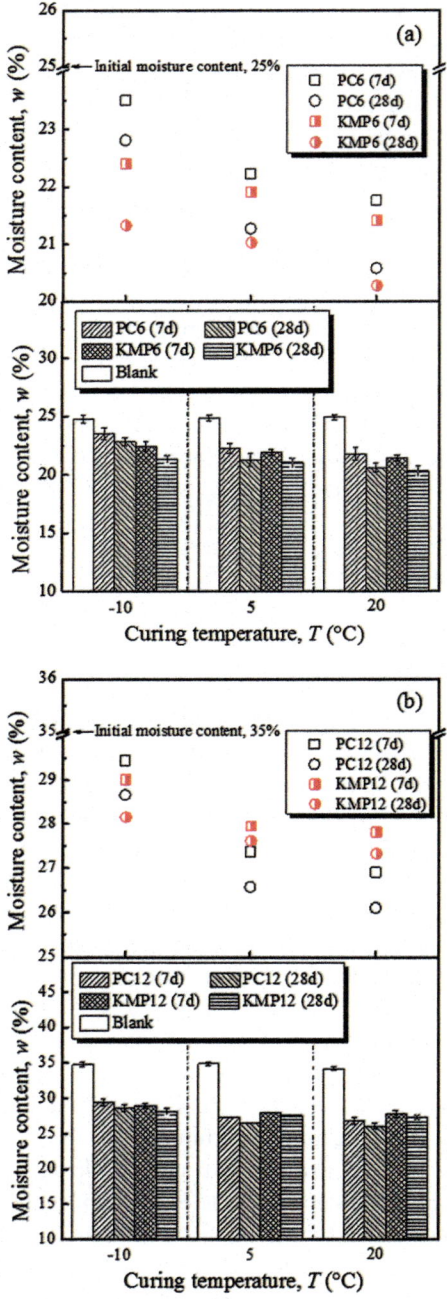

Fig. 2. Moisture content measured immediately after UCT of specimens with different curing temperature and IMC: (a) IMC is 25%; and (b) IMC is 35% (COV < 1.6%).

consume more water in the stabilized soil specimen with higher moisture content, and it is more notable at the curing temperature of 5 °C and 20 °C, as compared to PC stabilized soil. The possible reason for the result is that retardation degree of PC hydration is slightly alleviated by a lower heavy metals concentration in the pore water of stabilized soil specimen with higher moisture content and binder dosage.

Figure 3 depicts the moisture content variations of soil specimens with different curing temperature and IMC. The IMC for Fig. 3(a) and (b) are 25% and 35%, respectively. It can be observed that dry density of the stabilized soil rises remarkably with increasing curing temperature and curing time, as compared to the blank specimen (i.e., no binder). And the result obtained from dry density test is consistent with the moisture content test. The dry density of the blank soil specimen (1.59 g/cm^3) cured at

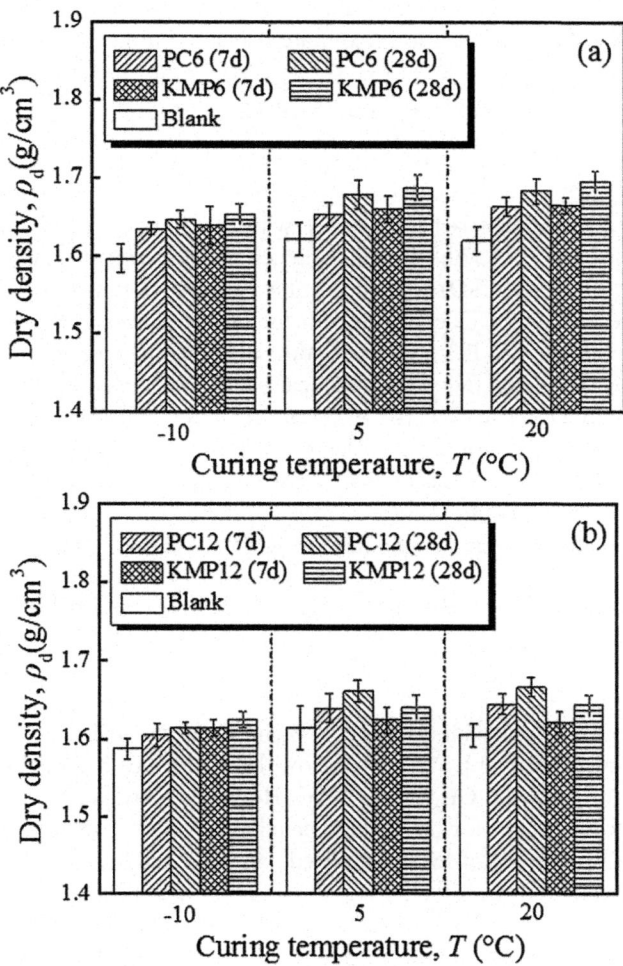

Fig. 3. Effects of curing temperature and IMC on the dry density: (a) IMC is 25%; and (b) IMC is 35% (COV < 2.5%).

temperature of $-10\,°C$ is slightly lower than other blank soil specimens (1.61–1.62 g/cm^3) due to frost heaving occurs in the soil specimen cured at temperature of $-10\,°C$. Figure 3(a) also illustrates that the dry density of the KMP stabilized soil is greater than the PC stabilized soil specimens with IMC of 25%, therefore the KMP stabilized soil can develops a denser internal structure, compared to PC stabilized soil.

Furthermore, in spite of the binder dosage increases to 12%, the dry density of the soil specimens slightly reduces when the IMC is 35%, as shown in Fig. 3(b). The dry density of the KMP stabilized soil is greater than the PC stabilized soil at $-10\,°C$. The difference indicates that the amorphous hydration products of KMP (i.e., $MgKPO_4$, $Mg(OH)_2$) are more than that of PC (i.e., CSH) in quantitative terms, thus KMP displays good cold-resistant properties when applied in the cold regions or cold seasons.

The effects of low temperature and high moisture content on the aforementioned soil pH and electrical conductivity of leachate as well as other environmental and mechanical properties including leachability of the heavy metals and unconfined compressive strength of the stabilized soil, are discussed in the following sections.

3.2   Soil PH and Electrical Conductivity of Leachate

Figure 4 presents the measured pH soil specimens with different curing temperature and IMC. The IMC for Fig. 4(a) and (b) are 25% and 35%, respectively. It can be observed that soil pH increases remarkably with the addition of binders due to the plenty product of OH^- after the hydration reaction of a certain quantity of alkaline substances (i.e., MgO and CaO) contained in the binders. The main reasons for the pH of the stabilized soil slightly decreases with the curing time are (a) the evolution of pozzolanic reaction in the PC stabilized soil, wherein the consumption of OH^- is needed, as expressed by the following chemical Eq. (1), the result is consistent with the previous results of Du et al. [4]; (b) A certain quantity of OH^- is consumed in the process of hydroxide precipitation of heavy metal, as expressed by the following chemical Eq. (2). The pH of the PC stabilized soil rises more significantly with the decrease of curing temperature, compared to KMP stabilized soil. This further indicates that the negative effect of low temperature to KMP hydration is more slightly.

$$m\mathrm{Ca(OH)}_2 + \mathrm{SiO}_2 + n\mathrm{H}_2\mathrm{O} \rightarrow m\mathrm{CaO} \cdot \mathrm{SiO}_2 \cdot (m+n)\mathrm{H}_2\mathrm{O} \tag{1}$$

$$\mathrm{M}^{2+} + 2\mathrm{OH}^- \rightarrow \mathrm{M(OH)}_2 \tag{2}$$

The PC stabilized soils with dosage of 6% and 12% have pH values of 9.94–10.89 and 10.77–11.69, respectively. Those are below the established hazardous classification threshold limit of 12.5 for GB 5085.1 [26]. However, they are beyond the stricter recommended value of 6-9 for the contaminated site remediation, i.e. the standard of DB 43T 1165 [27]. The pH of KMP stabilized soil is 8.13 to 8.91 at the curing time of 28 days, which can well meet the stricter remediation target.

Figure 5 depicts the EC of leachate with different curing temperature and IMC. The IMC for Fig. 5(a) and (b) are 25% and 35%, respectively. It can be observed that EC of the stabilized soil reduces remarkably with increasing curing temperature and curing time, as compared to the blank specimen. The decreased EC of the leachate

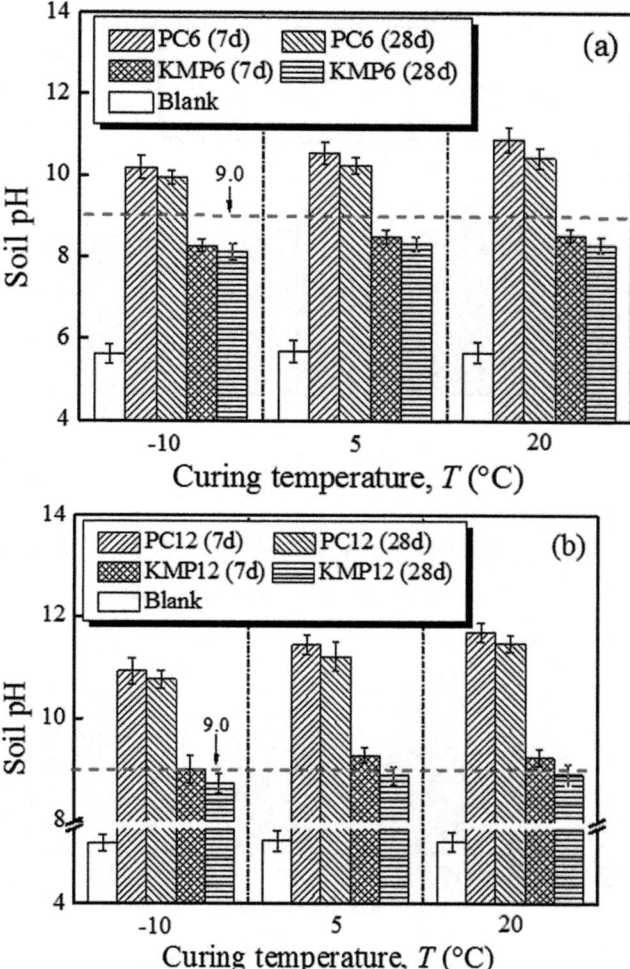

Fig. 4. Measured soil pH for different curing time and curing temperature: (a) IMC is 25%; and (b) IMC is 35% (COV < 2.5%).

suggests that a quality of water-soluble ions (containing metal cations, i.e., Pb^{2+}, Zn^{2+}, Cd^{2+}, Ca^{2+}, Mg^{2+}, K^+ and anions, i.e., PO_4^{3+}, OH^-) are involved in the reactions of the binder-contaminant-soil system, as shown by the chemical Eqs. (1), (2) and (3) for series of reactions.

$$3M^{2+} 2PO_4^{3+} \rightarrow M_3(PO_4)_2 \qquad (3)$$

Figure 5 also illustrates that the influence of curing temperature on the leachate EC of KMP stabilized soil is less than PC stabilized soil, and it further suggests that the KMP can well complete hydration even at a low temperature of $-10\ ^\circ C$, and the low curing temperature has much less interference with the hydration of KMP than PC.

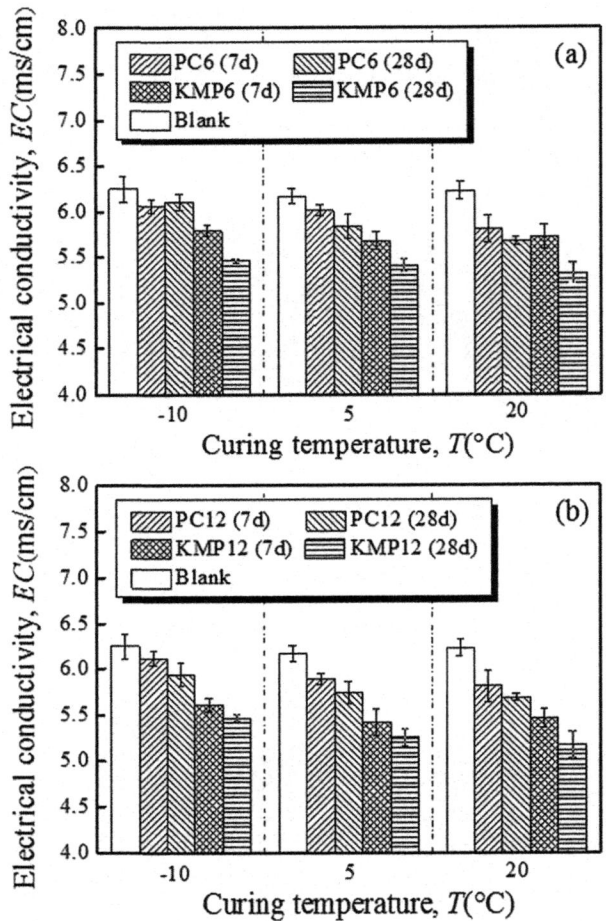

Fig. 5. Electrical conductivity of leachate with different curing temperature and IMC: (a) IMC is 25%; and (b) IMC is 35% (COV < 2.1%).

3.3   Leachability of the Heavy Metals

The leached Pb, Zn and Cd concentrations from original and stabilized soils with different IMC and curing temperature are compared in Figs. 6, 7 and 8. In all cases, the leached concentrations of Pb, Zn and Cd from the original soils obviously exceed the China MEP threshold limit of GB 5085.3 [28], emphasizing the need for remediation. The stabilized soils exhibit significantly lower concentrations of leached heavy metals, regardless of the curing time and curing temperature. It can be found that the effect of curing temperature is more significant on PC treated soil, compared to KMP stabilized soil. In addition, the leached Pb concentration of the KMP stabilized soil is much lower relative to the PC stabilized soil, as shown in the Fig. 6(a). It is evident from Fig. 6(a) that for all the curing temperature, the heavy metals leached concentrations of KMP stabilized soil can meet the China MEP threshold limit of GB 5085.3 at the curing time of 7 days. However, it needs 28 days for the PC stabilized soil to meet the same limit at the curing temperature of 20 °C, and it fails to meet the China MEP threshold limit for GB 5085.3 for the soil specimens cured at the temperature of −10 °C and 5 °C. As shown in the Fig. 6(b), heavy metals leached concentrations of the stabilized soils decreases with the increasing binder dosage and IMC, however, the leached concentrations of PC stabilized soil still cannot meet the China MEP threshold limit of GB 5085.3, except the soil specimen cured at the temperature of 20 °C.

The basic variation trends in the leachated concentration of Zn and Cd are similar to that of Pb, as shown in the Figs. 7 and 8. However, a minute difference is that the

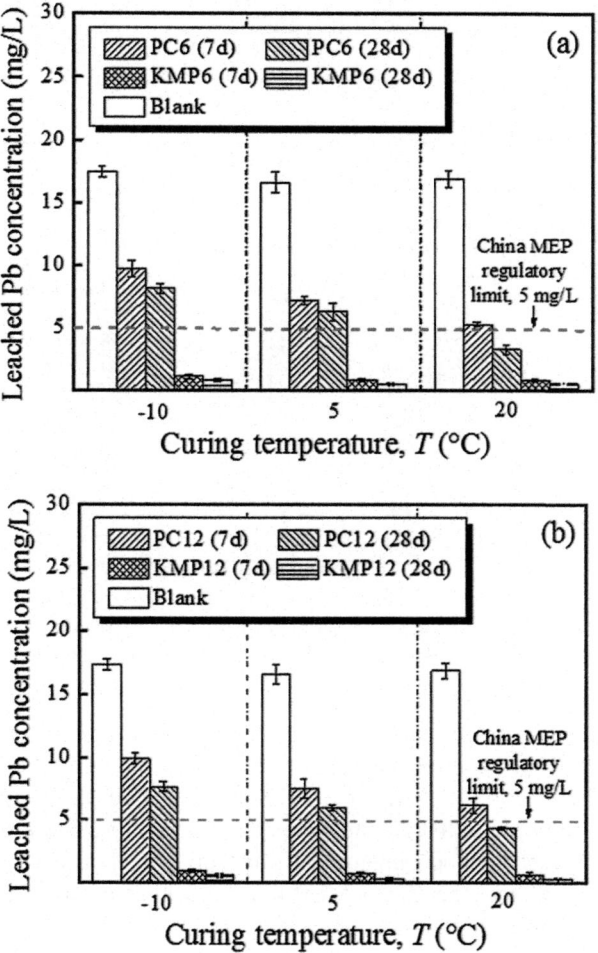

Fig. 6. Variations in the leachated Pb concentration of soil specimens with different curing temperature and IMC: (a) IMC is 25%; and (b) IMC is 35% (COV < 3.2%).

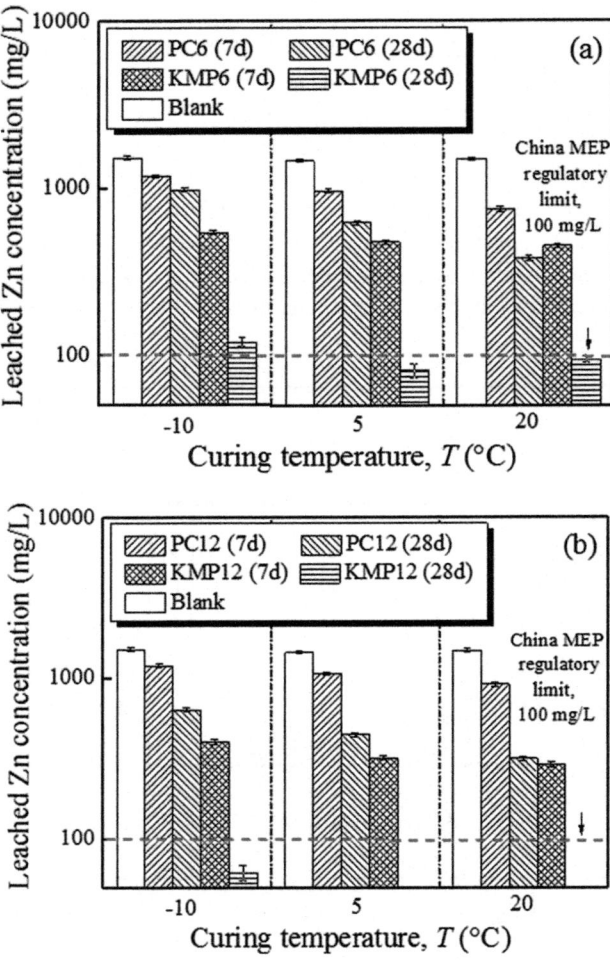

Fig. 7. Variations in the leachated Zn concentration of soil specimens with different curing temperature and IMC: (a) IMC is 25%; and (b) IMC is 35% (COV < 2.8%).

leachated concentration of Zn and Cd for all KMP and PC stabilized soil specimens with a low IMC (25%) cannot meet the China MEP threshold limit of GB 5085.3 under the curing temperature of −10 °C for cured 28 days. Overall, the results presented in Figs. 6, 7 and 8 demonstrate that the heavy metals leached concentrations of the KMP stabilized contaminated soil specimens, with low or high moisture content (25% or 35%), can both well meet the China MEP threshold limit when cured under a low temperature condition (−10 °C).

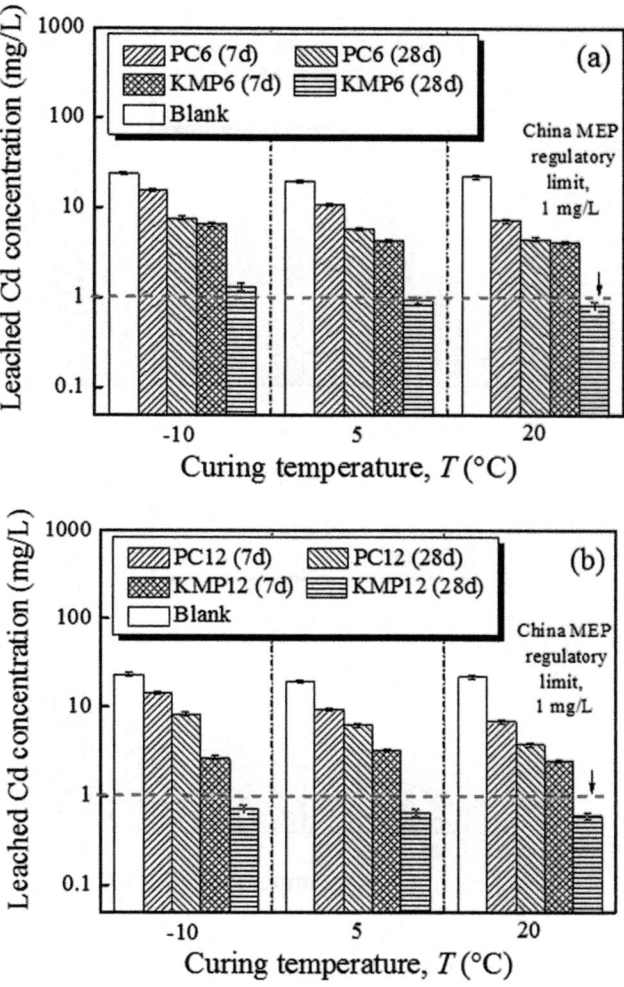

Fig. 8. Variations in the leachated Pb concentration of soil specimens with different curing temperature and IMC: (a) IMC is 25%; and (b) IMC is 35% (COV < 4.0%).

3.4 Unconfined Compressive Strength

Based on the measured UCS of the soil specimens with different curing temperature and IMC (Fig. 9), it can be seen that, similar to dry density, the UCS reduces with the decreasing curing temperature. The effect of low temperature cured condition to the UCS of PC stabilized soil is higher than KMP treated soil. For example, as presented in Fig. 9, the UCS of KMP stabilized soil cured at −10 °C is 0.78 times to that cured at 20 °C, while UCS of PC stabilized soil cured at −10 °C is only 0.48 times to that cured at 20 °C. USEPA suggested an UCS of 350 kPa at 28 days for materials to be disposed of to landfill [29]. As can be seen from Fig. 2(a), all the KMP stabilized soil specimens

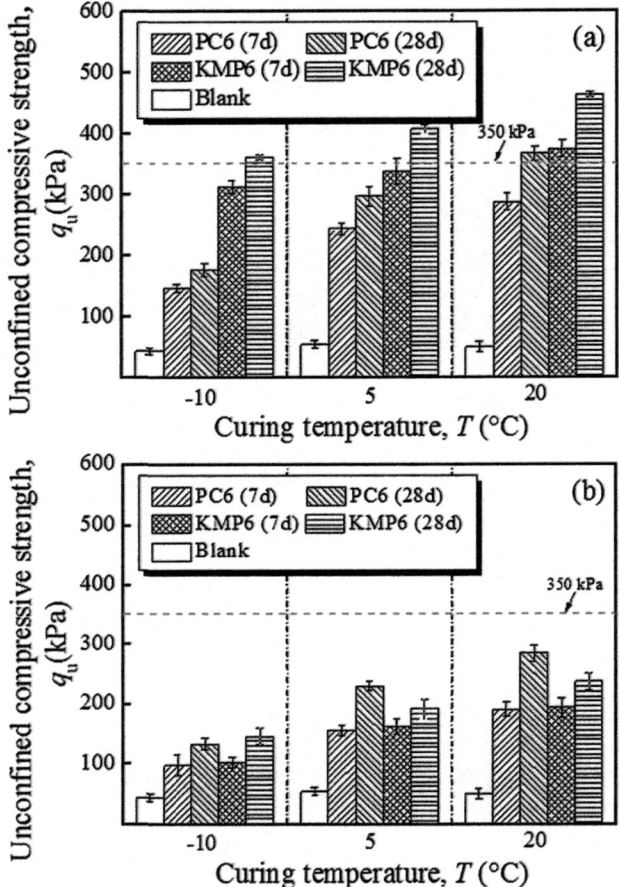

Fig. 9. Unconfined compressive strength with concentration of soil specimens with different curing temperature and IMC: (a) IMC is 25%; and (b) IMC is 35% (COV < 3.5%).

can fulfill the requirement in USEPA Guidance [29], while only the PC stabilized soil specimens at 20 °C fulfill the requirement.

By comparing the UCS of stabilized soils with different IMC (Fig. 9(a) and (b)), it is obvious that the soil specimen with increasing IMC of 35% shows a remarkable reduction in UCS relative to the IMC of 25%, in spite of the binder dosage increases to 12%, which can be attributed to more pores being formed inside the stabilized soil as a result of the larger moisture content. These results demonstrate that the synergistic effects of low temperature and high moisture content can adversely affect the UCS. As shown in the Fig. 9, the UCS of the soil specimens stabilized by KMP is greater than that by PC when the IMC is 25%. However, when the IMC is 35%, the UCS is higher for the PC stabilized soil specimens cured at 5 °C and 20 °C; meanwhile, the UCS of the KMP stabilized soil specimens is slightly greater than the PC stabilized soil specimens at −10 °C. In addition, no stabilized soil specimens can fulfill the requirement in USEPA Guidance [29] when the IMC of the soil specimens is 35%. Therefore, other necessary measures should be taken to dispose of the high moisture content soil if there is a special strength requirement (e.g., higher than 350 kPa as the USEPA Guidance suggested).

3.5 Discussion

From the results obtained from moisture content and UCS test (Figs. 2 and 9), it can be found that the PC stabilized soil shows poor hydration property and further reactions in the binder-contaminant-soil system at low temperature, and the similar behavior of PC at low temperature has been observed by earlier researcher in previous studies [15–18]. These results can be attributed to that low temperatures inhibit the dissolution of the anhydrous clinker phases and the PC hydration process [30]. Consequently, the amount of hydration products (e.g., calcium silicate hydrate (C-S-H), portlandite ($Ca(OH)_2$), calcite ($CaCO_3$)) decreases with temperature. This is adverse to the strength gain of stabilized soil since C-S-H is considered to be the major binding phase in hardened cement [31–33]. After the mixture of PC and contaminated soil, when cured at 5 °C (corresponding to the daylight temperature when practical construction), the hydration of PC occurs slowly. However, the hydration of PC nearly terminates when cured in a low temperature condition of −10 °C (corresponding to the low temperature at night in field environment). Furthermore, the high concentrations of heavy metals (e.g., Zn, Pb) in the pore water hinder hydration of PC to a far greater extent [4, 34, 35]. The coupled effects of temperature and heavy metals on the strength development of cemented contaminated soil are obviously demonstrated in the heavy metals leached results (Figs. 6, 7 and 8). The immobilization efficiency of PC to heavy metals is low with the decreased amount of hydration products.

Compared with PC, the setting time of the phosphate cement is quite short [20], which indicates that the hydration reaction rate of the phosphate cement is relatively higher than PC. When KMP are mixed with contaminated soil, the phosphate is dissolved instantaneously in the pore water to form an acidic aqueous medium, and the process is rapid. Previous study has reported that phosphate cement can be set and hardened within 1 h even though the temperature is −10 °C [19], indicating that the phosphate can sufficiently dissolve at low temperature and it is verified by the leached results (Figs. 6, 7 and 8) in our present study. Because the predominant immobilization mechanism of heavy metals with KMP is to form insoluble bivalent metals phosphate products (e.g., $Zn_3(PO_4)_2 \cdot 2H_2O$, $CaZn_2(PO_4)_2 \cdot 2H_2O$, $Pb_5(PO_4)_3F$) for the Pb and Zn contaminated clay [10], and the heavy metals will precipitate when the phosphate anion (PO_4^{3+}) abounds in the pore water even though at low temperature.

In the other hand, the dissolution of basic magnesia leads to the release of magnesium ions into pore solution. The strong acid-base reaction in the stabilized soil occurs, and the cementitious product of $MgKPO_4 \cdot 6H_2O$ is formed, as shown in the chemical Eq. (4). However, the UCS of the KMP stabilized soil is affected by the low temperature (as shown in Fig. 9). A possible explanation for this unexpected observation may be that the low temperature hinders the formation of $MgKPO_4 \cdot 6H_2O$ in some ways. A detail transformation process is observed by the Qiao [36] and Chau et al. [37], and the $MgHPO_4 \cdot 7H_2O$ and $Mg_2KH(PO_4)_2 \cdot 15H_2O$ are two reaction intermediates before the formation of the final product, $MgKPO_4 \cdot 6H_2O$, as shown in Eqs. (5), (6) and (7). At low temperature, the amount of the two reaction intermediates, whose mechanic strength are relative lower than final product, is possible more than those formed at relative high temperature. The quantitative phase analysis by XRD is needed to be further conducted.

$$Mgo + KH_2PO_4 + 5H_2O \rightarrow M_gKPO_4 \cdot 6H_2O \tag{4}$$

$$H_2PO_4^- + MgO + 7H_2O \rightarrow MgHPO_4 \cdot 7H_2O + OH^- \tag{5}$$

$$2MgHPO_4 \cdot 7H_2O + K^+ + OH^- \rightarrow Mg_2KH(PO_4)_2 \cdot 15H_2O \tag{6}$$

$$M_{g2}KH(PO_4)_2 \cdot 15H_2O + K^+ + OH^- \rightarrow 2MgKPO_4 \cdot 6H_2O + 4H_2O \tag{7}$$

4 Conclusions

This study presents results and analyses of the strength and heavy metals leached properties of KMP and PC stabilized soils. The influences of low-temperature and high moisture content to reactions in the binder-contaminant-soil system are evaluated. Based on the results of this research, the following conclusions can be drawn:

(1) The effect of low-temperature curing condition to the UCS of KMP stabilized soil specimens is lower relative to PC stabilized soil specimens. The KMP stabilized soil specimens with initial moisture content of 25% and cured at $-10\,°C$ can fulfill the requirement in USEPA Guidance (≥ 350 kPa), while the UCS of PC stabilized soil specimens with the same dosage content is far below the requirement.

(2) The effect of low-temperature curing condition to the heavy metals leached concentrations of KMP stabilized soil specimens are lower relative to PC stabilized soil specimens. The KMP stabilized soil specimens with initial moisture content of 35% and cured at $-10\,°C$ can fulfill the requirement in China MEP threshold limit; while the heavy metals leached concentrations of PC stabilized soil specimens with the same dosage content far exceed the threshold limit.

(3) The UCS of stabilized soil decreases with increasing initial moisture content. The UCS of KMP stabilized soil is lower than the PC stabilized soil specimen with high initial moisture content of 35%, while the UCS of KMP stabilized soil is higher than the PC stabilized soil with initial moisture content of 25%.

Acknowledgements. The authors are grateful for the support of Environmental Protection Scientific Research Project of Jiangsu Province (Grant No. 2016031), National High Technology Research and Development Program of China (Grant No. 2013AA06A206), the State Key Program of National Natural Science of China (Grant No. 41330641), National Natural Science Foundation of China (Grant No. 41472258), and Natural Science Foundation of Jiangsu Province (Grant No. BK2012022).

References

1. The World Bank. Overview of the current situation on brownfield remediation and redevelopment in China. The World Bank, sustainable development-East Asia and Pacific, Washington (2010)
2. United Nations. World urbanization prospects: the 2011 revision United Nations, Department of Economic and Social Affairs (DESA), Population Division, Population Estimates and Projections Section, New York (2012)
3. Harbottle, M.J., Al-Tabbaa, A., Evans, C.W.: A comparison of the technical sustainability of in situ stabilisation/solidification with disposal to landfill. J. Hazard. Mater. **141**(2), 430–440 (2007)
4. Du, Y.J., Jiang, N.J., Liu, S.Y., et al.: Engineering properties and microstructural characteristics of cement-stabilized zinc-contaminated kaolin. Can. Geotech. J. **51**(3), 289–302 (2013)
5. Spence, R.D., Shi, C.J.: Stabilization and Solidification of Hazardous, Radioactive, and Mixed Wastes. CRC Press, Boca Raton (2004)
6. Scrivener, K.L., Kirkpatrick, R.J.: Innovation in use and research on cementitious material. Cem. Concr. Res. **38**(2), 128–136 (2008)
7. Zha, F.S., Liu, J.J., Xu, L., et al.: Cyclic wetting and drying tests on heavy metal contaminated soils solidified/stabilized by cement. Chin. J. Geotech. Eng. **07**, 1246–1252 (2013)
8. Du, Y.J., Wei, M.L., Reddy, K.R., et al.: Effect of acid rain pH on leaching behavior of cement stabilized lead-contaminated soil. J. Hazard. Mater. **271**, 131–140 (2014)
9. Wei, M.L., Du, Y.J., Reddy, K.R., Wu, H.L.: Effects of freeze-thaw on characteristics of new KMP binder stabilized Zn and Pb contaminated soils. Environ. Res. Pollut. Res. **22**(24), 19473–19484 (2015)
10. Du, Y.J., Wei, M.L., Reddy, K.R., et al.: New phosphate-based binder for stabilization of soils contaminated with heavy metals: leaching, strength and microstructure characterization. J. Environ. Manag. **146**, 179–188 (2014)
11. Du, Y.J., Wei, M.L., Reddy, K.R., et al.: Effect of carbonation on leachability, strength and microstructural characteristics of KMP binder stabilized Zn and Pb contaminated soils. Chemosphere **144**, 1033–1042 (2016)
12. Chen, R., Kang, E., Ji, X., et al.: Cold regions in China. Cold Reg. Sci. Technol. **45**(2), 95–102 (2006)
13. Andersland, O., Anderson, D.: Geotechnical Engineering for Cold Regions. McGraw-Hill Book Co., New York (1978)
14. Freitag, D.R., McFadden, T.T.: Introduction to Cold Regions Engineering. ASCE Press, Reston (1997)
15. Lothenbach, B., Winnefeld, F., Alder, C., et al.: Effect of temperature on the pore solution, microstructure and hydration products of Portland cement pastes. Cem. Concr. Res. **37**(4), 483–491 (2007)
16. Gallucci, E., Zhang, X., Scrivener, K.L.: Effect of temperature on the microstructure of calcium silicate hydrate (CSH). Cem. Concr. Res. **53**, 185–195 (2013)
17. Myers, R.J., L'Hôpital, E., Provis, J.L., et al.: Effect of temperature and aluminium on calcium (alumino) silicate hydrate chemistry under equilibrium conditions. Cem. Concr. Res. **68**, 83–93 (2015)
18. Ogirigbo, O.R., Black, L.: Influence of slag composition and temperature on the hydration and microstructure of slag blended cements. Constr. Build. Mater. **126**, 496–507 (2016)

19. Yang, Q., Wu, X.: Factors influencing properties of phosphate cement-based binder for rapid repair of concrete. Cem. Concr. Res. **29**(3), 389–396 (1999)
20. Li, Y., Chen, B.: Factors that affect the properties of magnesium phosphate cement. Constr. Build. Mater. **47**, 977–983 (2013)
21. Abdelrazig, B.E.I., Sharp, J.H., El-Jazairi, B.: The chemical composition of mortars made from magnesia-phosphate cement. Cem. Concr. Res. **18**(3), 415–425 (1988)
22. Cai, G.H., Du, Y.J., Liu, S.Y., et al.: Physical properties, electrical resistivity, and strength characteristics of carbonated silty soil admixed with reactive magnesia. Can. Geotech. J. **52** (11), 1699–1713 (2015)
23. Shand, M.A.: The Chemistry and Technology of Magnesia. Wiley, Hoboken (2006)
24. Jin, F., Al-Tabbaa, A.: Evaluation of novel reactive MgO activated slag binder for the immobilisation of lead and zinc. Chemosphere **117**, 285–294 (2014)
25. Shackelford, C.D., Benson, C.H., Katsumi, T., et al.: Evaluating the hydraulic conductivity of GCLs permeated with non-standard liquids. Geotext. Geomembr. **18**(2), 133–161 (2000)
26. China MEP: Identification standards for hazardous wastes-Identification for corrosivity (GB5085. 1-2007). China Environmental Science Press, Beijing, China (2007)
27. Hunan EPD: Standards for Soil Remediation of Heavy Metal Contaminated Sites (DB43/T1165-2016). Environmental Protection Department of Hunan Province, China (2016)
28. China MEP: Identification standards for hazardous wastes-identification for extraction toxicity (GB5085. 3-2007). China Environmental Science Press, Beijing, China (2007)
29. USEPA: Prohibition on the disposal of bulk liquid hazardous waste in landfills–statutory interpretive guidance, office of solid waste and emergency response, EPA/530-SW-016, Washington D.C. (1986)
30. Fall, M., Pokharel, M.: Coupled effects of sulphate and temperature on the strength development of cemented tailings backfills: Portland cement-paste backfill. Cem. Concr. Compos. **32**(10), 819–828 (2010)
31. Gan, M.S.J.: Cement and Concrete. CRC Press, Boca Raton (1997)
32. Chew, S.H., Kamruzzaman, A.H.M., Lee, F.H.: Physicochemical and engineering behavior of cement treated clays. J. Geotech. Geoenvironmental Eng. **130**(7), 696–706 (2004)
33. Venkatarama Reddy, B.V., Lal, R., Nanjunda Rao, K.S.: Enhancing bond strength and characteristics of soil-cement block masonry. J. Mater. Civ. Eng. **19**(2), 164–172 (2007)
34. Olmo, I.F., Chacon, E., Irabien, A.: Influence of lead, zinc, iron (III) and chromium (III) oxides on the setting time and strength development of Portland cement. Cem. Concr. Res. **31**(8), 1213–1219 (2001)
35. Mellado, A., Borrachero, M.V., Soriano, L., et al.: Immobilization of Zn (II) in Portland cement pastes. J. Therm. Anal. Calorim. **112**(3), 1377–1389 (2013)
36. Qiao, F.: Reaction mechanisms of magnesium potassium phosphate cement and its application (2010)
37. Chau, C.K., Qiao, F., Li, Z.: Potentiometric study of the formation of magnesium potassium phosphate hexahydrate. J. Mater. Civil Eng. **24**(5), 586–591 (2012)

Author Index

© Springer Nature Singapore Pte Ltd. 2018
A. Farid and H. Chen (Eds.): GSIC 2018, *Proceedings of GeoShanghai 2018
International Conference: Geoenvironment and Geohazard*, pp. 595–597, 2018.
https://doi.org/10.1007/978-981-13-0128-5

Printed by Printforce, the Netherlands